-0. SEP. 1977

Benchmark Papers
in Microbiology

Series Editor: Wayne W. Umbreit
Rutgers—The State University

Published Volumes and Volumes in Preparation

MICROBIAL PERMEABILITY / John P. Reeves
CHEMICAL STERILIZATION / Paul M. Borick
MICROBIAL GENETICS / Morad Abou-Sabé
MICROBIAL PHOTOSYNTHESIS / June Lascelles
MICROBIAL METABOLISM / H. W. Doelle
ANIMAL CELL CULTURE AND VIROLOGY / Robert J. Kuchler
PHAGE / Sewell P. Champe
MICROBIAL GROWTH / P. S. S. Dawson
INFLUENCE OF TEMPERATURE ON MICROORGANISMS / J. L. Stokes
MOLECULAR BIOLOGY AND PROTEIN SYNTHESIS / Robert Niederman
ANTIBIOTICS AND CHEMOTHERAPY / M. Solotorovsky
INDUSTRIAL MICROBIOLOGY / Richard W. Thoma
MARINE MICROBIOLOGY / C. D. Litchfield
MICROBIAL NUTRITION / B. W. Koft
PHYSICAL FACTORS IN MICROBIOLOGY / D. W. Thayer

Benchmark Papers in Microbiology

———— A *BENCHMARK* ® Books Series ————

PHAGE

Edited by
SEWELL P. CHAMPE
Institute of Microbiology
Rutgers—The State University

Dowden, Hutchinson & Ross, Inc.
Stroudsburg, Pennsylvania

Library of Congress Cataloging in Publication Data

Champe, Sewell P   1932-   comp.
    Phage.

    (Benchmark papers in microbiology, v. 7)
    Bibliography: p.
    1. Bacteriophage--Addresses, essays, lectures.
I. Title. [DNLM: 1. Bacteriophage--Collected works.
QW160 C451p]
QR342.C48   576'.6482   74-13061
ISBN 0-87933-080-5

Acknowledgments
and Permissions

ACKNOWLEDGMENTS

GENETICS SOCIETY OF AMERICA—*Genetics*
  Genetic Studies of Lysogenicity in *Escherichia coli*

NATIONAL ACADEMY OF SCIENCES—*Proceedings of the National Academy of Sciences (U.S.)*
  A Bacteriophage Containing RNA
  A Change from Nonsense to Sense in the Genetic Code
  Chromosome Structure in Phage T4: I. Circularity of the Linkage Map
  Cohesion of DNA Molecules Isolated from Phage Lambda
  The Identification and Characterization of Bacteriophages with the Electron Microscope
  Isolation of the λ Phage Repressor
  Linkage Among Genes Controlling Inhibition of Lysis in a Bacterial Virus
  Morphogenesis of Bacteriophage T4 in Extracts of Mutant-infected Cells
  On the Topography of the Genetic Fine Structure

PERMISSIONS

The following papers have been reprinted with the permission of the authors and the copyright holders.
ACADEMIC PRESS, INC.
  Advances in Genetics
    Episomes
  Biochemical and Biophysical Research Communications
    The Linear Insertion of a Prophage into the Chromosome of *E. coli* Shown by Deletion Mapping
  Virology
    The Infection of *Escherichia coli* by T2 and T4 Bacteriophages as Seen in the Electron Microscope: I.
      Attachment and Penetration
    Some Unusual Properties of the Nucleic Acid in Bacteriophages S13 and φX174

ACADEMIC PRESS (LONDON) LTD.—*Journal of Molecular Biology*
  Density Alterations Associated with Transducing Ability in the Bacteriophage Lambda

AMERICAN SOCIETY OF BIOLOGICAL CHEMISTS, INC.—*Journal of Biological Chemistry*
  Virus-induced Aquisition of Metabolic Function: VII. Biosynthesis *de novo* of Deoxycytidylate Hydrox-
    ymethylase

ANNUAL REVIEWS, INC.—*Annual Review of Genetics*
  Developmental Pathways for the Temperate Phage: Lysis Versus Lysogeny

ASSOCIATED SCIENTIFIC PUBLISHERS (ELSEVIER-EXCERPTA MEDICA-NORTH HOLLAND)
    FOR THE CIBA FOUNDATION—*Ciba Foundation Symposium*
  Assembly in Biological Systems

AUSTRALIAN ACADEMY OF SCIENCE—*Replication and Recombination of Genetic Material*
  Role of Recombination in the Life Cycle of Bacteriophage T4

COLD SPRING HARBOR LABORATORY—*Cold Spring Harbor Symposia on Quantitative Biology*
  DNA Replication After T4 Infection
  The Mechanism of RNA Replication
  Physiological Studies of Conditional Lethal Mutants of Bacteriophage T4D
  Stages in the Replication of Bacteriophage ϕX174 DNA *in vivo*
  Structure and Function of DNA Cohesive Ends

FEDERATION OF AMERICAN SOCIETIES FOR EXPERIMENTAL BIOLOGY—*Federation Proceedings*
  Bacteriophage Assembly
  Virus-induced Acquisition of Metabolic Function

MACMILLAN JOURNALS LTD.—*Nature*
  Cleavage of Structural Proteins During the Assembly of the Head of Bacteriophage T4
  Co-linearity of the Gene with the Polypeptide Chain
  General Nature of the Genetic Code for Proteins

ROCKEFELLER UNIVERSITY PRESS—*The Journal of General Physiology*
  The Intracellular Growth of Bacteriophages: I. Liberation of Intracellular Bacteriophage T4 by Premature Lysis with Another Phage or with Cyanide

WISTAR INSTITUTE PRESS—*Journal of Cellular Physiology*
  The Rule of the Ring

Series Editor's Preface

Knowledge of microbial viruses has advanced so greatly over the past quarter century that it constitutes today a separate field of knowledge, with its own dogma, its own methodology, its own specialized language. As such, it becomes a formidable task to understand the recent publications in the field and to assess their significance, especially if one is not working on the subject. Yet it need not be so and a less difficult approach is possible. One might, for example, collect the significant papers, and this has indeed been done previously. There are plenty of collections of phage papers, but these have been designed mostly for workers in the field and mostly from the viewpoint of genetics. This collection is different; while in no way avoiding the technicalities of the field, Dr. Champe has brought together a group of first-rate significant papers (one hesitates to call a paper less than a decade old "classical") which define the field, classify the concepts, and provide the definitive experiments upon which our present concepts are based. One can emerge from a study of this book with a pretty fair idea of the science of bacteriophage. But perhaps even more important, before long even a novice can begin to understand "terminal redundancy," the mysteries of biological assembly, and even more exotic matters.

In addition to an excellent selection of papers, the comments and introductions of the editor perceptively point up some very fundamental principles of biological science and delineate, in a charming manner, ways of looking at living beings and of working with them which are of broad fundamental application. It may be that such approaches become particularly evident when one works with phages or it may be that particularly perceptive people were required before progress could be made in these studies, but at any rate these principles and approaches can well be of much broader significance. Indeed, one may learn much of the contemporary viewpoints and abiding principles of biological science from the editor's narrative with only a general examination of the individual papers.

Wayne W. Umbreit

Contents

I. ANATOMY OF A PHAGE

II. INSIDE *ESCHERICHIA COLI*

III. CISTRONS AND THE RULE OF THE RING

IV. GETTING IT ALL TOGETHER

V. THE VIRTUES OF TEMPERANCE

VI. THE SPARTANS

Contents by Author

Introduction

This collection of papers on bacteriophage spans the thirty-year period from 1942 to 1972 and attempts to document some of the highlights of this era, during which our understanding of the molecular basis of life saw monumental advances. In contrast to earlier collections, which emphasize primarily the role of phage as a tool of molecular genetics, I have tried to select material that pertains more to the biology of phage. A degree of redundancy was unavoidable since the two aspects are intimately related, but an attempt has been made to keep the overlap minimal.

Some of the papers are reports of original investigations that opened up new areas or decisively settled old questions; others are papers that, in my opinion, concisely review an aspect of the subject particularly well or develop unifying ideas. The final choice of papers was to a large extent arbitrary, determined by my own prejudices and by space limitations. The omission of many papers that are obviously as significant or incisive as those chosen was a necessity for which I apologize.

In the commentaries accompanying each section I have tried to put the selected papers in historical context and to provide some continuity to the ideas. For those in the field of phage research this history is elementary working knowledge, but for the beginning student or scientist in another area of biology the comments may prove useful as a guide. They are not intended to be exhaustive but do include references to current reviews.

The papers are arranged in six sections. The first four concern the general topics of phage structure, intracellular development, genetics, and assembly; the last two deal with the specialized aspects of temperate phage and the small single-stranded DNA and RNA phage. This hardly covers the vast world of phage but, I think, serves to underline the more important general principles and gives a glimpse of the diversity revealed by the best-studied prototypes.

In 1945 Demerec and Fano *(41)* collected the several well-behaved phage strains then in use in various laboratories, added a few of their own, and renamed the collection the T phage, T1 to T7. This proved to be a boon to phage research, since most investigators soon limited their studies to one or another of the T

1

phages and abandoned the former practice of each isolating his own phage strain. Later, a few mavericks again made new isolates and some of these revealed behavior unknown to the T phages, but in the early years the restriction to a few strains proved invaluable in comparing results from different laboratories. Another important advantage of working with the T phages was the fact that they could all multiply on *Escherichia coli*, a rapidly propagating, nonpathogenic genus of Enterobacteriaceae which can grow in a simple medium that has glucose as its sole carbon source.

Among the T phages, three strains — T2, T4, and T6 — proved to be very similar to one another and, because they were especially easy to propagate, assay, and store, were used predominantly in the early studies, from which gradually emerged our understanding of the concept of a virus. These studies with the T-even phages have thus been chosen to illustrate some of the important aspects of phage biology discussed in the first four sections, although in many respects the T-even phages are not representative of phage in general. There are differences also among the T-even phages, but, in general, what is true for one is true for all. In recent years most work has been confined to T4, but in earlier days the choice was often dictated by what was in the refrigerator at the time an idea for an experiment arose.

The absorbing account of the discovery of phage by Twort in 1915 and d'Herelle in 1917 and the ensuing controversy concerning the nature of this new creature can be found in Stent's previous collection *(170)* and in his book *Molecular Biology of Bacterial Viruses (171)*. d'Herelle's phage was defined solely by its bacteriolytic properties; he gave it its present-day name from the Greek word *phagein*, meaning "to eat," and indeed this is what a phage appeared to do. The cells of a turbid bacterial culture inoculated with a minute drop of the bacteriophagic principle would, after some hours, completely dissolve (lyse), leaving clear broth with little or no visible residue. Furthermore, and most importantly, this lysate contained the bacteriophagic principle at undiminished or even higher potency, for even a large dilution of it would induce the lysis of a second bacterial culture; a sample of the second lysate would induce the lysis of a third, and so on *ad infinitum*. It was thus apparent to d'Herelle that the phage must regenerate itself in the course of lysing the host cells. To dramatize this he pointed out that after only 22 passages, with a dilution factor of 10^4 per passage, the original sample would have been diluted into a volume comparable to that of the universe (quoted in *170*, p. 5). Or, stated in somewhat different terms, if the phage did not regenerate, there must have been at least 10^{88} particles in the original 0.001-ml sample for there to remain one particle after such a dilution. Clearly this could not be the case since such a sample could contain only some 10^{20} units of even a small molecule. Thus by 1926 d'Herelle had gleaned the essential nature of the phage as a lytic entity that replicates at the expense of the bacterial cell, but in spite of a surge of activity among microbiologists during the next decade, progress in understanding the details of the process was slow.

If a prize were given for the one technique that gave the greatest impetus to phage research, the winner would be the plaque assay, probably first performed by d'Herelle. The method is simplicity itself. If a small volume of liquid containing phage is mixed with a large number of sensitive bacteria (say 10^8) and spread onto solid agar, the surface of the agar, after overnight incubation, will be covered with a confluent opaque lawn of bacteria except at those places where a phage particle became immobilized when the liquid absorbed into the agar. At these places, clear round holes or plaques are visible; these are due to the growth of the phage and consequent lack of growth of the bacteria. Thus by spreading an appropriate dilution of a phage stock (so that the plaques are few enough to be counted), the concentration of phage or titer of the stock can be easily determined. Using this method a phage experiment was often a one-afternoon affair, giving results in time to plan the next afternoon's work.

In spite of the constant reassurance given by the faithful appearance of plaques, the early phage workers must have labored under the mild anxiety that the entities inferred to exist from their biological properties, but which were invisible under the best compound microscopes, might after all be illusions due to some improbable artifact. The physical existence of phage became more concrete when, between 1933 and 1936, Schlesinger *(157)* purified phage particles and showed them to have a size of the order of 0.1 micron and to be composed of protein and DNA. Most biologists are not really satisfied, however, until they can see the objects of their investigations. It was thus no doubt somewhat of a relief, as well as a remarkable revelation, when a phage particle was first brought to focus in the electron microscope by Luria and Anderson in 1942. As a point of departure their report has been chosen as the first paper in this collection.

I
Anatomy of a Phage

Editor's Comments on Papers 1 and 2

1 **Luria and Anderson:** *The Identification and Characterization of Bacteriophages with the Electron Microscope*

2 **Simon and Anderson:** *The Infection of* Escherichia coli *by T2 and T4 Bacteriophages as Seen in the Electron Microscope: I. Attachment and Penetration*

If Luria had been studying one of the small spherical phages such as S13, the microscopes of that day probably would have revealed little. But one of Luria's phages, called PC (later renamed T2) was large and the microscope revealed, as described in Paper 1, not a mere featureless sphere nor an amorphous blob but a particle morphologically resembling a sperm, though of course much smaller, with a distinctly differentiated head and tail. Luria and Anderson point out ". . . the extreme interest of the finding of such constant and relatively elaborate structural differentiation in objects of supposedly macromolecular nature."

The existence of a tail gave rise to speculation that the phage may be self-motile, and indeed many of the early micrographs of phage interacting with host bacterial cells showed a head-first attachment, as if the phage encountered the cell by swimming toward it. Anderson showed later, however, that this orientation was an artifact of drying and that attachment is actually by means of the tail *(4)*. There is no evidence that phage can swim nor any need to assume it; Brownian motion alone can account quite well for the collision frequency between phage and cells *(156)*.

Anderson's reminiscences *(26)* of his discussions with Luria and Delbrück during this period are fascinating to read. One disconcerting observation from these studies was that the phage particles never seemed to enter the cell but remained on the surface. Did the phage also replicate on the surface? The portentous solution to this puzzle had to await the attention of Hershey and Chase *(91)* ten years later.

With the development of better microscopes and better techniques of specimen preparation, additional details of the T2 morphology were revealed. The head was seen to be polyhedral, with fairly sharp vertices and edges *(204)*, and the tail itself had thin fiber-like appendages *(108)*, but it was not until the particle could be dissected into its component parts that the ultrastructure of the parts became discernible. Brenner *et al.* *(24)* in 1959 found that treating T2 phage with mild acid denatures the protein of the head but not of the tail components. The resulting precipitate could be redissolved by treatment with proteolytic enzymes, yielding morphologically intact but separated tail components. The tail was thus found to consist of an outer hollow cylinder (the sheath) surrounding another hollow inner cylinder (variously called the core or needle) to the end of which was attached a hexagonal baseplate from which the fibers emanate (see Fig. 19 of Paper 2). The isolated sheath was observed to be considerably shorter and fatter than the corresponding structure of the intact phage, from which it was inferred that the sheath is capable of contraction, a process shown later to occur during normal infection. Moody has recently proposed a molecular model of sheath contraction based on observations of the subunit structure of partially contracted sheaths *(138)*.

The isolation of pure fibers and sheath allowed the demonstration that the protein subunits of these substructures are different from each other and different from the subunit composing the head *(24)*. Further studies *(59, 120)* showed the T2 (or T4) head and sheath to be composed, respectively, of about 1000 and 150 identical subunits. Although these two proteins comprise the bulk of the protein mass of the phage, many additional minor but essential structural proteins have been identified. Recent studies by Laemmli (Paper 18), included in a later section, reveal 28 different proteins in T4, 11 of which are derived from the head and 17 from the tail. The most complex component appears to be the baseplate, which may consist of as many as 12 different kinds of protein. T4 also contains three internal proteins *(17, 86, 95)* and two small peptides *(49, 87)* which are released from the particle with the DNA when the head is ruptured.

The bonds that hold together the various phage components and their subunits appear for the most part to be noncovalent since almost all of the protein of the phage can be reduced to monomers by treatment with detergent. The involvement of some covalent bonds, however, has not been excluded.

In the light of what had become known about the structure of T4, Simon and Anderson undertook a careful reexamination of the interaction between the phage and host cell. Their 1967 paper reproduced here (Paper 2), from which the advances in electron microscopy since 1942 are apparent, reveals the remarkable complexity of the attachment and penetration process. The phage makes its first encounter with the cell wall by means of the distal tips of the long fibers in a manner reminiscent of the landing of a lunar module (Fig. 19). This is followed by a second contact between the cell wall and the pin structures of the baseplate. Finally, some unknown trigger causes the sheath to contract, pulling the baseplate away from the wall and pushing the needle through the wall. As a result, the pins, which apparently come under tension, uncoil to form the thinner structures called the short fibers. In a subsequent study Simon and Anderson *(162)* examined the baseplate of the phage and found that during the process of penetration this structure is transformed from a hexagon into a six-pointed star.

It should be pointed out that other phage types are quite different in morphology from the T-even group and must employ different mechanisms for attachment and penetration. Some phages are spherical (actually icosahedral) without visible tail structures, others have noncontractile tails lacking fibers, and still others are filamentous. A detailed description of the various types has been given by Bradley *(19)*.

Reprinted from *Proc. Natl. Acad. Sci. (U.S.)*, **28**(4), 127–130 (1942)

THE IDENTIFICATION AND CHARACTERIZATION OF BACTERIOPHAGES WITH THE ELECTRON MICROSCOPE

BY S. E. LURIA* AND THOMAS F. ANDERSON†

BACTERIOLOGICAL RESEARCH LABORATORIES OF THE DEPARTMENT OF SURGERY, COLLEGE OF PHYSICIANS AND SURGEONS, COLUMBIA UNIVERSITY, NEW YORK, AND RCA RESEARCH LABORATORIES, CAMDEN, N. J.

Communicated March 12, 1942

Bacteriophages, or bacterial viruses, are a group of viruses reproducing in the presence of living bacterial cells. Bacteriophages are particulate, and convincing evidence exists that (1) one particle of phage is sufficient to originate the lysis of a bacterial cell; in the lysis, a variable number of new phage particles (an average of 100 or more) are liberated per cell;[1] (2) the elementary particles of each phage strain seem to have a characteristic particle size as determined by any one of various indirect methods of investigation (ultrafiltration,[2] radiosensitivity,[3] diffusion,[4]) and diameters ranging from 10 to 100 mμ have been obtained for the various strains depending on the method of investigation, although diffusion experiments occasionally yield still smaller values.

The electron microscope has recently been applied with success to the study of viruses[5] and it therefore seemed desirable to attempt such a study of bacterial viruses, particularly since they offer favorable possibilities for the identification of the virus particles through a study of the reaction between the individual particles and the bacterial cell under the microscope. Indeed, a number of short reports have been published recently by German authors[6, 7] in which round particles have been described as corresponding to the phage particles, although Ruska[7] shows pictures of "sperm-shaped" particles from a phage suspension adhering to a bacterial membrane. From this evidence alone he is unable to decide whether these are particles of phage or bacterial products.

We have undertaken an investigation of the problems of phage structure, size, reproduction and lytic activity by means of the RCA electron microscope. Research on the last items is still in progress. The present report concerns itself with the identification and the morphological analysis of a number of strains of phage particles and their adsorption on sensitive bacterial cells. The results are illustrated by some of the electron micrographs (Plates I and II) which have brought to light many extremely interesting features. Details of the material and methods used will soon be published.

I.   Bacteriophage anti-coli PC (particle diameter by diffusion 44 mγ, Kalmanson and Bronfenbrenner[8]; by x-irradiation 50 mμ, Luria and Exner, unpublished).

Micrographs of high titer suspensions, figures 1, 2, 4, 5 and 6, show the constant presence of particles of extremely constant and characteristic aspect. They consist of a round "head," and a much thinner "tail," which gives them a peculiar sperm-like appearance. The "head" is not homogeneous but shows an internal structure consisting of a pattern of granules, distinguished by their higher electron scattering power. Deviations from the usual symmetrical internal pattern may be due to varying orientation of the particles or to other factors as yet unknown. The diameter of the head appears to be about 80 mμ; the tail is about 130 mμ long.

EXPLANATION OF PLATE
PLATE I

1. Electron micrograph of particles from a high titer suspension of bacteriophage anti-coli PC. $\times$ 38,000.

2. Particles from a high titer suspension of bacteriophage anti-coli PC. $\times$ 84,000.

3. *Escherichia coli* from suspension in distilled water. $\times$ 17,000.

4. *Escherichia coli* in suspension of bacteriophage anti-coli PC for ten minutes. $\times$ 17,500.

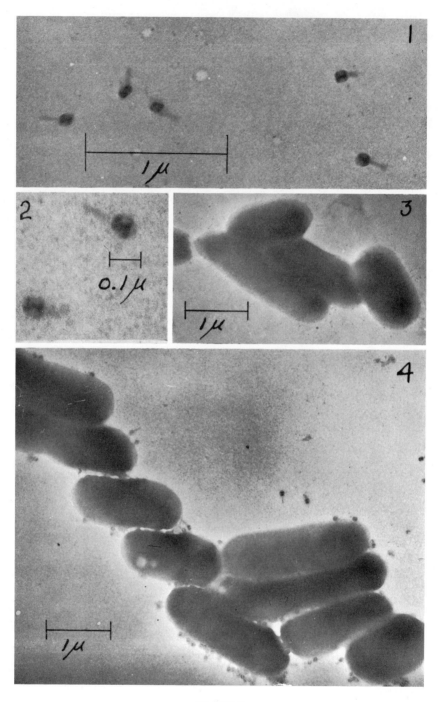

PLATE I

This gives a size which is in fair agreement with the figures deduced from the radiosensitivity method.   On the other hand, it is possible that the size as determined by x-rays corresponds more closely to the size of the granules.

When allowed to stand a few minutes in the presence of sensitive bacterial cells *Escherichia coli*, strain PC (Fig. 3), the particles described above are readily adsorbed (Figs. 4 and 5).   They appear to stick to the bacteria either by the head or by the tail.   Other conditions remaining constant, the number of particles adsorbed on a bacterium increases with the time of contact, although it is difficult at the present time to differentiate between adsorption and reproduction of the particles on the cell wall.   By allowing the phage to stay in contact with bacteria for a time of the order of the minimum time of lysis (21 minutes for PC phage, Delbrück and Luria[1]) it is possible to observe bacterial cells extensively damaged, surrounded by a very large number of particles, probably newly formed (Fig. 6).

II.   Bacteriophage anti-coli P 28, also active on *Escherichia coli* strain PC (particle size: irradiation, 36 mμ, Luria and Exner.[3]

Round particles are visible in the suspensions of this phage which are somewhat smaller than those described for phage PC (about 50 mμ in diameter).   An extremely thin tail, although difficult to demonstrate with certainty in the reproductions, seems to be visible in many instances.   In many micrographs the head is almost completely filled by a dense internal structure.   These particles, too, are readily adsorbed on sensitive bacterial cells.

III.   Bacteriophage anti-staphylococcus 3K (particle size: by ultrafiltration and ultracentrifugation 50–75 mμ, Elford;[2] by irradiation 48 mμ, Luria and Exner.[3]

Owing to technical reasons, the conditions for successful micrographing are here less favorable.   Nevertheless, the presence of approximately round particles of proper size has been established in preparations of this page also.

We are inclined to identify the particles described above with the actual particles of bacteriophage for the following reasons: (*a*) They are always present in highly active phage suspensions and missing in any control suspensions (media, bacterial cultures, bacterial filtrates, etc.); (*b*) they are readily adsorbed by the bacterial cells of the susceptible strain and fail to be adsorbed by other bacteria; (*c*) the size from a given strain is uniform and corresponds essentially to measurements by indirect methods; (*d*) the structure of both the "head" and the "tail" is characteristic of the strain of phage; (*e*) preliminary experiments on the lysis process seem to demonstrate the liberation of these particles from the lysing bacteria.

Conclusions.—We do not want to discuss here the bearing of the above described results on the problem of the nature of bacteriophage and of viruses in general.   We limit ourselves to pointing out the extreme interest of the finding of such constant and relatively elaborate structural differen-

11

tiation in objects of supposedly macromolecular nature. This result is of equal interest in the field of genetics, since genes, together with viruses, are currently supposed to be macromolecular entities.

The correspondence of the particle size as directly portrayed in the electron microscope with the results of indirect methods is also very remarkable. although it does not exclude the possibility of phage activity being sometimes associated with smaller particles. It is worth while emphasizing that the results of the present investigation, together with the recently published results of irradiation of bacteriophages, represent most desirable evidence for the validity of the so-called "hit theory" for the determination of the "sensitive volume" in sub-light-microscopic biological objects. This conclusion, too, seems to be interesting from the point of view of genetics, since the "hit theory," although widely criticized, has been used for calculating the approximate value of the dimensions of genes.

The authors are grateful to the National Research Council Committee on Biological Applications of the Electron Microscope for allocating time for this study, and to the RCA Laboratories for the use of their facilities, and to Dr. V. K. Zworykin for his interest and encouragement. The authors also thank Dr. Stuart Mudd for the use of the facilities of his laboratory for the preparation of material for study.

* Aided by a grant from the Dazian Foundation for Medical Research.

† RCA Fellow of the National Research Council.

[1] Ellis, E. L., and Delbrück, M., *Jour. Gen. Physiol.*, **22**, 365 (1939); Delbrück, M., and Luria, S. E., to be published.

[2] Elford, W. J., in Doerr and Hallauer, *Handbuch der Virusforschung*, vol. I, Julius Springer, Wien, 1938, p. 126.

[3] Luria, S. E., and Exner, F. M., *Proc. Nat. Acad. Sci.*, **27**, 370 (1941).

[4] Hetler, D. M., and Bronfenbrenner, J., *Jour. Gen. Physiol.*, **14**, 547 (1931).

[5] Stanley, W. M., and Anderson, T. F., *Jour. Biol. Chem.*, **139**, 325–538 (1941), and references given therein.

[6] Pfankuch, E., and Kausche, G. A., *Naturwiss.*, **28**, 46 (1940).

[7] Ruska, H., *Naturwiss.*, **29**, 367 (1941).

[8] Kalmanson, G. M., and Bronfenbrenner, Jr., *Jour. Gen. Physiol.*, **23**, 203 (1939).

EXPLANATION OF PLATE

PLATE II

5. *Escherichia coli* in suspension of bacteriophage anti-coli PC for 20 minutes. × 14,500.

6. *Escherichia coli* in suspension of bacteriophage anti-coli PC for 20 minutes. × 12,500.

7 and 8. Particles from a high titer suspension of bacteriophage anti-coli P28. × 38,000.

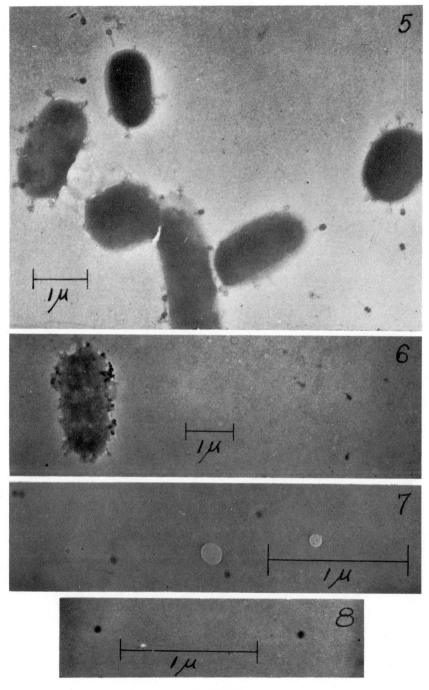

PLATE II

$$2$$

Reprinted from *Virology*, **32**(2), 279–297 (1967)

The Infection of *Escherichia coli* by T2 and T4 Bacteriophages as Seen in the Electron Microscope

I. Attachment and Penetration[1]

LEE D. SIMON[2] AND THOMAS F. ANDERSON

Department of Molecular Biology, The Institute for Cancer Research, Philadelphia, Pennsylvania 19111

Accepted March 2, 1967

The infection of *Escherichia coli* by T2 and T4 bacteriophages has been studied with the electron microscope using both thin-sectioning and negative-staining techniques. The initial attachment of T2 and T4 viruses to the bacterial surface is made by the distal ends of the phages' long tail fibers. The second major step in the adsorption process seems to be a repositioning of the bacteriophage so that its short tail pins which were at first more than 1000 A from the bacterial surface are brought into very close proximity to the cell wall. In the final step in the attachment of T2 and T4 bacteriophages the tail sheath contracts and the baseplate is pulled along the needle away from the cell wall. At this stage the tail pins are no longer visible, but short tail fibers about 370 A long connect the baseplate directly to the cell wall, while the long tail fibers are found attached to the wall at points apparently corresponding to their initial sites of attachment. The needle appears to penetrate about 120 A into the cell wall.

INTRODUCTION

The first step in the infection of *Escherichia coli* by T2 and T4 bacteriophages is the adsorption of the viruses to the bacteria. T-even viruses attach by their tails to sensitive cells (Anderson, 1953). The tails of the T-even phages are quite intricate, possessing several morphologically, chemically, and serologically different types of structural components (Brenner *et al.*, 1959; Edgar and Lielausis, 1965). The normal tail consists of two coaxial hollow tubes, the

inner now being referred to as the needle, and the outer as the sheath. At their distal tips the needle and the sheath terminate in the baseplate. Long tail fibers, extending from the baseplate, seem to play an important role in the specific attachment of the phage to sensitive bacteria (Williams and Fraser, 1956; Kellenberger and Sechaud, 1957; Wildy and Anderson, 1964).

After the attachment of the phage to its host cell, the phage's DNA is transferred into the bacterium (Hershey and Chase, 1952) apparently through the phage's hollow needle. However, electron microscopic examination has failed to determine whether or not the needle penetrates the bacterial cell wall (Stouthamer *et al.*, 1963). Thus, it has not been clear whether the phage DNA is injected directly into the bacterial cytoplasm, into the space between the cell wall and cytoplasmic membrane, or released on the surface of the bacterium where the DNA might immediately enter

[1] This work was supported in part by predoctoral training grant NIH-5-T-1-GM 000658 from the U.S. Public Health Service to the University of Rochester, and by grants GB-982 and GB-4640 from the National Science Foundation and CA-06927 from the U.S. Public Health Service to The Institute for Cancer Research.

[2] This paper is based on a thesis submitted to the Biology Department of the University of Rochester in partial fulfillment of the requirements for the Ph.D. degree.

the cell in a manner similar to that of transforming DNA.

The roles of the tail fibers and other phage components in the attachment process and the penetration of the cell wall by the phage's needle will be illustrated and discussed in this paper.

MATERIALS AND METHODS

Bacteriophage strains. The phages used in the experiments described in this paper were a cofactor-requiring T4 strain (obtained from D. Brown), another T4 strain (obtained from E. Kellenberger), and a T2 strain (obtained from M. Bayer). Phages were grown according to the usual procedures either in a tryptone broth or in a synthetic medium containing ammonium lactate, the so-called "F" medium (Adams, 1959).

Bacterial strains. The bacterial strains used were our laboratory culture of *E. coli* B and *E. coli* CR-63 (obtained from R. S. Edgar). They were grown with aeration either in tryptone broth or in F medium at 37°.

Preparation of phage stocks. Phage stocks used for electron microscopic observation of negatively stained unadsorbed phages were prepared as follows: the phage in crude lysates were purified by differential centrifugation; each phage pellet obtained was allowed to resuspend overnight in 1 ml of 1% ammonium acetate at 4°. A few phage stocks were further purified by density gradient centrifugation in cesium chloride. The fraction containing the T2 or T4 phage was dialyzed against 1% ammonium acetate at 4° to remove the cesium chloride.

Phage stocks for adsorption experiments were purified by low speed centrifugation to remove bacterial debris. These stocks were stored in the medium in which they were grown (tryptone broth or F medium) at 4°.

Adsorption of phage on bacteria. All adsorption experiments were carried out in tryptone broth or in F medium since T-even phages will not adsorb to *E. coli* cells in otherwise more desirable media such as ammonium acetate solutions (Williams and Fraser, 1956) or distilled water. The multiplicities of infection (moi) were usually 50 or 200 to make it easier to find adsorbed phages in sections studied in the electron microscope. The use of such high multiplicities will be discussed in more detail later.

Fixation, embedding, and sectioning of specimens. Fixation and dehydration were the same as described by Kellenberger et al. (1958) except for the addition of 0.5% (w/v) uranyl acetate to the 1% OsO_4 fixative solution.

All specimens intended for fixation were grown in aerated tryptone broth. Uninfected *E. coli* cells were fixed during the logarithmic phase of growth. *E. coli* cells which had been lysed from without were fixed 10 minutes after the infecting phages (moi ~200)' had been added. *E. coli* cultures which were infected with phages (moi ~50) but which had not been lysed from without were fixed 1 hour after the addition of phages.

The fixed specimens were embedded in Vestopal W (purchased from Henley and Co., New York), "activated" before use by the addition of 1% (w/w) tertiary-butyl perbenzoate (purchased from K and K Laboratories, New York) and 0.05% (w/w) benzoin. The fixed specimens were dehydrated in a graded acetone series; they were then placed in open weighing bottles containing 25 ml of 10% (v/v) "activated" Vestopal in acetone. After the acetone had been allowed to evaporate overnight at 20°, the specimens were transferred to gelatin capsules filled with fresh "activated" Vestopal. The capsules were covered and then placed in an oven at 60–65° where the Vestopal was allowed to polymerize for at least 48 hours.

Thin sections were cut using a diamond knife (purchased from E. I. duPont de Nemours and Co., Wilmington) on an LKB Ultrotome microtome. Unless otherwise stated, the section thickness was less than 300 A (according to the thermal feed meter on the Ultrotome). Sections were picked up on copper grids covered with thin carbon-coated Formvar films.

Sections were stained with both uranyl acetate and lead hydroxide as follows: grids were floated overnight, section side down, on 2% aqueous uranyl acetate. The grids were then rinsed with distilled water and allowed to dry on filter paper. Sub-

15

sequently, the sections were stained for 3–7 minutes with lead hydroxide according to Karnofsky (1961).

Negative staining of specimens. For negative staining, an unfixed specimen droplet on the surface of a grid was mixed with a droplet of 2% silicotungstate (adjusted to pH 6.8–7.2 with NaOH). The excess liquid was removed with filter paper, and the grid was quickly placed into the column of the electron microscope to dry (Anderson, 1962).

Electron microscopy. Specimens were examined either in a Siemens Elmiskop I electron microscope operated at 80 kV and

equipped with a 50-μ objective aperture and a pointed filament, or in an RCA EMU 3 E electron microscope operated at 100 kV and equipped with a 50-μ objective aperture.

RESULTS

Initial Steps in the Adsorption of T2 and T4 Bacteriophages to the E. coli Cell Wall

Attachment of the long tail fibers to the cell wall. As a control experiment, a stock of tryptophan-requiring T4 bacteriophages (Anderson, 1948) was suspended in F type medium (Adams, 1959) which contained no tryptophan. A drop of this phage suspension

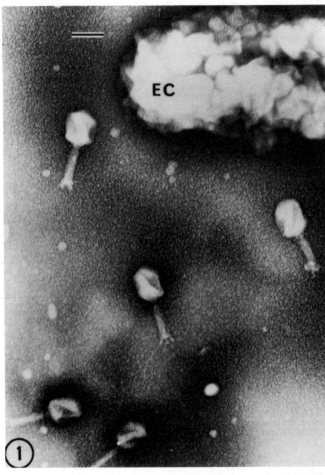

Fig. 1. An electron micrograph of cofactor-requiring T4 phages suspended with *Escherichia coli* B cells in F medium without tryptophan. A droplet of cofactor-requiring phages in F medium was mixed on the surface of a grid with a droplet of *E. coli* B also in F medium. After 1 minute the mixture was examined by negative staining with 2% silicotungstic acid at pH 6.8–7.2. Without tryptophan very few of these phages attach to the *E. coli* cells (EC). The length of the bar in this and in subsequent micrographs represents 1000 A.

was mixed on the surface of an electron microscope grid with a drop of *E. coli* B cells also suspended in F medium without tryptophan. After 1 minute, a small amount of 4 % silicotungstic acid was added to the bacteria and phage mixture on the grid. The excess fluid was then drawn off with filter paper (Anderson, 1962) and the grid was immediately placed into the vacuum of the electron microscope column to dry. Figure 1 is an electron micrograph of such a preparation. This picture shows that in the

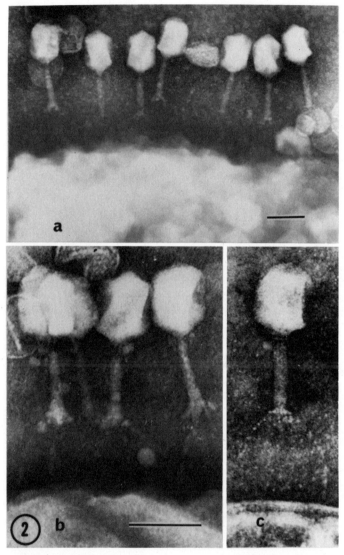

FIG. 2. Early stages in the attachment of T4 phages to *E. coli* B cells as seen in negatively stained preparations. Cofactor-requiring T4 phages were mixed on the surface of a grid with *E. coli* B cells in F medium. Tryptophan was then added, and about 15 seconds later the preparation was negatively stained. Within 1 minute it was placed into the column of the electron microscope and dried. (a) The phages are oriented with their tails toward the *E. coli* cell wall. The baseplates of the viruses are about 1000 A from the bacterial surface. (b) Long tail fibers are visible, connecting the baseplates to the *E. coli* wall. (c) The long tail fibers extending from the phage's baseplate to the cell wall are bent, meeting the wall about 900 A from the point where a perpendicular from the baseplate would intersect the cell wall.

absence of tryptophan, cofactor-requiring T4 phages do not seem to attach to or to orient with respect to the bacteria.

In the presence of tryptophan, though, cofactor-requiring T4 phages show a striking orientation with respect to the bacterial cell wall. The electron micrographs in Fig. 2a, b, c are of preparations similar to that of Fig. 1, except that tryptophan had been added to the preparations only 15 seconds before they were stained and 1 minute before they were dried in the microscope (as suggested by Mr. Dennis Brown). Figure 2a shows many phages oriented with their tails pointing toward the bacterial cell wall; the distance from their baseplates to the cell wall is about 1000 A. In Fig. 2b,c long fibers are seen linking the phage tails to the cell wall. These long tail fibers are characteristically bent near their centers so that their distal tips touch the cell wall at points some 700–900 A from the point where a line drawn perpendicular to the baseplate would intersect the cell wall.

Similar observations were made on thin sections like that of Fig. 3a showing a T4 particle apparently in the process of attaching to an *E. coli* CR-63 cell; Fig. 3b is a similar micrograph of a T2 bacteriophage attaching to an *E. coli* B cell. The orientation of the long fibers and the positions of the phages correspond to the observations made using negative staining. It is of interest to note in Fig. 3a that the tail pins extend only 100 A from the baseplate toward the cell wall.

Movement of the phage closer to the cell wall. After the attachment of the phage's long tail fibers to a sensitive bacterium, the main body of the phage seems to move closer to the bacterial surface. Figure 4 is an electron micrograph of a section showing a T4 particle apparently attached to the cell wall. Here the phage's baseplate is only 100 A from the bacterial surface.

It has not been possible to confirm this observation using negative staining techniques because the stain immediately surrounding intact bacteria is too thick. However, negatively stained phages occasionally can be seen in close association with small cell wall fragments. Figure 5 shows a negatively stained T4 phage and a bacterial fragment which were found in a

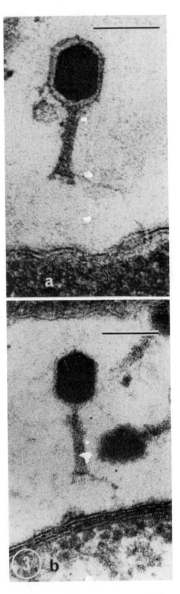

FIG. 3. Attachment of the long tail fibers of T2 and T4 phages to *E. coli* cells as seen in thin sections. (a) A T4 phage is seen attached to an *E. coli* CR-63 cell by a long tail fiber which is bent in the same manner as those observed in negatively stained preparations. The tail pins extend toward the cell wall. (b) A T2 phage appears to be attached to an *E. coli* B cell by a long tail fiber. The long fiber and tail pins are oriented similarly to those of the T4 phage.

lysate purified by several cycles of differential centrifugation and then resuspended in 1 % ammonium acetate. The piece of

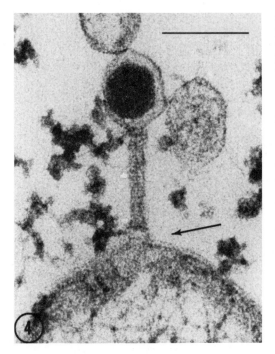

FIG. 4. A section through a T4 phage with its baseplate about 100 A from an *E. coli* CR-63 wall fragment. The phage appears to be attached to the cell wall by its tail pins (arrow).

bacterial wall is about 125 A from the baseplate, apparently touching one of the short tail pins. Neither in negatively stained preparations nor in thin sections have we ever seen the baseplate closer to the cell wall than 100 A during any phase of the adsorption process.

Contraction of the Tail Sheath

The position of the baseplates of adsorbed phages with regard to the bacterial cell wall. After the baseplate has moved to about 100 A from the cell wall, a series of events takes place which seems to involve the contraction of the tail sheath. The baseplate remains attached to the distal end of the sheath but releases itself from the tip of the needle so that it slides up the needle and away from the cell wall as the sheath contracts. In Fig. 6, which shows many T4 bacteriophages attached to the wall of an *E. coli* B cell, the baseplates have moved to positions 300–370 A away from the cell wall. However (as will be described in detail later), the occurrence

of superposition effects in negatively stained preparations and the results of experiments using thin sections suggest that the most likely estimate of the distance from the baseplate to the bacterial surface after contraction of the sheath is about 370 A.

It seemed possible that the baseplates might be in intimate contact with an invisible surface layer 370 A thick which forms around each bacterium after infection. All attempts, however, to observe such a layer failed. If such a layer exists it is not made visible by negative staining with uranyl acetate or silicotungstate, or by positive staining of thin sections with heavy metals; nor does it exclude colloidal palladium added like a negative stain to infected cells.

Short tail fibers. In thin sections through

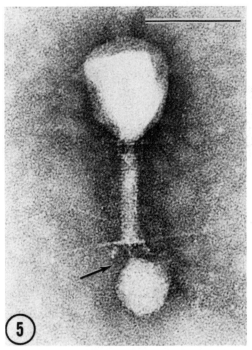

FIG. 5. T4 phage in close association with an *E. coli* B cell wall fragment, as observed by negative staining. This micrograph is of a lysate purified by several cycles of differential centrifugation and resuspended in 1% ammonium acetate. The bacterial wall fragment is about 125 A from the phage's baseplate; it appears to be in contact with one of the tail pins (arrow).

19

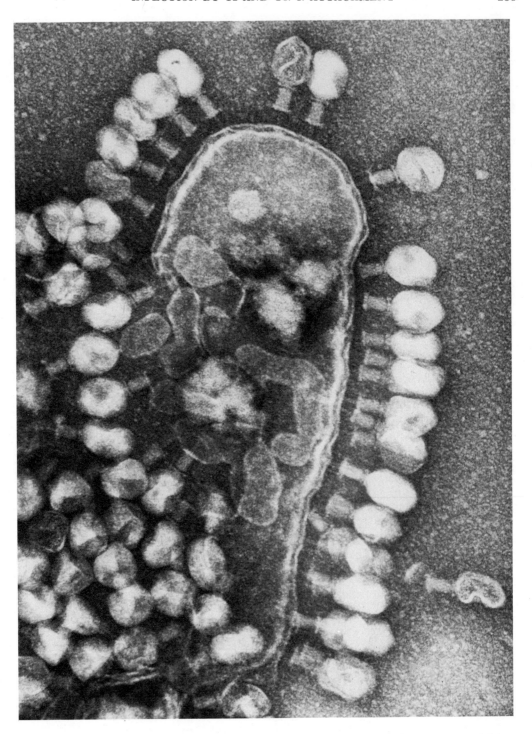

FIG. 6. T4 phages adsorbed on an *E. coli* B cell. Cofactor-requiring phages were suspended with bacteria in F medium containing tryptophan. After 10 minutes at 37°, a droplet of the suspension was negatively stained and examined in the électron microscope. The phages' sheaths are contracted, and their baseplates are 300–400 A from the cell wall.

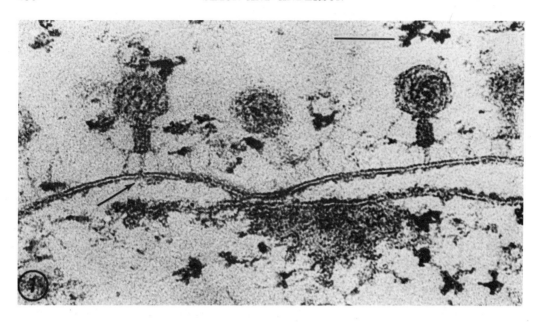

FIG. 7. T4 phages adsorbed on the wall of an *E. coli* cell which has lysed from without. *E. coli* B cells were heavily infected with T4 phages (moi ~200). The cells lysed from without. Ten minutes later they were fixed and subsequently embedded. The T4 phages visible in this section are attached to the cell wall by short tail fibers; their long tail fibers extend laterally from the tips of the baseplates. The cell wall appears as two continuous electron-opaque lines separated by an electron-transparent space. The arrow indicates where the needle of an adsorbed phage may have penetrated the cell wall.

T4 particles adsorbed on *E. coli* cells which they have lysed from without (Fig. 7), it is clear that the phages are attached to the bacterial wall by short tail fibers. Since short tail fibers have not been reported before, it seems desirable to give them special attention here. These short fibers extend directly from the baseplates to the wall. The serial sections shown in Fig. 8a,b,c suggest that there are *at least* four short fibers emanating from points about 420 A apart on the baseplate.

Short tail fibers are also visible in negatively stained preparations of *E. coli* B cells lysed from within by T4 phages as shown in Fig. 9. As in sections, the bacteriophage baseplates are seen to be attached to cell wall fragments by short fibers about 370 A long and separated by about 400 A. The kinked long tail fibers on these phages are pointing away from the pieces of cell wall to which the phages are attached by their short fibers.

T2 phages appear to have similar tail fiber arrangements to those observed on T4.

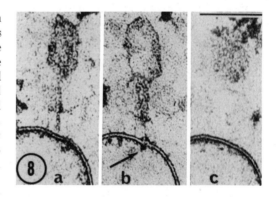

FIG. 8. Consecutive serial sections through an adsorbed T4 phage. The section thickness is under 300 A. (a) Two short tail fibers extend from the baseplate to the cell wall. The distal tip of the needle is not visible in the plane of this section. The outer layers of the cell wall in the immediate vicinity of the phage's needle are continuous and unbroken. (b) In this section two short tail fibers again connect the baseplate to the wall. The arrow indicates where the needle appears to have penetrated the broken cell wall. (c) The cell wall in this section is unbroken. Only part of the phage's head is visible.

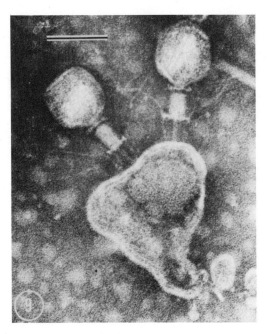

FIG. 9. Adsorbed phages in broth lysates from T4-infected *E. coli* CR-63 cells seen by negative staining. An *E. coli* CR-63 culture was infected with T4 phages (moi ∼5) in tryptone broth. After 2 hours at 37°, chloroform was added, and then the suspension was centrifuged at low speed to remove large bacterial debris. Droplets of the supernatant fluid were negatively stained, after which they were examined in the electron microscope. Two phages are attached by their short tail fibers to a small bacterial fragment. Their long tail fibers extend laterally from their baseplates. Material resembling negatively stained DNA is visible within the wall fragment.

Figure 10 shows a negatively stained T2 phage which was purified by differential centrifugation and was suspended in 1% ammonium acetate. This phage, though unadsorbed, has a contracted sheath and it has both long and short tail fibers. The short fibers extend from the baseplate away from the head, parallel to the needle, while the long fibers extend in lateral directions.

Occasionally the long tail fibers of unattached T2 and T4 bacteriophages with contracted sheaths are sharply kinked, making acute angles of 40–50° near their centers. The T4 virus negatively stained in Fig. 11a shows several acutely bent long fibers. Figure 11b is an electron micrograph

showing a negatively stained T4 phage attached to a bacterial fragment. Short fibers go directly from the baseplate to the wall. The proximal parts of the long tail fibers extend away from the cell wall, forming included angles of 130–150° with the baseplate; near their centers these long fibers bend, making acute angles of 40–50°; the distal parts of the long fibers attach to the wall.

Similar observations that the distal ends of the phage's long tail fibers may remain attached to the cell wall after contraction of the sheath have been made on relatively thick (silvery gold colored) sections of *E. coli* B cells infected with T2 phages (moi ∼50). When a section happened to contain a T2 phage and the phage's long tail fibers, electron micrographs such as those in Fig. 12 were obtained. In all these micrographs short fibers are seen in addition to the long tail fibers. The long tail fibers observed in Fig. 12 connect the baseplates to the cell

FIG. 10. The tail fiber arrangement on an unadsorbed T2 phage with a contracted sheath. A T2 phage stock, purified by differential centrifugation and then resuspended in 1% ammonium acetate, was examined by negative staining. Occasionally, phages were observed which, though unadsorbed, had contracted sheaths. On the phage shown here both long and short tail fibers are visible. The T2 tail fiber arrangement appears identical to that seen on T4 phages: the short tail fibers extend from the baseplate away from the head, parallel to the needle; the long tail fibers extend laterally.

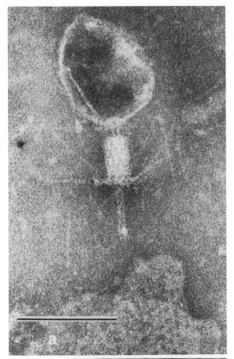

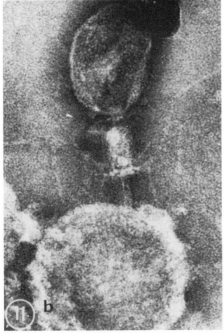

walls. The proximal parts of these fibers extend from the baseplates in directions away from the bacteria; near their centers, the long fibers bend sharply so that their distal halves are almost perpendicular to the bacterial surfaces. The long fibers usually meet the cell walls about 700–900 A from the points of intersection of the phages' needles with the walls. Similar observations have been made on thin sections through adsorbed T4 particles.

In an attempt to determine whether the short tail fibers are present only after sheath contraction, negatively stained T2 and T4 bacteriophages with extended sheaths were carefully examined in the electron microscope. A number of observations pertaining to the short tail fibers were made. (1) Frequently, particles are observed whose tail pins appear ill-defined, and occasionally structures resembling short tail fibers can be seen on these phages (Fig. 13). These short fibers are about 375 A long, the same length as the short tail fibers observed on adsorbed phages. (2) Pairs of T4 phages (Fig. 14) are sometimes observed linked to each other by 600 A fibers connecting their baseplates in positions where tail pins are usually seen. (3) In several high resolution electron micrographs, employing both thin-sectioning and negative staining techniques, the tail pins appear to have a periodic structure with a repeat distance of about 40 A (Fig. 15).

Attachment of T4 to cyanide-poisoned E. coli. To reduce metabolic changes in the cell wall after infection, *E. coli* B cells in phosphate buffer containing 0.01 *M* potassium cyanide were infected with high multiplicities of T4 and examined 10 minutes later. As seen in Fig. 16, the cyanide had no

FIG. 11. T4 phages with contracted sheaths and acutely bent long tail fibers. The phages shown in these electron micrographs are from a lysate purified only by low speed centrifugation. (a) The long fibers on this negatively stained T4 particle are bent making acute angles of 40–50° near their centers. (b) This adsorbed T4 phage is attached to the cell wall by both its long and short tail fibers. Short fibers connect the baseplate directly to the *E. coli* B wall fragment. The proximal parts of the long tail fibers extend away from the cell wall; near their centers, these fibers bend sharply; their distal halves extend toward the cell wall and in some cases appear attached to it. The long fibers meet the cell wall over 700 A from the point of intersection of the needle with the wall.

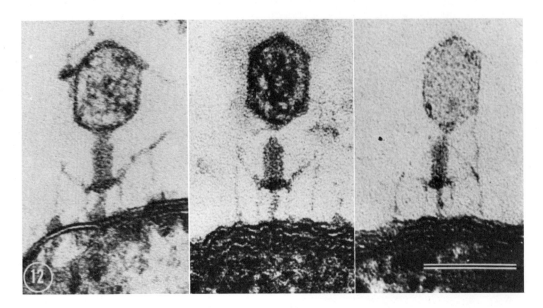

Fig. 12. Sections through adsorbed T2 phages attached to intact *E. coli* B cells or to large wall fragments by both their long and short tail fibers. *E. coli* B cells were infected with T2 phages (moi ∼50);
after 1 hour they were fixed and then embedded. Rather thick, silvery gold-colored sections were cut
and subsequently examined in the electron microscope. The long tail fibers extend laterally from the
baseplates; they flex near their centers and their distal tips are attached to the cell walls. The long tail
fibers attach to the walls at points over 800 A from the intersections of the needles with the walls. The
short tail fibers go directly from the baseplates to the walls.

visible effect: baseplates of adsorbed T4
particles are still separated by 400 A from
the cell wall and are connected to it by
short tail fibers and the central needle,
while the long tail fibers have lifted away.

Penetration of the Wall by the Needle

Several micrographs of thin sections
through T4 bacteriophages attached to *E.
coli* cell wall fragments show T4 needles
which have penetrated through the bacterial
wall. In Figs. 8 and 17, the needles appear to
have penetrated just through the cell walls
(about 120 A) but not further. Thin threads,
apparently of viral DNA, extend from the
tips of the needles of several of the adsorbed
phages shown in Fig. 17. In some cases this
DNA may be helically supercoiled. Frequently, the contrast between the needle
and the surrounding bacterial components
is not great enough to permit direct visualization of the extent to which a phage
needle has penetrated a bacterial cell.
Some cell walls are obviously interrupted at
their junctures with T4 needles. Micrographs

of serial sections (section thickness ∼ 250
A) through a single adsorbed T4 phage
(Fig. 8a,b,c) demonstrate that the outer
layers of the *E. coli* cell wall are broken and
penetrated only at their points of intersection with the phage's needle. Since phage
lysozyme does not seem to attack the
lipoprotein and lipopolysaccharide layers
of the *E. coli* wall (Salton, 1964), it is likely
that these soft layers are pierced mechanically by the bacteriophage's needle.

The depth to which T2 and T4 needles
penetrate bacterial walls has also been
estimated by comparing the lengths of the
needles of attached phages remaining
outside of the cells (extramural lengths)
with the average length of the needles of
unattached phages with contracted sheaths.
In all cases, the lengths of the needles were
measured from their most distal points to
their points of intersection with the baseplates; the magnifications of the electron
micrographs were calculated using the
dimensions given for T4 particles by
Kellenberger *et al.* (1965). The data pre-

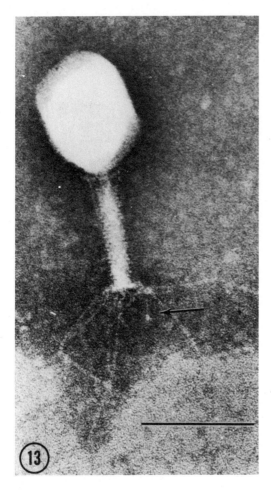

FIG. 13. Short tail fibers visible on a negatively stained T4 phage with an extended sheath. The phages in a T4 lysate were purified by differential centrifugation, after which they were resuspended in 1% ammonium acetate. Droplets of the suspension were then negatively stained and examined in the electron microscope. Six long tail fibers are evident on the phage in this micrograph; tail pins cannot be seen but a short fiber (indicated by the arrow) is visible. The short fiber is about 375 A long.

sented in Fig. 18 show that the distance from the tips of the needles to the baseplates for unadsorbed phages is 490 ± 30 A. Of the 25 attached phages examined, none had a needle with an extramural length greater than 380 A. The largest group of attached phages had needles with extramural lengths of 360–380 A; the other attached phages appeared to have needles whose extra-

mural lengths varied from 320 A to 359 A. Therefore, it seems that the length of an attached T2 or T4 phage's needle visible outside of the bacterium is at least 110–130 A shorter than the average length of an unattached phage's needle.

It was also observed that the distance from the proximal end of the sheath to the head is constant (about 120 A) whether or not the sheath is contracted and whether or not the bacteriophage is attached. This observation indicates that the proximal end of the sheath does not move along the needle during the adsorption process.

DISCUSSION

Brief Summary of the Events Involved in the Attachment of T2 and T4 Bacteriophages to E. coli Cells

The steps in the attachment of T2 and T4 bacteriophages to E. coli cells are diagrammatically illustrated in Fig. 19. Unadsorbed T2 and T4 phages (a) attach to the bacterial surface initially by the distal ends of the phages' long tail fibers (b). Subsequently, a repositioning of the bacteriophage brings the tail pins, which were over 1000 A from the bacterial surface, into close proximity to the cell wall (c). Finally, the sheath contracts and the baseplate is pulled along the needle away from the cell wall (d). At this stage the tail pins are no longer visible, but short tail fibers, probably under tension and about 370 A long, connect the baseplate directly to the cell wall. The needle, apparently under compression, penetrates the lipoprotein and lipopolysaccharide layers of the wall. It then penetrates through the rigid layer of the cell wall which may have been weakened by the phage's lytic enzyme, as will be discussed in a later paper. Throughout the process the long tail fibers remain attached to the wall at points corresponding to their initial sites of attachment, 700–900 A from the point where the needle has penetrated.

Attachment of Long Tail Fibers

Several authors (Williams and Fraser, 1956; Wildy and Anderson, 1964) have shown that isolated T-even tail fibers attach to E. coli cells. Therefore, the observation pre-

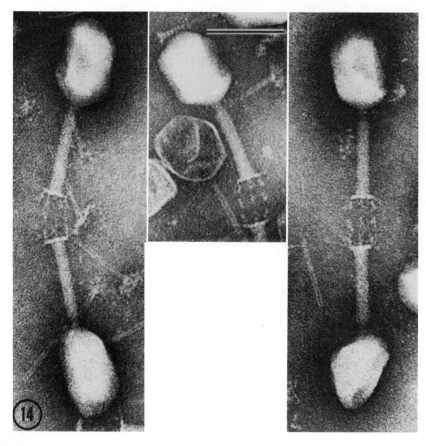

FIG. 14. T4 phages aggregated end to end. T4 phages were purified by 3 cycles of differential centrifugation after which they were resuspended in F medium without tryptophan at 4°. About 3 weeks later, samples from the suspension were negatively stained and inspected in the electron microscope. The phages in these micrographs are associated end to end in pairs. The two baseplates of each pair are about 600 A apart, and they appear to be linked by short fibers. Therefore, each fiber is probably at least 300 A long. These fibers might be extended tail pins. In addition, many phages were either partially or completely disrupted. Numerous free needles, isolated baseplates and other degradation products are seen here.

sented in this paper that the initial attachment of T2 and T4 phages to the bacterial surface is by the long tail fibers is not surprising. But the discovery that only the distal tip of each long tail fiber on intact phages binds to the cell wall is unexpected because purified fibers clump sensitive bacteria as though tail fibers formed bridges between them. From this observation, Wildy and Anderson concluded that isolated tail fibers have at least *two* binding sites. Further work with more highly purified tail fibers should clear up this anomaly.

The long tail fibers on phages bound to bacterial cell walls by their short tail fibers may remain attached to the bacterial surface (Fig. 12) or they may become detached (Figs. 9 and 16). This observation suggests that the long tail fibers bind reversibly to the bacterial wall during the adsorption process. The "surface reaction theory" of phage attachment described by Stent and Wollman (1952) is in agreement with this suggestion that the initial, long fiber attachment of phages to bacteria is a reversible step followed by the irreversible one of adsorption of the short tail fibers and contraction of the sheath (cf. Stent, 1963).

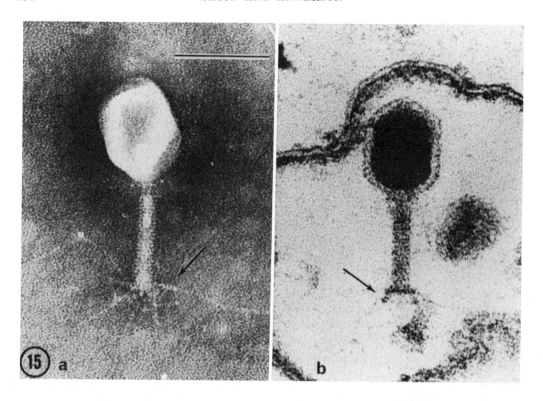

Fig. 15. T4 particles with tail pins which exhibit periodic substructures. (a) This negatively stained T4 phage is from a stock purified by density gradient centrifugation followed by dialysis overnight against 1% ammonium acetate. A tail pin (arrow) about 110 A long and about 60 A in diameter, extends laterally from an ill-defined baseplate. The pin appears to have a 3-repeat periodic substructure with a center-to-center spacing of about 40 A. (b) This T4 particle, viewed in thin sections, also has a tail pin (arrow) which seems to have a 3-repeat periodic substructure. Its dimensions are similar to those seen on the negatively stained phage.

Movement of the Bacteriophage into Close Proximity to the E. coli Surface

The attachment of the phage's long tail fibers to the bacterial surface positions the phage's baseplate about 1000 A from the cell wall. The virus subsequently comes to lie with its tail pins touching the cell wall and its baseplate about 100 A from it. The factors responsible for the phage's movement toward the bacterial surface have not been experimentally determined. However, it seems likely that Brownian motion could randomly agitate the attached virus until by chance it moves close enough to the cell wall for the next steps in the adsorption process to occur: the appearance of the short tail fibers and the contraction of the tail sheath.

Short Tail Fibers and Their Relationship to the Tail Pins

Several lines of evidence indicate that the short tail fibers are derived from the tail pins. (1) The tail pins appear to be brought into intimate contact with the cell wall before the sheath contracts. The pins, therefore, are in excellent positions to bind directly to the cell wall in the same spatial arrangement shown by the short tail fibers. (2) Occasionally, bacteriophages with extended sheaths are observed without visible tail pins. Frequently, these phages have 350–400 A fibers extending from their baseplates (Fig. 13). This observation suggests that the tail pins can exist in either a contracted (100 A long) or an extended (350–400 A long) state. (3) The tail pins

27

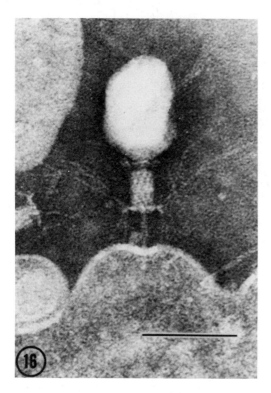

FIG. 16. Attachment of T4 to cyanide-poisoned *E. coli* B. *E. coli* B cells were suspended in phosphate buffer containing tryptophan and 0.01 *M* potassium cyanide. About 3 minutes later T4 phages were added (moi ∼200). The bacterial culture cleared, indicating that the cells had lysed from without. Ten minutes after addition of the phage, droplets of the suspension were negatively stained and examined in the electron microscope. The phage in this micrograph is attached to a small wall fragment by its short tail fibers; its long tail fibers extend laterally from the tips of the baseplate. The cyanide has caused no apparent change in the morphology of attachment.

have a diameter, *c*, equal to about 60 A. They seem to have a periodic structure with a repeat distance, *d*, of 40 A (Fig. 15). If each tail pin were actually a single thin fiber with a thickness $t = 20$ A, helically coiled to form a cylinder, the length of the fiber, *L*, could be estimated from the relation:

$$L = \sqrt{n\pi^2(c - t)^2 + d^2}$$

where $n = 3$ is the number of gyres made by the fiber. The length *L* of the helically

coiled fiber would then be 396 A which is in good agreement with the length of a short tail fiber, 370 A. (4) Pairs of phages with extended sheaths frequently have been observed to be associated with each other through fibers positioned similarly to the tail pins (Fig. 14). These fibers lie nearly perpendicular to the planes of the baseplates, extending in directions away from the phage heads. Since the two baseplates of each pair of phages are separated by 600 A, each short fiber must be at least 300 A long. (5) F. A. Eiserling (personal communication) while studying amber mutants lacking long tail fibers noted that these phages usually possessed tail pins. However, he occasionally found phages without tail pins; short tail fibers, 350–400 A in length, were visible on these phages. Since the phages did not possess long tail fibers, Eiserling concluded that the short tail fibers were derived from the tail pins. The preceding observations support the hypothesis that during the second stage of the adsorption process, when the phage's baseplate is brought into very close proximity to the bacterial surface, the tail pins attach to the cell wall; when the sheath contracts, pulling the baseplate away from the wall, the tail pins uncoil producing the short tail fibers which extend directly from the baseplate to the bacterial surface.

The results presented in this paper show that the adsorption of T2 and T4 phages to *E. coli* involves binding by two different phage components, an initial attachment by long tail fibers and a subsequent attachment by short tail fibers.

This dichotomy may explain the observations of Franklin (1961) and of Edgar and Lielausis (1965) that some of the neutralizing activity of anti-T4 serum is directed against phage components other than the long tail fibers. Some of the neutralizing activity of this antiserum may be directed against tail pins or short tail fibers, and further experiments may confirm this possibility.

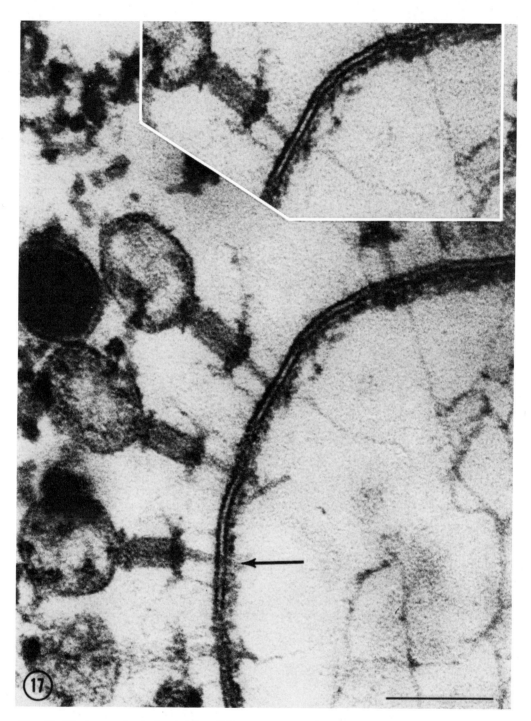

FIG. 17. T4 phages adsorbed on an *E. coli* B cell wall, as seen in a thin section. The phages are bound to the bacterial surface by short tail fibers extending directly from their baseplates to the cell wall. The baseplates have "boat-type" profiles; the central regions of the baseplates lie parallel to the cell wall; the tips of the baseplates extend away from the wall. The needle of one of the phages can be seen to penetrate just through the cell wall (arrow). Thin 30 A wide fibrils probably of phage DNA extend on the inner side of the cell wall from the distal tips of the T4 needles. The inset shows a second micrograph of the same field taken 0.3 μ underfocus. In both micrographs the phage's DNA appears to have a 40 A periodicity. This fact may be observed best by visually fusing the two micrographs as if they were a stereo pair.

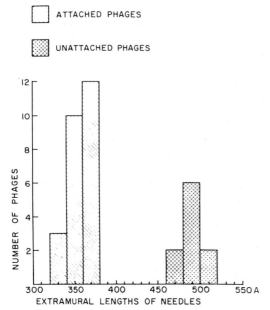

FIG. 18. Histograms of the lengths of needles of unadsorbed particles with contracted sheaths.

Penetration of the E. coli Cell Wall by the Needles of Infecting T2 and T4 Bacteriophages

Electron micrographs of thin sections (Figs. 8 and 17) show that the *E. coli* cell wall is penetrated about 120 A by the needles of infecting T2 and T4 bacteriophages. In agreement with this observation, the extramural lengths of the needles of attached phages are at least 110–130 A shorter than the average length of needles of unattached

Lengths of the needles of unattached phages were measured from the phages' baseplates to the distal tips of the needles; lengths of the needles of adsorbed phages were measured from the phages' baseplates to the cell walls on which they had adsorbed. The average distance from the baseplates to the tips of the needles of the unattached phages was 490 A. The longest distance from the baseplate to the cell wall measured on adsorbed phages was 380 A. This observation suggests that the needles of infecting phages extend at least 110 A into their bacterial host cells.

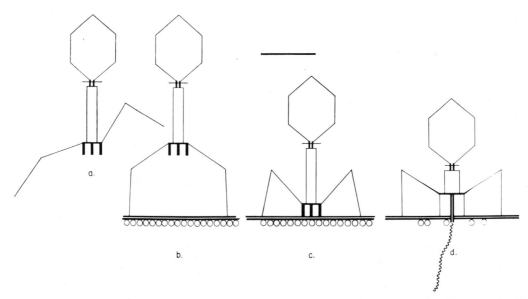

FIG. 19. Schematic illustration of the major steps in the attachment of T2 and T4 phages to the *E coli* cell wall. (a) An unattached phage is depicted. A few of its long tail fibers and tail pins are seen. (b) The long tail fibers have attached to the cell wall. Opposite long fibers attach to the wall about 1800 A apart. They are bent slightly near their centers. The baseplate of the phage is over 1000 A from the cell wall. (c) The phage has moved closer to the wall; its tail pins are in contact with the wall. Long tail fibers are shown in this stage since they have been observed in the preceding and in the following stages. However, direct evidence about the morphology of the long fibers during this phase of the attachment process has not yet been obtained. (d) The tail sheath has contracted. Short tail fibers and long tail fibers link the baseplate to the cell wall. The needle, through which the phage's DNA is ejected, has penetrated the wall.

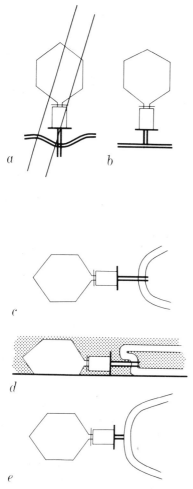

phages (Fig. 18). Extramural measurements *for adsorbed particles* are probably biased toward shorter values because the tips of the needles tend to be hidden by the wall in negatively stained preparations. Also, in thin sections (cf. Fig. 8a), part of the tip of the needle may not be included in the section. In both cases the extramural lengths of the needles of attached phages would appear shorter than they really are (Fig. 20). Among the measurements of the apparent extramural lengths the maximum value would be expected to correspond to the actual extramural lengths of the needles. The data on attached phages presented in Fig. 18 show a maximum limit of 360–380 A. It appears likely, therefore, that the actual distance from the *E. coli* cell wall to the baseplate on an adsorbed T2 or T4 bacteriophage is about 370 A.

Since the average distance from the baseplates to the distal tips of the needles on unattached phages is about 490 A, it seems that a length of about 120 A disappears from the distal tips of the needles during the process of bacteriophage adsorption to *E. coli* cell walls. Thus, the evidence from micrographs of individual and serial thin sections and from the data presented in Fig. 18 indicates that the needles of adsorbed phages penetrate about 120 A into the bacterial wall.

The thickness of the intact cell wall is about 120 A. Therefore, it appears that the needles of adsorbed T2 and T4 phages penetrate just through the *E. coli* cell wall and not through the 60–70 A thick protoplasmic membrane immediately surrounding the bacterial cell. It would thus seem that the phage DNA is injected between the cell wall and the cytoplasmic membrane. From there it might be expected that the phage DNA infects the cell in a manner analogous to that in which pure DNA extracted from λ phages (Meyer *et al.*, 1961) and from ΦX 174 phages (Guthrie and Sinsheimer, 1960) infects protoplasts.

FIG. 20. Schematic diagram showing the apparent shortening of the length of the needle on an adsorbed phage caused by artifacts in thin sectioning (a and b) and in negative staining (c, d, and e). (a) A phage is attached to an *E. coli* cell wall; its needle has penetrated the wall. The two diagonal lines indicate the limits of a thin section, seen in side view, through the adsorbed phage. The section does not contain the distal end of the phage's needle. If this section were examined in the electron microscope, an image such as that shown in part (b) would be seen. (b) The needle and the cell wall are superimposed so that the needle appears as though its distal tip were in contact with the wall; the apparent distance from the baseplate to the tip of the needle is less than the normal distance. (c) A phage is seen attached to an *E. coli* wall fragment. (d) This figure shows a side view of an electron microscope grid on which the phage has been placed. The phage is surrounded by electron-dense negative staining material. The cell wall has folded back on itself over the needle. (e) The diagram shows the electron micrograph which would be obtained from this preparation. The cell wall is superimposed over the needle, causing the distal end of the needle to be obscured. The needle, therefore, appears shorter than it actually is.

ACKNOWLEDGMENTS

The authors wish to thank Dr. W. Schreil for his helpful suggestions regarding fixation and embedding procedures.

REFERENCES

ADAMS, M. (1959). "Bacteriophages." Wiley (Interscience), New York.

ANDERSON, T. F. (1948). The activation of bacterial virus T4 by L-tryptophan. *J. Bacteriol.* **55**, 637–649.

ANDERSON, T. F. (1953). The morphology and osmotic properties of bacteriophage systems. *Cold Spring Harbor Symp. Quant. Biol.* **18**, 197–203.

ANDERSON, T. F. (1962). Negative staining and its use in the study of viruses and their serological reactions. *In* "Symposia of the International Society for Cell Biology, V. 1, The Interpretation of Ultrastructure." (R. J. C. Harris, ed.), pp. 251–262. Academic Press, New York.

BRENNER, S., STREISINGER, G., HORNE, R. W., CHAMPE, S. P., BARNETT, L., BENZER, S., and REES, M. W. (1959). Structural components of bacteriophage. *J. Mol. Biol.* **1**, 281–292.

EDGAR, R. S., and LIELAUSIS, I. (1965). Serological studies with mutants of phage T4 D defective in genes determining tail fiber structure. *Genetics* **52**, 1187–1200.

FRANKLIN, N. C. (1961). Serological study of tail structure and function in coliphages T2 and T4. *Virology* **14**, 417–429.

GUTHRIE, G. D., and SINSHEIMER, R. L. (1960). Infection of protoplasts of *Escherichia coli* by subviral particles of bacteriophage ΦX 174. *J. Mol. Biol.* **2**, 297–305.

HERSHEY, A. D., and CHASE, M. (1952). Independent functions of viral protein and nucleic acid in growth of bacteriophage. *J. Gen. Physiol.* **36**, 39–56.

KARNOFSKY, M. (1961). Simple methods for "staining with lead" at high pH in electron microscopy. *J. Biophys. Biochem. Cytol.* **11**, 729–732.

KELLENBERGER, E., BOLLE, A., BOY DE LA TOUR, E., EPSTEIN, R. H., FRANKLIN, N. C., JERNE, N. K., REALE-SCAFATI, A., SÉCHAUD, J., BENDET, I., GOLDSTEIN, D., and LAUFFER, M. A. (1965). Functions and properties related to the tail fibers of bacteriophage T4. *Virology* **26**, 419–440.

KELLENBERGER, E., RYTER, A. and SÉCHAUD, J. (1958). Electron microscope study of DNA-containing plasms. II. Vegetative and mature phage DNA as compared with normal bacterial nucleoids in different physiological states. *J. Biophys. Biochem. Cytol.* **4**, 671–678.

KELLENBERGER, E. and SÉCHAUD, J. (1957). Electron microscopical studies of phage multiplication. II. Production of phage-related structures during multiplication of phages T2 and T4. *Virology* **3**, 256–274.

MEYER, F., MACKAL, R. P., TAO, M., and EVANS, E. A., JR. (1961). Infectious deoxyribonucleic acid from λ bacteriophage. *J. Biol. Chem.* **236**, 1141–1143.

SALTON, M. R. J. (1964). "The Bacterial Cell Wall." Elsevier, Amsterdam.

STENT, G. S. (1963). "Molecular Biology of Bacterial Viruses." W. H. Freeman, San Francisco.

STENT, G. S., and WOLLMAN, E. L. (1952). On the two-step nature of bacteriophage adsorption. *Biochim. Biophys. Acta* **8**, 260–269.

STOUTHAMER, A. H., DAEMS, W. T., and EIGNER, J. (1963). Electron microscope studies of bacteriophage adsorption with negative and positive staining. *Virology* **20**, 246–250.

WILDY, P., and ANDERSON, T. F. (1964). Clumping of susceptible bacteria by bacteriophage tail fibers. *J. Gen. Microbiol.* **34**, 273–283.

WILLIAMS, R. C., and FRASER, D. (1956). Structural and functional differentiation in T2 bacteriophage. *Virology* **2**, 289–307.

II
Inside *Escherichia coli*

Editor's Comments on Papers 3 Through 6

On the scale of biological complexity, the simpler an organism, the greater is the proportion of its functions relegated directly to reproduction. This must be so because reproduction is the one activity that an organism cannot do without. For a complex organism, such as a fish, one can arrive at a fair understanding of the object by studying the anatomy and physiology of the adult individual with little regard to the mechanism of reproduction. A virus, on the other hand, is at the extreme end of the scale, for almost all its functions are involved in reproduction. We are used to thinking of the "mature" virus particle as the goal of the replication process, and it is useful to do so. But if we drop our anthropomorphic prejudices and accept that a virus life cycle has no goal other than to repeat itself, then it is clear that no one stage of the cycle has preeminence over another in its teleological claim. The concept of virus must thus include equally the replication process as well as the virus particle. A virus is not merely the morphologically defined object seen in the electron microscope but is the sum of its different states of existence. We shall now examine the state of existence of the virus which follows the interaction of the virus particle with the host cell. It will be seen that this state, called the vegetative virus, is a dynamic one whose nature changes from moment to moment.

Although, even before 1930, d'Herelle *(84)* and Burnet *(25)* had concluded that phage are released from infected cells in a burst that occurs a certain time after infection, there were reports to the contrary which suggested that phage production is a continuous process. Early studies on the dynamics of phage production were clouded by the fact that the phage particles added to a culture of bacteria do not all adsorb to the cells at the same time. Furthermore, the newly released progeny phage can readsorb to either uninfected cells or to infected cells that had not yet lysed. These factors often obscured the true dynamics of phage production. In 1939 Ellis and Delbrück *(55)* devised a way to overcome this difficulty by taking advantage of the fact that the fraction of phage adsorbed in a given time interval depends on the concentration of bacteria. If phage and bacteria are mixed and after a period of time, when say 90 percent of the phage have adsorbed, are diluted by a large factor, no additional phage will be adsorbed. The progeny phage that are released will likewise remain as free phage particles because the bacteria are now too dilute to adsorb a significant fraction over the period of the experiment. This was the basis for the one-step growth experiment, which conclusively proved that phage-infected cells do not release their progeny in a continuous manner but only after a well-defined time interval, called the latent period, which for their phage was about 30 minutes at 37°C.

The experiments of Ellis and Delbrück, however, said nothing about the existence of intracellular phage before lysis of the cell, since as Doermann states in the introduction of Paper 3: ". . . the cell wall of the infected bacterium presents a closed door to the investigator in the period between infection and lysis." Doermann, as he describes in this paper, devised a way to open this door and thus to examine the cell contents for the presence of infectious phage particles. These experiments showed no infectious phage particles present during the first half of the latent period; not even the infecting phage particle could be recovered. During the latter half of the latent period intracellular phage appeared rapidly until, at about 10 minutes after infection, there were on the average one new phage particle per infected cell, a time that marks the end of what later came to be known as the eclipse period. Doermann concluded that his results were ". . . in agreement with the gradually emerging concept that a profound alteration of the infecting phage particle takes place before reproduction ensues."

Within 6 months of the appearance of Doermann's paper, the nature of the "profound alteration" became clear. Hershey and Chase *(91)* demonstrated that, upon adsorption to the cell surface, the phage dissociates into its two principal parts. The DNA moves to the interior of the cell, while the outer protein coat remains fixed to the cell surface, from which it could be removed without affecting the replication process. Thus the eclipse period discovered by Doermann was explained: The integrity of the phage particle is destroyed shortly after attachment of the phage to the host cell. More importantly, this experiment showed that, at least during the initial period after infection, the part of the infecting phage particle which provides genetic continuity to its progeny is its DNA.

The significance of this conclusion to genetics was immediately apparent and spurred attempts to determine the structure of DNA. During the following year Watson and Crick *(197, 198)* deduced the double-helical structure of the molecule from the X-ray diffraction data of Wilkins and pointed out its biological implications. The idea of a genetic role for DNA did not, of course, originate with Hershey and Chase. It had been proclaimed by the DNA transformationists for years but had somehow met with resistance, probably because as Hershey says *(26)*: ". . . some redundancy of evidence was needed to be convincing and that diversity of experimental materials was often crucial to discovery."

Thus by 1953 the following picture of phage replication had emerged: The parental phage DNA somehow directs the synthesis of more phage DNA (autocatalytic synthesis) and, in addition, directs the synthesis of the phage proteins (heterocatalytic synthesis). The intricacy of the biochemical processes leading from the newly injected phage DNA to the finished phage particle was not even imagined at the time, but at least the problem had been stated in molecular terms.

The next several years were marked by a surge of biochemical studies aimed at elucidating the intracellular events in phage replication. Cohen *(36)* had already measured the course of macromolecular synthesis in T2-infected cells and found that although DNA and RNA synthesis ceases immediately after infection, protein synthesis, curiously, proceeds without interruption at the preinfection rate (see Fig. 1 of

Paper 4). The proteins synthesized immediately after infection were not likely to be the products of host genes since it was known that T2 infection quickly shuts off the synthesis of host enzyme activities. Nor were these proteins the ones that would eventually form the outer coat of the progeny phage particles, since it had been found that phage structural proteins do not begin to appear until some 10 minutes after infection. Thus, it was reasoned, these early proteins must be some phage-specific enzymes required for the early steps of replication. But what functions did they perform? A clue was provided by the finding that suppression of protein synthesis by an antibiotic immediately after phage infection prevents subsequent synthesis of phage DNA, which normally commences at about 8 minutes after infection. This finding, in turn, raised a further question: Why is the host's metabolic machinery insufficient for the synthesis of phage DNA?

Part of the answer came with the discovery by Wyatt and Cohen *(209)* that the T-even phage possess a unique kind of DNA in which the usual pyrimidine–cytosine is totally replaced by hydroxymethyl cytosine (HMC). Since there was no evidence that uninfected *E. coli* ever made HMC, it was concluded that the enzyme (or enzymes) necessary for its biosynthesis is induced by phage infection. Flaks and Cohen *(61)* proceeded to demonstrate an enzyme activity, dCMP-hydroxymethylase, which converts the nucleotide deoxycytidylate (dCMP) to deoxyhydroxymethyl cytidylate (dHMCMP) in extracts of T2-infected cells. This was followed by the discovery by Kornberg and his collaborators *(117)* of a whole battery of new phage-induced enzyme activities which were involved in the synthesis of the components of T-even phage DNA. Among them was an enzyme (dCTPase) which degrades dCTP to dCMP. Since the nucleoside triphosphate dCTP is a precursor of *coli* DNA synthesis and the nucleoside monophosphate dCMP is the substrate for the hydroxymethylase, the lack of cytosine in phage DNA was neatly explained: The phage-induced enzyme dCTPase depletes the supply of the normal host nucleotide and the phage uses the product to synthesize its own unique nucleotide. Paper 4 by Cohen summarizes these studies, an overview of which is given in Fig. 7 of this paper. For a more detailed and up-to-date account of this subject, Cohen's monograph *(37)* is an excellent source.

There were two possible ways to interpret the acquisition of new metabolic functions by phage-infected cells. Either the new enzymes are synthesized *de novo* after infection, or they preexist in a latent inactive state in the cell and are activated by infection. To decide between these alternatives, Mathews, Brown, and Cohen (Paper 5) purified dCMP hydroxymethylase from phage-infected cells that had been labeled with radioactive amino acids *before* infection. Their purified enzyme was found to contain essentially no radioactivity, proving that the enzyme was synthesized from its constituent amino acids after infection.

In general, this kind of experiment still does not preclude the possibility that the host cell has the genetic potential for making the phage-induced protein, but that the potential remains completely unexpressed until after phage infection. In the case of the T-even phage, which abolish host-protein synthesis, this seemed unlikely. A rigorous proof, however, requires the identification of a phage gene that codes for the protein.

In a number of instances, including the hydroxymethylase, this has been shown to be the case *(135)*.

As we have seen, a bacterial cell infected with a T-even phage can be characterized as one in which the introduction of a new set of genes results in the sudden and exclusive expression of these new genes. It was soon apparent that this situation provided a unique experimental system to answer a fundamental question of molecular biology: Does the injected phage DNA (or any DNA) act directly as a template on which proteins are synthesized, or is the information carried by the DNA first transferred to an intermediate that acts as the template?

By 1958 several lines of reasoning had already suggested that RNA rather than DNA is the direct template *(195)*. This notion, together with the identification of an RNA-containing particle — the ribosome — as the organelle of protein synthesis *(214)*, led naturally to the idea that the ribosomal RNA, which comprises the bulk of the cellular RNA, is the template for protein synthesis. This view, however, was open to several objections. First, very little net RNA synthesis could be detected in phage-infected cells, certainly not enough to be anywhere near equivalent to the amount of ribosomal RNA present before infection. Second, studies on the kinetics of induced enzyme synthesis in uninfected cells by Jacob and Monod *(97)* suggested that the informational intermediate is unstable, in contrast to ribosomal RNA, which had been found to be highly stable *(40)*. In perspective the paradox was that, on the one hand, there was evidence for the involvement of "RNA" as a template for protein synthesis, but, on the other hand, the only kind of RNA known (aside from the relatively low-molecular-weight soluble RNA) was ribosomal RNA, for which the implicatory evidence was negative.

The resolution of the problem came from a more careful examination of the small net amount of RNA synthesized after phage infection. If, indeed, the template RNA was unstable, a low level of accumulation would not be inconsistent with a high rate of synthesis. Volkin and Astrachan *(191)* examined this RNA and found it to be unlike the bulk of host RNA, but rather had a base ratio more like that of T2 DNA. Nomura, Hall, and Spiegelman *(141)* showed further that this RNA had a sedimentation coefficient and electrophoretic mobility different from *coli* RNA. The decisive experiment was performed by Brenner, Jacob, and Meselson *(21)*, who demonstrated in 1961 that this unstable phage-specific RNA becomes associated with preexistent ribosomes.

The quest for messenger RNA would have been so much easier if Nature had not chosen to use the same kind of macromolecule both as a structural component of the protein-synthesizing machinery and as the template for protein synthesis. Undoubtedly there is a method to this madness, but as yet no one has provided any convincing insight as to why a ribosome could not do what we think it does without RNA.

It has been mentioned previously that the phage replicative process can be divided into two distinct periods according to the kind of protein synthesized. The proteins that appear during the early period function predominantly to replicate the phage DNA. The commencement of DNA replication ushers in the late period, during which the structural proteins of the phage particle are made. Furthermore, the synthesis of early proteins ceases at about the time that mature phage particles first appear. Thus the synthesis of the two classes of protein is regulated in a way that makes these proteins available specifically when they are needed. In addition to this temporal regulation, the synthesis of the various proteins is also regulated quantitatively; some proteins are synthesized in several hundredfold greater quantity than others.

Since there are two steps of information transfer in gene expression — DNA to RNA (transcription) and RNA to protein (translation) — regulation could conceivably be effected by stimulation or inhibition of either or both of these steps for particular genes or groups of genes. Presently, by far the bulk of evidence supports a transcriptional mechanism as the major means by which the regulation of phage gene expression is accomplished, although some degree of translational control cannot be ruled out.

Just how this control is exerted on the T-even phage transcriptional process is far from clear. One fact which stands out is that the transcription of late genes requires continuous synthesis of the phage DNA. If DNA synthesis is blocked even after some phage DNA has accumulated, the subsequent transcription of late genes is blocked. This finding led Riva, Cascino, and Geiduschek *(152)* to propose a model whereby newly replicated DNA, unlike parental DNA, is competent for late transcription but is continuously modified to an incompetent state. Continuing replication replenishes the competent DNA, but if DNA replication is arrested the competent DNA is depleted. Their studies further suggested *(153)* that the relevant physical differences between competent and incompetent DNA is that the former has single-stranded interruptions whereas the latter does not. Thus the modification that regenerates the incompetent state could be the sealing of the interruptions by a DNA ligase. The interruptions in the newly synthesized competent DNA could be due either to the action of an endonuclease or as a result of replication itself.

Although this model satisfactorily explains the coupling of late gene expression to DNA replication, it is but one aspect of the regulatory strategy. Another way in which the transcription process might be controlled is by phage-controlled modifications of the RNA polymerase. The RNA polymerase of *E. coli* is a complex enzyme that consists of five polypeptide subunits. One of these subunits—the σ subunit—enables the enzyme to recognize sites (promotors) on the DNA for the initiation of transcription. Bautz and Dunn *(8)* found that the *coli* polymerase loses this σ subunit after T4 infection. Provocative but still indecisive evidence suggests that the phage may replace the host σ subunit by one or more σ-like proteins of its own making *(173, 187)*.

After infection by phage T7, a totally new and much simpler RNA polymerase is made which is the product of one of the phage genes *(35, 177)*. The phage polymerase gene, and a few others, are transcribed by the *coli* polymerase. The emerging picture is that different phages have evolved different mechanisms of regulation. This subject has been recently reviewed by Calendar *(28)*.

Finally, we turn to the question of the structure of the replicating phage DNA. The early blackboard model of DNA replication imagined the separation, either locally or totally, of the two strands of the DNA duplex followed by or concurrent with synthesis of the two complementary strands, resulting in two duplex molecules, each identical to the original. As far as it goes, this picture is accurate enough for the first step of phage DNA replication but, as it turns out, is only one phase of a more complex series of events. In 1963 Frankel *(63)* discovered that replicating T4 DNA sediments much more rapidly than DNA isolated from phage particles, implying that the replicating DNA is of higher molecular weight than mature phage DNA. In a later paper reproduced here (Paper 6), Frankel further documents this finding and suggests two possible mechanisms by which this apparently multichromosomal structure might be generated (Fig. 11). More recently, Bernstein and Bernstein *(15)* have shown by direct autoradiographic visualization that replicating T4 DNA exists as large circular and branched concatenates as much as 21 times the length of mature phage DNA. Obviously, such structures must be reduced to phage-size pieces when the DNA is packaged into individual phage particles. We shall return to this point and its profound implications in Part III.

The processes thus far described lead, in a regulated way, to the accumulation in the infected cell of copies of the phage DNA and of the specific proteins that will eventually envelope the DNA. The question as to how the parts of the phage particle are assembled will be dealt with later, but first we shall digress to consider the genetics of phage—what phage has taught us about genetics and what genetics has taught us about phage.

3

Reprinted from *J. Gen. Physiol.*, **35**(4), 645–656 (1952)

THE INTRACELLULAR GROWTH OF BACTERIOPHAGES

I. Liberation of Intracellular Bacteriophage T4 by Premature Lysis with Another Phage or with Cyanide

By A. H. DOERMANN[*][‡]

(From the Department of Genetics, Carnegie Institution of Washington, Cold Spring Harbor)

(Received for publication, August 20, 1951)

Direct studies of bacteriophage reproduction have been handicapped by the fact that the cell wall of the infected bacterium presents a closed door to the investigator in the period between infection and lysis. As a result it was impossible to demonstrate the presence of intracellular phage particles during this so called latent period, and, much less, to estimate their number or to describe them genetically. This barrier has now been penetrated. It is the purpose of the first two papers of this series to describe two methods for disrupting infected bacteria in such a way that the intracellular phage particles can be counted and their genetic constitution analyzed.

The first method used to liberate intracellular bacteriophage depends on the induction of premature lysis in infected bacteria by "lysis from without" which occurs when a large excess of phage particles is adsorbed on bacteria (1). It was found by nephelometric tests that T6 lysates are efficient in disrupting cells when moderately high multiplicities are used (2). The further observation was made that the addition of a large number of T6 particles to bacteria previously infected with T4, would, under some conditions, cause liberation of T4 particles before the expiration of the normal latent period of these cells. It therefore seemed hopeful that a method of reproducibly disrupting infected bacteria could be developed on the basis of this preliminary knowledge.

* The experiments described here were carried out while the author was a fellow of the Carnegie Institution of Washington. The author is indebted to Dr. M. Demerec and the staff of the Department of Genetics of the Carnegie Institution of Washington for providing facilities for this work. In particular the stimulating discussions with Dr. Barbara McClintock are gratefully acknowledged. The manuscript was prepared while the author held a fellowship in the Department of Biology of the California Institute of Technology. He is grateful to Dr. G. W. Beadle and the staff of that department for their interest in this work, and especially to Dr. Max Delbrück for criticism of the manuscript.

‡ Present address: Biology Division, Oak Ridge National Laboratory, Oak Ridge, Tennessee.

645

40

The first experiments in devising a method of this kind were made with phage T5 (3). It was found that T5 is liberated before the end of the latent period if the infected cells are exposed to a high excess of T6. However, the extremely low rate of adsorption of T5 coupled with difficulties in inactivation of unadsorbed phage by specific antisera indicated that this phage was a poor choice. Hence T4 was chosen because of its fast rate of adsorption and because of the availability of high titer antisera against it. The first experiments with T4, along with the T5 results, showed conclusively that, by itself, lysis from without is not sufficiently rapid for the purpose of this investigation. It is likely that phage growth continues after the addition of the lysing agent T6. Therefore the attempt was made to stop phage growth while T6 was allowed to accomplish lysis from without. Low temperature could not be used for this purpose

TABLE I

Composition of the Growth Medium

Material	Amount
	gm. per liter
KH_2PO_4	1.5
Na_2HPO_4(anhydrous)	3.0
NH_4Cl	1.0
$MgSO_4 \cdot 7HOH$	0.2
Glycerol	10.0
Acid-hydrolyzed casein	5.0
dl-Tryptophan	0.01
Gelatin*	0.02
Tween-80	0.2

* To reduce surface inactivation of free phage particles (4).

since it also inhibits lysis from without. A search for a suitable metabolic inhibitor was therefore undertaken, and cyanide was eventually chosen as the most suitable one.

Materials and Methods

The experiments described here were carried out with the system $T4r_{48}$[1] growing at 37°C. in *Escherichia coli*, strain B/r/1. The latter is a T1 resistant, tryptophan-dependent mutant of B/r obtained from Dr. E. M. Witkin.

Two media were used in these experiments, namely the growth medium and the lysing medium. The composition of the growth medium used for both bacterial and phage cultures, is given in Table I. The lysing medium consists of growth medium with

---

[1] The subscript refers to a particular *r* mutant of T4 which arose by mutation and was numbered after the system of Hershey and Rotman (12) using the high subscript number to avoid confusion with the mutants already described by Hershey and Rotman.

the addition of one part in ten of a high titer T6 phage filtrate (concentration of T6 in lysing medium was *ca.* 4×10^9 particles per ml.) and cyanide brought to a final concentration of 0.01 M. Specially designed experiments showed that at this concentration the cyanide does not inactivate free phage particles, nor does the amount which reaches the plate affect titration by interference with plaque development.

T6 was used as the lysing phage because in several experiments it proved to be a more effective lysing agent than any of the other T phages tested. Since only single stocks of the phages were compared in the early experiments, the superiority of T6 over the other phages may have been due to a difference in the particular stock used, and not to an inherent difference among the phages. In fact, later experiments with different T6 stocks showed marked differences in lysing efficiency, and phage titer proved to be a poor criterion of lysing ability. The experiments described here were made with T6 stocks selected for their ability to induce lysis from without. The selections were made on the basis of nephelometric comparisons.

Platings were made in agar layer (0.7 per cent agar) poured over nutrient agar plates (1.3 per cent agar), and in order to assay T4 in the presence of high titer T6, the indicator strain, B/6, was used. B/6 is completely resistant to T6 (no host range mutants have so far been found which will lyse the strain used here) and gives full efficiency of plating (compared to B) with T4.

EXPERIMENTAL

Experiments with the Standard Lysing Medium.—The experimental procedure used consisted essentially of a one-step growth experiment (5) with certain modifications. B/r/1 cells in the exponential growth phase were concentrated by centrifugation to about 10^9 cells per ml. To these concentrated bacteria $T4r_{48}$ was added and this adsorption mixture was incubated for 1 to 2 minutes with aeration, allowing at least 80 per cent of the phage to be adsorbed to the bacteria. Then a 40-fold or larger dilution was made into growth medium containing anti-T4 rabbit serum. The serum inactivated most of the residual unadsorbed phage. After several minutes' incubation in the serum tube, a further dilution was made to reduce the serum concentration to one of relative inactivity. The resulting culture will be referred to as the *source culture* (SC). The entire experiment was carried out with the infected bacteria from SC. The titer of infected B/r/1 in this tube was approximately 10^5 cells per ml.

Simultaneously with the dilution into the tube containing serum, another dilution from the adsorption tube was made. From the latter an estimation of the unadsorbed phage was made by assaying the supernatant after sedimentation of the cells. This step permits calculation of the multiplicity of infection (5).

From SC a further dilution of 1:20 was made at some time before the end of the latent period. The resulting culture, containing approximately 5×10^3 infected bacteria per ml., was used for determining the normal end of the latent period and for estimating the average yield of phage per infected cell. It will be called the *control growth tube* (GT). In addition, a number of precisely timed 20-fold dilutions were made from SC into lysing medium. These were titrated after they had been incubated in the lysing medium for 30 minutes or longer. Serial platings from the lysing medium cultures over a longer period of time have shown that the phage titer remained constant after 30 minutes' incubation. The titer calculated from these platings, divided by the titer of infected bacteria given by the preburst control platings, gives the average yield per infected cell. As a working hypothesis, this yield was considered to be the average number of intracellular phage particles per bacterium at the time of dilution into the lysing medium. Dividing these numbers by the control burst size gives the fraction of the control yield found in the experimental lysing medium tubes.

The results of several typical experiments are shown in Fig. 1 in which the data are plotted on semilogarithmic coordinates. The fraction of the control yield found in a given experimental culture is plotted against the time at which the dilution is made into lysing medium. Curve 1 shows the results from a single experiment in which the bacteria were infected with an average of 7 phage particles each. Curve 2 is the composite result of four experiments in which the bacteria were infected with single phage particles. Curve 3 is the control one-step growth curve derived from the control growth tube platings in the four experiments of curve 2.

Several striking results can be seen in these experiments. First, it is clearly seen that during the early stages of the latent period the virus-host complex is inactivated by the cyanide-T6 mixture, and that not even the infecting particles are recovered. Even when 7 phage particles were adsorbed on each bacterium, less than two are recovered per cell at the earliest stage tested, and the shape of the curve suggests that if earlier stages had been tested, still fewer would have been recovered. In experiments with singly infected bacteria, the earliest tests indicated that less than one infected bacterium in 80 liberated any phage at all. A second point to be noted is that the multiplicity of infection appears to influence slightly the time at which phage particles can be recovered from the cell, and it continues to affect the fraction found in the bacteria at a given time. That this difference is a real one seems clear from the consistency among the points of curve 2. This result has been observed in each experiment, although the effect appeared to be less pronounced in some experiments made at the lower temperature of 30°C. Attention should also be drawn to the fact that the shape of the curves is clearly not exponential. In fact, it parallels with a delay of several minutes the approximately linear DNA increase observed in this system (6).

In connection with the preceding experiments a test was made to establish whether the cyanide concentration chosen was maximally effective in inhibiting phage synthesis. Using the described technique, but changing the cyanide concentration of the lysing medium from 0.01 M to 0.004 M and 0.001 M in parallel aliquots, no difference was detectable in the three lysing media. Thus 0.01 M cyanide is well beyond the minimum concentration necessary and was considered to be adequate for these experiments.

Action of Cyanide in the Absence of the Lysing Agent T6.—Cohen and Anderson (7) reported a loss of infectious centers when infected bacteria were incubated in the presence of the antimetabolite 5-methyltryptophan. Although the details

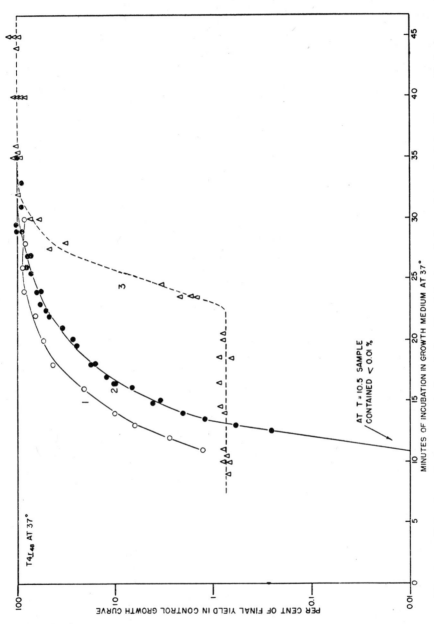

Fig. 1. The intracellular bacteriophage population during the latent period, as determined by the cyanide-lysis procedure. Curve 1 represents a single experiment in which the bacteria were infected with 7 phage particles per cell. Curve 2 is derived from four single infection experiments. Curve 3 is the control one-step growth curve from the single infection experiments.

of their experiments differed somewhat from those presented here, the loss of infectious centers in their experiments suggested testing whether cyanide could cause a similar loss of infected bacteria in the present procedure. An experiment was made which was identical with the standard cyanide lysis experiment

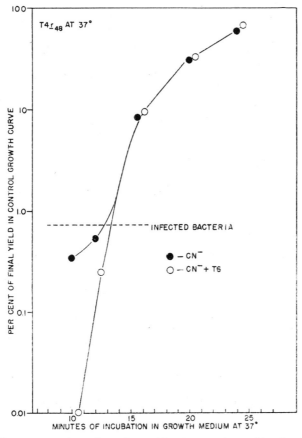

FIG. 2. The comparative effect of cyanide alone and cyanide plus T6 on singly infected bacteria at various stages in the latent period.

except that T6 was omitted from one set and included in a parallel set of lysing medium cultures (Fig. 2). As in the case of 5-methyltryptophan, it is seen that cyanide alone caused a loss of infectious centers when added in the early stages of phage growth, although the loss is less than that produced by cyanide and T6 together. Furthermore, in the second half of the latent period, comparison of the two media showed clearly and surprisingly that a definite *rise* in titer of infective centers occurred even when the lysing agent, T6, was omitted from the lysing medium. In fact, during the second half of the latent period, phage liberation is identical in the two media.

In order to see whether lysis is actually occurring and can account for the liberation of phage, a nephelometric experiment was made introducing CN⁻ at two points in the latent period. Three cultures of B/r/1 growing exponentially in growth medium were infected with T4r_{48} (*ca.* fivefold multiplicity). One culture served as a control for normal lysis. To the second culture cyanide (0.01 M final concentration) was added 7.5 minutes after addition of the virus and to

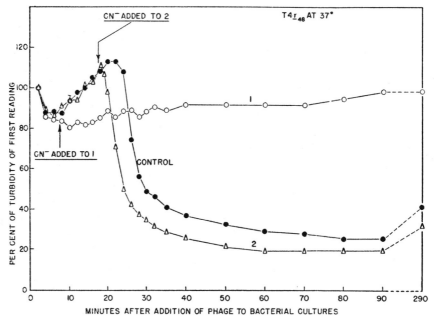

FIG. 3. The turbidity of T4-infected bacterial cultures as affected by addition of cyanide at two stages in the latent period.

the third tube 17.5 minutes after addition of the T4r_{48}. The turbidities of these three cultures were followed with a nephelometer designed like that described previously by Underwood and Doermann (8), but with four separate units which permit independent readings on the four tubes without removing any of them from the instrument. The results indicate that CN⁻ added to infected bacteria early during the latent period does not induce lysis (Fig. 3). From the plaque count experiment (Fig. 2) it is seen that a loss of infective centers does occur. This loss must therefore be due to some cause other than lysis of these cells. In the later stages of the latent period, the turbidimetric experiment indicates that lysis occurs promptly upon the addition of CN⁻ to the culture

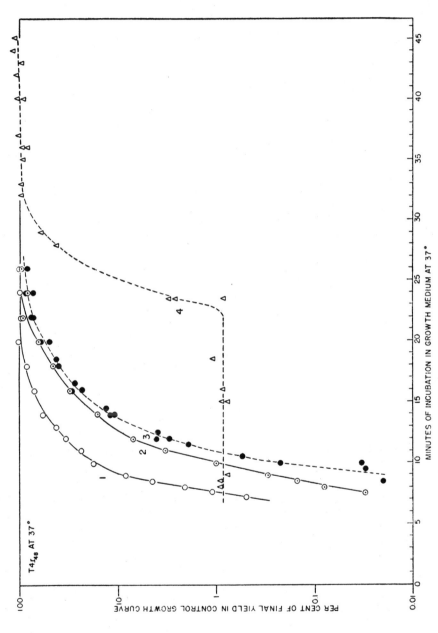

Fig. 4. The intracellular bacteriophage population during the latent period, as determined by premature lysis in the presence of 5-methyltryptophan. Curve 1 represents a multiple (fivefold) infection and curve 2 a single infection experiment with a different T4r mutant than that used in the other experiments described here. Curve 3 is a composite of three single infection experiments made with the same phage as was used in the cyanide experiments. Curve 4 is the standard one-step growth curve from the experiments which gave the data for curve 3.

(Fig. 3). The increase of infective centers in comparable cultures (Fig. 2) at these later stages is probably brought about by liberation of phage particles concurrent with this lysis.

Experiments Using 5-Methyltryptophan as the Metabolic Inhibitor.—In trying to find a suitable metabolic inhibitor for instantaneously stopping phage growth, a large number of experiments was done using the antimetabolite 5-methyltryptophan (5MT)[2] whose bacteriostatic action is blocked by tryptophan (9). The technique used was similar to the cyanide lysis procedure except that tryptophan was omitted from the lysing medium and 5MT was used in place of cyanide. The results (Fig. 4) are quite similar to the cyanide results in all respects except one. They are similar in failure to recover any phage particles during the early stages of the latent period, in the difference between single and multiple infection, and in the shapes of the curves. They are different, however, in that both the single and the multiple infection curves are moved to the left along the time scale by 3 to 4 minutes. This indicates that more phage is liberated per cell if lysis is induced in the presence of 5MT than if it is brought about in the presence of CN^-. This difference may be interpreted on the basis of two alternative hypotheses.

First, it might be suspected that CN^- penetrates the cell and reaches its site of inhibition more quickly than 5MT. This would allow more phage reproduction to go on between the time of exposure to the 5MT and the time at which the cell breaks open. In this event, a higher concentration of 5MT would enable penetration of an inhibitory amount in a shorter period of time, thus reducing the amount of phage found. To test this, the concentration of 5MT in the lysing medium was increased fivefold. No difference in the amount of phage liberated was found, suggesting that the rate of penetration of the poison is not limiting its effectiveness.

A second hypothesis is that the reaction blocked by 5MT may be one of the earlier ones involved in the synthesis of phage constituents. At the time of addition of 5MT many individual phages may already have acquired these constituents and thus be able to go on to maturity before lysis disperses the enzyme equipment of the infected cell. Cyanide, on the other hand, may block one of the terminal reactions in phage production, with the result that at a given time fewer individuals will have passed this reaction than will have passed the 5MT-inhibitable step. Consequently, fewer particles will be liberated when using cyanide than when 5MT is used.

DISCUSSION

Earlier experiments (2) and tests made of the lysing efficiency of the T6 stocks used here indicate that rapid lysis occurs when T6 stocks are added in

[2] Obtained through the courtesy of Dr. M. L. Tainter, Sterling-Winthrop Research Institute, Rensselaer, New York.

sufficient concentration to bacterial cultures. The very first experiments with bacteria infected with T5 (3) left no doubt that lysis from without by T6 will liberate T5 particles prematurely from infected bacteria. From the evidence contained in the present paper it cannot be definitely established whether the combined action of T6 and cyanide liberates all of the mature phage present in the cells. However, the fact that, during the terminal stages of intracellular development, the cyanide-lysis method yields as much phage as does spontaneous lysis, suggests that the cyanide method liberates all of the mature phage. Furthermore, during the second half of the latent period, exactly the same amount of phage is liberated by cyanide alone as by cyanide plus T6. This suggests that cyanide acts promptly in arresting phage growth. Otherwise one would expect to find a consistently higher number of phage particles in the cyanide medium than in the medium in which cyanide and T6 are combined. The experiments presented here therefore warrant the working hypothesis that mature intracellular phage is effectively liberated by the treatment described, and that the method gives a true picture of the intracellular phage population. The validity of this working hypothesis will be conclusively demonstrated for the phage T3 in the second paper of this series (10).

The bearing of the present experiments on our concept of phage reproduction might be discussed here. The finding that the original infecting particles are not recoverable from the cells during the first stages of the latent period appears at first sight surprising. Nevertheless some indirect evidence indicates that this is to be expected. The discovery that yields from mixedly infected bacteria may contain new combinations of the genetic material of the infecting types (11–13) suggests that some alteration of the infecting particles may occur. Furthermore, in mixed infections of bacteria with unrelated phages only one type is reproduced. The other type, although adsorbed on the cells, it not only prevented from multiplying but the infecting particle of that type is lost (5, 14). On the basis of multiple infection experiments with ultraviolet-inactivated phage particles, Luria (15) has proposed that reproduction of phage occurs by reproduction of subunits which are at some later stage assembled into complete virus particles. The failure to find infective phage particles within the infected cell in the early stages of reproduction agrees with what would have been predicted from these experiments.

The results of our experiments agree quite well with the scheme which Latarjet (16) suggested on the basis of x-ray inactivation studies of phage inside infected bacteria. Latarjet differentiated three segments of the latent period of phage growth. Using T2 he found that during the first segment of 6 to 7 minutes' duration, singly infected bacteria show the same inactivation characteristics as do unadsorbed phage particles. In the second period, from time 7 to time 13 minutes, the phage in infected cells became more resistant to x-rays, even during the first 2 minutes of this segment in which the inactivation curves still retain a single hit character. During the last 4 minutes of this period the curves take on a multiple hit character. In the final segment, from time 13 minutes to the end of the latent period, the curves retain the multiple hit character, but gradually regain the original x-ray sensitivity characteristic of free phage. These x-ray experiments suggest again that a rather drastic alteration occurs to the infecting particle, and that particles with the original characteristics are not found in the cell until the second half of the latent period. This is precisely what is observed in the results presented here. Our experiments were done with T4, but comparison seems legitimate since the two viruses are quite closely related (17).

Results of a similar nature to those discussed here were published by Foster (18). In studying the effect of proflavine on the growth of phage T2, Foster found that the time at which this poison was added influenced the amount of phage liberated by the bacteria. No phage is liberated from T2-infected cells (latent period 21 minutes) if proflavine is added during the first 12 minutes after infection even though lysis of the cells does occur at the normal time. When proflavine is added at later points in the latent period, lysis yields phage particles, the number depending on the time of proflavine addition. When the results of these single infection experiments are compared to the cyanide single infection experiments (Fig. 1) the results are seen to be quite similar. From other experiments Foster concluded that proflavine inhibits one of the final stages in the formation of fully infective phage. These facts, taken together, suggests that proflavine experiments were, in fact, measuring intracellular phage.

SUMMARY

A method is described for liberating and estimating intracellular bacteriophage at any stage during the latent period by arresting phage growth and inducing premature lysis of the infected cells. This is brought about by placing the infected bacteria into the growth medium supplemented with 0.01 M cyanide and with a high titer T6 lysate. It was found in some of the later experiments that the T6 lysate is essential only during the first half of the latent period. Cyanide alone will induce lysis during the latter part of the latent period.

Using this method on T4-infected bacteria it is found that during the first half of the latent period no phage particles, not even those originally infecting the bacteria, are recovered. This result is in agreement with the gradually emerging concept that a profound alteration of the infecting phage particle takes place before reproduction ensues. During the second half of the latent period mature phage is found to accumulate within the bacteria at a rate which is parallel to the approximately linear increase of intracellular DNA in this system. However, the phage production lags several minutes behind DNA production.

When 5-methyltryptophan replaced cyanide as the metabolic inhibitor, similar results were obtained. The curves were, however, displaced several minutes to the left on the time axis.

The results are compared with Latarjet's (16) data on x-radiation of infected bacteria and with Foster's data (18) concerning the effect of proflavine on infected bacteria. Essential agreement with both is apparent.

BIBLIOGRAPHY

1. Delbrück, M., *J. Gen. Physiol.*, 1940, **23,** 643.
2. Doermann, A. H., *J. Bact.*, 1948, **55,** 257.
3. Doermann, A. H., *Ann. Rep. Biol. Lab.*, *Long Island Biol. Assn.*, 1946, 22.
4. Adams, M. H., *J. Gen. Physiol.*, 1948, **31,** 417.
5. Delbrück, M., and Luria, S. E., *Arch. Biochem.*, 1942, **1,** 111.
6. Cohen, S. S., *Bact. Rev.*, 1949, **13,** 1.
7. Cohen, S. S., and Anderson, T. F., *J. Exp. Med.*, 1946, **84,** 525.
8. Underwood, N., and Doermann, A. H., *Rev. Scient. Instr.*, 1947, **18,** 665.
9. Anderson, T. F., *Science*, 1945, **101,** 565.
10. Anderson, T. F., and Doermann, A. H., *J. Gen. Physiol.*, 1952, **35,** 657.
11. Delbrück, M., and Bailey, W. T., Jr., *Cold Spring Harbor Symp. Quant. Biol.*, 1946, **11,** 33.
12. Hershey, A. D., and Rotman, R., *Proc. Nat. Acad. Sc.*, 1948, **34,** 89.
13. Hershey, A. D., and Rotman, R., *Genetics*, 1949, **34,** 44.
14. Delbrück, M., *J. Bact.*, 1945, **50,** 151.
15. Luria, S. E., *Proc. Nat. Acad. Sc.*, 1947, **33,** 253.
16. Latarjet, R., *J. Gen. Physiol.*, 1948, **31,** 529.
17. Adams, M. H., *in* Methods in Medical Research, (J. H. Comroe, editor), Chicago, The Yearbook Publishers, Inc., 1950, **2,** 1.
18. Foster, R. A. C., *J. Bact.*, 1948, **56,** 795.

4

Reprinted from *Fed. Proc.*, **20**(2), 641–649 (1961)

Virus-induced acquisition of metabolic function

SEYMOUR S. COHEN[1]

Department of Biochemistry, University of Pennsylvania School of Medicine, Philadelphia, Pennsylvania

THE DETAILED chemical study of the course of bacterial-virus multiplication has been in progress slightly less than 15 years, antedating the chemical study of the multiplication of animal and plant viruses by about a decade. This pioneering role for the bacteriophage systems stems from their exceptionally favorable attributes in such matters as ease of cultivation and assay of host and virus, the establishment of the time and multiplicity of infection, and so forth. Of some historical interest is the fact that the bacterial viruses subjected to the most detailed chemical exploration, i.e., the T-even phages T2, T4, and T6, are biochemical freaks. The unique properties of these phages have contributed to their extreme virulence, which, in itself, has facilitated the study of these viruses.

Initially we posed our problems in the somewhat trivial terms of the molecular mechanism of the extremely virulent activity of the T-even phages on *E. coli*. Many other laboratories have developed their problems in the more general terms of the molecular bases of inheritance for these same phages. In the past three years, the problems of virulence in these systems have been more or less solved in terms of the discovery of a series of rapid syntheses of many enzymes and proteins controlled by the insertion of viral DNA into the infected bacterium. Thus, the problems of the virulence of these phages can now be posed in the terms and problems of molecular genetics. Our progress in this area has now catapulted us into the same pot of molecular biology in which almost everyone else is stewing. In this paper, we summarize the unifying studies, of the past few years, of virus-controlled syntheses of enzymes and some other proteins in infected *E. coli*.

EARLY DATA ON POLYMER SYNTHESIS
IN INFECTED CELLS

Until 1952, it was thought that the T-even viruses might merely redirect the existing functional metabolic machinery of the bacterium to the production of virus by replacing a set of critical bacterial templates by viral templates. Thus infected cells could no longer divide nor could they be induced to produce β-galactosidase (1).

Nevertheless, respiration and assimilation into polymeric substance appeared to continue at the same rate after infection as before infection (2). If one concentrated on the fate of assimilated phosphorus, for example, this now was directed almost entirely toward the production of viral DNA, while the net synthesis of RNA—the nucleic acid present in largest amount in the bacterium—stopped. The discovery of the new and unique pyrimidine, 5-hydroxymethyl cytosine (HMC) (3, 4), for the first time led to the suspicion that the virus might be contributing another, qualitatively new element, i.e., the ability to control a metabolic function which, perhaps, did not exist at all in the uninfected bacterium.[2] This function was one essential to viral duplication in providing a new base, a pyrimidine without which viral DNA could not be made.

Almost from the beginning of chemical study, it had been appreciated that there was a mystery in protein synthesis in infected cells. As shown in Fig. 1, protein synthesis continued from the inception of infection (8); however, very little of the protein which was formed before viral DNA synthesis began, appeared in the virus which was eventually liberated (9, 10). Was this bacterial protein which continued to be synthesized? For example, it is known that infection causes a leak in the permeability barrier of the cell, and that there is a subsequent repair (11). It was possible that this repair involved protein synthesis.

It was demonstrated—first, by means of the specific analogue, 5-methyl tryptophan (12, 13), and, subsequently, with chloramphenicol (14, 15), that the early synthesis of protein was essential to the synthesis of viral DNA. Indeed, once this protein was made, the inhibition of further protein synthesis did not block the production of functionally active viral DNA (13, 15). It could, therefore, be asked what function was fulfilled by this early protein, which was not itself a component of virus, but was nevertheless essential to the production

---

[1] The work performed in the author's laboratory has been aided greatly by grants from the Commonwealth Fund, New York, N. Y. and the Upjohn Company, Kalamazoo, Mich.

[2] Although it had been known that deoxyribonuclease is markedly increased in amount after infection with certain viruses, and that this increase is inhibited by chloramphenicol and thienylalanine (5, 6), this effect was most frequently attributed to the activation of pre-existing enzyme, rather than to a *de novo* synthesis of the enzyme (7, 6). This preference probably reflected, in large part, the intellectual climate (at the time) relative to these problems, although evidence for both possibilities had been obtained. Whether one or the other, or both, is correct has not yet been established.

641

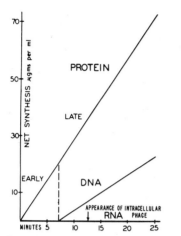

FIG. 1. Time course of polymer syntheses in *E. coli* infected by T-even phages.

of viral DNA. One obvious suggestion was that the early protein was involved in the formation of HMC. Having conceded the possibility of the expansion of the metabolic machinery, it would also be suggested that other enzymes might be made, particularly those which might account for the enormous stimulation of DNA synthesis observed on infection.

In Fig. 1, it can be seen that there is no net synthesis of RNA. The use of P^{32} then showed an incorporation of this isotope into the RNA fraction which was only 2% that of the incorporation into DNA (8). This result has been confirmed many times, but, although it was initially concluded that this amount of isotope in the RNA fraction reflected a contamination of the RNA, it has been shown rigorously that this P^{32} is, indeed, in RNA (16). This newly synthesized material is a chemically unique type of RNA, whose base composition mimics that of viral DNA. In addition, this RNA has an active turnover, in which the ribose nucleotides are converted to deoxyribonucleotides of viral DNA (16). It has been difficult to demonstrate a function for this RNA, but it has now been shown that the early steps of phage multiplication, which exclude DNA synthesis, require pyrimidines and adenine, presumably for RNA (17, 18).

It can be seen in Fig. 1, that DNA synthesis in infected cells stops for some minutes. DNA synthesis resumes at a markedly stimulated rate which, in some media (e.g., lactate), appears to compensate for the RNA that no longer accumulates (8). The new DNA, made after infection, is entirely viral, containing HMC and lacking cytosine (19, 20).

SIGNIFICANCE OF HYDROXYMETHYL CYTOSINE

The existence of the new base in viral DNA and the lack of cytosine, found normally in bacterial RNA and DNA, suggested that hydroxymethylation of cytosine to HMC converted the former base to a form unsuitable for normal bacterial syntheses. As we shall see, deoxycytidine triphosphate (dCTP), which is essential to the synthesis of bacterial DNA is, indeed, converted very efficiently in infected cells to deoxycytidylate (dCMP), the precursor of the HMC deoxyribonucleotide, hydroxymethyl deoxycytidylate (dHMP). Virulence and parasitism in this system are, thus, exhibited at the molecular level by appropriating a normal essential nucleotide for the formation of a unique viral metabolite. It should be noted that the host DNA is degraded in infection, and the dCMP therein contained is freed and converted to viral pyrimidine (21, 22).

In studies of the isolation of HMC derivatives from viral DNA, it was soon observed that nucleotides containing this base were present in a structure which resisted enzymatic release (23). Several workers (24–26) then showed that, in viral DNA, HMC was glucosylated at the hydroxymethyl group. It appears that the presence of glucose does, indeed, protect—to a considerable extent—phosphodiester bonds involving dHMP. Thus, viral DNA carries a significant molecular mechanism for self-protection when present in the DNase-rich medium of the infected cell. However this may not be the sole reason for the survival of viral DNA.

In addition, we can note that, although the base composition of the HMC viruses—T2, T4, and T6—are essentially identical (4), the glucose contents of these viruses differ considerably (26). Thus, T6 contains mainly large amounts of diglucosyl derivatives of HMC plus nonglucosylated HMC, T4 contains only the monoglucosyl HMC, while T2 contains mainly monoglucosylated and nonglucosylated HMC. HMC has thereby been implicated in mechanisms of parasitism and virulence, survival, and speciation among the T-even viruses.

BIOSYNTHESIS OF THYMINE AND HYDROXYMETHYL CYTOSINE

In exploring the biosynthesis of viral pyrimidine, it was found that exogenous orotic acid and pyrimidine of host DNA could be converted into viral HMC and thymine. The —CH₂OH and —CH₃ of the latter bases were derived from the β-carbon of serine (27). The methyl group of thymine did not come from the methyl of methionine (28). A series of nutritional experiments with bacterial mutants and competition experiments with infected cells then revealed that the addition of the one-carbon fragments to form viral bases did not occur at the level of free bases or nucleoside (29). In order to test the syntheses at the nucleotide level, it was necessary to study these possible reactions in cell-free extracts. Friedkin and Kornberg (30) demonstrated the presence of an enzyme, thymidylate synthetase, in extracts of normal *E. coli* which produced thymidylate (dTMP) from deoxyuridylate (dUMP), formaldehyde (HCHO), and tetrahydrofolic acid (THFA) (Fig. 2). Testing for

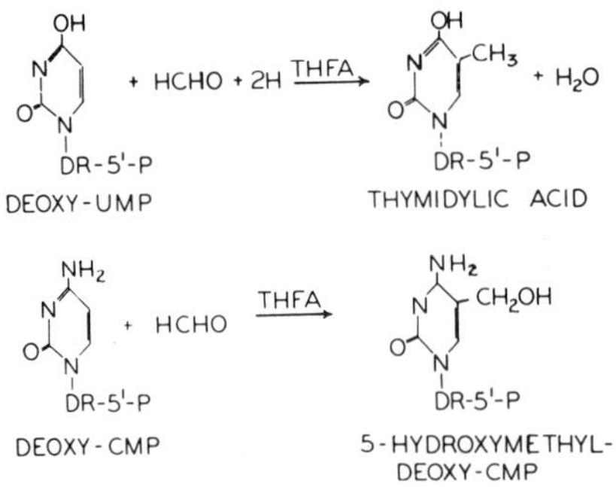

FIG. 2. Utilization of formaldehyde in the enzymatic synthesis of thymidylate and 5-hydroxymethyl deoxycytidylate.

similar reactions with dCMP in normal and T2-infected *E. coli*, Flaks and I showed that, with dCMP as a substrate, only extracts from infected cells were capable of fixing C^{14}-labeled formaldehyde in an acid stable form to yield dHMP, as in Fig. 2 (31, 32).

Although thymidylate synthetase was, indeed, present in uninfected *E. coli* strain B, the amount was shown to be markedly increased in phage-infected bacteria (31, 32). The mechanism of the reaction forming thymidylate does not appear to involve a free hydroxymethyl derivative, unlike the reaction which produces dHMP (30, 33). However, it was observed that hydroxymethyl deoxyuridylate and thymidylate were generated to a small extent during the conversion of dCMP to dHMP (33). This has been shown to be due to the appearance of a third enzyme, dCMP deaminase, in infected cells (Flaks, unpublished results) which not only deaminated dCMP but also deaminated dHMP. The appearance of this enzyme in viral infection has also been described recently in another laboratory (34).

PROPERTIES OF dCMP HYDROXYMETHYLASE

Three assays have been developed for the enzyme based on the fixation of C^{14}-labeled HCHO to dCMP to form radioactive dHMP (35). The ion-exchange method, reported in 1959 by Somerville et al. (36), is the most simple and accurate of the three, and is the method of choice, at present, in following the purification and other properties of the enzyme. The hydroxymethylase is completely inactive on cytosine ribonucleotide; however, cytosine arabinonucleotide appears to possess of the order of 0.6 % of the substrate activity of dCMP (37). Using a purified enzyme preparation, the following Km values have been determined: 6×10^{-4} M for dCMP, 1.5×10^{-3} M for formaldehyde, and 1×10^{-4} M for THFA.

The separation of the hydroxymethylase and the thymidylate synthetase may be effected easily; the first enzyme is far more stable (33). A much greater degree of purification of the hydroxymethylase was accom-

plished on a diethylaminoethyl (DEAE) cellulose column, as described by Kornberg et al. (38) and, recently, in somewhat greater detail by Pizer and myself (35). Fig. 3 presents the elution from the column of the protein and enzyme derived from infected cells in 40 litters of Biogen culture containing 10^9 infected bacteria per milliliter. Phage multiplication was stopped, at 15 min, by addition of chloramphenicol to the culture, which was expelled onto ice; the cells were collected by centrifugation. The infected cells (40–50 g, wet wt) were frozen and thawed, and extracted twice in the Waring Blendor, for 1 min, in 250 ml of 0.05 M glycylglycine buffer, at pH 7.2 (0.001 M with respect to the gluta-thione). The extract, containing about 4 g protein, was centrifuged (at 30,000 rpm) for 30 min and, after removal of nucleic acids by precipitation with streptomycin, the fractionation of DEAE cellulose was effected by using two linear gradient elutions with phosphate buffer.

As can be seen in Fig. 3, a peak of enzyme activity is obtained at a low point in the protein elution, and the central tubes of this peak have a constant specific activity. The values presented in the graph were obtained for the purified fractions in this peak by reanalysis about 16 days after infection, and about 9 days after separation on the column. Specific activities of 2,200 were obtained by the column assay for key tubes 8 days before the time for the assays indicated on the chart. Extrapolation to the earliest point in the fractionation suggests specific activities, for the enzyme, of the order of 5,000—a value slightly lower than that obtained by Kornberg et al. A personal communication, i.e., 6,000—using our original, less reliable assay (32).

The peak tubes were pooled, concentrated, and examined by analytical ultracentrifuge and electrophoresis in 0.02 M KPO_4 buffer, pH 7.4, 0.1 M with respect to KCl. Two components of 85 % and 15 % were observed in both analyses. The major component had an $S_{20,w}$ of 4.4 and a mobility of -5.4×10^{-5} cm^2 volts^{-1} sec^{-1}. The minor component had an $S_{20,w}$ of 5.3 and a mobility of -9.3×10^{-5} cm^2 volts^{-1} sec^{-1}. Analysis of the specific activities in the peak of enzyme elution indicates that the enzyme would probably be a major protein component. From the $S_{20,w}$ for this component, assuming a frictional ratio of 1, a molecular weight of 49,000, and a turnover number of the order of 125 moles dCMP per mole of enzyme per minute have been calculated.

With these numbers, we may estimate the number of enzyme molecules per cell. In the Biogen experiment, 10^{10} infected cells gave rise to about 1 mg of extract protein capable of converting 0.3 μmole dCMP in 20 min. It can be calculated that an infected cell contains about 8,300 molecules of hydroxymethylase in 15 min, or produces about 9 molecules of enzyme per second over a 15-min interval. Since the appearance of enzyme occurs at almost a constant rate, it can be calculated that—at the time DNA synthesis begins (about 7 min)—the hydroxymethylase content of the cells can maintain

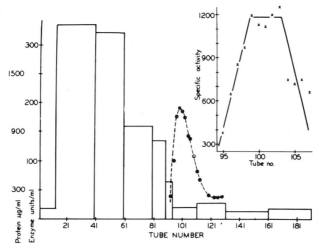

FIG. 3. Isolation of the deoxycytidylate hydroxmethylase by elution from a DEAE column. Dotted line = enzyme.

a rate of dHMP synthesis sufficient to produce 120 T2-virus particles per cell, if continued for 14 min. This value is close to the usual yield of a T-even phage per cell in a one-step growth experiment in the media used.

Pizer and I have now asked if any active enzyme is to be found in uninfected cells. To test this it has been necessary to run assays with approximately 2×10^7 cpm per assay tube or 2.4 μmoles C^{14} H_2O at 7 μc per micromole. Extracts of uninfected cells were prepared, and amounts of protein (1.6–4.7 mg) were used derived from 8.2×10^9 to 2.6×10^{10} cells. Among the blanks were used the reaction mixture minus protein, and the reaction mixture with extract lacking added dCMP. In another control, the reaction mixture was run with an extract of infected cells. The expected activity of enzyme was obtained despite the unusual radioactivity of the HCHO used. After 20 min, 0.55 μmoles DHMP was added as carrier, and this substance was purified initially on Dowex 50-H+ and subsequently by paper chromatography in isobutyrate-NH$_4$OH. The nucleotide was hydrolyzed, and the free HMC was purified by paper chromatography in butanol-NH$_4$OH and isopropanol-HCl. The purification steps for carrier HMC are summarized in table 1. It can be seen that after the final chromatography, the activity present in carrier HMC in all assays was far less than could be expected on the hypothesis that the cells contained at least 1 molecule of dCMP hydroxymethylase per cell. Furthermore, there was no suggestion of a trend in fixation as the amount of cell protein was increased. It may be remarked that the exclusion of activity for 1 molecule of enzyme in any auxotroph or otherwise "deficient" bacterium does not seem to have been reported previously.[2a]

[2a] Note added in proof: The following experiments have been performed with Dr. Pizer since the submission of the manuscript.

BIOSYNTHESIS OF dCMP HYDROXYMETHYLASE

Enzyme cannot be detected in uninfected cells, but is found in large amount in all infected cells (8,300 molecules/cell, or 0.7 % of the protein extracted at 15 min, since each infected cell can give rise to virus containing HMC. Several possibilities can be visualized concerning the appearance of the enzymatic activity.

1) The virus injects the enzyme. In addition to the

Cells were grown in the presence of S^{32} and infected in the presence of S^{35}. Fifteen minutes after infection cells were harvested and fractionated, as in Fig. 3, with numerous estimations of protein, radioactivity, and enzyme activity. The fractions containing enzyme possessed the highest radioactivity per milligram of protein. However, methionine isolated from this enzyme-containing fraction possessed a specific radioactivity of one-third that of the S^{35} in the medium. In addition to the possibility implied by this result that there is a protein or polypeptide precursor to the dCMP hydroxymethylase, it appears more likely that our enzyme preparation was not pure and was contaminated with unlabeled host protein, since the enzyme preparation reacted strongly with an antiserum prepared against protein isolated from normal cells by the method used in isolating the hydroxylase.

As a result of the observations of heterogeneity in the enzyme preparation, we have undertaken sedimentation studies of enzymatic activity in a partition cell and diffusion measurements with a porous disc method. The analyses have revealed an $S_{20.w}$ of 4.0 S and $D_{20.w}$ of 6.1×10^{-7} cm²/sec. The molecular weight of the enzyme has been calculated to be $64,000 \pm 9,000$. It is obvious that if the specific enzyme activity can be increased by a factor of 3, and the molecular weight is only slightly changed, the turnover member will approximately triple, and the total number of enzyme molecules per cell and the rate of enzyme synthesis will fall to about one-third that indicated in the text. On the other hand, the calculation of turnover number from an observed specific activity of 1,200 and a molecular weight of 64,000 leads to a turnover number of about one-third that used in Table 1. In either extreme the presence of active dCMP hydroxymethylase in uninfected cells appears to be excluded by the data in Table 1.

TABLE 1. *Exclusion of Active dCMP Hydroxymethylase in E. coli*

Assay Tube	Chromatography			Expected cpm for 1 Molecule Enzyme/Cell†
	Nucleotide	Free base		
	Isobutyrate-NH₄OH	Butanol-NH₄OH	Isopropanol-HCl	
*mg protein**	*cpm per 0.55 μmole*			
0	9900	890	22	0
1.56	5050	440	15.4	251
3.12	4450	288	3.7	502
4.68	2800	332	6.4	753
3.12 + 0 dCMP	1430	120	0.5	0

* 1 mg protein = 5.6 × 10⁹ cells. One molecule of enzyme (TN 125)/cell would produce 2.3 × 10⁻⁵ μmoles dHMP per mg protein per 20 min. Formaldehyde-C¹⁴ was used at 7 × 10⁶ cpm per μmole.

† See footnote 2a.

fact that we have been unable to detect enzyme in disrupted virus preparations (39), it can be calculated that the weight of the enzyme found in infected cells (7 × 10⁻¹⁶ g) approximates the entire weight of a T-even phage (5 − 7.5 × 10⁻¹⁶ g). This hypothesis does not appear reasonable.

2) The enzyme is present in uninfected bacteria in an inhibited state. We have disrupted bacteria in a variety of ways without revealing enzymatic activity. In addition, we have mixed and incubated extracts of infected and uninfected cells without inhibiting enzymatic activity or producing an increase in this activity (39). This shows the absence of excess inhibitor in uninfected cells and the absence of a system in extracts of infected cells capable of activating the hypothetical, inhibited complex in uninfected cells.

3) The enzyme is synthesized after insertion of the phage contents into the bacterium. The conditions required for the appearance of the dCMP hydroxymethylase bear on this hypothesis.

a. Infection must be effected with a T-even phage; enzyme is not formed during infection with other T phages, such as T1 or T5, or after induction of a lysogenic system such as K12-λ (39). Thus, infection with HMC phage is required for the development of the activity.

b. Osmotically shocked ghosts of T-even phages, although capable of adsorbing to the bacteria and affecting bacterial metabolism, cannot induce enzyme. The contents of the phage head are therefore essential for the appearance of enzyme. These contents consist mainly of DNA and small amounts of a specific internal protein (40), a polypeptide (41), and polyamines (42). The polyamines may be replaced by other cations in the phage without affecting phage viability (43).

c. The contents of the phage head do not contain free pyrimidine deoxyribonucleotides (39), which might be imagined to induce the biosynthesis of the hydroxymethylase.

d. As presented in Fig. 4, ultraviolet irradiated phages induce the appearance of the enzyme in normal amount under conditions (15 hits, virus:cell = 4) in which virus cannot multiply[3] or induce the formation of DNA (39). This work has been extended to a study of single infection with irradiated phage using the column assay. It appears that the site, in the phage, which controls the appearance of enzyme has a sensitivity to ultraviolet light about one-twentieth that of all the sites which control virus multiplication. (N. Delihas, in a personal communication). Assuming that the target for ultraviolet inactivation is phage DNA, all the major portions of which are equally sensitive to radiation, one may suggest that the portion of DNA controlling hydroxymethylase synthesis is a small fraction (perhaps about 1/20) of the total DNA but, nevertheless, this site is quite large (about 10,000 nucleotide pairs, or a molecular weight of about 6 × 10⁶).

It is essential, therefore, to insert a large viral polynucleotide into the bacterium. Once the DNA has entered the cell, enzyme appears at an almost constant rate for 10–15 min. We have shown that protein synthesis is essential for enzyme production (39). Thus, if protein synthesis is blocked by 5-methyl tryptophan, as in Fig. 5, or by deprivation of an amino acid requiring mutant, enzyme production does not occur. Enzyme synthesis can be begun by a later supply of the appropriate amino acid to the inhibited or deprived cell. Thus, the prerequisite of protein synthesis prior to synthesis of viral DNA is indeed explained, in part, by the requirement of protein synthesis for the appearance of the dCMP hydroxymethylase. Nevertheless, it can still be supposed that the protein which is synthesized is not the enzyme itself but, rather, a protein necessary to free the pre-existing inhibited enzyme or otherwise to activate an inactive or incomplete enzyme. Although we do not think that this is likely, we are currently attempting to disprove this by demonstrating a *de novo* synthesis of the enzyme from metabolites supplied exogenously after infection.

BIOSYNTHESIS OF THYMIDYLATE SYNTHETASE

Our studies with this enzyme originated, in part, as a result of the similarities of the early methods of assay for synthetase and hydroxymethylase (31). From the fixation of acid-stable C¹⁴H₂O, it was shown that *1)* this enzyme was normally present in *E. coli* capable of synthesizing thymine; *2)* the enzyme rapidly increased sixfold to sevenfold during infection of strain B with T2; *3)* the increase did not occur on infection with ghosts; and *4)* the increase of enzyme required protein synthesis (33).

Extending this work, Barner and I explored thymine-requiring mutants, which had been observed to syn-

---

[3] *E. coli* strain W will adsorb T2 and be killed by this phage, without giving rise to DNA and virus (23). Despite the abortive quality of the infection, this strain will produce dCMP hydroxymethylase (Pizer and Cohen, unpublished results).

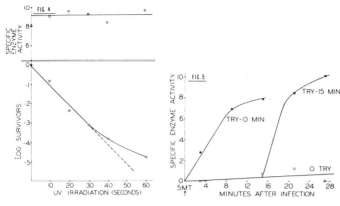

FIG. 4. Effect of ultraviolet irradiation of virus on hydroxymethylase formation. Cells were multiply infected with virus irradiated to varying levels of survival. The enzyme contents of such infected cells were determined.

FIG. 5. Effect of 5-methyl tryptophan on formation of hydroxymethylase. 5 MT = 5 methyl tryptophan. Try = tryptophan. Cells were infected in the presence of 5 MT and tryptophan was added in two systems at different times and omitted in a third system. The development of the hydroxymethylase was determined.

thesize thymine after infection with T2 (44) and T5 (45). Using a sensitive carrier assay, it was shown that although the extracts of thymine-requiring strains lacked the synthetase, infection with T2 or T5 resulted in the extensive production of the synthetase (46). It was found that in infections with T5, a virus which does not contain HMC, production of the synthetase occurs without a concomitant production of the dCMP hydroxymethylase. Indeed, production of the synthetase in T5-infected cells exceeds that in T2-infected cells.

We have obtained two highly active inhibitors for this enzyme, which is crucial in the economy of most cells. In the absence of thymidylate synthesis and exogenous thymine, a normal synthesis of DNA cannot occur. The unbalanced cell growth, which then ensues, results in the irreversible loss of the ability to multiply or "thymineless death" (44). Such a situation can be provoked by 5-fluorouracil deoxyriboside in many cells, including tumors, since the nucleotide, fluorodeoxyuridine-$5'$-phosphate, is a potent inhibitor for thymidylate synthetase (47). Fluorouridine-$5'$-phosphate is essentially inactive.

In the search for a fluorouracil nucleoside which is less readily degraded by the mammal, Dr. J. Fox of the Sloan-Kettering Institute has synthesized D-arabinofuranosyl 5-fluorouracil, and we have prepared the $5'$-phosphate of this nucleoside by an enzymatic method. This new nucleotide (F-aUMP) is also an inhibitor of thymidylate synthetase, although one-hundredth as potent as F-dUMP. In the study of compounds which can provoke thymineless death by this mechanism, T2 or T5 virus-infected bacteria are easily the best source of the thymidylate synthetase.

The production of thymidylate synthetase can be controlled by three chemically different DNA's. These are the cytosine-containing DNA of strain B, the HMC-containing DNA of T-even phages used to infect strain B_T^-, and the cytosine-containing DNA of T5.[4] It was,

therefore, of interest to see if the three forms of the enzyme were different, and the two inhibitors, F-dUMP and F-aUMP, were used to compare the sensitivities of the active sites of the enzyme in extracts of strain B and of virus-infected B_T^-. Miss H. D. Barner has recently improved the assay greatly for these experiments. As can be seen in Fig. 6, the active sites of the bacterial enzyme and two viral-induced types of thymidylate synthetase are indistinguishable by this test.

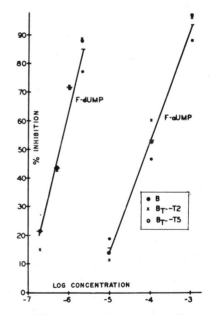

FIG. 6. Inhibition of two fluorouracil nucleotides on thymidylate synthetases induced by chemically different DNAs. F-dUMP = Fluorodeoxyuridine-$5'$-phosphate. F-aUMP = D-arabinofuranosyl fluorouracil-$5'$-phosphate.

[4] Of course, we cannot yet state that there are chemical differences between the fractions of *E. coli* DNA and T5 DNA which control the production of this enzyme, but the over-all compositions of the total DNA's of these forms are quite different (4).

OTHER VIRUS-INDUCED ENZYME
AND PROTEIN SYNTHESES

In Fig. 7 is presented a summary of new and stimulated syntheses induced in infection by a T-even phage. After the initial description of the appearance of the three systems presented earlier—dCMP hydroxymethylase, thymidylate synthetase, and dCMP deaminase—a number of other laboratories undertook to study a cellular system in which the insertion of a chemically unique viral DNA induced the rapid production, in all cells, of many unique enzymes essential to the rapid multiplication of the viral DNA and many other viral-specific proteins. Many questions could still be posed concerning the exclusion of cytosine, and the steps involving purine nucleotides, DNA polymerase, glucosylation, and so forth, in the synthesis of virus DNA.

In testing to see whether the DNA polymerase of T_2-infected cells was specific for HMC nucleotides, it was observed that infected cells elaborated a specific dCTPase, which removed pyrophosphate from dCTP to form dCMP (36, 38, 48). Thus, a compound essential for the formation of bacterial DNA was eliminated, i.e., dCTP, providing the dCMP essential for the formation of dHMP via the hydroxymethylase. The dCTPase is not produced in T_5 infection.

According to Koerner (in a personal communication), the activity of dCTPase in T_2-infected cells is ten times as much as is necessary to generate the dCMP as a substrate for hydroxymethylation. It may be pointed out, as well, that this is also consistent with the possibility that all deoxyribose is derivable via cytosine nucleotides.

Several laboratories (36, 38) have also described the phosphorylation of dHMP and the production of dHDP and dHTP. Although I have presented this as two new steps, in this reaction and, indeed, in the activation of all the deoxyribonucleotides to the triphosphate level, it is not certain that two separate, specific enzymes are involved. The dCMP kinase of *Azotobacter vinelandii* forms only the diphosphate (49), and the conversion of

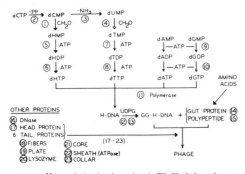

FIG. 7. New and stimulated reactions in T6. Circled numbers = new or stimulated units. Filled in arrows = new proteins. Hollow arrows = either stimulated or new enzyme.

dCDP to dCTP can be effected to dCTP by phosphoenol pyruvate and pyruvic kinase. It would be of interest to know whether pyruvic kinase fulfills this role in infected cells for other deoxynucleoside diphosphates. No dCMP kinase is produced in *E. coli* after T_2, T_3, and T_7 infection; however, a marked increase in activity appears in T_5 infection (38, 50).

Tenfold to 20-fold increased activities have also been observed in T_2, T_4, T_6, and T_5 infection for dGMP kinase and dTMP kinase, but not for dAMP kinase. The last appears to be present in uninfected cells in excess (38, 50). The increase in thymidylate kinase is blocked by chloramphenicol, suggesting the synthesis of protein (50). As in the case of the thymidylate synthetase, described earlier, it may be asked if the increase in these kinases represents more of the old enzymes or of new types of polypeptides having similar active sites. Bessman and van Bibber (51) have reported that the dGMP kinase formed after infection no longer has the requirement for K^+ that is characteristic of the kinase formed before infection.

Given the dHTP and the other normal triphosphates, several groups of investigators (38, 48) have studied the polymerase present after infection and its activity in the synthesis of a DNA containing HMC. About a tenfold increment has been seen in the DNA polymerase, and the purified enzyme of both uninfected and infected cells appears to be able to handle dCTP as well as dHTP. In these tests heated DNA derived from T_2 or calf's thymus can be used interchangeably.

Koerner et al. (48) have prepared a monoglucosyl dHTP, and have observed that this does not participate in the reaction with DNA polymerase. On the other hand, according to Kornberg and his collaborators (38), DNA containing HMC, generated from dHTP and DNA polymerase, can be monoglucosylated in part by uridine diphosphate glucose (UDPG) in the presence of a new enzyme formed in bacteria infected by T_2. Extending this important observation (A. Kornberg, in a personal communication), this group has found that two separable enzymes are formed in T_4-infected cells, one which adds a small amount of glucose to T_2 DNA, and a second which converts T_2 DNA to a form in which all HMC is completely monoglucosylated. Two separable enzymes are also present in T_6-infected cells, one like that in T_2-infected cells and a second that does not monoglucosylate, but adds a second glucose to T_2 DNA and T_4 DNA.

With the apparent synthesis of 13 enzymes in the first minutes after infection, we are finally in a position to account, in outline at least, for the stimulated DNA synthesis first observed so many years ago (8). These phage-infected bacteria are evidently a source par excellence for the study not only of the individual enzymes, since they are present in extraordinarily high concentration, but also of the specific control and kinetics of protein synthesis by specific DNA. However, in addition there are many other problems which remain in this system, e.g., when are the other viral proteins produced, and how are these syntheses related to the production of viral DNA?

Of the viral proteins, the internal "gut" protein of Levine et al. (40) is unique in appearing with the enzymes of DNA synthesis. With an immunochemical method, this specific protein is detectable 2–3 min after infection (52). As with the hydroxymethylase, infection with ultraviolet-irradiated phage induces its formation despite the prevention of DNA synthesis. This synthesis of internal protein is not blocked by proflavine, but is prevented by chloramphenicol. It appears to be significant that if this inhibitor is added at various times early in infection, a correlation appears to exist between DNA-synthesized and internal protein formed prior to addition of the agent. When the inhibitor was added 10 min after infection, the rate of DNA synthesis remained high, until the phage equivalents of DNA approached that of internal protein. At this point, the rate of DNA synthesis fell (52).

According to Hershey (41), the formation of the polypeptide, but not the polyamines, is also blocked by chloramphenicol. The recent studies of Kellenberger et al. (53), showing the late formation of a chloramphenicol-sensitive principle essential to the condensation of elements of the DNA pool, raises the problem of the possible relation of this polypeptide to the condensing principle and its mechanism of action.

Following the condensation of phage DNA, between 9 and 11 min after infection, it appears that centers are produced on which may be constructed the numerous structures (at least seven) of the phage head and tail. Of these, at least two possess enzymatic activity essential for the penetration of phage DNA into the bacteria. These are lysozyme (54) and an ATPase (55). It has not been reported when these enzymes are elaborated or whether or not the appearance of the activity and the phage structure take place simultaneously.

CONCLUDING REMARKS

The phenomena of the appearance of new enzymes, described above will certainly not be confined to the T phages. The development of a polysaccharidase in Klebsiella infected by a phage was described some years ago by Adams (56) and by Park (57). In this interesting case, the polysaccharidase, which hydrolyzed the bacterial capsule, was found both in a soluble form in the lysate and associated with the phage. The enzyme, thus, appears to facilitate attachment of the phage to the bacterium.

Several similar phenomena have been described in the animal-virus system. The appearance of neuraminidase in myxovirus-infected cells is well known, although inadequately explored. More recently, Rogers (58) has described the appearance of arginase in rabbit epithelium infected by rabbit papilloma virus. Thus, the acquisition or increase, or both, of metabolic function as a result of the addition of the viral genome to the infected cell will have the widest significance in explaining not only parasitic mechanisms, but the possible mode of action of tumor viruses as well. We have indicated the existence of 23 proteins whose production is controlled by the insertion of T-even phage DNA. Unquestionably a number of other enzymes will be found to be increased, e.g., the formation of deoxyribose, and so forth. On the other hand, other systems, such as respiration and amino acid activation, appear not to be increased, and these can be conceived as possibly rate-determining in the system as a whole.

Since bacterial DNA is degraded very soon after infection, it is reasonable to suppose that the code determining the amino acid sequence in each new protein is contained within the nucleotide sequence of viral DNA. Although HMC and its glucosylated derivatives can readily be conceived as equivalent to cytosine in this sequence, it is unlikely that all elements for this sequence controlling a single protein will be functionally identical for all amino acids; it is also quite reasonable to imagine that bacterial and virus-induced proteins will not be entirely identical in amino acid sequence. Whether this is true, remains to be determined. The solution of this question is only the first step pointing to one of the fundamental questions of molecular biology, i.e., the correlation of the nucleotide sequence in DNA with amino acid sequence in the protein whose structure is controlled by the DNA.

The problem of the intermediate steps in the control of protein synthesis by viral DNA is also one of the utmost importance. It can be inferred, although proof can scarcely be considered adequate, that the early RNA synthesis is functionally related to the synthesis of the new enzymes and other proteins. It is a general phenomenon, in these early protein syntheses, that they come to a halt at about 15 min after infection. Why do they stop? Is there a competition for amino acids between enzymes and viral coat proteins, or is there a destruction of the specific RNA templates by conversion of component ribonucleotides to the deoxyribonucleotides of viral DNA, or both? Is the turnover of RNA templates and competitive relation to protein and DNA a general phenomenon, which may account for the fact that differentiated cells elaborating specific proteins do not normally produce DNA and duplicate chromosomes?

Although, in a technical sense, these systems are admirable for the exploration of these questions, it is now evident that the T-even phages contain too much DNA and too much information, that too many proteins are produced, and that multiplication in cells infected by these phages requires too many parallel events. This system is too complicated to permit us to relate a twentieth of the genome in chemical terms to one of the new proteins. Of course, if we could insert a defined portion of the genome at will, we would be in a much better position. However, *E. coli* has never been transformed nor has the DNA of a T-even phage ever been reproducibly fractionated.

It may be suggested that the pursuit of these problems requires an infectable cell, transformable at high efficiency, and a tiny phage, possessing a bare minimum of DNA necessary for the inheritable transmission of virulence. Is the bare minimum of DNA that in $\phi \times 174$, and how many proteins does the DNA complement of this phage control? If only a single polypeptide chain is controlled in infections by $\phi \times 174$, we may cease to worry about the requirement for transformability.

However, if this DNA determines the biosynthesis of several different polypeptides, we shall wish to fractionate the DNA and introduce the determinants separately (as in bacterial transformation). We are at the stage, therefore, at which the study of the molecular relations of DNA as a genetic determinant and the protein produced as the expression of this determinant will be markedly facilitated by finding new and more simple biological systems to replace our old friend, T_2.

REFERENCES

1. MONOD, J., AND E. WOLLMAN. *Ann. inst. Pasteur* 73: 937, 1947.
2. COHEN, S. S., AND T. F. ANDERSON. *J. Exp. Med.* 84: 511, 1946.
3. WYATT, G. R., AND S. S. COHEN. *Nature* 170: 1072, 1952.
4. WYATT, G. R., AND S. S. COHEN. *Biochem. J.* 55: 774, 1953.
5. PARDEE, A. B., AND I. WILLIAMS. *Arch. Biochem. Biophys.* 40: 222, 1952.
6. KUNKEE, R. E., AND A. B. PARDEE. *Biochim. et Biophys. Acta* 19: 236, 1956.
7. KOZLOFF, L. *Cold Spring Harbor Symposia Quant. Biol.* 18: 209, 1953.
8. COHEN, S. S. *Cold Spring Harbor Symposia Quant. Biol.* 12: 35, 1947.
9. HERSHEY, A. D., A. GAREN, D. K. BRASER AND J. D. HUDIS. Carnegie Institution of Washington Year Book No. 53, 1954, p. 210.
10. WATANABE, I. *Biochim. Biophys. Acta* 25: 665, 1957.
11. PUCK, T. T., AND H. H. LEE. *J. Exp. Med.* 101: 151, 1955.
12. COHEN, S. S., AND C. B. FOWLER. *J. Exp. Med.* 85: 771, 1947.
13. BURTON, K. *Biochem. J.* 61: 473, 1955.
14. TOMIZAWA, J., AND S. SUNAKAWA. *J. Gen. Physiol.* 39: 553, 1956.
15. HERSHEY, A. D., AND N. E. MELECHEN. *Virology* 3: 207, 1957.
16. VOLKIN, E., AND L. ASTRACHAN. *The Chemical Basis of Heredity*, edited by W. D. MCELROY AND B. GLASS. Baltimore, Md.: Johns Hopkins Press, 1957, p. 686.
17. PARDEE, A. B., AND L. S. PRESTIDGE. *Biochim. Biophys. Acta* 37: 544, 1959.
18. VOLKIN, E., *Fed. Proc.* 19: 305, 1960.
19. HERSHEY, A. D., J. DIXON AND M. CHASE. *J. Gen. Physiol.* 36: 777, 1953.
20. VIDAVER, G., AND L. M. KOZLOFF. *J. Biol. Chem.* 225: 335, 1957.
21. KOZLOFF, L. M., K. KNOWLTON, F. W. PUTNAM AND E. A. EVANS, JR. *J. Biol. Chem.* 188: 101, 1951.
22. WEED, L. L., AND S. S. COHEN. *J. Biol. Chem.* 192: 693, 1951.
23. COHEN, S. S. *Cold Spring Harbor Symposia Quant. Biol.* 18: 221, 1953.
24. VOLKIN, E. *J. Am. Chem. Soc.* 76: 5892, 1954.
25. SINSHEIMER, R., *Science* 120: 551, 1954.
26. JESAITIS, M. A., *J. Gen. Physiol.* 106: 233, 1957.
27. COHEN, S. S., AND L. L. WEED. *J. Biol. Chem.* 209: 789, 1954.
28. GREEN, M., AND S. S. COHEN. *J. Biol. Chem.* 225: 387, 1957.
29. COHEN, S. S., J. LICHTENSTEIN, H. D. BARNER, AND M. GREEN. *J. Biol. Chem.* 228: 611, 1957.
30. FRIEDKIN, M. *The Kinetics of Cellular Proliferation.* New York City: Grune & Stratton, Inc., 1959.
31. FLAKS, J. G., AND S. S. COHEN. *Biochim. Biophys Acta* 25: 667, 1957.
32. FLAKS, J. G., AND S. S. COHEN. *J. Biol. Chem.* 234: 1501, 1959.
33. FLAKS, J. G., AND S. S. COHEN. *J. Biol. Chem.* 234: 2981, 1959.
34. KECK, K., H. R. MAHLER AND D. FRASER. *Arch. Biochem. Biophys.* 86: 85, 1960.
35. PIZER, L. I., AND S. S. COHEN. New York City: Academic Press, Inc.
36. SOMERVILLE, R., K. EBISUZAKI, AND G. R. GREENBERG. *Proc. Nat. Acad. Sci. U. S.* 45: 1240, 1959.
37. PIZER, L. I., AND S. S. COHEN. *J. Biol. Chem.* 235: 2387, 1960.
38. KORNBERG, A., S. B. ZIMMERMAN, S. R. KORNBERG AND J. JOSSE. *Proc. Nat. Acad. Sci. U. S.* 45: 772, 1959.
39. FLAKS, J. G., J. LICHTENSTEIN AND S. S. COHEN. *J. Biol. Chem.* 234: 1507, 1959.
40. LEVINE, L., J. L. BARLOW AND H. VAN VUNAKIS. *Virology* 6: 702, 1958.
41. HERSHEY, A. D. *Virology* 4: 237, 1957.
42. AMES, B. N., D. T. DUBIN AND S. M. ROSENTHAL. *Science* 127: 814, 1958.
43. AMES, B. N., AND D. T. DUBIN. *J. Biol. Chem.* 235: 769, 1960.
44. BARNER, H. D., AND S. S. COHEN. *J. Bact.* 68: 80, 1954.
45. CRAWFORD, L. V. *Virology* 7: 359, 1959.
46. BARNER, H. D. AND S. S. COHEN. *J. Biol. Chem.* 234: 2987, 1959.
47. COHEN, S. S., J. G. FLAKS, H. D. BARNER, M. LOEB AND J. LICHTENSTEIN. *Proc. Nat. Acad. Sci. U. S.* 44: 1004, 1958.
48. KOERNER, J. F. M. S. SMITH AND J. M. BUCHANAN. *J. Am. Chem. Soc.* 81: 2594, 1959.
49. MALEY, F., AND S. OCHOA. *J. Biol. Chem.* 233: 1538, 1958.
50. BESSMAN, M. J. *J. Biol. Chem.* 234: 2735, 1959.
51. BESSMAN, M. J., AND M. J. VANBIBBER. *Biochem. Biophys. Research Commun.* 1: 101, 1959.
52. MURAKAMI, W. T., H. VAN VUNAKIS AND L. LEVINE. *Virology* 9: 624, 1959.
53. KELLENBERGER, E., J. SECHAUD AND A. RYTER. *Virology* 8: 478, 1959.
54. KOCH, G., AND W. J. DREYER. *Virology* 6: 291, 1958.
55. DUKES, P. P., AND L. M. KOZLOFF. *J. Biol. Chem.* 234: 534, 1959.
56. ADAMS, M. H., AND B. H. PARK. *Virology* 2: 719, 1956.
57. PARK, B. H., *Virology* 2: 711, 1956.
58. ROGERS, S. *Nature* 183: 1815, 1959.

5

Copyright © 1964 by the American Society of Biological Chemists, Inc.

Reprinted from J. Biol. Chem., 239(9), 2957–2963 (1964)

Virus-induced Acquisition of Metabolic Function

VII. BIOSYNTHESIS *DE NOVO* OF DEOXYCYTIDYLATE HYDROXYMETHYLASE*

CHRISTOPHER K. MATHEWS,† FRED BROWN,‡ AND SEYMOUR S. COHEN

From the Department of Therapeutic Research, University of Pennsylvania School of Medicine, Philadelphia 4, Pennsylvania

(Received for publication, March 11, 1964)

In 1957, Flaks and Cohen (1) discovered that infection of *Escherichia coli* with T-even bacteriophages leads to the appearance of a new enzymatic activity, the hydroxymethylation of deoxycytidylate (Reaction 1).

$$\text{Deoxycytidylate} + 5,10\text{-methylenetetrahydrofolate} \rightarrow \\ \text{5-hydroxymethyldeoxycytidylate} + \text{tetrahydrofolate} \quad (1)$$

This reaction apparently accounts for the formation of the unique pyrimidine, 5-hydroxymethylcytosine, which appears in the deoxyribonucleic acid of T-even phages (2). Subsequent reports from this and several other laboratories (3–8) have shown that at least 12 enzymatic activities are either created or stimulated in phage-infected cells. More recently, similar phenomena have also been observed in virus-infected animal cells (9–12), indicating that the acquisition of new synthetic capabilities may be a common feature of virus-infected cells.

In those cases of phage-induced enzyme synthesis involving an increase in the level of an enzyme preexisting in the host cell, such as thymidylate synthetase (1, 8, 13, 14), several investigators have shown that the host cell enzyme and the corresponding phage-induced enzyme can be distinguished by their different catalytic and physical properties (5, 7, 15, 16). These observations, plus the fact that the appearance of phage-induced enzymatic activity requires protein synthesis (7, 17, 18), suggest that the virus directs the synthesis *de novo* of new enzyme molecules. However, alternate explanations can be envisaged. For example, the addition to or deletion from a preformed polypeptide chain of a peptide fragment could conceivably change the properties of an enzyme enough to account for the results which have been observed. In the case of an enzyme such as deoxycytidylate hydroxymethylase, in which activity is undetectable in uninfected cells (8), a similar mechanism could account for the appearance of enzymatic activity after infection. Such a mechanism would postulate the existence in uninfected cells of a preformed specific inactive enzyme precursor.

Evidence for enzyme synthesis *de novo* could be obtained by infecting in nonradioactive medium cells whose proteins had been prelabeled with a radioactive isotope (19, 20). If the enzyme molecules are assembled entirely after infection, then they should

be nonradioactive. To perform such an experiment, one must be able to isolate the enzyme in essentially pure form. Pizer and Cohen (21) attempted this experiment with dCMP hydroxymethylase. However, impurities in their enzyme preparation prevented them from obtaining conclusive evidence, although their results were consistent with the concept of synthesis *de novo*. We have now purified the same enzyme so that it is practically free of host cell protein. With this procedure we have carried out a direct test of synthesis *de novo;* the results of these experiments are described in this paper.

Even if we were to show complete synthesis *de novo* after infection, it might be argued that although the host cell may not contain any preformed enzyme precursor, it does contain the genetic information for enzyme synthesis and that infection activates this specific process. This type of inducibility could be designated as derepression, but is considered to be completely inapplicable to the appearance of T-even phage-induced enzymes, particularly that of dCMP hydroxymethylase, for the following reasons.

It should be noted that no evidence of derepression has ever been observed in T-even phage development. Unlike inducible systems, such as β-galactosidase, in which a basal level of enzyme is always detectable, no trace of dCMP hydroxymethylase can be detected in uninfected cells (8). Furthermore, phage mutants incapable of inducing the enzyme have been reported (22); indeed, a mutant is known which induces a temperature-sensitive dCMP hydroxymethylase (23). This evidence is similar then to observations with phage-induced thymidylate synthetase, an enzyme separable from the normal host enzyme (15) and noninducible on infection by certain phage mutants (24). Furthermore, the catalytic behavior in reactions with fluordeoxyuridylate of thymidylate synthetase induced by different phages differs as a function of the virus rather than of the host (25). Thus it appears that all these activities, whether qualitatively new or merely increments in preexisting levels, nevertheless reflect syntheses of new specific polypeptides, genetically determined by the viral genome rather than by the host. Such an inference still does not exclude the need to determine whether there is any specific host contribution of amino acids to the primary polypeptide structure whose synthesis may be induced by the virus. The chemical evidence of synthesis *de novo* presented below is thus an essential complement to the genetic data and inferences adumbrated in this discussion.

EXPERIMENTAL PROCEDURE

Biological Materials—*Escherichia coli* strain B and wild-type T6r⁺ bacteriophage were used for most of these studies. In

* This investigation was supported by Grant E-3963 from the National Institute of Arthritis and Metabolic Diseases of the United States Public Health Service.

† Postdoctoral Fellow of the United States Public Health Service. Present address, Department of Biology, Yale University, New Haven, Connecticut.

‡ Kellogg Foundation Fellow. Permanent address, Research Institute (Animal Virus Diseases), Pirbright, Surrey, England.

2957

61

some experiments we used *E. coli* strain B$_{4s}$, a methionine auxotroph of *E. coli* strain B (26). A large batch of phage was prepared as previously described (21), by infection of a 40-liter culture of bacteria. After purification by differential centrifugation, the virus preparation (2×10^{12} plaque-forming units per ml) was stored at 4° as a suspension in 0.85% sodium chloride. Under these conditions the preparation retained nearly full infectivity for 1 year.

Media—Medium M-9 was used throughout (3), except in experiments involving sulfate limitation. In this case a modified medium, designated M-9S, was used. This contained, per liter, 6.0 g of Na$_2$HPO$_4$, 3.0 g of KH$_2$PO$_4$, 1.0 g of NH$_4$Cl, 200 mg of MgCl$_2$, 0.5 mg of FeCl$_3$, and Na$_2$SO$_4$ added to the desired concentration. Glucose, sterilized separately as a 20% solution, was added to a concentration of 5.0 g per liter.

Chemicals—Sodium sulfate-^{35}S and methionine-methyl-^{14}C were purchased from Volk Radiochemical Company. Formaldehyde-^{14}C was obtained from Isotope Specialties Company and was standardized according to MacFadyen (27). Streptomycin sulfate was purchased from Nutritional Biochemicals Corporation. Deoxycytidylic acid and folic acid were obtained from the California Corporation for Biochemical Research. Tetrahydrofolic acid was prepared by catalytic hydrogenation of folic acid (28) and was stored in vacuum-sealed tubes in the dark at room temperature.

DEAE-cellulose was purchased from Distillation Products Industries. The dry powder was washed on a filter with 1 M Na$_2$HPO$_4$ until the filtrate was colorless, and it was then washed with water until the pH of the suspension was approximately 7. The product was washed repeatedly with water by decantation to remove fine particles. CM-Sephadex C-50, fine grade, was obtained from Pharmacia Fine Chemicals, Inc.

Hydroxylapatite was prepared as described by Tiselius, Hjérten, and Levin (29). Calcium phosphate was prepared by a modification of the procedure of Main and Cole (30). Just before use, sufficient material for 1 column was made as follows. To 40 ml of 0.5 M Na$_2$HPO$_4$ were added, simultaneously and dropwise, 200 ml of 0.5 M CaCl$_2$ and 200 ml more of the same Na$_2$HPO$_4$ solution. The mixture was stirred continuously during the addition. The precipitate was washed four times by decantation with distilled water and was then suspended in the starting buffer for preparation of the column.

Assays for dCMP hydroxymethylase activity were carried out as described by Pizer and Cohen (21). Protein was determined by the biuret method (31) in crude fractions, while the more highly purified fractions were assayed by the spectrophotometric procedure (31). The radioactive counting techniques used have been described previously (14). In the experiments involving ^{35}S, all values reported are uncorrected for radioactive decay, since each experiment was carried out within a time period such that 5% or less of the isotope had decayed.

Purification of Enzyme

Growth and Infection of Cells—Cells were grown as described previously (21), in a Biogen culture apparatus. Growth was followed turbidimetrically in a Klett colorimeter (filter No. 420). When the optical density of the suspension was 200 Klett units, corresponding to a titer of about 1×10^9 cells per ml, 2 g of DL-tryptophan were added as adsorption cofactor, and infection was initiated by the addition of 3 phage particles per bacterium. Aliquots of the phage stock suspension were diluted to 2 liters

with 0.85% NaCl before addition to the culture. After 20 minutes the infection was stopped by expulsion of the cells onto ice, and all subsequent operations were carried out at 0–4°. The addition of chloramphenicol at this point, done in the previous study (21), was found to be unnecessary. The cells were collected in a Sharples continuous flow centrifuge at a flow rate of 165 ml per minute. This procedure yielded between 50 and 70 g of cells (wet weight). The cell mass was suspended in 300 ml of 0.05 M potassium phosphate buffer, pH 6.5, and the resulting mixture was homogenized for 1 minute in a Waring Blendor. The homogenate was stored frozen overnight.

Preparation of Extract—The frozen homogenate was thawed and homogenized again for 1 minute. It was then centrifuged for 45 minutes at 10,500 r.p.m. in a Lourdes high speed centrifuge. The supernatant fluid was decanted, and the precipitate was resuspended in 200 ml of the same potassium phosphate buffer and homogenized. The centrifugation was repeated, and the supernatant fluids were combined.

Streptomycin Sulfate Treatment—To the crude extract was added 0.3 volume of 5% streptomycin sulfate solution with rapid stirring. After 5 minutes the mixture was centrifuged for 10 minutes at 5000 r.p.m. and the precipitate, containing most of the nucleic acid of the extract, was discarded.

Ammonium Sulfate Fractionation—To every 100 ml of the above supernatant fluid was added 27.7 g of solid ammonium sulfate. The mixture was stirred for 20 minutes and then centrifuged for 30 minutes at 6000 r.p.m. The precipitate was discarded, and to every 100 ml of the supernatant fluid was added 21.0 g of ammonium sulfate. Stirring and centrifugation were repeated as before, except that in this case the supernatant fluid was discarded. The precipitate was dissolved in a minimal volume (about 40 ml) of 0.02 M potassium phosphate buffer, pH 6.7. The solution was dialyzed for 16 hours against 8 liters of the same buffer. The protein in this solution represented that precipitating between 45 and 75% of saturation.

DEAE-cellulose Column Chromatography—The dialyzed ammonium sulfate fraction was clarified by brief centrifugation at 5000 r.p.m. The phosphate concentration was adjusted to 0.2 M by the addition of 1 M potassium phosphate, pH 6.5, and the solution was adsorbed on a column (3 × 20 cm) of DEAE-cellulose which had been equilibrated with 0.2 M phosphate buffer, pH 6.5. The column was washed with about 400 ml of the same buffer until the absorbance of the eluate at 280 mμ fell below 0.100, and the washings, containing no enzyme activity, were discarded. The enzyme was eluted from the column with a linear gradient. The mixing chamber contained 350 ml of 0.3 M phosphate buffer, pH 6.5, and the reservoir contained an equal volume of 1.0 M phosphate, pH 6.0. Fractions of 15 ml were collected. The enzymatically active fractions (tubes 10 to 24) were pooled and concentrated by dialysis against a suspension of polyethylene glycol (Carbowax 4000, Union Carbide Company). When the volume had been reduced approximately 5-fold, the dialysis bag was tied off, and the solution was dialyzed for an additional 16 hours against 8 liters of 0.05 M potassium phosphate, pH 6.5.

Calcium Phosphate Column Chromatography—The pooled and concentrated material from the previous step was applied to a column (2 × 36 cm) of calcium phosphate, which had been equilibrated with 0.05 M potassium phosphate, pH 6.5. After the enzyme solution had been adsorbed on the column, 20 ml of the latter buffer was added, and a linear gradient was applied, with 120 ml of 0.05 M phosphate, pH 6.5, in the mixing chamber, and 120 ml of 0.50 M phosphate buffer at the same pH in the reservoir. Fractions of 3 ml were collected. Enzyme activity migrated with the principal protein peak (see below). The most active fractions were combined and dialyzed for 4 hours with vigorous stirring against 2 liters of 0.05 M potassium phosphate, pH 6.0.

CM-Sephadex Column Chromatography—The dialyzed mixture from the previous step was adsorbed on a column (1.5 × 20 cm) of CM-Sephadex C-50, fine grade, which had been thoroughly washed with 0.05 M potassium phosphate, pH 6.0. To the top of the column was added 15 ml of the same buffer, and a linear potassium phosphate gradient at pH 6.0 was applied, with 120 ml of 0.05 M buffer in the mixing chamber, and 120 ml of 0.50 M buffer in the reservoir. Fractions of 3 ml were collected. The second major protein peak contained most of the activity (see below), and these fractions were pooled to provide a highly purified preparation of dCMP hydroxymethylase.

RESULTS AND DISCUSSION

Results of a typical enzyme fractionation are summarized in Table I. The method of cell extraction reported herein gave enzyme recoveries about twice those reported in the earlier study from this laboratory (21). Fractionation with ammonium sulfate served to concentrate the preparation while effecting a 2-fold rise in specific activity. The elution schedule for the DEAE-cellulose column is similar to that used previously (21). However, the recovery of enzyme activity and the degree of purification were both improved considerably in the present study. The chromatographic step with calcium phosphate, summarized in Fig. 1, was substituted for a similar step with hydroxylapatite which was used in the early part of this work. Calcium phosphate is easier to prepare and gives much higher flow rates than those obtained with hydroxylapatite. Elution

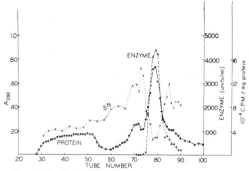

FIG. 1. Distribution of enzyme activity, protein, and radioactivity in the eluate from a calcium phosphate column. Details are given in the text. ○——○, absorbance at 280 mμ; △- - -△, enzyme activity; ×····×, specific radioactivity.

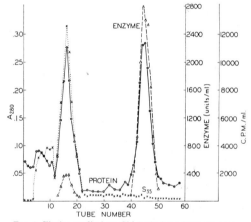

FIG. 2. Elution profile from CM-Sephadex column chromatography. Details are given in the text. ○——○, absorbance at 280 mμ; △- - -△, enzyme activity; ×····×, radioactivity.

profiles are quite similar, with resolution actually somewhat better with calcium phosphate. The final step in the procedure, chromatography on CM-Sephadex, is illustrated in Fig. 2. At this stage the enzyme preparation may approach homogeneity. Within experimental error, the specific activities of the most active fractions are equal. About 10% of the enzyme activity was found in the first major peak eluted from the column. This observation is discussed below in greater detail.

The active fractions from any of the three columns could be stored at 4° for up to 10 days with only minor losses in activity. At an earlier stage in the purification, the dialyzed ammonium sulfate fraction on one occasion was stored frozen for 2 weeks with a 30% loss in activity. Infected cells could be stored frozen for at least 1 month without a noticeable effect on the amount or activity of extractable enzyme.

Total recoveries of enzyme activity in the last two steps were higher than those reported in Table I, for only the most active

TABLE I

Purification of dCMP hydroxymethylase and its separation from labeled host cell protein

Fraction	Volume	Protein concentration	Specific enzyme activity	Total enzyme activity	Specific radioactivity per mg of protein
	ml	*mg/ml*	*units/mg*	*units × 10⁻³*	*c.p.m. × 10⁻⁴*
Extract	540	13.3	36	258	10.75
Streptomycin supernatant	670	10.2	47	258	7.96
45 to 75% ammonium sulfate	90			237	
Ammonium sulfate after dialysis	147	13.5	97	193	11.29
Pooled DEAE-cellulose fractions (tubes 9 to 23)	220			128	
DEAE-cellulose pool after concentration and dialysis	69	0.58	2850	113	9.12
Pooled calcium phosphate fractions (tubes 76 to 82)	19	0.54	7540	76.9	4.32
Calcium phosphate pool after dialysis	19	0.51	6120	59.1	
Pooled CM-Sephadex column fractions (tubes 42 to 48)	21	0.25	8900	46.5	0.15

fractions were pooled for further use. Between 70 and 90% of the enzyme applied to each column was usually recovered. Dialysis of the preparation at any stage in the purification resulted in activity losses of the order of 10 to 20%. This appeared to be a result of denaturation in solution rather than enzyme adsorption on the dialysis membrane, since the amount of protein in the preparation did not decrease during dialysis. Because of the possible existence of inactivated enzyme in even the most highly purified preparations, however, the actual specific activity of the pure enzyme might be considerably higher than the value of 8,900 units per mg of protein, listed in Table I for the pooled fractions from the final column. Specific activities of 9,500 units per mg have been recorded for individual fractions from the column.

The purified enzyme is colorless with a single light absorption peak at 280 mμ, indicating the absence of bound cofactors. It passes through a column of Sephadex G-200 slightly more slowly than bovine serum albumin (see below). This behavior on Sephadex G-200 is consistent with a molecular weight of 68,000, the value determined by Pizer and Cohen (21) from measurements of the sedimentation and diffusion of the enzymatic activity. With this molecular weight and the maximal activity for the enzyme, the turnover number is 320 molecules of substrate per minute per molecule of enzyme.

Separation of dCMP Hydroxymethylase from Host Cell Protein—As mentioned above, a demonstration that dCMP hydroxymethylase is synthesized entirely after infection depends upon being able to obtain a very highly purified enzyme. If not homogeneous, the preparation must at least be substantially free of contamination by host cell protein. To test whether our purification procedure met the latter criterion, the following experiment was carried out. A 2-liter culture of *E. coli* strain B growing in a sulfate-deficient medium (2.5 × 10⁻⁵ M) was allowed to incorporate 1 mc of sodium sulfate-³⁵S. When growth had ceased, owing to exhaustion of the sulfate of the medium, the cells (1.9 g) were centrifuged and washed. They were then mixed with a normal batch of phage-infected cells grown in the Biogen (71 g). This mixture, containing labeled host cell protein and nonradioactive enzyme, was used as a starting material for purification of the hydroxymethylase. Enzyme activity and radioactivity were determined at each step. Fig. 1 and 2 show the radioactivity elution profiles from the calcium phosphate column and from the CM-Sephadex column, respectively. Fig. 1 records specific radioactivity in counts per minute per mg of protein, while Fig. 2 shows observed radioactivity in counts per minute per ml of eluate. Results of the experiment are summarized in the last column of Table I. The specific activity of the pooled CM-Sephadex column fractions was reduced to 1.4% of that of the starting material. One of the most active enzyme fractions (tube 45) had a specific radioactivity of 800 c.p.m. per mg of protein, about 0.75% of that of the extract. The elution profile suggests that this radioactivity is "background," and is not specifically associated with enzyme molecules. From this evidence it seemed that similar results in an actual test situation would provide satisfactory evidence for enzyme synthesis *de novo*.

Enzyme Synthesis in Cells Prelabeled with Sulfate-³⁵S—In planning an experiment to demonstrate synthesis *de novo*, we were initially guided by earlier investigations designed to show that the enzyme β-galactosidase is synthesized *de novo* following induction of a bacterial culture (19, 20). The work of Hogness, Cohn, and Monod (19) seemed particularly applicable, since

these workers also labeled cellular proteins by growth in the presence of sulfate-³⁵S. In the experiments of Hogness *et al.* (19), cells were grown in the presence of limiting sulfate (radioactive), and when growth had ceased due to exhaustion of the sulfate, they were starved for an additional period to deplete intracellular pools of protein precursor. At that point, excess nonradioactive sulfate was added to the medium, and the cells were induced. Cells were harvested, the β-galactosidase purified, and its specific radioactivity determined.

At least three possible sources of error are inherent in the above design, especially in its application to a study of phage-induced protein synthesis. These are (*a*) the presence of radioactive material in the medium at the time of infection, (*b*) the possible failure of the internal pools of labeled protein precursors to be depleted completely during sulfur starvation, and (*c*) the possible breakdown of host cell protein and its reutilization in phage-directed protein synthesis.

Regarding the first point, Roberts *et al.* (32) had reported that when a culture of *E. coli* received sulfate as its sole sulfur source at limiting concentration, between 25 and 40% of the total sulfur is returned to the medium in a nonutilizable form, mostly as oxidized glutathione. Our work confirmed this finding; when cultures grown under conditions similar to those of Roberts *et al.* ceased growing because of sulfur starvation, approximately one-third of the total radioactivity was in the medium. Chemical analysis (33) showed that the labeled material was not inorganic sulfate. Moreover, it was apparently not utilized for growth, for even after nonradioactive sulfate had been added to the starved cultures and growth had resumed, there was no *net* uptake of radioactivity. This, of course, does not rule out the possibility of equal rates of uptake and release of labeled material by the cells. Therefore, the presence of a large amount of possibly utilizable radioactive material in the medium after starvation must be construed as a possible complicating factor.

With regard to the second possible source of error, Roberts *et al.* (32) have also shown that sulfate incorporated by *E. coli* is taken into two "internal reservoirs" which can be utilized for protein synthesis when the sulfate in the medium is consumed. These reservoirs are the glutathione pool and an alcohol-soluble protein fraction. Sulfur from these sources can support cell growth, although at a much reduced rate, for 6 to 8 hours. Thus, a relatively short period of sulfur starvation, such as that used by Hogness *et al.* (19), would not suffice to eliminate these pools. The presence of these reservoirs and their possible utilization even after the addition of nonradioactive sulfate must also be considered a disadvantage of this type of experiment.

The third source of error which must be considered is the possible breakdown and reutilization of the protein components of the host cell. Earlier reports (19, 20, 34) indicated that the rate of protein turnover in logarithmic cultures of *E. coli* is so low as to be virtually undetectable. However, more recent investigations (35, 36) have shown that in nongrowing cultures of *E. coli*, protein turnover is significant, amounting to 4 to 5% per hour. To our knowledge, studies of protein turnover in phage-infected bacteria have not been described. However, it has been reported (37) that bacteriophage T6r⁺ derives 6 to 12% of its protein nitrogen from host cell protein. On the other hand, the purity of the phage was not established in these studies, so that the host protein components present in the phage may represent contamination by cellular material. However, if the early phage-induced enzymes are partially synthesized from degraded host cell protein, then it might be impossible to demonstrate enzyme synthesis *de novo* by this type of experiment.

In spite of the above difficulties, we proceeded with an experiment quite similar in concept to that of Hogness *et al.* (19). A 40-liter culture of *E. coli* B was grown in limiting sulfate concentration (3.0×10^{-5} M). To this culture was added 1 mc of carrier-free sodium sulfate-^{35}S. When growth had ceased, owing to exhaustion of the sulfate, the cells were incubated further for 60 minutes to partially deplete the labeled internal pools of protein precursors. A separate experiment showed that this treatment did not affect the ability of the cells to produce enzyme upon subsequent infection. Following the starvation period, nonradioactive sodium sulfate was added to 1.0×10^{-3} M, and the cells resumed growth. When the culture had attained an optical density of 200, it was infected, and an extract was prepared from the infected cells. This extract was used as starting material for purification of the hydroxymethylase. Fig. 3 illustrates the results of chromatography on CM-Sephadex, the final step in the preparation. A small but significant amount of radioactivity migrated with the peak of enzyme activity. That this radioactivity was associated specifically with enzyme molecules was indicated by the fact that the specific radioactivity of the enzyme-containing fractions was constant throughout the peak. Moreover, radioactivity migrated with enzyme activity upon rechromatography on CM-Sephadex or upon chromatography on Sephadex G-200. The specific radioactivity of the pooled enzyme fractions was 2030 c.p.m. per mg of protein, approximately 4% of the specific radioactivity of the initial extract. Thus, it became necessary to redesign the experiment in such a way that the three sources of error discussed above were either minimized or eliminated.

Enzyme Synthesis in Cells Prelabeled with Methionine-methyl-^{14}C—The above experiment was carried out again, with two major modifications. First, methionine labeled with ^{14}C in the methyl position was substituted for sulfate-^{35}S as the source of radioactivity, and a methionineless strain of *E. coli* B was used. For this purpose we used *E. coli* B$_{45}$, a strain which requires methionine or vitamin B$_{12}$ for growth (26). In this strain, synthesis of the hydroxymethylase does not occur following phage infection unless methionine is present (18), suggesting that the enzyme contains methionine. The substitution of methionine-^{14}C for sulfate-^{35}S eliminated the difficulties associated with the accumulation of radioactivity in various pools of intracellular protein precursors. Moreover, if radioactive methionine is released from cellular protein after infection, the nonradioactive methionine added after starvation is an effective competitor with this material for entry into protein biosynthesis.

The second modification of the experimental design was to grow the labeled cells in a small volume of medium, from which they could be centrifuged and washed before infection. This enabled us to remove the radioactivity (15 to 20% of the total label applied as methionine-methyl-^{14}C) that was excreted into the medium in a form other than methionine. The small batch of highly radioactive cells was mixed with a normal large-scale batch of unlabeled infected cells which were used as carrier for the enzyme purification.

The experiment was carried out as follows. A 4-liter culture of *E. coli* B$_{45}$ was grown in a 20-liter carboy on a horizontal rolling apparatus at a speed of 165 r.p.m. These growth conditions are similar to those in the Biogen apparatus and the cells grow at approximately the same rate. The cells were grown in a limiting

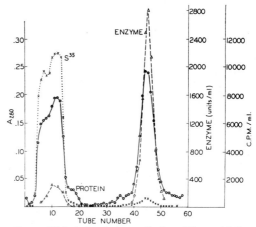

Fig. 3. CM-Sephadex elution profile from ^{35}SO$_4^-$ prelabeling experiment. O——O, absorbance at 280 mμ; $\triangle$-- -$\triangle$, enzyme activity; $\times\cdots\times$, radioactivity.

concentration of L-methionine (7 μg per ml) containing 0.5 mc of methionine-methyl-^{14}C. Growth ceased at an optical density of 148. The cells were incubated for an additional 30 minutes and then centrifuged and washed. They were suspended in 4 liters of methionine-free medium and incubated further for 15 minutes. At this point nonradioactive methionine was added to a concentration of 10 μg per ml. Growth resumed immediately and when the optical density reached 200, the cells were infected and harvested. This procedure yielded 6.5 g of infected cells (wet weight). A large batch (52 g) of unlabeled infected cells was prepared simultaneously, and separate extracts were prepared from the two batches of cells. Both extracts were shown to be enzymatically active, each having 55 units per mg of protein. The extracts were then mixed and used as starting material for purification of the hydroxymethylase. The specific radioactivity of the mixed extract was 34,700 c.p.m. per mg of protein; this was of the same order of magnitude as the specific activity of the starting extract from the sulfate-^{35}S experiment (52,900 c.p.m. per mg of protein). In the following discussion, specific activities are referred to the specific activity of the mixed extract as the starting point for the experiment. Results would be the same if we used the specific activity of the extract of radioactive cells as the starting point and corrected all subsequent activities for the dilution occurring when the radioactive and nonradioactive cells were mixed.

Results of the CM-Sephadex chromatographic step are shown in Fig. 4. In this experiment the enzyme in the second peak was practically nonradioactive. The pooled enzyme fractions had a specific radioactivity of 208 c.p.m. per mg of protein, or about 0.6% of the specific radioactivity of the mixed extract.

As mentioned earlier, a small portion of the enzyme activity applied to the CM-Sephadex column was eluted in the first peak, which contained highly radioactive material. This phenomenon was observed even in the mixing control experiment, where we knew that all of the hydroxymethylase in the preparation was nonradioactive. The reason for this observation is not clear. However, we do know that the enzyme molecules in the first peak

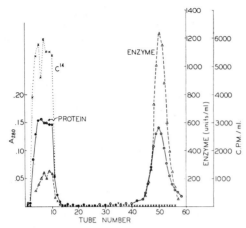

Fig. 4. CM-Sephadex elution profile from methionine-^{14}C pre-labeling experiment. O——O, absorbance at 280 mμ; △---△, enzyme activity; ×····×, radioactivity.

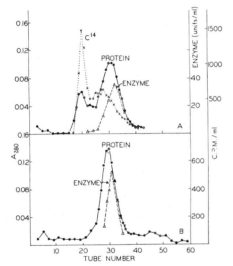

Fig. 5. Chromatography on Sephadex G-200 of the two peaks of enzyme activity from a CM-Sephadex column. *A*, first peak; *B*, second peak. Bovine serum albumin was added as a marker in each instance before chromatography. O——O, absorbance at 280 mμ; △---△, enzyme activity; ×····×, radioactivity.

are not specifically associated with radioactive material because chromatography of the pooled fractions from the first peak on a column of Sephadex G-200 in 0.1 M phosphate, pH 6.5, gave a partial but definite separation of enzyme activity from radioactivity (Fig. 5*A*). Moreover, the enzyme in the first peak was eluted at approximately the same position as was the enzyme from the second peak when the latter was chromatographed on Sephadex G-200 under identical conditions (Fig. 5*B*).

In order to correct for any large difference in the proportions of methionine in the purified enzyme and in the cellular protein, it was necessary to determine the specific radioactivity of the methionine in the mixed extract and in the isolated hydroxymethylase. Samples of the purified enzyme and the mixed extract (2 to 3 mg of protein in each) were hydrolyzed in sealed tubes with 6 N HCl at 110° for 18 hours. The hydrolysates were taken to dryness under a stream of warm air, and the residues were dissolved in a small volume of water. This process was repeated to remove HCl completely.

Methionine was determined in the hydrolysates by a microbiological method, with *E. coli* B$_{45}$. One-milliliter samples containing protein hydrolyzate or a methionine standard and Medium M-9 plus 5 mg of glucose per ml were inoculated with 0.05 ml of a suspension of *E. coli* B$_{45}$ which had been grown in limiting methionine. For each assay, three different levels of each hydrolysate were run, along with standards containing 1, 2, or 3 μg of L-methionine. The samples were incubated at 37° for 18 hours with occasional shaking, and the optical densities were measured at 420 mμ in the Beckman model DU spectrophotometer. Growth response was linear, and the standard curves extrapolated back to zero. A hydrolysate of tobacco mosaic virus protein, which contains no methionine, did not stimulate growth (38) and neither enhanced nor inhibited growth in this assay when it was added to reaction mixtures in the presence of methionine standards. A hydrolysate of pancreatic ribonuclease (Worthington) was assayed by this method. Two separate determinations gave results of 4.1 and 4.3% for the proportion of methionine in the hydrolysate. This compares favorably with the value of 4.4% calculated on the basis of 4 moles of methionine per mole of ribonuclease.[1]

When a hydrolysate of the mixed extract was subjected to paper chromatography with 1-butanol-acetic acid-water (40:10:50, upper phase), all of the radioactivity migrated with the same R_F value as methionine, suggesting that most, if not all, of the radioactivity of the hydrolysates was due to methionine. In view of the specific character of the isotopic label and the biological assay, we carried out radioactivity determinations and methionine assays directly on the hydrolysates of the mixed extract and the purified enzyme, without prior isolation of the methionine. The specific activity of the methionine of the extract was 422 c.p.m. per μg of methionine, compared with a value for the enzyme hydrolysate of 3 c.p.m. per μg. From this we can calculate that the methionine of the enzyme was at most 0.7% as radioactive as the methionine of the extract.

Because of the 20-minute period of phage-directed protein synthesis occurring in nonradioactive medium, the specific activity of the methionine of the enzyme is actually less than 0.7% of the specific activity of the methionine of the cellular protein *at the time of infection*. Moreover, at least part of this low radioactivity is probably due to a failure of the purification method to separate host cell protein completely from the enzyme. It will be recalled that in the control experiment where a nonradioactive extract of infected cells was added to an extract of ^{35}S-labeled uninfected cells, the most active enzyme fraction had 0.75% of the specific activity of the starting extract. Even disregarding the above factors, however, we can say that no more than 0.7% of the methionine of the enzyme came from material which was

[1] E. Brand, cited in J. H. Northrup, M. Kunitz, and R. M. Herriott, *Crystalline enzymes*, Columbia University Press, New York, 1948, p. 26.

present in the cells before infection.[2] This effectively rules out any theory of enzyme synthesis which postulates the uniform deletion or addition of a peptide to a preformed but inactive precursor.

If we assume (a) that the specific activity of the pure enzyme is 9,500 units per mg of protein (the highest value we have obtained for any individual fraction) and (b) that the 258,000 units recovered in the extract of the ^{14}C-labeled methionine experiment represents the total amount of enzyme synthesized in a 40-liter culture of infected cells, then we can calculate that the total amount of enzyme protein is 27 mg. For a molecule of molecular weight 68,000, this weight of enzyme is equivalent to 2.4 × 10[17] molecules. Assuming that every cell in the culture is infected and knowing that the cell density at the time of infection is 10[9] cells per ml, each infected cell must contain about 6,000 molecules of enzyme 20 minutes after infection. If all the precursor methionine were in a few molecules of enzyme, each uninfected cell can contain no more than 0.7% as many enzyme equivalents as an infected cell, or about 42 molecules. Pizer[3] has increased the sensitivity of the hydroxymethylase assay procedure to the point where it can detect an amount of enzyme activity equivalent to 1 molecule per cell. With this technique he showed that there is less than 1 molecule of active hydroxymethylase per uninfected cell.

Our initial control experiment, in which we mixed ^{35}S-labeled cell proteins with unlabeled enzyme, permitted the isolation of enzyme containing the same level of radioactivity (0.7% of the radioactivity of the host) as we obtained in the ^{14}C-labeled methionine experiment. A correction for this degree of contamination would eliminate the possibility of even 42 inactive precursor molecules.

SUMMARY

Deoxycytidylate hydroxymethylase has been purified from extracts of *Escherichia coli* strain B infected with bacteriophage T6r[+]. The preparation is essentially free of host cell protein.

The purification procedure has been used to provide direct evidence that the enzyme is synthesized *de novo* after infection. Cells prelabeled by growth in the presence of methionine-methyl-^{14}C were infected in nonradioactive medium. Hydroxymethylase purified from these cells was virtually nonradioactive. The methionine of this enzyme was less than 0.7% as radioactive as the methionine of the cellular protein at the time of infection.

Acknowledgment—The authors are grateful to Dr. Lewis I. Pizer for helpful discussions throughout the course of this study.

REFERENCES

1. FLAKS, J. G., AND COHEN, S. S., *Biochim. et Biophys. Acta*, **25**, 667 (1957).
2. WYATT, G. R., AND COHEN, S. S., *Biochem. J.*, **55**, 774 (1953).

[2] The methionine content of the enzyme is about 5%, which is equivalent to approximately 22 residues of methionine in each molecule of enzyme. This means that an average of only 0.15 molecule of radioactive methionine could be present in 1 molecule of an inactive precursor of the enzyme. Since the smallest number of precursor methionine molecules in a single molecule of enzyme is 1, at least 85% of the hydroxymethylase molecules would be lacking even 1 precursor molecule of the 22 methionine molecules required.

[3] Unpublished results.

3. KORNBERG, A., ZIMMERMAN, S. B., KORNBERG, S. R., AND JOSSE, J., *Proc. Natl. Acad. Sci. U. S.*, **45**, 772 (1959).
4. SOMERVILLE, R., EBISUZAKI, K., AND GREENBERG, G. R., *Proc. Natl. Acad. Sci. U. S.*, **45**, 1240 (1959).
5. BESSMAN, M. J., AND BELLO, L. J., *J. Biol. Chem.*, **236**, PC72 (1961).
6. OLESON, A. E., AND KOERNER, J. F., *Federation Proc.*, **22**, 406 (1963).
7. MATHEWS, C. K., AND COHEN, S. S., *J. Biol. Chem.*, **238**, PC853 (1963).
8. COHEN, S. S., *Federation Proc.*, **20**, 641 (1961).
9. KIT, S., PIEKARSKI, L. J., AND DUBBS, D. R., *J. Molecular Biol.*, **6**, 22 (1963).
10. BALTIMORE, D., FRANKLIN, R. M., EGGERS, H. J., AND TAMM, I., *Federation Proc.*, **22**, 407 (1963).
11. GOTTSCHALK, A., *Physiol. Revs.*, **37**, 66 (1957).
12. GREEN, M., *Cold Spring Harbor symposia on quantitative biology, Vol. XXVII*, Long Island Biological Association, Cold Spring Harbor, Long Island, New York, 1962, p. 219.
13. FLAKS, J. G., AND COHEN, S. S., *J. Biol. Chem.*, **234**, 2981 (1959).
14. BARNER, H. D., AND COHEN, S. S., *J. Biol. Chem.*, **234**, 2987 (1959).
15. GREENBERG, G. R., SOMERVILLE, R., AND DEWOLF, S., *Proc. Natl. Acad. Sci. U. S.*, **48**, 242 (1962).
16. APOSHIAN, H. V., AND KORNBERG, A., *J. Biol. Chem.*, **237**, 519 (1962).
17. FLAKS, J. G., LICHTENSTEIN, J., AND COHEN, S. S., *J. Biol. Chem.*, **234**, 1507 (1959).
18. PIZER, L. I., AND COHEN, S. S., *Biochim. et Biophys. Acta*, **53**, 409 (1961).
19. HOGNESS, D. S., COHN, M., AND MONOD, J., *Biochim. et Biophys. Acta*, **16**, 99 (1955).
20. ROTMAN, B., AND SPIEGELMAN, S., *J. Bacteriol.*, **68**, 419 (1954).
21. PIZER, L. I., AND COHEN, S. S., *J. Biol. Chem.*, **237**, 1251 (1962).
22. WIBERG, J. S., DIRKSEN, M., EPSTEIN, R. H., LURIA, S. E., AND BUCHANAN, J. M., *Proc. Natl. Acad. Sci. U. S.*, **48**, 293 (1962).
23. DIRKSEN, M., HUTSON, J. C., AND BUCHANAN, J. M., *Proc. Natl. Acad. Sci. U. S.*, **50**, 507 (1963).
24. SIMON, E. H., AND TESSMAN, I., *Proc. Natl. Acad. Sci. U. S.*, **50**, 526 (1963).
25. MATHEWS, C. K., AND COHEN, S. S., *J. Biol. Chem.*, **238**, 367 (1963).
26. GREEN, M., AND COHEN, S. S., *J. Biol. Chem.*, **225**, 387 (1957).
27. MACFADYEN, D. A., *J. Biol. Chem.*, **158**, 107 (1945).
28. HATEFI, Y., TALBERT, P. T., OSBORN, M. J., AND HUENNEKENS, F. M., in H. LARDY (Editor), *Biochemical preparations, Vol. 7*, John Wiley and Sons, Inc., New York, 1960, p. 89.
29. TISELIUS, A., HJERTÉN, S., AND LEVIN, Ö., *Arch. Biochem. Biophys.*, **65**, 132 (1956).
30. MAIN, R. K., AND COLE, L. J., *Arch. Biochem. Biophys.*, **68**, 186 (1957).
31. LAYNE, E., in S. P. COLOWICK AND N. O. KAPLAN (Editors), *Methods in enzymology, Vol. III*, Academic Press, Inc., New York, 1957, p. 447.
32. ROBERTS, R. B., COWIE, D. B., ABELSON, P. H., BOLTON, E. T., AND BRITTEN, R. J., *Studies of biosynthesis in Escherichia coli*, Carnegie Institution of Washington, Publication 607, Washington, D. C., 1955, pp. 318–402.
33. JONES, A. S., AND LETHAM, D. S., *Analyst*, **81**, 15 (1956).
34. KOCH, A. L., AND LEVY, H. R., *J. Biol. Chem.*, **217**, 947 (1955).
35. BOREK, E., PONTICORVO, L., AND RITTENBERG, D., *Proc. Natl. Acad. Sci. U. S.*, **44**, 369 (1958).
36. MANDELSTAM, J., *Bacterial Revs.*, **24**, 289 (1960).
37. KOZLOFF, L. M., *Cold Spring Harbor symposia on quantitative biology, Vol. XVIII*, Long Island Biological Association, Cold Spring Harbor, Long Island, New York, 1953, p. 209.
38. TSUGITA, A., GISH, D. T., YOUNG, J., FRAENKEL-CONRAT, H., KNIGHT, C. A., AND STANLEY, W. M., *Proc. Natl. Acad. Sci. U. S.*, **46**, 1463 (1960).

6

Reprinted from *Cold Spring Harbor Symp. Quant. Biol.*, **33**, 485–493 (1968)

DNA Replication after T4 Infection

Fred Robert Frankel

Department of Microbiology, School of Medicine, University of Pennsylvania, Philadelphia, Pennsylvania

VEGETATIVE INTERMEDIATE

Watanabe, Stent, and Schachman (1954) were the first to demonstrate that the vegetative DNA in cells infected with labeled T2 phage possessed macromolecular properties. They found it to be indistinguishable by sedimentation analysis from purified DNA extracted from phage particles. It is clear, however, that the conditions to which the DNA had been subjected had severely reduced its molecular weight. Several groups using extraction and analysis procedures more suitable for fragile macromolecules then found that the intracellular DNA was indistinguishable from undegraded DNA molecules from phage particles (Pouwels et al., 1963; Kozinski et al., 1963). By contrast, we found that little or none of the newly synthesized DNA (Frankel, 1963) and only part of the parental DNA (Frankel, 1966a) in infected cells resembled the DNA derived from the mature phage. A large fraction of the intracellular DNA sedimented much more rapidly than mature phage DNA and consequently must differ from it (Frankel, 1966b). However, the basis for this difference is still not entirely clear.

Figure 1 shows the results of sedimentation analysis of both the parental and newly synthesized DNA at several times after infection. Almost immediately after infection part of the parental DNA began to sediment rapidly, and there was little subsequent alteration of the ratio of normal to rapid sedimentation throughout the infection. It has not been possible to distinguish the slower sedimenting parental component from mature phage DNA. The heterogeneous sedimentation of the other parental component marked also the sedimentation position of the newly synthesized DNA during the first 15 min of infection. After this time there was little increase in the amount of the rapidly sedimenting material, and the bulk of the newly synthesized DNA appeared in phage particles which in these analyses remained undetected at the bottom of the centrifuge tube. The small amount of DNA at the position of normal phage DNA molecules may represent DNA released from weak incomplete phage structures, rather than a true intermediate between the rapidly sedimenting DNA and DNA in phage. Thus, there appears to be no

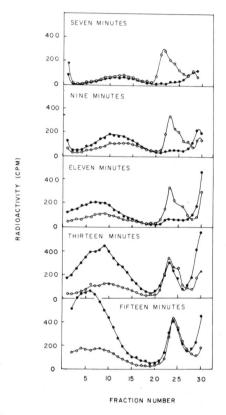

Figure 1. Sedimentation of ^{32}P-labeled T4 DNA of parental origin (open circles) and ^{3}H-labeled newly synthesized DNA (filled circles) in neutral sucrose. Lysates were prepared at the times indicated according to a procedure described previously (Frankel, 1968). Sedimentation was for 35 min at 32,000 rpm, 17°, through 5% to 20% sucrose gradients; sedimentation was from right to left.

accumulation of non-particle phage DNA molecules (Frankel, 1966a).

The newly synthesized DNA that sediments rapidly is a precursor of phage DNA, as shown in the following experiment. Infected cells were grown in medium containing ^{3}H-thymidine from 10 to 12 min, and the labeled precursor was then removed. Samples of the culture were then lysed under

485

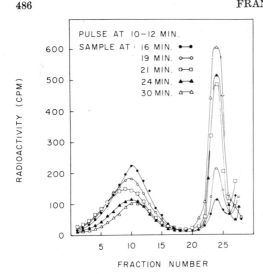

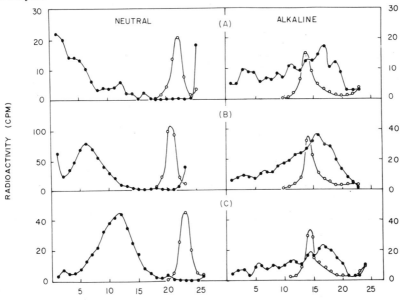

FIGURE 2. Sedimentation of ^{3}H-labeled newly synthesized DNA in neutral sucrose. The two-minute labeling period was terminated by filtration of the culture and resuspension in medium containing 100-fold excess thymidine. Lysates were prepared at the times indicated by a procedure described previously (Frankel, 1966b). Sedimentation was for 50 min at 22,600 rpm, 11°. The figure is a composite of five separate analyses.

conditions that also caused phage particles to dissolve. This permitted a single analysis for both the rapidly sedimenting DNA and phage DNA. The results are shown in Fig. 2. It is clear that labeled phage DNA was produced at the expense of the prelabeled rapidly sedimenting DNA.

PROPERTIES OF THE INTERMEDIATE

Previous studies indicated that little or no protein was associated with the rapidly sedimenting material (Frankel, 1966b). This was shown mainly by our inability to label the DNA with radioactive amino acids and the absence of an effect of pronase on its sedimentation. It is possible that cell polysaccharide is associated with the DNA. Our density analyses which showed the DNA to have the normal density of phage DNA would not have eliminated this possibility. Thus it is not yet possible to say whether the unusual sedimentation property of the vegetative DNA is exclusively an attribute of DNA alone.

That the vegetative DNA does differ from mature phage DNA molecules can be shown in other ways. The left half of Fig. 3 shows the re-analysis of three selected fractions of newly synthesized DNA. It

FIGURE 3. Recentrifugation of fractions of ^{3}H-labeled newly synthesized DNA (closed circles) and admixed ^{32}P-labeled reference DNA (open circles) in neutral (left) or alkaline (right) sucrose. The three successive fractions (A, B, and C) were obtained from a 25 ml, neutral, 5% to 20% sucrose gradient after centrifugation for 45 min at 20,800 rpm. The sample for this preparative run was a lysate prepared at 24 min after infection with T4amB17 (gene 23). Left: 32,000 rpm, 30 min. Right: 32,000 rpm, 90 min.

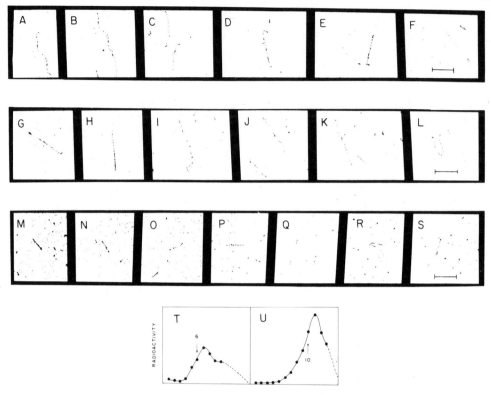

FIGURE 4. Autoradiographs of rapidly sedimenting strands [(A) to (L)] and mature strands [(M) to (S)] from intracellular T4 DNA.

Frames (A) to (L) were derived from the fraction of the alkaline sucrose gradient indicated in (T). The sample for this run was a culture infected with T4amB17, grown in synthetic medium containing ^{3}H-thymidine (22.1 C/mmole) until 18 min, and lysed and denatured as in Frankel, 1968. Sedimentation was for 162 min at 23,000 rpm on a 25 ml sucrose column. Frames (A), (B), and (C) are consecutive photographs of an apparently single filament.

Frames (M) to (S) were derived from the fraction of the alkaline sucrose gradient indicated in (U). The sample for this run was a culture infected with T4$^+$, grown as above for 26 min, and lysed and denatured. Sedimentation as above. These autoradiographs show a high background because of the proximity of the selected fraction to the unincorporated ^{3}H-thymidine at the top of the sucrose gradient.

The sucrose gradient fractions were immediately diluted with water or buffers containing denatured T4 DNA at 5 μg/ml, drawn across clean glass slides, and drained. Some slides were coated with chrome-gelatin before this procedure, and others, after. The slides were overlaid with Kodak autoradiographic stripping film, AR10, and stored at 4° over silica gel. Exposure time for most samples was 74 days. The scale shows 50 μ. The procedures follow those of Cairns (1961).

emphasizes both the considerable sedimentation heterogeneity of the material and its relative stability during isolation and re-analysis. When the DNA of these fractions was denatured and the DNA strands analyzed on alkaline sucrose gradients, the results shown on the right half of Fig. 3 were obtained. The vegetative DNA clearly contains few strands that sediment like those from phage particles. The most rapidly sedimenting fraction of vegetative DNA contained strands with a sedimentation rate up to two times greater than that of phage DNA strands. This corresponds to a strand length of about seven times that of phage DNA, if the sedimentation equation of Abelson and Thomas (1966) holds for these molecules. The shorter

strands may have arisen during the isolation procedure from enzymatic cleavage or radiation effects.

Are the strands as long as the sedimentation equation indicates? We have previously presented evidence that they may be (Frankel, 1968). The most compelling argument was that the rapidly sedimenting strands show fragility properties analogous to those of native DNA molecules of known molecular length. In addition, we have preliminary autoradiographic evidence that suggests that rapidly sedimenting single strands from vegetative DNA are longer than phage DNA strands. These autoradiographs are shown in Fig. 4. Mature intracellular phage DNA showed an average length of 38 μ. This is somewhat less than the value obtained by Cairns (1961), probably because of extensive folding of the single stranded DNA. The rapidly sedimenting strands showed an average length of 82 μ.

All vegetative DNA does not contain long strands. When a rapidly sedimenting fraction of vegetative DNA was isolated at 9 min after infection, its strands sedimented more slowly than phage DNA strands. Kozinski has also found that parental DNA that has replicated at least one time and probably contains strands no longer than phage DNA strands shows rapid sedimentation (Kozinski et al., 1967). This rapid sedimentation, at least in early samples, apparently results either from an altered DNA conformation or from its association with cell components.

KINETIC PROPERTIES OF LONG STRANDS

When the DNA of infected cells grown at 36° was examined from the onset of DNA synthesis, strands of gradually increasing length were observed, approaching the length of phage DNA strands at about 10 min (Frankel, 1968). This apparent rate of strand growth is comparable to or somewhat lower than other measurements of the rate of DNA synthesis in infected cells (Werner, 1968). Only after 10 min did longer strands appear. This explains why rapidly sedimenting DNA isolated at 9 min after infection contains mainly short single strands. The short strands that are synthesized at an early time after infection are precursors for the longer strands, as shown in Fig. 5A and 5B.

Previous experiments showed that DNA synthesized during a short pulse at 15 min showed only rapid sedimentation in neutral sucrose gradients, with no labeled material detectable at the position of normal molecules (Frankel, 1966b). This suggested that normal molecules are not precursors which aggregate to form the rapidly sedimenting DNA but that new synthesis occurs directly onto the rapidly sedimenting structure. The nature of this pulse-labeled DNA was examined further by alkaline sedimentation. DNA labeled at 15 min for 30 sec and 10 sec is shown in Fig. 5C and 5D, respectively. The strands formed a rather continuous distribution presumably resulting from synthesis at growing points that had proceeded to different extents from the ends of molecules. The longest strands may have been produced by growing points that had progressed beyond the length of normal phage DNA molecules at the time of the pulse, or perhaps by recombination. This question will be considered further below. The shortest strands in Fig. 5D may be similar to those described by Okazaki et al. (1968).

Are the long strands precursors of phage DNA? Although Fig. 2 shows that the rapidly sedimenting vegetative DNA is a precursor of phage DNA, the ambiguous nature of the material makes it useful to ask this more direct question. DNA of infected cells was labeled until the 13th min after infection, and the fate of the labeled strands examined after removal of the isotope. The results of this experiment are shown in Fig. 6. They

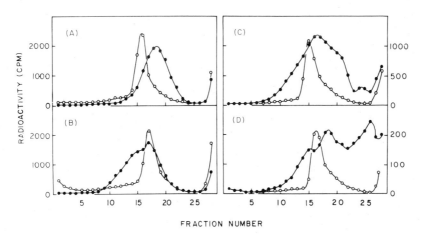

FIGURE 5. A. Sedimentation of newly synthesized DNA labeled with ³H-thymidine from 4 min to 7 min after infection (filled circles) and admixed ³²P-labeled reference DNA (open circles) in alkaline sucrose. Sedimentation was for 90 min at 32,000 rpm, 15°.

B. Effect of a 6 min 'chase' in culture shown in A. At 7 min, a 100-fold excess of thymidine was added to the above culture and a sample removed at 13 min.

C., D. Incorporation of ³H-thymidine at 15 min after infection (36°) for 30 sec (C) or 10 sec (D).

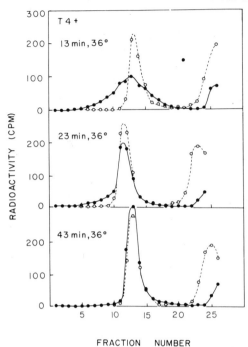

RADIOACTIVITY (CPM)

T 4 +

13 min, 36°

23 min, 36°

43 min, 36°

FRACTION   NUMBER

FIGURE 6. The effect of a 'chase' on DNA labeled with ³H-thymidine from 4 min to 13 min after infection with T4⁺ (filled circles); reference DNA (open circles). At 13 min the isotope was removed by filtration and the culture grown in medium containing 100-fold excess of thymidine. Samples were taken at the times indicated. Sedimentation was in alkaline sucrose as in Fig. 5.

suggest that long strands are transformed to phage length strands. It was also possible to examine this question by use of phage mutants unable to form normal heads. Cells infected with these mutants do not convert the rapidly sedimenting vegetative DNA to normal phage DNA, as shown in Fig. 7 for mutants in four different cistrons (Frankel and Mayne, 1965). Likewise, cells infected with these mutants accumulate long DNA strands at late times after infection. By use of temperature sensitive mutations in these genes, it was possible to accumulate at a restrictive temperature a large pool of DNA containing long strands, and then to test the effect of further incubation at a permissive temperature. The results of such an experiment are shown in Fig. 8. Similar results were obtained with mutants in genes 20, 22, and 31. Within 50 min at 30° about half of the long strands synthesized at 42° had been reduced specifically to phage length strands. Maintenance of the culture at the restrictive temperature resulted only in nonspecific degradation of the strands. It is difficult to prove, however, that the appearance of DNA at the position

of phage length strands is not due to breakdown of the DNA and resynthesis, although a newly synthesized product would have appeared initially as heterogeneous, slowly sedimenting strands, and this was not observed.

We conclude that the long strands synthesized in T4-infected cells are not an abnormal terminal product but a direct precursor of phage DNA, and that their conversion to phage length strands depends on the proper functioning of the head cistrons as well as other genes (Frankel and Batcheler, in prep).

DNA REPLICATION AND STRAND LENGTH

DNA strands probably grow by the stepwise accretion of mononucleotides at or near the terminus of a polynucleotide chain. However, if T4 vegetative DNA contains strands that are longer than strands of phage DNA, one may question the nature of the process that produces these exceptional molecules. The recombinational events described by Tomizawa (1967) and Kozinski and Kozinski (1964) that occur following infection of E. coli by T4 could lead to the formation of long DNA strands. Replication alone or replication and recombination could also explain their appearance. We have examined the possible requirement of DNA replication in the process.

DNA synthesis does not occur in cells infected by T4amN122 (gene 42), a mutant unable to synthesize normal deoxycytidylate hydroxymethylase. We could detect no increase in sedimentation rate of the parental DNA during such an infection although an increase would have occurred after T4⁺ infection. A similar result was obtained with T4amB17 (gene 23), a mutant unable to synthesize head protein, when 40 µg/ml FUdR was included in the medium, and with T4td8, a thymidine-requiring mutant, when thymidine was absent. It is likely that a slight increase in strand length would have been obscured by traces of DNA breakdown, which in fact were evident in some samples at late (30 min, 60 min) times after infection. Under similar conditions Tomizawa and Kozinski were able to detect some covalent association of parental DNA molecules.

We have also tested for the elongation of newly synthesized DNA during inhibition of DNA synthesis. A temperature sensitive mutant in gene 42 (T4tsLB1) is able to promote DNA synthesis at 30°, but incorporation of ³H-thymidine is severely arrested at 42°. Figure 9 shows an experiment in which DNA synthesis was allowed to proceed for 9 min at 30°, followed by a further 10-min incubation at 42°. The arrest of DNA synthesis was incomplete in this experiment, and there was

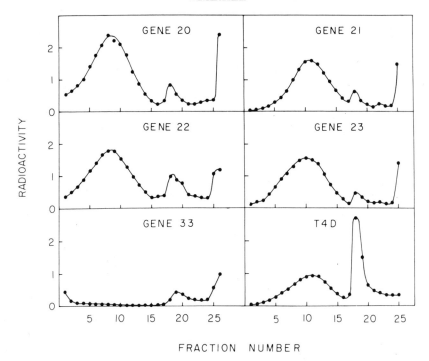

FIGURE 7. Sedimentation in neutral sucrose of newly synthesized DNA labeled with ^{3}H-thymidine until 20 min after infection with phages bearing amber mutations in the head genes indicated. Sedimentation as in Fig. 2.

extensive strand lengthening. However, strand elongation was halted completely by simultaneous temperature shift and transfer to medium containing FUdR. A similar result was obtained using FUdR alone after infection with T4amB17 (gene 23). Figure 10 shows the result of an experiment in which DNA synthesis was inhibited by removal of thymidine from *E. coli* B3 infected with T4td8.

All of these results suggest that DNA replication is required in order to observe an increase in strand length of either the parental or newly synthesized DNA. It is known that arrest of DNA synthesis has either no effect or a positive effect on both genetic and molecular recombination in T4. Therefore, recombination by breakage and rejoining of molecules may occur under these conditions without readily being detected by sedimentation analysis alone. Conversely, extensive chain elongation seems to require replication.

CONCLUSIONS

Presented in Fig. 11 are two models that might account for the formation of long, phage-specific DNA strands in T4-infected cells. The lower is a variation of the one suggested by Streisinger and

co-workers (Streisinger et al., 1964, 1967). According to this model, the parental and progeny strands elongate as a result of recombination at terminal regions of homology. Mixed parental recombinants, produced by recombination at an end of one molecule and an internal homologous region of another, could also form. Both of these processes should occur without extensive DNA synthesis. The formation of long progeny strands would also occur by further replication of the recombinant molecule; this process would be directly affected by inhibitors of DNA synthesis. We could detect no elongation of parental or newly synthesized DNA strands by sedimentation analysis when DNA synthesis was inhibited. Since recombination is known to occur under these conditions, these results suggest either that hydrolysis of DNA strands is obscuring elongation without showing an adverse effect on the formation of recombinants, or that recombination and strand elongation are not necessarily equivalent phenomena and that the model may not adequately account for the latter.

Long DNA strands may also be synthesized by the novel mechanism shown in the upper part of the figure. This mechanism was suggested by Fulton (1965) to explain the transfer of more than a single

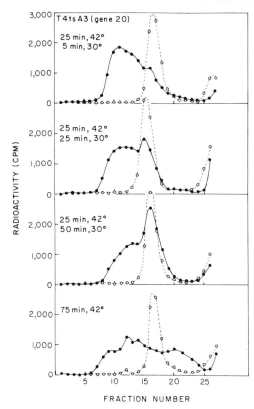

FIGURE 8. The effect of a 'chase' on DNA labeled with ³H-thymidine until 25 min after infection with T4tsA3 at 42° (filled circles); reference DNA (open circles). At 25 min the isotope was removed by filtration and the culture grown at 30° in medium containing 100-fold excess of thymidine. Samples were taken at the times indicated. Sedimentation was in alkaline sucrose as in Fig. 5.

According to the model, the synthesis of a multichromosomal DNA molecule occurs from a growing point that travels without interruption around a circular parental template. Additional growing points may either move along the displaced daughter molecule (as shown) or along the circular parental strand following the initial growth point (as suggested by Werner, 1968 and this volume). Inhibition of DNA synthesis will cause immediate cessation of strand growth. Recombination in this model might occur between two parental circles creating double length circles (Rush and Warner, this volume) or between long daughter molecules.

The circular model of replication allows several predictions. First, if the initial growing point moving around the circle is to travel beyond the point of its origin, the circularizing nascent strand must become covalently linked to the distal end of the parental strand that is being displaced. Otherwise, the first molecule may be released from the structure. Our results show clearly that at an early time after infection no newly synthesized strands are linked to the parental DNA (Frankel,

unit of Hfr chromosome to a recipient cell during mating. We proposed (Frankel, unpubl., 1965; 1968) a similar model for T4 DNA replication to account for our observation that during the eclipse period after infection, normal, mature phage DNA molecules cannot be detected. Furthermore the model can explain the continuous linear rate of DNA synthesis at late times, and the minor effect of chloramphenicol on this rate (see Amati et al., this volume). Some of our original reasons for invoking the model are not very persuasive; for example, simply the condition that all molecules in the pool exist in some stage of replication may explain their abnormal sedimentation properties. A model similar to this one is being presented to explain other modes of DNA replication (Gilbert and Dressler, this volume).

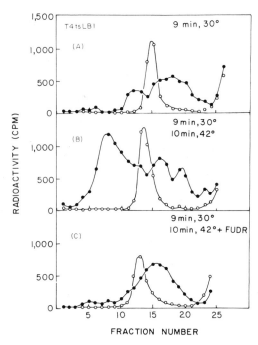

FIGURE 9. Sedimentation in alkaline sucrose of ³H-labeled DNA (filled circles) synthesized after infection with T4tsLB1 for (A) 9 min at 30°, (B) 9 min at 30° followed by 10 min at 42°, or (C) 9 min at 30° followed by filtration and growth for 10 min at 42° in medium containing fluorodeoxyuridine. Sedimentation as in Fig. 5.

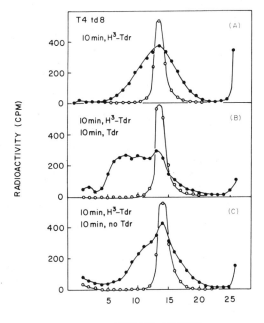

FIGURE 10. Sedimentation in alkaline sucrose of ³H-labeled DNA synthesized after infection of *E. coli* B3 (thy⁻) with T4*td*8 (filled circles), and admixed reference DNA (open circles).

A. Growth in medium containing ³H-thymidine from 4 min to 10 min.

B. Same as A, followed by transfer to and growth for 10 min in medium containing 5 µg/ml of thymidine.

C. Same as B except that second medium contained no thymidine. Sedimentation as in Fig. 5.

1968). The same conclusion can probably be inferred from the work of others (Kozinski and Kozinski, 1964). It is possible that the covalent linkage occurs at a later time, but this is difficult to test because of extensive recombination at late times. A second prediction is that each of the parental strands will experience a different fate

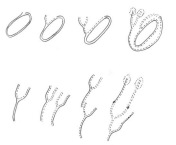

FIGURE 11. Two models for the replication of T4 DNA.

depending on its polarity, one remaining in the pool of vegetative DNA and the other being passed to progeny phage. Parental transfer experiments do not support this idea (Maaløe and Watson, 1951; Hershey and Burgi, 1956), although an examination of the strandedness of the transferred parental DNA has not been made.

As both models are drawn, the formation of mature phage DNA molecules is a late function controlled not by DNA replication but by the process of head condensation, as suggested by Streisinger and co-workers. It is compatible with this idea that strand length continues to increase without the appearance of phage length molecules if head formation is prevented by chloramphenicol or by mutations in the head cistrons. This interaction of the DNA termini with head proteins could act to prevent the initiation of new growth points, which in fact do not increase in number at late times (Werner, 1968). The limitation of the number of growth points, while compatible with the circular model, would seem to lead ultimately to the arrest of DNA synthesis according to other models. This does not occur.

It is clear that more information is necessary before a choice of models describing the replication process in T4-infected cells can be made.

ACKNOWLEDGMENTS

The author thanks C. Clark for excellent assistance in some of the experiments described above and D. Frankel for discussions and other help. Financial support for this research was provided by an award from the National Institutes of Health (A1-05722).

REFERENCES

ABELSON, J. and C. A. THOMAS, JR. 1966. The anatomy of the T5 bacteriophage DNA molecule. J. Mol. Biol. *18:* 262.

CAIRNS, J. 1961. An estimate of the length of the DNA molecule of T2 bacteriophage by autoradiography. J. Mol. Biol. *3:* 756.

FRANKEL, F. R. 1963. An unusual DNA extracted from bacteria infected with phage T2. Proc. Nat. Acad. Sci. *49:* 366.

——. 1966a. The absence of mature phage DNA molecules from the replicating pool of T-even-infected *Escherichia coli.* J. Mol. Biol. *18:* 109.

——. 1966b. Studies on the nature of replicating DNA in T4-infected *Escherichia coli.* J. Mol. Biol. *18:* 127.

——. 1968. Evidence for long strands in the replicating pool after T4 infection. Proc. Nat. Acad. Sci. *59:* 131.

FRANKEL, F. R. and S. H. MAYNE. 1965. Maturation of T4 phage DNA. Abstracts, 66th Meeting of the American Society of Microbiology.

FULTON, C. 1965. Continuous chromosome transfer in *Escherichia coli.* Genetics, *52:* 55.

HERSHEY, A. D. and E. BURGI. 1965. Genetic significance of the transfer of nucleic acid from parental to offspring phage. Cold Spring Harbor Symp. Quant. Biol. *12:* 91.

KOZINSKI, A. W. and P. KOZINSKI. 1964. Replicative fragmentation in T4 bacteriophage DNA, II. Biparental molecular recombination. Proc. Nat. Acad. Sci. *52:* 211.

KOZINSKI, A. W., P. KOZINSKI, and R. JAMES. 1967. Molecular recombination in T4 bacteriophage deoxyribonucleic acid. J. Virol. *1:* 758.

KOZINSKI, A. W., P. KOZINSKI, and P. SHANNON. 1963. Replicative fragmentation in T4 phage: inhibition by chloramphenicol. Proc. Nat. Acad. Sci. *50:* 746.

MAALØE, O. and J. D. WATSON. 1951. The transfer of radioactive phosphorus from parental to progeny phage. Proc. Nat. Acad. Sci. *37:* 507.

OKAZAKI, R., T. OKAZAKI, K. SAKABE, K. SUGIMOTO, and A. SUGINO. 1968. Mechanism of DNA chain growth, I. Possible discontinuity and unusual secondary structure of newly-synthesized chains. Proc. Nat. Acad. Sci. *59:* 598.

POUWELS, P. H., G. VELDHUISEN, M. JANZ, and J. COHEN. 1963. The molecular fate of deoxyribonucleic acid of bacteriophage T4 after infection of *Escherichia coli.* Biochem. Biophys. Res. Commun. *13:* 83.

STREISINGER, G., R. EDGAR, and G. DENHARDT. 1964. Chromosome structure in phage T4. I. Circularity of the linkage map. Proc. Nat. Acad. Sci. *51:* 775.

STREISINGER, G., J. EMRICH, and M. STAHL. 1967. Chromosome structure in phage T4. III. Terminal redundancy and length determination. Proc. Nat. Acad. Sci. *57:* 297.

TOMIZAWA, J. 1967. Molecular mechanisms of genetic recombination in bacteriophage: joint molecules and their conversion to recombinant molecules. J. Cell Physiol. *70:* 201.

WATANABE, I., G. S. STENT, and H. K. SCHACHMAN. 1954. On the state of the parental phosphorus during reproduction of bacteriophage T2. Biochim. Biophys. Acta *15:* 38.

WERNER, R. 1968. Distribution of growing points in DNA of bacteriophage T4. J. Mol. Biol. *33:* 679.

DISCUSSION

A. KOZINSKI: We have confirmed the appearance of a rapidly sedimenting moiety in alkaline sucrose gradients. Controlled nicking of this material assures us that these structures are indeed linear. We cannot isolate any appreciable amount of this material prior to 10 min, at which time 20–25 phage equivalent units of DNA were formed. We were unable to observe the covalent addition of newly synthesized progeny DNA to parental strands at the early stages of replication. I have two questions. (1) Do you ever observe covalent addition of progeny DNA at the initiation of replication? (2) Are you able to detect a sizable amount of a fast sedimenting material when 5–10 phage equivalent units of DNA have been synthesized?

F. FRANKEL: As I indicated in my discussion, we do *not* find a covalent association of parental and progeny DNA shortly after the initiation of DNA synthesis. Nor do we find rapidly sedimenting DNA strands before 10 min, under the conditions of our infection.

D. BOTSTEIN: How does the circular model of replication account for polarized segregation?

F. FRANKEL: I believe that Dr. Mosig has discussed this point in her presentation.

III
Cistrons and the Rule of the Ring

Editor's Comments on Papers 7 Through 15

Today, biochemical techniques have advanced to the stage where, in certain cases, the nucleotide sequences of single genes or parts of genes can be determined. We can even see individual DNA molecules in the electron microscope and thereby determine length and gross structural features. In general, however, our knowledge of the number, kind, and arrangement of genes in an organism is still best obtained from genetic analysis, that is, by mating mutant individuals and observing dominance relationships, complementation ability, and segregation patterns. Genetic analysis can also provide unique insights into the physiology of an organism. There is no better way to determine the physiological role of a compound, enzyme, or structure than to eliminate or qualitatively change that entity by mutation.

The idea that the hereditary mechanisms of a virus bear any relationship to those of higher organisms was a new one in 1948 when, as described in Paper 7, Hershey and Rotman first began infecting cells with two different phage mutants. Their discovery that mixedly infected cells yield recombinant phage particles having characteristics inherited from both parents, and that, just as with *Drosophila,* a genetic map could be constructed from the recombinant frequencies, were revolutionary findings prompting the cautionary note that ". . . the linkage system discerned in the virus may yet prove to have little in common with the more familiar linkage systems in cellular organisms." Among the mutants of T2 which Hershey and Rotman studied were those with an altered plaque morphology. The mutants gave a larger and sharper plaque than the wild-type and were called r mutants because in liquid culture they induced lysis of the host more rapidly than the wild-type. Hershey and Rotman found that almost every independently isolated r mutant would yield wild-type recombinants when crossed with any other r mutant.

According to genetic dogma of the time, recombination could not occur within a gene. If this were true, it would be necessary to conclude that every r mutant corresponds to a different phage gene. Hershey and Rotman were apparently not happy with this interpretation and suggested that ". . . the different r mutants might represent alterations of the same biochemical property." Although the precise definition of a "biochemical property" was left open, this statement bore the seeds of truth.

Benzer *(11)* then made the discovery that one group of *r* mutations, which were located in a cluster on Hershey's map and called *r*II, had one property in common: These *r*II mutants, unlike the wild-type and *r* mutants located elsewhere, would not replicate and thus would not form plaques on strains of *E. coli* lysogenic for phage λ. Benzer quickly recognized the technical advantage which this finding provided to genetic studies with *r*II mutants. The progeny of a cross between two *r*II mutants could be plated on a λ lysogen to detect and measure the wild-type recombinants *selectively* since only such recombinants would form plaques. Whereas, previously, to measure a recombination frequency of 0.01 percent required the screening of 10^6 plaques on 10,000 plates, Benzer could now do this on only one plate.

With this ability to detect low frequencies of recombination, the fine structure of the *r*II region was revealed to consist of several hundred independently mutable and recombinable sites. The idea that each of these sites represented a separate gene thus became further suspect since, even though T4 is a complex virus, it was not reasonable that it would contain hundreds of genes all concerned with the time of lysis. The paradox hinged, of course, on what was meant by a gene. To this question Benzer provided an operational answer based on the finding that *r*II mutants are separable into two groups A and B. Mixed infections of a λ lysogen with any two A-group mutants or any two B-group mutants failed to yield progeny, but mixed infection with any A-group mutant and any B-group mutant yielded progeny at a level similar to wild-type. This classification of mutants into A and B groups was based, not on recombination, but on functional complementation; the function missing in A mutants could be supplied by B mutants and vice versa. When, however, the map positions of the A and B mutations were compared, they were found to be located in two distinctly defined adjacent nonoverlapping regions. Benzer named these regions cistrons *(12)* and proposed that a cistron is the segment of the genetic material which determines the amino acid sequence of a polypeptide chain. The gene was thus redefined in terms of function, and the notion that the parts of a gene are inseparable by recombination was dropped. The terms "gene" and "cistron" are now used interchangeably, although the latter is often reserved for genetic loci with demonstrated functional unity.

Benzer found the cistron to be a strictly linear structure *(13)* the components of which are induced to mutate by agents that either interact with DNA or that mimic the normal DNA bases. The genetic properties of the cistron thus affirmed its identification with DNA, the structure of which had been determined several years earlier. A summary of much of the work on the *r*II region is given in Paper 8, which includes a detailed genetic map.

The paramount question of molecular genetics in the decade following 1953 was the nature of the code relating the amino acid sequence of a polypeptide to the nucleotide sequence of its cistron. Although the ultimate deciphering of the code was accomplished biochemically *(140)*, phage genetics provided several key parts of the solution. One of the questions concerned the existence of nonsense in the code. It had been reasoned *(39)* that since there are only four nucleotides in DNA but 20 amino acids in proteins, the coding unit of the DNA, called the codon, must be a group of nucleotides. The size of this group must be larger than two since four letters arranged in groups of two still can form only 16 words. The minimum codon size therefore must be three, from which 64 words can be formed. The 44 extra codons might serve as alternates for certain of the amino acids, but, it was argued, some might code for no amino acid (nonsense codons). In Paper 9, Benzer and Champe describe studies with *r*II mutants which showed that nonsense codons can be created by mutation. It was further shown that these nonsense mutations exhibit an effect on the *r*II gene function in one host strain but not in another, implying that the genetic code is not universal with respect to its nonsense elements. In genetic terminology, the host in which the nonsense mutation is not expressed is said to harbor a suppressor mutation. Suppression of nonsense mutations is now thoroughly understood in terms of mutational changes of specific transfer RNA molecules, thereby altering their codon recognition specificity *(69, 74, 166)*. The class of nonsense mutations studied by Benzer and Champe was later to be called *amber* (see below). Two additional nonsense classes (*ochre* and *opal*) were subsequently identified *(20, 23)*. With the eventual complete biochemical deciphering of the genetic code *(140)* it was found that these three codons (UAG, UAA, and UGA) are the only three of the 64 which do not, in general, code for an amino acid.

In early speculations on possible coding schemes *(39)* it was recognized that the process of translating a sequence of nucleotides must include some mechanism that allows the correct nucleotide groups to be read while ignoring overlapping groups formed from adjacent codons. This could be accomplished most simply if the message were translated sequentially, codon by codon, starting at a fixed origin. The correct codons would thus automatically be read if the phase of reading or the reading frame is preserved as translation proceeds down the message. The proof that this mechanism is the correct one was provided by genetic studies with *r*II mutants and was stimulated by the suggestion of Brenner et al. *(22)* that acridines, a class of compounds that intercalate between the bases of the DNA helix, induce mutations in T4 by causing deletions and insertions of single nucleotides. Crick and colleagues argued that if acridines act in this manner, an acridine-induced mutation should throw the reading out of phase from the point of the mutation to the terminating end of the cistron and thus cause a drastic alteration of the polypeptide product. If this mutation is, for example, a single nucleotide deletion, a second mutation, which is a single nucleotide insertion, would be expected to restore the reading frame except in the region between the mutations. In support of this hypothesis it was found, in fact, that revertants of acridine-induced *r*II mutants were often composed of the original mutation plus a closely linked second mutation.

Crick saw that by extending this argument the size of the codon could be determined, for if the codon were n-nucleotides in length, then the combination of n-mutations, all of which are single nucleotide deletions or all of which are single nucleotide insertions, should result in a normal reading frame outside the region of the cistron encompassed by the mutations. How Crick applied this to show that $n = 3$ is the subject of Paper 10. The complete report of this work, a monumental opus not recommended for casual reading, can be found in the *Philosophical Transactions (7)*.

The simplest kinds of code used by cryptographers are those in which the elements of the cryptogram and the corresponding elements of the translated message have the same linear sequence; i.e., they are collinear. More complex codes have been devised for which collinearity is not the case. One of the first questions asked by genetic cryptographers was whether the gene and its polypeptide product are collinear, as might be naively expected, or have some more intricate topological relationship. An experimental proof of collinearity would be the demonstration that a series of genetically ordered missense mutations in a cistron produce altered polypeptide products in which the positions of the amino acid changes are in the same linear order as the sites of the mutations. Although such a proof is simple in concept, when the question was first posed there was no system sufficiently developed to allow analysis of both the gene and its protein.

Sarabhai et al. (Paper 11) devised an ingenious approach to the problem. They reasoned that a nonsense mutation should result in premature termination of a growing polypeptide chain at the site of the mutation, yielding an N-terminal fragment of the chain. If then the gene and its protein product are collinear, a series of nonsense mutations in a gene should produce a corresponding series of polypeptide fragments with lengths proportional to the genetic distance of the mutational site from the N terminus of the gene. This reduces the problem from one of protein sequence determination to one of length determination. The major experimental requirement for this method to be successful is an easy means of identifying the functionless nonsense polypeptide fragment among the myriad of other proteins of the cell. Again T4 proved to be the material of choice, owing to the fact that, during the latter part of the latent period, most of the protein synthesized is the phage-head protein. The gene for this protein had recently been identified and the necessary nonsense mutants had been isolated. The results of these studies, described in Paper 11, showed that, for once at least, Nature had not conspired to confound us: The gene and its protein proved to be collinear.

Although such excursions in the interior of the gene provided insights into the nature and function of the genetic material, they did little to illuminate the replicative processes of the phage. Until the early 1960s only a handful of genes had been identified for T4, and the genetic map was best characterized as being composed of empty spaces. This could be understood if the genes populating these uncharted regions were indispensable genes. For such genes, a mutation that destroys the gene function would be lethal to the phage and the mutant would not be recoverable. We have seen already, however, that the "indispensability" of a phage gene can depend on the host in which the phage is growing. The *r*II function is indispensable in a host carrying prophage λ (a restrictive host) but is dispensable in a host lacking the prophage (a permissive host). Epstein, Edgar, and their colleagues reasoned that perhaps other phage genes might exhibit a similar conditional host-dependent lethality. They thus selected two different strains of *E. coli* independently isolated from nature and undertook to isolate phage mutants that would form plaques on one strain but not on the other. After collecting a number of such mutants, which they called *amber*, they proceeded to map them and found, to their surprise, not a cluster of closely linked mutational sites as had been found for *r*II mutants, but sites distributed throughout the uncharted regions of the map defining many different genes. Although it was not recognized until later. Epstein and Edgar had isolated nonsense mutants of the same type that had previously been identified among *r*II mutants (Paper 9). These *amber* mutants could be recovered even though the affected gene was indispensable because one of the host strains they used, the permissive host, fortuitously carried a suppressor specific for this kind of nonsense. Another less serendipitous approach to isolating conditional lethal mutants had been employed years earlier for other organisms but had been overlooked by phage geneticists. The idea was to seek mutants that can replicate at one temperature (usually 30°C) but not at another temperature (usually 42°C). Edgar and Epstein obtained many such temperature-sensitive mutants for T4 and found that the mutations occurred in many of the same genes defined by *amber* mutants *(50, 53)*.

The genetic and physiological studies by Epstein, Edgar, et al. with these conditional lethal mutants are summarized in Paper 12. Some mutants proved to be defective in DNA synthesis while others were found to be blocked in the assembly of the pieces of the phage particle. Among the assembly-defective mutants a number of different types could be identified, the study of which has contributed greatly to our understanding of the assembly process (see Part IV). One striking feature of the T4 chromosome revealed by these studies is the strong tendency for genes of related function to be clustered. The partitioning is not perfect, but the distribution is far from random. In the decade since these initial studies were reported, a number of additional genes have been discovered. To date there are 64 known genes defined by conditional lethal mutations and an additional 25 or so nonessential genes (see the Appendix).

At about the same time that the T4 genome was being unveiled, Streisinger and his colleagues began to reconsider the phenomenon of phage heterozygosity, the nature of which had been a puzzle since its discovery by Hershey and Chase *(90)* in 1951. Heterozygotes in T4 can be observed among the progeny of cells mixedly infected with, for example, *r* and *r*+ phage. Most of the progeny will form plaques that are either pure *r* or pure *r*+ (aside from rare mutants), but a small proportion (about 1 percent) produce plaques that contain both *r* and *r*+ phage in about equal numbers. After ruling out trivial explanations, it was concluded that the phage particles initiating these rare plaques of mixed genotype must have contained two copies of the *r* gene—one copy being the mutant allele and the other copy the wild-type allele. Further work showed that, although any genetic locus of the phage could be heterozygous, a phage particle that is heterozygous for one locus is, in general, not heterozygous for other loci. This suggested that the region of heterozygosity is a small part of the total genome.

By 1960 the accumulated genetic evidence *(46)* strongly suggested that the regions of heterozygosity were located near a physical terminus of a DNA strand. This, in turn, implied that the DNA molecule could not be covalently continuous but must somehow be internally interrupted, for otherwise only those loci near the two ends of the DNA molecule could be heterozygous, contrary to the evidence that any locus could be heterozygous. This model thus predicted that if T4 DNA were melted it would fall into a number of pieces, but when the measurements were performed no such fragmentation was found *(14)*.

When a paradox based on trusted facts arises in science it is often a signal that something is incorrect in the basic assumptions. A basic assumption of the model predicting interruptions in the DNA is that the sequence of genes is the same for each DNA molecule derived from a population of phage. In Paper 13 Streisinger, Edgar, and Denhardt challenge this assumption and propose instead that (1) the sequence of genetic elements at one end of the linear chromosome is always repeated at the other end, and (2) the ends of a population of chromosomes are distributed over the genome; i.e., the population consists of chromosomes whose sequence of elements are circular permutations of one another. To illustrate this more clearly let us define the T4 genome as the full complement of elements a . . . z (26 elements). Then according to their assumptions the chromosome of one phage particle looks like this: abc . . . zabc; a second like this: bcd . . . zabcd; a third like this: cde . . . zabcde; and so on, each chromosome having the three terminal elements repeated. As bizarre as they seem, these assumptions resolve the paradox, for they place the repeated genes always at the ends of the chromosome and also allow all genes to be at the ends. One obvious prediction of this model is that the genetic map of T4 should be circular. The mapping experiments in Paper 13 test this prediction and show it to be true.

In a later paper *(174)*, Streisinger, Emrich, and Stahl argued that if a phage chromosome is terminally redundant, head-to-tail pairing and recombination between two replicas should generate double-length molecules and reiteration of the process should produce long chains of repeated genomes (concatemers). During maturation of the phage particle this long chain must be cut into phage-size pieces the length of which must be determined by some mechanism extrinsic to the DNA. Streisinger proposed that the length determinant is the volume of the phage head (from which comes the obvious sounding but not trivial statement that T4 contains a headful of DNA). If this is true, a genetic deletion in T4 should not change the amount of DNA in the phage particle but rather should increase the length of the region of terminal redundancy. Again this prediction was upheld by experiment *(174)*.

At first sight Streisinger's two assumptions—terminal redundancy and circular permutation—appear to be unrelated to one another. As Frankel *(64)* pointed out, however, these two aspects of the model have an appealing unity since both can be generated by a single mechanism. If replicating phage DNA consists of long chains of repeated 26-element genomes—a . . . za . . . za . . . za . . . z . . .—and if these are cut, starting at an end, into 29-element lengths during maturation, the resulting pieces will be both terminally redundant and circularly permuted. A lucid account of the development of these ideas is given in Paper 14 by Stahl, who has been one of the principal contributors to the concept of genetic circularity and its implications.

Direct confirmation of the proposed structure of the T4 chromosome was provided by Thomas and collaborators *(133, 184, 185)* using electron microscopic visualization of DNA which had been denatured and reannealed. If the DNA molecules derived from a population of phage have circularly permuted elements and contain repetitious terminals, such a procedure should yield circles of duplex DNA with appended regions of single-stranded DNA ("bushes") representing the repetitious terminals that cannot find complementary partners, as illustrated in Fig. 2 of Paper 15. In this paper Thomas summarizes the results of application of this method, and variations of it, to other phages (P22, T3, T5, T7, and λ). In all cases terminal repetitions were found ranging in length from about 10 nucleotide pairs for λ up to 6000 nucleotide pairs for T2 and T4. For some phages, however, (T3, T5, T7, λ) the DNA was found to have a unique nonpermuted sequence. For these phages it must be supposed that, unlike T4, the concatenated DNA is cut by two staggered sequence-specific breaks, one in each chain of the duplex. These breaks must be located on each side of the terminally repetitious sequence and so oriented that, of the four ends so generated, the two interior ends are 5'. DNA polymerase, which synthesizes only in the 5' to 3' direction, could then elongate the 3' ends generated by the breaks, thereby displacing the 5' ends from the duplex until finally the fragments are physically separated. In the special case of phage λ whose mature DNA has 5' single-stranded ends (see Part V, Paper 25), packaging apparently occurs without elongation of the 3' ends.

To date all known double-stranded phage DNAs have been found to be terminally repetitious. Watson *(196)* has recently proposed a theory that may explain why this is so. He argues that if a linear DNA molecule replicates bidirectionally, starting at some internal point, and proceeds according to the scheme proposed by Okazaki *(142)* (or a similar scheme characterized by 5' to 3' growth with bridge formation at the replicating fork followed by "backward" growth on the opposite template strand), chain growth should reach the 5' end of the template without impediment but should be stopped short of the 3' end because once the 5' end has been reached the parental strands separate and bridge formation is prevented. The result will be a pair of daughter helices each having a 3' single-stranded tail, one member of the pair with the tail on the left end and the other member with the tail on the right end. Further replication of these daughter molecules would lead progressively to shorter and shorter molecules, with catastrophic consequences. This obviously does not happen. The reason it does not, Watson suggests, is that the daughter molecules form end-to-end hydrogen-bonded dimers by virtue of the fact that their single-stranded tails were derived from the repetitious ends of the duplex and thus have complementary sequences. This dimer could then be converted to a covalently continuous double-stranded form by the action of appropriate known enzymes (polymerase, exonuclease, ligase), and the process could be repeated to produce larger multimers. According to this model terminal redundancy is thus necessary to avoid what might be called "the single-strand catastrophy" that would otherwise result from the replicative process.

This model was developed for phage T7, whose DNA is known to replicate as a linear molecule *(206)*. For T4 it has been proposed that DNA replication occurs, at least in its later stages, by a rolling-circle mechanism *(72)* as illustrated in Fig. 11 of Paper 6, and indeed evidence has been presented for circular structures in replicating T4 DNA *(15)*. In this case end completion does not pose a problem, and concatemers could be formed merely by continuous replication around the circle. The rolling-circle model for T4, however, has recently been put in doubt, at least as the primary mechanism *(45)*. If it occurs at all, concatemers, formed in the early stages of replication by the same mechanism proposed for T7, might be required for efficient circle formation, since concatemers would have greater internal homology than monomers. In any case, whether concatemers arise by circular replication or by fusion of 3'-tailed monomers, repetitious terminals are a prerequisite. This strategy has an appealing circularity to its logic: Terminal redundancy results in concatemer formation, and the cutting of a concatemer produces a molecule with terminal redundancy. For cases in which the concatemer is not cut at specific sequences, terminal redundancy makes biological sense in another respect. If, as seems likely, the length-determining mechanism is not completely precise, packaging of complete phage genomes would be better assured by cutting off a bit more than a complete genome. As Thomas concludes in Paper 15, "To repeat one's self only when one has reached the end, that may be *The Rule of the Ring*."

7

Reprinted from *Proc. Natl. Acad. Sci. (U.S.)*, **34**(3), 89–96 (1948)

LINKAGE AMONG GENES CONTROLLING INHIBITION OF LYSIS IN A BACTERIAL VIRUS*

BY A. D. HERSHEY AND RAQUEL ROTMAN†

DEPARTMENT OF BACTERIOLOGY AND IMMUNOLOGY, WASHINGTON UNIVERSITY SCHOOL OF MEDICINE, ST. LOUIS, MISSOURI

Communicated by M. Demerec, December 5, 1947

Experiments in which bacterial cells are simultaneously infected with two different viruses of the T2 serological group have suggested that both can grow in the same cell,[1] and that interaction between them produces new genetic types of virus.[2, 3] Further experiments, in which the mixed clones of virus coming from single bacterial cells were examined, have confirmed these inferences, and indicate that the new types arise only in those bacteria in which both viruses succeed in growing freely. These experiments are being continued in an effort to learn something about the mechanism of genetic interaction, and will not be reported in further detail at this time. In this paper we wish to describe a new series of mutations in the bacterial virus T2H,[1] discovered in connection with the experiments mentioned above, which point to the operation in this virus of a system of genetic linkage.

The virus T2H, like all its relatives, exists in the form of a lysis-inhibiting wild type (r^+) and an r (for rapidly lysing) mutant phenotype, which has lost the lysis-inhibiting character.[1] The principal new finding to be reported here is that all the independently arising r mutants of T2H that we have examined are genetically different. The basic observations may be summarized as follows.

A genetic difference between two r mutants is shown when bacteria infected with both liberate viral progeny containing a measurable proportion of lysis-inhibiting types, whereas bacteria infected with the same number of particles of either mutant alone yield only r viral progeny. The experimental procedure used is the one-step growth experiment of Delbrück and Luria, with mixed multiple infection, carried out as described earlier.[1] For simplicity of language, we shall speak of this experiment as a cross, described as heterologous when new types appear among the yield of virus, and homologous when only the parent types are recovered.

Certain details of the method require mention. First of all, one examines by it not the progeny of a genetic cross strictly speaking, but simply a mixed population of virus growing in the same community. The crosses reported in this paper were so arranged that the yield of virus examined came from about 40,000 mixedly infected bacteria. Variations in yields from individual bacteria[4] are thus left out of account. Care was taken to ensure adequate and equal multiplicity of infection, with respect both to titers of stocks used and to adsorption to bacteria. Fortunately, all mutants were found to be equally well adsorbed in sensitive tests with admixed wild type (table 2). The effect of unequal multiplicity on the yield of segregants was also tested. The largest proportion of wild type resulted from equal multiplicity of infection with the two parent types, but the yield was not markedly affected by variations between 1 : 2 and 2 : 1. Variations in total multiplicity above about 3 of each type per bacterium had no measurable effect on the result. A multiplicity of about 5 of each type per bacterium was habitually used. Attention was also paid to the total yield of virus in crosses. In parallel experiments, the yields seldom differed by more than 15 per cent, and fell between 300 and 600 per bacterium, depending in part on variations in the nutrient broth. Because of these variations, difficult to control, important comparative results were always checked by parallel experiments.

The *r* mutants so far examined, among which 15 have been named *r*1, *r*2, etc., in order of isolation, are all identical with respect to type of plaque. Their behavior in intercrosses places them in two clearly defined classes, called A and B. Crosses between any pair belonging to the two classes yield about 15 per cent of wild type among the total yield of virus, no significant differences having been established. Crosses within classes yield proportions of wild type characteristic for each pair, falling between 0.5 and 8 per cent. The first eleven *r* mutants to be examined were intercrossed in all combinations, yielding only one example (*r*1) of class A, and no homologous pairs. Six others were examined less completely in a search for a second member of class A; all proved to be class B, and at least two differed from all the others. The second member of class A was finally obtained from an *r* mutant of the distinct virus T4, by crossing with T2H.[2] This mutant is called *r*14. Three examples of class B were discarded without numbering.

The results of the quantitative intercrosses, summarized in table 1, present three significant features. (1) The amounts of wild type appearing in the three possible crosses among sets of three mutants of class B (e.g., *r*2, *r*5 and *r*7) are approximately additive, but generally fall noticeably short, so that no linear order valid for several loci can be discerned. (2) The results of intercrosses among sets of four mutants (e.g., *r*2, *r*4, *r*5 and *r*7) cannot be interpreted in terms of four independent transfer frequencies, but indicate some type of linkage. (3) If linkage is assumed, the classes A and B are probably to be regarded as independent linkage groups. However, the fact that *r*1 of class A gives equal amounts of wild type with all members of class B is not adequate proof of this independence, for the same thing is seen at a lower level in the crosses between *r*13 and the other members of class B.

The different *r* mutants also differ with respect to rate of spontaneous

87

TABLE 1

PER CENT OF WILD TYPE IN VIRUS YIELDS FROM BACTERIA INFECTED WITH PAIRS OF
r MUTANTS

←class A→						class B				
r1	r14	r13	r2	r4	r8	r9	r3	r5	r7	r6
r1	0.5	14	17	16	(15)	(13)	14	15	15	15
	r14	(13)	X	X	X	X	X	X	15	X
		r13	6	6	(7)	X	(7)	(6)	7	8
			r2	1.1	1.7	1.5	2.0	3.9	3.0	2.4
				r4	1.1	(1.5)	2.6	2.9	3.5	3.2
					r8	(0.8)	(2.1)	(2.3)	(2.3)	2.7
						r9	(0.9)	(1.8)	(2.2)	(2.4)
							r3	(1.2)	(1.0)	1.4
								r5	0.8	(1.1)
									r7	(0.5)

The crosses indicated by X have not been made.   Results in parentheses are from
single experiments; the others are averages of several.   Self-crosses yield less than 0.1%
wild type, usually none.

reversion to lysis-inhibitor during serial passage of the virus.[1]   In two or
more independent tests starting from different single plaques of each of
several r mutants, the number of passages required to yield lysates containing 10 to 20 per cent of lysis-inhibiting virus varied as follows: for r9, two or
three; for r5, r6, r11, five; for r2, r3, r4, r8, r10, r14, six or seven; for r7,
r13, eight; for r1, ten; and for r15, about twenty-two.   The lysis-inhibitor
recovered from r2, r13, and r15 differed from authentic wild type in appearance of plaques, being of intermediate character between r and wild
type.   On further serial passage, variants apppeared which resembled wild
type more closely.   Both types of lysis-inhibiting virus obtained from these
three mutants were genetically different from wild type, for when either
was crossed with wild type, excessive numbers of r mutants were found
among the progeny.   This was true of seven independently arising reversions from one or another of the mutants named.   Presumably the lysis
inhibiting virus selected in these cases arose by suppressor mutations,
rather than by back mutation at the respective r loci.   On the other hand,
lysis-inhibiting viruses recovered from each of the 11 remaining r mutants
mentioned above were indistinguishable from wild type both phenotypically
and in back cross.   We have not, however, done the careful experiments
which might reveal suppressor mutations closely linked to an r locus.

Two different r loci were found to mutate independently in a double-
mutant virus obtained as described in the following section of this paper.
In four independent serial passages starting from different single plaques of
the double mutant r1r5, the reversion passed through the more stable type
r1 in each case.   This suggests that the difference in the rate of reversion of
the two single mutants is due to differences in stability of loci, rather than to

different selection rates among mutants growing in competition with wild type.   Direct estimations of selection rates have not been made, however.

In contrast with the spontaneous reversions, lysis-inhibiting virus arising in crosses between *r* mutants is always homologous with wild type.   Tests of this point included segregants from the cross *r2* × *r13*, two mutants whose spontaneous reversions are heterologous with wild type.   There is, furthermore, no correlation between the individual stability of two *r* mutants and the amount of wild type they throw in crosses.

A few of the mutants, including two lines each of *r2* and *r13* mentioned above, were carried through the transformations *r* → *hr* → *r* by crossing with a host-range mutant (*h*), followed by a cross with wild type.   The resulting stocks showed the same behavior in crosses as the originals.   It is unlikely, therefore, that these are multiple-locus mutants.

In order to compare the ability of the different *r* mutants to grow competitively in the bacterial cells, mixed infections with *r* mutant and wild type were made.   In only two cases did the proportion of wild type in the

TABLE 2

COMPETITION BETWEEN *r* MUTANTS AND WILD TYPE IN MIXED INFECTION

| MUTANT | % WILD TYPE | | | MUTANT | % WILD TYPE | | |
	INPUT	ADSORBED	YIELD		INPUT	ADSORBED	YIELD
r1	54	59	53	*r4r7*	52	55	53
r5	50	57	54	*r2*	68	77	64
r1r5	53	55	37	*r3*	58	58	60
r13	51	48	60	*r6*	60	60	57
r4	60	64	65	*r2r3*	63	70	67
r7	59	54	61	*r2r6*	60	56	64
r4r7r13	59	54	54	*r3r6*	66	71	69

yield differ appreciably from the proportion in the infecting mixture, or from the proportion in the fraction adsorbed to bacteria.   With the mutant *r13* the yield contained a proportion of wild type in excess of that adsorbed amounting to about 10 per cent of the total virus (3 experiments).   The significance of this is not clear.   The other exception is easily explained.   The cross between the double mutant *r1r5* and wild type yielded an increased proportion of *r* mutant amounting to about 15 per cent of the total virus (2 experiments), owing to the segregation of the two *r* loci described below.   For other multiple mutants studied, all of class B, the loss of wild type was too small to be detected, presumably because of linkage between the loci.   The data for individual experiments are summarized in table 2.

Segregation of Loci.—The cross between any two *r* mutants yields two new viral types: wild type and an *r* mutant distinguishable from others only by back cross to the parental types, in which it proves homologous with both.   This behavior supports the idea that each mutant represents a modification at a different genetic locus among many controlling inhibition of lysis.

For any pair X and Y, the segregation may be written: $(X^+) \times (^+Y) \rightleftharpoons (^{++}) \times (XY)$. In testing this hypothesis, it is necessary to prepare single-plaque stocks from a number of the r progeny of the cross, and test each one by back cross to each of the parent types. Tests of this kind have confirmed the hypothesis in every case. It was also found, within the limits imposed by the small scale of the experiments, that in each cross the two recombination types appeared in equal numbers, and the two parental types in the same proportion with which the bacteria were initially infected. These data are shown in table 3. In one experiment, virus yields from single bacteria[4] infected with $r4$ and $r7$ were found to contain a small proportion of wild type in nearly every burst, the average being about 4 per cent. From one of the yields, tests of a number of r progeny revealed the three additional viral types expected.

TABLE 3

SEGREGATION OF r LOCI IN INTERCROSSES

CROSS	WILD TYPE (% OF YIELD)	DISTRIBUTION OF TYPES AMONG r PROGENY		
$r1 \times r5$	15	4 $r1$	4 $r5$	2 $r1r5$
$r1 \times r7$	15	9 $r1$	5 $r7$	6 $r1r7$
$r1 \times r13$	15	5 $r1$	3 $r13$	4 $r1r13$
$r4 \times r13$	6	12 $r4$	17 $r13$	1 $r4r13$
$r7 \times r13$	7	7 $r7$	7 $r13$	1 $r7r13$
$r2 \times r3$	3.3	36 $r2$	47 $r3$	3 $r2r3$
$r2 \times r6$	3.4	33 $r2$	41 $r6$	1 $r2r6$
$r3 \times r6$	1.9	63 $r3$	23 $r6$	1 $r3r6$
$++ \times r1r5$	37	2 $r1$	2 $r5$	6 $r1r5$
$r1r5 \times r1r13$	0.0	13 $r1r5$	13 $r1r13$	2 $r1r5r13$ 2 $r1$
$r4r7 \times r7r13$	0.0	12 $r4r7$	15 $r7r13$	3 $r4r7r13$ 0 $r7$

The equality of numbers of the two recombination progeny, only suggested by the data of table 3, has been confirmed on a statistically adequate scale for crosses between host-range and r mutants to be reported elsewhere.

Multiple Factor Tests.—In an effort to learn something about the nature of the linkage, multiple factor crosses have been made. These are of three types.

The first type consists of three factor crosses involving the loci $r1$, $r4$, and $r7$, the two latter being rather closely linked but probably independent of the first. If this is correct, one would expect the factor $r1$ to distribute itself equally between the two types of virus resulting from the enchange of $r4$ with respect to $r7$. The effect should be an equal segregation of wild type from the crosses $r1r4 \times r7$ and $r1r7 \times r4$, amounting in either case to about half that segregating in the cross $r4 \times r7$. For the cross $r1 \times r4r7$, the yield of wild type should be nearly as large as that for $r1 \times r7$, owing to the linkage. The data shown in table 4 confirm these expectations very well.

TABLE 4

SEGREGATION OF WILD TYPE IN MULTIPLE FACTOR INTERCROSSES

CROSS	% WILD TYPE	CROSS	% WILD TYPE
r1r4 × r7	1.4 (3)	r7r13 × r4r13 × r4r7	0.16 (2)
r1r7 × r4	1.4 (3)	r4 × r7 × r13	7.3 (1)
r1 × r4r7	11 (2)	r2r3 × r2r6 × r3r6	0.02 (7)
r1 × r7	15 (4)	r2 × r3 × r6	3.0 (2)
r4r13 × r7	1.8 (4)	r2r3 × r6	0.8 (7)
r7r13 × r4	0.7 (3)	r2r6 × r3	0.3 (6)
r4r7 × r13	5.0 (3)	r3r6 × r2	1.4 (5)
r4 × r13	6.3 (5)	r2 × r3	2.0 (6)
r7 × r13	7.0 (8)	r2 × r6	2.4 (6)
r4 × r7	3.5 (9)	r3 × r6	1.4 (7)

The results shown are averages of the number of experiments indicated in parentheses. The 22 measurements, including all crosses r4 × r13, r7 × r13 and r4 × r7 made during several months, show a coefficient of variation of 20%.

The second type of cross corresponds to the conventional three factor test for linkage order. We have used two sets of three factors all belonging to the class B. The sets include (r2, r3 and r6) and (r13, r4 and r7), the former being more closely linked than the latter. The data for these crosses, also included in table 4, tend to place the factors in the order listed above, though the correspondence with expectations for random crossing over between linear structures is rather poor. The data for r2, r3 and r6, particularly, are compatible with the possibility that a crossover in one region tends to be accompanied by a second in the adjacent region.

The third type of cross, suggested by analogous experiments with bacteria,[5] was tried for the purpose of estimating the amount of repeated exchange (i.e., exchange between recombination progeny and progeny of the parental types) taking place under the conditions of the experiments. The combinations used were such that segregation of wild type would require the interaction (simultaneous or successive) of three different mutants. The sets chosen were (r4r13, r7r13, r4r7) and (r2r3, r2r6, r3r6). In different experiments, multiplicities of about three and five per bacterium of each of the three mutants of the set were tested, with similar results. The data for these crosses, and for the corresponding sets of three single mutants, are shown in table 4. It will be seen that repeated exchanges can be detected by this method; that they are less frequent for the more closely linked factors; and that their number is too small to be of quantitative importance in the linkage tests already described.

It should be noted that repeated exchange of the type detected above should not have any effect in two factor crosses. Repeated exchange between sister recombination progeny, if such arise, presents a more serious difficulty. Until effects from this cause can be evaluated, no great quantitative significance can be attached to the results of intercrosses.

Discussion.—Differences among the *r* mutants described in this paper indicate that there are an indefinite number of distinct loci controlling lysis inhibition in the bacterial virus T2H. The suggestion from the genetic data is that the mechanism of lysis-inhibition[6] may be exceedingly complex. On the other hand, the fact that identical mutants are not encountered shows that the different *r* mutations occur with similar frequency. This remarkable circumstance might be interpreted as an indication that the different mutations result from a single type of chemical change occurring at different points in the genetic material of the viral particle. If this were so, the different *r* mutants might represent alterations of the same biochemical property. This argument must be qualified in view of our ignorance of the number of possible *r* mutations.

The existence of suppressor mutations noted in this paper, and of intermediate mutant types,[3] complicates the system somewhat further.

It is evident that the rather high frequency of *r* mutations, about 10^{-4} per viral duplication,[1] is at least partly due to the large number of loci involved. It is probably not feasible to measure the stability of any given *r* locus. For this reason the test of equal rates of *r* mutation in wild type and in the identical phenotype obtained by reversion from *r*,[3] is of no value as a test of reversibility of mutation. The test by crossing with wild type, reported in this paper, shows that some reversions are not back mutations, but that most of them probably are.

The situation described for the *r* mutants does not prevail among viral mutants in general. We have examined six independently arising host-range mutants of the h^c phenotype,[3] which proved to be homologous in intercrosses. This locus can be assigned a position in the linkage class B however, and it is for their usefulness in mapping that the *r* mutants are likely to prove most valuable.

The linkage system itself we shall discuss only briefly, pending the completion of experiments of another type. It is clear that the clases A and B are independent, or nearly so, and that some kind of linkage exists within the classes. Regarding the linkage structure, some of our data suggest linear arrangement with crossing over, but others are somewhat contradictory. Among difficulties which cannot be discussed very intelligently at present, the necessity for reconciling our data with the results of Luria[7] should be mentioned. It is probably necessary to add that the linkage system discerned in the virus may yet prove to have little in common with the more familiar linkage systems in cellular organisms.

Summary.—Different *r* mutants of the virus T2H differ (*a*) in rate of spontaneous reversion to lysis-inhibiting types; (*b*) in the kind of reversion they undergo, either to a form genetically like, or to forms unlike, the wild type; (*c*) in the amount of wild type they throw in intercrosses. With respect to (*c*) there are two independent linkage groups of distinct *r* loci.

Analysis of one of these groups by single and multiple factor intercrosses has failed to establish or refute the idea of crossing over between linear structures. The existence of some kind of linkage system conditioning segregation and reassortment of genetic factors among viral particles seems quite clear, but it remains to be learned whether reciprocal, or even material, exchanges are involved.

* Aided by a grant from the United States Public Health Service.

† On leave from the Instituto Bacteriologico de Chile, Santiago, Chile.

[1] Hershey, A. D., *Genetics*, **31**, 620–640 (1946).

[2] Delbrück, M., and Bailey, W. T., Jr., *Cold Spring Harbor Symp. Quant. Biol.*, **11**, 33–37 (1946).

[3] Hershey, A. D., op. cit., 67–76.

[4] Delbrück, M., *J. Bact.*, **50**, 131–135 (1945).

[5] Lederberg, J., "Gene Recombination and Linked Segregations in *Escherichia coli*," *Genetics*, **32**, 505–525 (1947).

[6] Doermann, A. H., see "Discussion" in reference 3, and complete paper to appear in *J. Bact.* **55**, 257–276 (1948).

[7] Luria, S. E., these PROCEEDINGS, **33**, 253–264 (1947).

8

Reprinted from *Proc. Natl. Acad. Sci. (U.S.)*, **47**(3), 403–415 (1961)

ON THE TOPOGRAPHY OF THE GENETIC FINE STRUCTURE

By Seymour Benzer

DEPARTMENT OF BIOLOGICAL SCIENCES, PURDUE UNIVERSITY

Read before the Academy, April 27, 1960*

In an earlier paper,[1] a detailed examination was made of the structure of a small portion of the genetic map of phage T4, the rII region. This region, which controls the ability of the phage to grow in *Escherichia coli* strain K, consists of two adjacent cistrons, or functional units. Various rII mutants, unable to grow in strain K, have mutations affecting various parts of either or both of these cistrons. The topology of the region; i.e., the manner in which its parts are interconnected, was intensively tested and it was found that the active structure can be described as a string of subelements, a mutation constituting an alteration of a point or segment of the linear array.

This paper is a sequel in which inquiry is made into the topography of the structure, i.e., local differences in the properties of its parts. Specifically, are all the subelements equally mutable? If so, mutations should occur at random throughout the structure and the topography would be trivial. On the other hand, sites or regions of unusually high or low mutability would be interesting topographic features.

The preceding investigation of topology was done by choosing mutants showing no detectable tendency to revert. This avoided any possible confusion between recombination and reverse mutation, so that a qualitative (yes-or-no) test for re-

Editor's Note: A foldout map that accompanied the original article has been omitted, owing to limitations of space.

combination was possible. The class of non-reverting mutants automatically included those marked by relatively large alterations, which will be referred to as "deletions." Such a mutant is defined for the present purposes as one which inter-

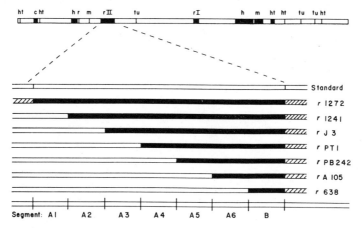

Fig. 1a.—At the top, the *r*II region is shown compared with the entire genetic map of the phage. This map is a composite[15] of markers mapped in T4 and the related phage T2. Seven segments of the *r*II region are defined by a set of "deletions" beginning at different points and extending to the right-hand end (and possibly beyond, as indicated by shading).

sects (fails to give recombination with) two or more mutants that do recombine with each other. Deletions provided overlaps of the sort needed to test the topology and to divide the map into segments.

The present investigation of topography, however, is concerned with differentiation of the various points in the structure. For this purpose mutants which do revert are of the greater interest, since they are most likely to contain small alter-

Fig. 1b.—Mapping a mutation by use of the reference deletions. If mutant x has a mutation in segment 1, it is overlapped by *r*1272, but not by *r*1241. Therefore, standard-type recombinants (as indicated by the dotted line) can only arise when x is crossed with *r*1241.

ations. As a rule (there are exceptions) an *r*II mutant that reverts behaves as if its alteration were localized to a point. That is to say, mutants that intersect with the same mutant also intersect with each other. In a cross, recombination can be scored only if it is clearly detectable above the spontaneous reversion noise of the mutants involved. Therefore, the precision with which a mutation can be mapped is limited by its reversion rate. The detailed analysis of topography can best be done with mutants having low, non-zero reversion rates.

Some thousands of such *r*II mutants, both spontaneous and induced, have been analyzed and the resultant topographic map is presented here.

Assignment of Mutations to Segments.—To test thousands of mutants against one another for recombination in all possible pairs would require millions of crosses. This task may be greatly reduced by making use of deletions. Each mutant is first tested against a few key deletions. The recombination test gives a negative result if a deletion overlaps the mutation in question and a positive result if it does not overlap. These results quickly locate a mutation within a particular segment of the map. It is then necessary to test against each other only the group of mutants having mutations within each segment, so that the number of tests needed is much smaller. In addition, if the order of the segments is known, the entire set of point mutations becomes ordered to a high degree, making use of only qualitative tests.

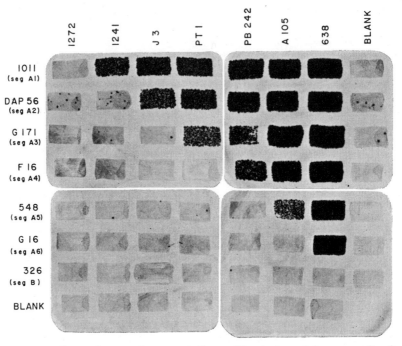

FIG. 2.—Crosses for mapping rII mutations. The photograph is a composite of four plates. Each row shows a given mutant tested against the reference deletions of Figure 1a. Plaques appearing in the blanks are due to revertants present in the mutant stock. The results show each of these mutations to be located in a different segment.

Procedure for crosses—The broth medium is 1% Difco bacto-tryptone plus 0.5% NaCl. For plating, broth is solidified with 1.2% agar for the bottom layer and 0.7% for the top layer. Stocks are grown in broth using *E. coli* BB which does not discriminate between rII mutants and the standard type. To cross two mutants, one drop of each at a titer of about 10^9 phage particles/ml is placed in a tube and cells of *E. coli* B are added (roughly 0.5 ml of a 1-hour broth culture containing about 2×10^8 cells/ml). The rII mutants are all able to grow on strain B and have an opportunity to undergo genetic recombination. After allowing a few minutes for adsorption, a droplet of the mixture is spotted (using a sterile paper strip) on a plate previously seeded with *E. coli* K. If the mutants recombine to produce standard type progeny, plaques appear on K. A negative result signifies that the proportion of recombinants is less than about 10^{-3}% of the progeny.

Within any one segment, however, the order of the various sites remains undetermined. This order can still be determined, if desired, by quantitative measurements of recombination frequencies.

In order to facilitate this project many more deletions have been mapped than were described in the previous paper. These suffice to carve up the structure into 47 distinct segments. By virtue of the proper overlaps, the order of almost all of these segments is established.   Observe first the seven large mutations in Figure 1a. These are of a kind which begin at a particular point and extend all the way to one end.   Thus, they serve to divide the structure into the seven major segments shown.

Consider a small mutation located in the segment A1, as indicated in Figure 1b. It is overlapped by r1272 and therefore when crossed with it cannot give rise to standard type recombinants.   It will, however, give a positive result with r1241 or any of the others, since, with them, recombinants can form as indicated by the

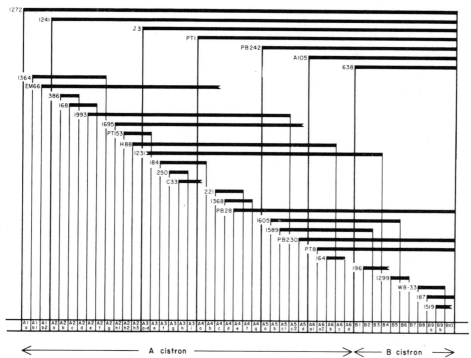

Fig. 3.—Deletions used to divide the main segments of Figure 1 into 47 smaller segments. (Some ends have not been used to define a segment, and are drawn fluted.)   The A and B cistrons, which are defined by an independent functional test, coincide with the indicated portions of the recombination map.   Most of the mutants are of spontaneous origin.   Possible exceptions are EM66, which was found in a stock treated with ethyl methane sulfonate, and the PT and PB mutants, which were obtained from stocks treated with heat at low pH.   The PT mutants were contributed by Dr. E. Freese.

dotted line.   A point mutation located in the second segment will give zero with mutants r1272 and r1241 but not with the rest, and so on.   Thus, if any point mutant is tested against the set of seven reference mutants in order, the segment in which its mutation belongs is established simply by counting the number of zeros. Figure 2 shows photographs of the test plates for seven mutants, each having its mutation located in a different segment.

Only these seven patterns, with an uninterrupted row of zeros beginning from the left, have ever been observed for thousands of mutants tested against these seven deletions.   The complete exclusion of the other 121 possible patterns confirms the linear order of the segments.

Now a given segment can be further subdivided by means of other mutations having suitable starting or ending points.   Figure 3 shows the set used in this study and the designation of each segment.   Each mutant is first tested against the seven which have been chosen to define main segments.   Once the main segment is known,

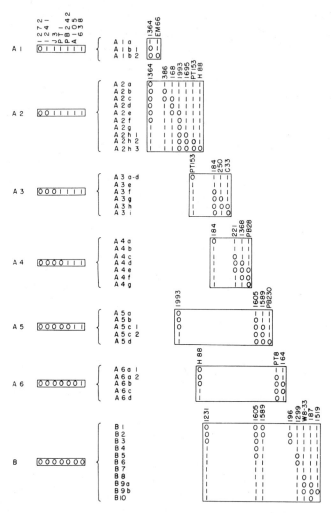

Fig. 4.—The test pattern which identifies the location of a point mutation in each of the segments of Figure 3.   The test is done in two stages.   An unknown mutant is first crossed with the "big seven" of Figure 1 in order.   *Zero* signifies no detectable recombination and *one* signifies some, and the number of zeros defines the major segment.   Once this is known, the mutant is crossed with the pertinent selected group of deletions to determine the small segment to which it belongs.

the mutant is tested against the appropriate secondary set.   Figure 4 shows the pattern which identifies the location of a point mutation within each of the small segments.   Thus, in two steps, a point mutation is mapped into one of the 47 segments.

The order of the first 42 segments, Ala through B6, is uniquely defined. Unfortunately, there remains a gap between $r1299$ and $rW8\text{-}33$. Therefore the order of segments B8 through B10, although fixed among themselves, could possibly be the reverse of that shown.[17] Also if there exists space to the right of segment B10, a mutation in that segment might map as if it were in segment B7, so that the latter segment must be tentatively regarded as a composite.

In the previous topology paper, the possibility that the structure contains branches was not eliminated. As pointed out by Delbrück, the existence of a branch would not lead to any contradiction with a linear topology if loss of a segment containing the branch point automatically led to loss of the entire branch.

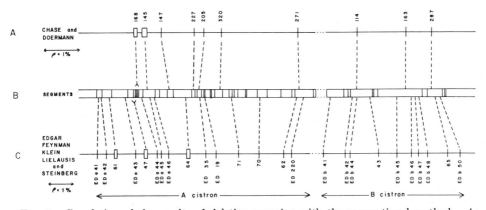

FIG. 5.—Correlation of the results of deletion mapping with the conventional method. *A:* The map constructed by Chase and Doermann[2] for ten rII mutants of phage T4B, using quantitative measurements of recombination frequency. The interval between adjacent mutations is drawn proportional to the frequency of recombination in a cross between the two. *C:* The map constructed in similar fashion by Edgar *et al.* (personal communication) with rII mutants of the very closely related phage T4D. The procedure used by Edgar *et al.* gives higher recombination frequencies. Therefore, the scales of the two maps are adjusted in the figure to produce a good over-all fit. Some of the mutations cover several sites and are drawn as having a corresponding length. A gap is left between the two cistrons because crosses between mutations in different cistrons give abnormally high frequencies due to the role of heterozygotes[16].

All of these mutations have also been mapped by the deletion method, and dotted lines indicate their locations in the various segments (*B*). The length of each segment is drawn in proportion to the number of distinct sites that have been found within it.

To show that a given segment is *not* a branch, it is required to find a mutation which penetrates it partially. From the mutations shown in Figure 3, it can be concluded that no branch exists that contains more than one of the 47 segments.

Comparison of Deletion Mapping by Recombination Frequencies.—The conventional method of genetic mapping makes use of recombination frequency as a measure of the distance between two mutations and requires careful quantitative measurements of the percentage of recombinant type progeny in each cross. By the method of overlapping deletions the order of mutations can be determined entirely by qualitative yes-or-no spot tests. Maps obtained independently by the two methods are compared in Figure 5. The upper part of the figure (*A*) shows the order obtained by Chase and Doermann[2] for a set of ten mutants, the distance between adjacent mutations being drawn proportional to the percentage of standard-type recombinants occurring among the progeny of a cross between the two. The central part of the figure (*B*) shows the rII region divided into the segments of

Figure 3, with the size of each segment drawn in proportion to the number of distinct sites which have been discovered within it (see below). As indicated by the dotted lines, there is perfect correlation in the order. In the lower part of the figure, a similar comparison is made for a set of rII mutations in the closely related phage strain T4D, which have been mapped, using recombination frequencies, by Edgar, Feynman, Klein, Lielausis, and Steinberg. Again the order agrees perfectly with that obtained by the use of deletions.

Topography for Spontaneous Mutations.—We now proceed to map reverting mutants of T4B which have arisen independently and spontaneously. The procedure is exactly as in Figure 4: first localizing into main segments, then into smaller segments. Finally mutants of the same small segment are tested against each other. Any which show recombination are said to define different sites. If two or more reverting mutants are found to show no detectable recombination with each other, they are considered to be repeats and one of them is chosen to represent the site in further tests. A set of distinct sites is thus obtained, each with its own group of repeats.

This procedure is based on the assumption that revertibility implies a point mutation. While this is a good working rule for rII mutants, a few exceptions have been found which appear to revert (i.e., give rise to some progeny which can produce detectable plaques on stain K) yet fail to give recombination with two or more mutants that do recombine with each other. If a mutant chosen to represent a "site" happens to be of this kind, mutations it overlaps will appear to be at the same site. Therefore, a group of "repeats" remains subject to splitting into different groups when they are tested against each other. This has not yet been done for all of the sites described here. It is, of course, in the nature of the recombination test that it is meaningful to say that two mutations are at different sites, while the converse conclusion is always tentative.

Figure 6 shows the map obtained for spontaneous mutants, with each occurrence of a mutation at a site indicated by a square. Within each segment the sites are drawn in arbitrary order. Other known sites are also indicated even though no occurrences were observed among this set of spontaneous mutants.

That the distribution is non-random leaps to the eye. More than 500 mutations have been observed at the most prominent "hotspot," while, at the other extreme, there are many sites at which only a single occurrence, or none, has so far been found.

To decide whether a given number of recurrences is significantly greater than random, the data may be compared with the expectation from a Poisson distribution. Figure 7 shows a distribution calculated to fit the least hot of the observed spontaneous sites, i.e., those at which one or two mutations have occurred, on the assumption that these sites belong to a uniform class of sites of low mutability. Comparing the observations with this curve, it would seem that if a site has four occurrences, there is a two-thirds probability that it is truly hotter than the class of sites of low mutability. Those having five or more are almost certainly hot. It can be concluded that at least sixty sites belong in a more mutable class than the coolest spots. Whether the hot sites can be divided into smaller homogeneous groups, assuming a Poisson distribution within each class, is difficult to say. Each of the two hottest sites is obviously unique.

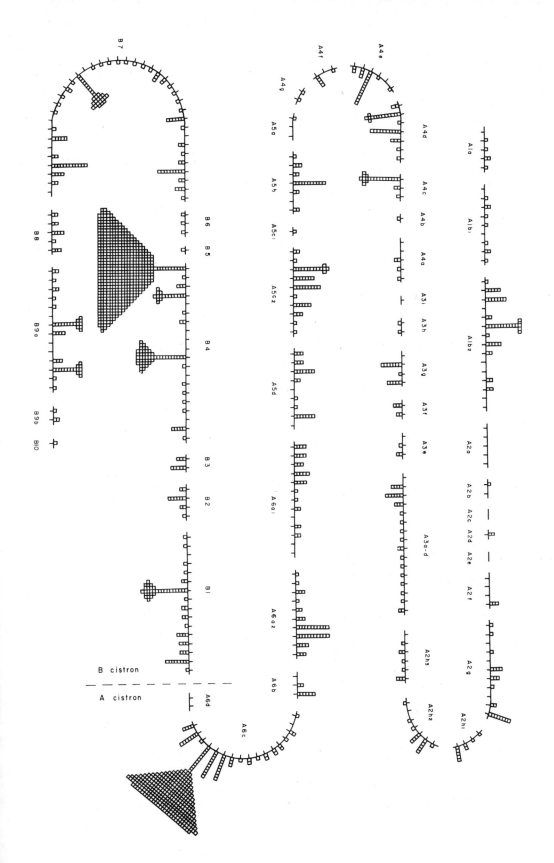

From the distribution it can be predicted that there must exist at least 129 spontaneous sites not observed in this set of mutants. This is a minimum estimate since it is calculated on the assumption that the 2-occurrence sites are no more mutable than the 1-occurrence sites. If this is not correct, the predicted number of

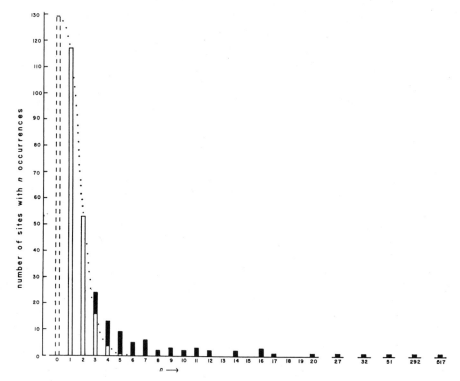

Fig. 7.—Distribution of occurrences of spontaneous mutations at various sites. The dotted line indicates a Poisson distribution fitted to the numbers of sites having one and two occurrences. This predicts a minimum estimate for the number of sites of comparable mutability that have zero occurrences due to chance (dashed column at n = 0). Solid bars indicate the minimum numbers of sites which have mutation rates significantly higher than the one- and two-occurrence class.

0-occurrence sites will be larger. Also, of course, there could exist a vast class of sites of much lower mutability. With 251 spontaneous sites identified and at least 129 more to be found, the degree of saturation of the map achieved with this set of 1,612 spontaneous mutants can be no greater than 66 per cent.

←

Fig. 6.—Topographic map of the *r*II region for spontaneous mutations. Each square represents one occurrence observed at the indicated site. Sites with no occurrences indicated are known to exist from induced mutations and from a few other selected spontaneous ones. The order of the segments is known for A1a through B7, but is only tentative for B7 through B10.[17] The arrangement of sites within each segment is arbitrary.

Each mutant arose independently in a plaque of either standard-type T4B or, in somewhat less than half of the cases, revertants of various *r*II mutants. All revertants (except F) gave results very similar to T4B. The pattern for *r*II mutants isolated from revertant F differs noticeably only in a reduced rate at the hotspot 117 (the site of its original *r*II mutation) and therefore does not significantly alter the topography. All the data for mutants isolated from standard type and from revertants are pooled in this figure.

Topography for Induced Mutations.—By the use of specific mutagens, new topographic features are revealed. This has been shown for *r*II mutants induced during reproduction of the phage inside the bacterial host cell with 5-bromouracil (Benzer and Freese[3]), proflavine (Brenner, Barnett, and Benzer[4]), and 2-aminopurine (Freese[5]). Other effective mutagens are 2,6-diaminopurine (Freese[5]) and 5-bromodeoxycytidine (Gregory, personal communication). Mutations may also be induced *in vitro*, i.e., in extracellular phase particles, by ethyl methane sulfonate (Loveless[6]) and nitrous acid (Vielmetter and Wieder;[7] Freese;[8] Tessman[9]).

*r*II mutants induced by all of these mutagens have now been mapped with respect to each other and spontaneous ones, and the results are given in Figure 8 (facing page 416) which shows the locations of over 2,400 induced and spontaneous mutations. Only *r*II mutants that have low reversion rates and are not too "leaky" on K have been included.

Each "spectrum" differs obviously from the spontaneous one. While the specificities of the various mutagens overlap in many respects, each differs significantly from the others at specific points. In making the comparison it must be borne in mind that the total number of mutants mapped is not the same for each mutagen and also that each induced set inevitably includes some proportion of spontaneous mutants. (An upper limit to this background can be set from the number of occurrences at the hottest spontaneous sites.) Also, none of the spectra are "saturated." Therefore, even if two mutagens act similarly upon a given site, it is possible, due to chance, that a few occurrences would be observed in one spectrum and not the other. Within these limitations, the map shows the comparative response at each site to each mutagen as well as the locations of various kinds of hotspots in various segments of the *r*II region.

The study of the induced mutations has added 53 new sites to the 251 identified by the spontaneous set alone, bringing the total to 304. (Four sites more are shown in Figure 8, but they come from a selected group of mutants outside this study.) Thus, a closer approach toward saturation of all the possible sites must have been made. By lumping together all the data, both spontaneous and induced, one can again make an estimate of the number of sites which must be detectable if one were to continue mapping mutants in the same proportion for the same mutagens. The result is that there must exist still a minimum of 120 sites not yet discovered. This appears discouragingly similar to the estimate based on spontaneous mutations alone. However, it need not be surprising if the use of mutagens brings into view some sites which have extremely low spontaneous mutability. With 308 sites identified and at least 120 yet to be found, the maximum degree of saturation of the map is 72 per cent.

Discussion.—One topographic feature, non-random mutability at the various sites, is obvious. Another question is whether mutable sites are distributed at random, or whether there exist portions of the map that are unusually crowded with or devoid of sites. The mapping technique used here defines only the order of sites from one segment to another (but not within a given segment). The distance between sites remains unspecified. However, all mutations in a segment more distal to a given point must be farther away than those in a more proximal segment. If the number of sites in a segment is used as a measure of its length, as in Figure 5, it can be seen that there is no major discrepancy between these distances and those defined in terms of another measure of distance, recombination frequency. On a gross scale, therefore, there is no evidence for any large portion of the rII region that is unusually crowded or roomy with respect to sites. This does not necessarily mean that some other measure of distance would not reveal such regions, since it is at least conceivable that mutable sites coincide with points highly susceptible to recombination. The distribution of sites on a finer scale, within a small segment, remains to be investigated.

The number of points at which mutations can wreck the activity of a cistron is very large. This would be expected if a cistron dictates the formation of a poly peptide chain and "nonsense" mutations[10] are possible which interrupt the completion of the chain. Such mutations would be effective at any point of the structure, whereas ones which lead to "missense," i.e., the substitution of one amino acid for another, might be effective at relatively special points or regions which are crucial in affecting the active site or folding.

It would be of interest to compare the number of genetic sites to the material embodiment of the rII region in terms of nucleotides. Unfortunately, the size of the latter is not well known. Estimates based upon its length, in units of recombination frequency compared to the length of the entire genetic structure, are uncertain. A more direct attempt has been made using equilibrium sedimentation in a cesium chloride gradient and looking for a change in density of mutants known by genetic evidence to have portions of the rII region deleted (Nomura, Champe, and Benzer, unpublished). This technique has been successful in characterizing defective mutants of phage λ (Weigle, Meselson, and Paigen[11]) and is sufficiently sensitive to detect a decrease of 1 per cent in the amount of DNA per phage particle, but has so far failed with rII mutants. Although other explanations are possible, this result may suggest that the physical structure corresponding to the rII region represents less than 1 per cent of the total DNA of the phage particle, or less than 2,000 nucleotide pairs. If this is so, the number of possible sites would be of the order of at least one-fifth of the number of nucleotide pairs.

The data show that, if each site is characterized by its spontaneous mutability and response to various mutagens, the sites are of many different kinds. Some response patterns are represented only once in the entire structure. According to the Watson-Crick model[12] for DNA, the structure consists of only two types of elements, adenine-thymine (AT) pairs and guanine-hydroxymethylcytosine (GC) pairs. This does not mean, however, that there can only be two kinds of mutable sites, even if a site corresponds to a single base pair. Considering only base pair substitutions, a given AT pair can undergo three kinds of change: AT can be replaced by GC, CG, or TA. Certain of these changes may lead to a mutant phenotype, but some may not. The frequency of observable mutations at a particular AT pair will be determined by the sum of the probabilities for each type of change, each multiplied by a coefficient (either one or zero) according to whether that specific alteration at that particular pair does or does not represent a mutant type. Thus, if the probability that a base pair will be substituted is independent of its neighbors, the various AT sites may have seven different mutation rates. Similarly, there are seven rates possible for the various GC sites, so that it would be possible to account for fourteen classes by this mechanism. Some of these may have (total) spontaneous mutation rates that are similar. If a mutagen induces only certain substitutions, it will facilitate further discriminations between sites but there should still be no more than fourteen classes.

If one allows for interactions between neighbors, the number of possible classes increases enormously. Such interactions are to be expected. As an example, consider the fact that AT pairs are held together much less strongly than are GC pairs.[13] If several AT pairs occur in succession, this segment of the DNA chain will be relatively loose, making it easier to consummate an illicit base pairing during replication. Thus, guanine and adenine, which make a very satisfactory pair of hydrogen bonds but require a larger than normal separation between the backbones, could be more readily accommodated. This would lead, in the next replication, to a replica in which one of the AT pairs has been substituted by a CG pair, with the orientation of purine and pyrimidine reversed. Thus, a region rich in AT pairs will tend to be more subject to substitution. If the same (standard-type) phenotype can be achieved by alternative sequences, the ones containing long stretches of AT pairs would tend to be lost because of their high mutability. In other words, cistrons ought to have evolved in such a way as to eliminate hotspots. The spontaneous hotspots that are observed would be remnants of an incomplete ironing-out process. In fact, a map of the *r*II region of the related phage T6 (Benzer, unpublished) also shows hotspots at locations corresponding to *r*131 and *r*117. However, while the first of these has a mutability similar to that in T4, the second is lower by a factor of four.

This point is emphasized by the data on reverse mutations. It is not uncommon for an *r*II mutant to have a reverse mutation rate that is greater than the total forward rate observed for the composite of at least 400 sites. That some of these high-rate reverse mutations represent true reversion (and not "suppressor" mutations) has been established in several cases by the most stringent criteria, including the demonstration that the revertant has exactly the same forward mutation rate at the same site as did the original standard type (Benzer, unpublished). It would therefore appear that certain kinds of highly mutable configurations are systematically excluded from the standard form of the *r*II genetic structure, and a mutation may recreate one of these banned sequences.

In the attempt to translate the genetic map into a nucleotide sequence, the detection of the various sites by forward mutation is necessarily the first step. By studies on the specificity of induction of reverse mutations,[14] one site at a time can be analyzed in the hope of identifying the specific bases involved.

Summary.—A small portion of the genetic map of phage T4, the two cistrons of the *r*II region, has been dissected by overlapping "deletions" into 47 segments. If any branch exists, it cannot be larger than one of these segments. The overlapping deletions are used to map point mutations and the map order established by this method is consistent with the order established by the conventional method that makes use of recombination frequencies. Further dissection has led to the identification of 308 distinct sites of widely varied spontaneous and induced mutability. The distributions throughout the region for spontaneous mutations and those induced by various chemical mutagens are compared. Data are included for nitrous acid and ethyl methane sulfonate acting *in vitro*, and 2-aminopurine, 2,6-diaminopurine, 5-bromouracil, 5-bromodeoxycytidine, and proflavine acting *in vivo*. The characteristic hotspots reveal a striking topography.

It is a pleasure to thank Mrs. Karen Sue Supple, Mrs. Joan Reynolds, and Mrs. Lynne Bryant for their indefatigable assistance in mapping mutants and bookkeeping. I am indebted to Dr. Robert S. Edgar and his associates for permission to make use of their unpublished data in Figure 5 and to Dr. Ernst Freese for several deletions as well as mutants induced with 2-aminopurine and 5-bromodeoxyuridine. This research was supported by grants from the National Science Foundation and the National Institutes of Health.

* Given by invitation of the Committee on Arrangements for the Annual Meeting as part of a Symposium on Genetic Determination of Protein Structure, Robley C. Williams, Chairman.

[1] Benzer, S., these Proceedings, **45**, 1607 (1959).

[2] Chase, M., and A. H. Doermann, *Genetics*, **43**, 332 (1958).

[3] Benzer, S., and E. Freese, these Proceedings, **44**, 112 (1958).

[4] Brenner, S., L. Barnett, and S. Benzer, *Nature*, **182**, 983 (1958).

[5] Freese, E., *J. Molec. Biol.*, **1**, 87 (1959).

[6] Loveless, A., *Nature*, **181**, 1212 (1958).

[7] Vielmetter, W., and C. M. Wieder, *Z. Naturforsch*, **14b**, 312 (1959).

[8] Freese, E., *Brookhaven Symposia in Biol.*, **12**, 63 (1959).

[9] Tessman, I., *Virology*, **9**, 375 (1959).

[10] Crick, F. H. C., J. S. Griffith, and L. E. Orgel, these Proceedings, **43**, 416 (1957).

[11] Weigle, J., M. Meselson, and K. Paigen, *J. Molec. Biol.*, **1**, 379 (1959).

[12] Watson, J. D., and F. Crick, *Cold Spring Harbor Symposia Quant. Biol.*, **18**, 123 (1953).

[13] Doty, P., J. Marmur, and N. Sueoka, *Brookhaven Symposia in Biol.*, **12**, 1 (1959).

[14] Freese, E., these Proceedings, **45**, 622 (1959).

[15] Brenner, S., in *Advances in Virus Research* (New York: Academic Press, 1959), pp. 137–158.

[16] Edgar, R. S., *Genetics*, **43**, 235 (1958).

[17] The terms topology and topography are used here in the following senses (Webster's New Collegiate Dictionary, 1959)—*topology:* the doctrine of those properties of a figure unaffected by any deformation without tearing or joining; *topography:* the art or practice of graphic and exact delineation in minute detail, usually on maps or charts, of the physical features of any place or region.

[18] *Note added in proof.* Recent data have established that the orientation shown for segments B8 through B10 is the correct one.

Reprinted from *Proc. Natl. Acad. Sci. (U.S.)*, **48**(7), 1114–1121 (1962)

A CHANGE FROM NONSENSE TO SENSE IN THE GENETIC CODE

By Seymour Benzer and Sewell P. Champe

DEPARTMENT OF BIOLOGICAL SCIENCES, PURDUE UNIVERSITY

Communicated May 28, 1962

For the genetic information in a cistron to be translated into a polypeptide chain, each coding unit in the nucleotide sequence must correspond to one of the twenty or so amino acids. If not every possible sequence corresponds to an amino acid, mutations that substitute one base for another could, in certain cases, cause the continuity of the information to be interrupted, and such "nonsense" mutations would block completion of the polypeptide chain.

By virtue of a mutant with special properties, it is possible to identify nonsense mutations within the A cistron of the rII region of phage T4. In this paper the criterion for nonsense is applied to certain "ambivalent" rII mutants,[1] i.e., ones whose phenotypes can be reversed by suppressor mutations in the bacterial host, *Escherichia coli*. The results show that an ambivalent mutation that behaves like nonsense in one bacterial host may nevertheless make sense in a second (suppressor containing) host. This suggests that a suppressor mutation in the bacterium can result in addition to the cell's dictionary of a new sensible coding unit, constituting a change in the genetic code of the bacterial cell.

The r1589 System as a Genetic Test for Nonsense Mutations.—The rII genetic region of phage T4 consists of two contiguous regions, A and B, each behaving as a separate functional unit or cistron.[2] For the phage to grow in *E. coli* strain KB, both the A and B activities are needed. This requirement may be satisfied by infecting the cell with two mutants, one damaged in the A, the other damaged in the B, in which case the mutants are said to complement one another. Full complementation occurs when one puts together any A mutant and any B mutant having point mutations or deletions of any size within the cistron boundaries. Thus, ordinarily, a defect in one cistron has no effect upon the functioning of the other. As illustrated in Figure 1, each cistron can be thought to produce a specific messenger RNA molecule that is in turn translated into a polypeptide chain.

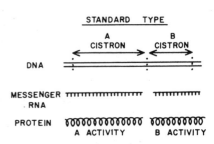

Fig. 1.—Scheme illustrating the independent functioning of the two cistrons of the *r*II region of standard type phage T4.

An exceptional mutant (*r*1589) has been described in which the two cistrons are effectively joined together.[3] In *r*1589, a segment is deleted that includes the divide between the cistrons and a portion of each. The A function is thereby lost, but the B function remains, as shown by the fact that *r*1589 will complement with any mutant that provides an intact A cistron. Apparently, the deleted tip of the B cistron is nonessential. However, due to the absence of the cistron-dividing element, the B fragment in *r*1589 no longer functions independently of the A. Crick *et al.*[4] showed that the B function of *r*1589 can be turned off by crossing certain deletions into the A fragment.. The same effect was produced by some mutations induced by proflavine and believed to be single nucleotide deletions or additions. Their explanation of this effect was that the nucleotide sequence is read in successive coding units starting from a fixed point, so that a deletion in the A cistron of length not equal to an integral multiple of the coding ratio would cause a shift of the reading frame, with disastrous effect on the translation of the B fragment. Thus, the properties of *r*1589 can be understood if the A and B fragments are transcribed into a single messenger RNA which is then translated into a single polypeptide chain, as shown in Figure 2*A*.

As illustrated in Figure 2*B*, the B function of *r*1589 might also be impaired, without a shift in the reading frame, by a substitution mutation changing a coding unit into one that does not correspond to any amino acid. In such a case, protein synthesis could not continue beyond that point. Substitution mutations of the "transition" type,[5] i.e., such that one base pair is exchanged for another and the orientation of purine and pyrimidine with respect to the two DNA chains remains unchanged, can be identified by the fact that they are inducible to revert by DNA base analogues such as 2-aminopurine. To test whether an analogue-revertible A cistron mutation is nonsense, it is inserted (by genetic recombination) in series with *r*1589 and the double mutant tested for B cistron activity. If the mutation in question is nonsense, the B activity should be cut off. If, on the other hand, the mutation is "missense," i.e., if the coding unit corresponds to a different amino acid from the original, the B activity should not be cut off, as illustrated in Figure 2*C*.

The ambivalent phage mutants and bacterial host strains: Certain *r*II mutants of

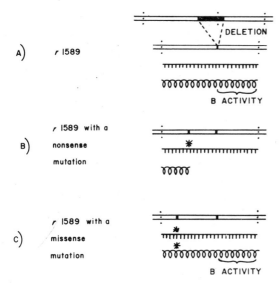

A) r 1589

B) r 1589 with a nonsense mutation

C) r 1589 with a missense mutation

Fig. 2.—Scheme illustrating the properties of the mutant r1589. In this mutant, a deletion has occurred so that fragments of the two cistrons are joined, the B fragment retaining its activity. A nonsense mutation in the A cistron cuts off the B activity, but a missense mutation has no effect.

Although the illustration shows protein synthesis proceeding from left to right, this assumption is not necessary. An incomplete protein, blocked by a nonsense mutation, might never be released from the template.

phage T4 are "ambivalent," i.e., inactive in one strain of *E. coli* but active in another.[1] Modification of the bacterial host (by mutation) can lead to activation of an entire group of mutants with defects located at various points in the *r*II region of the phage genome. The altered bacterium is thus said to contain a suppressor mutation that suppresses a specific ambivalent subset of phage mutations. The ambivalent subset relevant to the following discussion is subset 1, and is defined as those mutants which are active on strain KB-3 but not on KB. All of the known members of subset 1 located in the genetic map to the left of r1589 are shown in Figure 3. The effect of the suppressor in KB-3 on the subset 1 *r*II mutants is shown by the data in Table 1.

All the subset 1 mutants listed in Table 1 are inducible to revert by 2-aminopurine and thus presumably arose by substitutions of the transition type. The specificity of reversion induction indicates that, in all five cases, the transition involved in

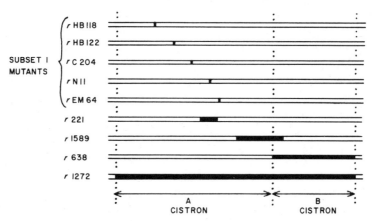

Fig. 3.—Genetic map of the *r*II region showing the locations of the mutations cited in this paper. The first five mutations are revertible by base analogues. The last four are "deletions," indicated by solid black lines. In crosses of the deletion r1589 with r638, or with other mutants located in the tip of the B cistron shown covered by r1589, no standard type recombinants have ever been detected in as many as 10^5 progeny. By the method of shrinkage of distance between outside markers,[21] r1589 has been shown to be a true deletion.[22]

TABLE 1

Activities of Various rII Mutants on Two Different $E.$ $coli$ Strains

		Burst size	
		$E.$ $coli$ KB	$E.$ $coli$ KB-3
Standard type	r^+	229.	190.
Deletions	r1589	0.01	0.01
	r221	0.03	0.01
	r1272	0.04	0.01
Analogue-revertible mutants of ambivalent subset 1	rHB118	0.05	157.
	rHB122	0.1	150.
	rC204	0.08	187.
	rN11	0.04	175.
	rEM64	0.09	168.
Analogue-revertible mutants not of subset 1	rAP129	1.8	2.5
	rNA27	0.09	0.04
	rN74	0.2	0.05
	rAP218	0.1	0.1
	r1814	0.3	0.1
	rH221	0.1	0.08
	r607	0.00	0.01
	rEM114	0.3	0.3

The procedure for measurement of burst sizes, as well as the complementation measurements of Table 2, has been previously described.[3] KB-3, derived from KB, contains a suppressor mutation that specifically reverses the phenotype of ambivalent mutants of subset 1. In this respect, it is similar to strain KT previously used.[1]

the original mutations is GC → AT. All except rHB122 are phenotypically reversible by 5-fluorouracil.[6]

Characterization of the ambivalent mutations: Each mutant to be tested was first crossed with r1589. The progeny were plated on $E.$ $coli$ B, on which all produce plaques, and from them the double mutants were isolated. This was done with the help of replica plating, testing each progeny plaque for recombination against the two parent phages. A suspected double mutant was replated on B, a pure stock made from a single plaque, and tested to assure that it gave no recombination (less than 10^{-2} per cent) with either parent type, or with deletions overlapping either end of the r1589 deletion. The latter test ruled out the possibility of a false double mutant due to a newly arising point mutation within the r1589 segment. All the double mutants were also tested to assure that they gave recombination with mutants located at sites neighboring the original point mutation. This ruled out the possibility that an apparent double mutant might be due to a newly arising deletion overlapping the original site. These checks are necessary, as spurious double mutants sometimes do appear.

In Table 2, the various mutants are analyzed to determine which ones represent nonsense mutations. Note first the properties of mutant r1589 by itself. The burst size on strain KB is negligible, but when the cells are simultaneously infected with r638 (to supply the A function), a large burst occurs, showing that the B function in r1589 is intact. Thus, r1589 behaves in the same way as r221, a deletion restricted to the A cistron. For the large deletion r1272, covering both cistrons, no B activity is present. All three mutants give the same results regardless of whether KB or KB-3 is used as host.

When the deletion r221 is crossed into r1589, it cuts off the B function of the latter. This is the effect described by Crick *et al.*[4] due to the interdependence of the A and B fragments in r1589, and suggests that the length of r221 is a nonintegral number of coding units.

The application of the test to the subset 1 point mutants shows that all cut off

110

TABLE 2

DETERMINATION OF NONSENSE OR SENSE CHARACTER OF VARIOUS MUTATIONS

	Burst Size			Burst Size		
	E. coli KB	E. coli KB(+r638)	Conclusion	E. coli KB-3	E. coli KB-3(+r638)	Conclusion
Standard type						
r^+	112.	135.		167.	127.	
Deletions						
$r1589$	0.01	77.		0.01	136.	
$r221$	0.03	109.		0.01	51.	
$r1272$	0.04	0.01		0.0	0.02	
Double mutant of $r221$ and $r1589$						
$r221-r1589$	0.02	0.025		0.0	0.05	
Double mutants of $r1589$ with analogue-revertible mutants of ambivalent subset 1						Conclusion
$rHB118-r1589$	0.00	0.00	Nonsense	0.01	100.	Sense
$rHB122-r1589$	0.00	0.02	Nonsense	0.01	1.5	Sense
$rC204-r1589$	0.00	0.00	Nonsense	0.00	37.	Sense
$rN11-r1589$	0.00	0.01	Nonsense	0.01	18.	Sense
$rEM64-r1589$	0.00	0.00	Nonsense	0.02	79.	Sense
Double mutants of $r1589$ with analogue-revertible mutants not of subset 1						
$rAP129-r1589$	0.00	58.	Sense			
$rNA27-r1589$	0.08	84.	Sense			
$rN74-r1589$	0.08	73.	Sense			
$rAP218-r1589$	0.07	85.	Sense			
$r1814-r1589$	0.1	28.	Sense			
$rH221-r1589$	0.09	76.	Sense			
$r607-r1589$	0.02	121.	Sense			
$rEM114-r1589$	0.07	80.	Sense			

If a mutation in the A cistron cuts off the B activity of $r1589$, no burst is obtained in the complementation test with $r638$ (second and fourth columns). For mutants of ambivalent subset 1, the burst size is greatly increased by using KB-3 as the host in the complementation test (fourth column), indicating that the suppressor mutation in KB-3 has changed the reading of these mutations from nonsense to sense. No distinction is made here between "sense" and "missense," since it should be immaterial whether a correct or incorrect amino acid is inserted. Burst size differences less than a factor of about two are not significant.

the B activity of r1589 when KB is the host, i.e., they behave as if they were nonsense mutations. However, on switching to the suppressor strain, KB-3, the B function is expressed. In other words, a mutation that behaves as nonsense in strain KB becomes sense by virtue of a suppressor mutation in the host. One of the five mutants (rHB122) gives a considerably weaker response. This mutant also differs from the others in not responding to 5-fluorouracil.[6]

For the other eight analogue-reverting mutants listed in Tables 1 and 2, the B cistron function (of the r1589 double mutant) is unimpaired in the KB host. Thus, these mutations do not interrupt the reading of the genetic information and are interpreted as missense mutations in which the nucleotide change is such that the normal amino acid is replaced by a different one. Among seven other analogue-revertible mutants that have been tested qualitatively, three were of the missense type while four were nonsense.[7] The latter were not members of ambivalent subset 1, so that nonsense substitutions are not restricted to that group.

Discussion.—The above analysis is based on the assumption, which also underlies the work of Crick *et al.*,[4] that the A and B cistrons of the rII region control the amino acid sequences of polypeptide chains. Although there is as yet no direct evidence for this, a property common to certain rII mutations and to certain mutations affecting the enzyme alkaline phosphatase gives support to this assumption. That is, the same suppressor that acts on a subset of mutations in the structural cistron of *E. coli* for the phosphatase enzyme acts concomitantly on subset 1 rII mutations of phage T4.[8] Furthermore, as shown by Garen and Siddiqi in the accompanying paper,[9] the phosphatase mutants in question produce no detectable phosphatase protein in the absence of the suppressor, as would be expected if nonsense mutations were involved. Thus, mutations in unrelated cistrons, one in the phage and one in the bacterium, that appear, by quite independent criteria, to be nonsense, are also related by sensitivity to the same suppressor. Recent evidence by Brody and Yanofsky[10] in the tryptophan synthetase system shows that suppressors can act by modifying protein structure. The properties of rII mutants are therefore readily understandable on the assumption that the products of the rII cistrons are proteins.

A second assumption on which our criterion for nonsense is based is that the product of the A cistron fragment of r1589 containing the mutation in question does not modify the B activity except in determining whether or not translation into a polypeptide chain is possible. It might be argued that some altered forms of the A (protein) fragment could interact with the B fragment in such a way that the B activity would be affected. It seems most unlikely, however, that the present data could be due to such an effect, since it is implausible that the effects would be the same for the interaction of the A with the B fragment in rII and for the interaction of the phosphatase molecule with itself.

To characterize a mutation as "nonsense," as distinguished from the "gibberish" of Crick *et al.*,[4] produced by a shift of the reading frame, the mutation must have arisen by a base pair substitution. We have used as a criterion for a substitution of the transition type that the mutation be revertible by the base analogue 2-aminopurine. This is consistent with many of the facts of mutagenic specificity.[5, 6] However, it cannot be ruled out that base analogues may have effects in addition to causing substitutions.[11]

112

Within these limitations, the genetic analysis indicates that certain mutations give rise to a coding unit that does not correspond to an amino acid, so that the genetic code in *E. coli* must not be completely degenerate. Further, the same coding unit, in the phage genome, that is nonsense in strain KB of *E. coli* becomes sense by virtue of a suppressor mutation in strain KB-3. Thus, unless KB-3 can use unusual amino acids, this strain translates at least two coding units into the same amino acid. This implies, in agreement with other evidence, [4, 12, 13] that the genetic code is partially degenerate.

As previously suggested, [1, 14–16] the action of external suppressors can be understood in terms of the mechanism of protein synthesis. The translation of genetic information from messenger RNA into amino acid sequences takes place via the mediation of sRNA adaptors having the dual function of accepting a specific amino acid and attaching to a specific nucleotide sequence on the messenger RNA template.[17, 18] The role of sRNA as an adaptor has recently been demonstrated experimentally.[19] Since the attachment of each amino acid to the correct sRNA depends upon a specific enzyme, it is the set of sRNA adaptors and activating enzymes that determines the genetic code. If the structure of each activating enzyme and each sRNA adaptor is specified by a cistron in the organism, then the genetic code is under the genetic control of the organism itself and could change by mutation.

Thus, the action of the suppressor mutation in KB-3 could be explained by the appearance of a new or modified sRNA adaptor that fits a coding unit absent from the KB dictionary. Alternatively, an adaptor might have always been present but unable to accept an amino acid with any of the activating enzymes, in which case the suppressor mutation might act by creating a new activating enzyme (or modifying the structure of one already present). It has recently been shown that the degeneracy of leucine observed *in vitro*[12, 13] is due to the existence of two sRNA adaptors having different coding specificities.[23]

The changes in the code involved in suppressor mutations may be relatively minor ones having to do with appearance or disappearance of degeneracies. Since the available evidence[20] favors a close similarity of the codes in various organisms, strongly selective factors must force the set of adaptors and activating enzymes, in spite of changes in their individual structures,[24] to maintain a quasi-universal code.

Summary.—Genetic analysis indicates that certain *r*II mutations of phage T4 give rise to coding units that do not correspond to an amino acid, i.e., they are nonsense mutations. Mutations that behave like nonsense in one bacterial host nevertheless make sense in another host containing a suppressor mutation. The suppressor mutation thus constitutes, in a limited sense, an hereditary alteration of the genetic code.

We wish to thank Mmes. Daine Auzins, Judith Berry, and Barbara Williams for assistance in isolation of the double mutants. This research was supported by grants from the National Science Foundation and the National Institutes of Health.

[1] Benzer, S., and S. P. Champe, these PROCEEDINGS, **47**, 1025 (1961).
[2] Benzer, S., these PROCEEDINGS, **41**, 344 (1955).
[3] Champe, S. P., and S. Benzer, *J. Mol. Biol.*, **4**, 288 (1962).
[4] Crick, F. H. C., L. Barnett, S. Brenner, and R. J. Watts-Tobin, *Nature*, **192**, 1227 (1961).

[5] Freese, E., *J. Mol. Biol.*, **1**, 87 (1959).

[6] Champe, S. P., and S. Benzer, these Proceedings, **48**, 532 (1962).

[7] Crick, F. H. C., personal communication, for four of the mutants.

[8] Benzer, S., S. P. Champe, A. Garen, and O. H. Siddiqi, manuscript in preparation.

[9] Garen, A., and O. H. Siddiqi, these Proceedings, **48**, 1121 (1962).

[10] Brody, S., and C. Yanofsky, manuscript in preparation.

[11] Trautner, T. A., M. N. Swartz, and A. Kornberg, these Proceedings, **48**, 449 (1962).

[12] Martin, R. G., J. H. Matthaei, O. W. Jones, and M. W. Nirenberg, *Biochem. Biophys. Res. Comm.*, **6**, 410 (1962).

[13] Speyer, J. F., P. Lengyel, C. Basilio, and S. Ochoa, these Proceedings, **48**, 441 (1962).

[14] Lieb, M., and L. A. Herzenberg, unpublished.

[15] Yanofsky, C., and P. St. Lawrence, *Ann. Rev. Microbiol.*, **14**, 311 (1960).

[16] Yanofsky, C., D. R. Helinski, and B. Maling, in *Cellular Regulatory Mechanisms*, Cold Spring Harbor Symposia on Quantitative Biology, vol. 26 (1961), p. 11.

[17] Hoagland, M. B., *Brookhaven Symposia in Biol.*, **12**, 40 (1959).

[18] Crick, F. H. C., "The Biological Replication of Macromolecules," in *Symposium Soc. Expt. Biol.*, **12**, 138 (1958).

[19] Chapeville, F., F. Lipmann, G. von Ehrenstein, B. Weisblum, W. Ray, and S. Benzer, these Proceedings, **48**, 1086 (1962).

[20] von Ehrenstein, G., and F. Lipmann, these Proceedings, **47**, 941 (1961).

[21] Nomura, M., and S. Benzer, *J. Mol. Biol.*, **3**, 684 (1961).

[22] Bode, W., doctoral dissertation, University of Cologne (1962).

[23] Weisblum, B., S. Benzer, and R. Holley, manuscript in preparation.

[24] Benzer, S., and B. Weisblum, these Proceedings, **47**, 1149 (1961).

Reprinted from *Nature*, **192**(4809), 1227–1232 (1961)

GENERAL NATURE OF THE GENETIC CODE FOR PROTEINS

By Dr. F. H. C. CRICK, F.R.S., LESLIE BARNETT, Dr. S. BRENNER
and Dr. R. J. WATTS-TOBIN

Medical Research Council Unit for Molecular Biology,
Cavendish Laboratory, Cambridge

THERE is now a mass of indirect evidence which suggests that the amino-acid sequence along the polypeptide chain of a protein is determined by the sequence of the bases along some particular part of the nucleic acid of the genetic material. Since there are twenty common amino-acids found throughout Nature, but only four common bases, it has often been surmised that the sequence of the four bases is in some way a code for the sequence of the amino-acids. In this article we report genetic experiments which, together with the work of others, suggest that the genetic code is of the following general type:

(a) A group of three bases (or, less likely, a multiple of three bases) codes one amino-acid.

(b) The code is not of the overlapping type (see Fig. 1).

(c) The sequence of the bases is read from a fixed starting point. This determines how the long sequences of bases are to be correctly read off as triplets. There are no special 'commas' to show how to select the right triplets. If the starting point is displaced by one base, then the reading into triplets is displaced, and thus becomes incorrect.

(d) The code is probably 'degenerate'; that is, in general, one particular amino-acid can be coded by one of several triplets of bases.

The Reading of the Code

The evidence that the genetic code is not overlapping (see Fig. 1) does not come from our work, but from that of Wittmann[1] and of Tsugita and Fraenkel-Conrat[2] on the mutants of tobacco mosaic virus produced by nitrous acid. In an overlapping triplet code, an alteration to one base will in general change three adjacent amino-acids in the polypeptide chain. Their work on the alterations produced in the protein of the virus show that usually only one amino-acid is changed as a result of treating the ribonucleic acid (RNA) of the virus with nitrous acid. In the rarer cases where two amino-acids are altered (owing presumably to two separate deaminations by the nitrous acid on one piece of RNA), the altered amino-acids are not in adjacent positions in the polypeptide chain.

Brenner[3] had previously shown that, if the code were universal (that is, the same throughout Nature), then all overlapping triplet codes were impossible. Moreover, all the abnormal human hæmoglobins studied in detail[4] show only single amino-acid changes. The newer experimental results essentially rule out all simple codes of the overlapping type.

If the code is not overlapping, then there must be some arrangement to show how to select the correct triplets (or quadruplets, or whatever it may be) along the continuous sequence of bases. One obvious suggestion is that, say, every fourth base is a 'comma'. Another idea is that certain triplets make 'sense', whereas others make 'nonsense', as in the comma-free codes of Crick, Griffith and Orgel[5]. Alternatively, the correct choice may be made by starting at a fixed point and working along the sequence of bases three (or four, or whatever) at a time. It is this possibility which we now favour.

Experimental Results

Our genetic experiments have been carried out on the B cistron of the r_{II} region of the bacteriophage $T4$, which attacks strains of *Escherichia coli*. This is the system so brilliantly exploited by Benzer[6,7]. The r_{II} region consists of two adjacent genes, or 'cistrons', called cistron A and cistron B. The wild-type phage will grow on both *E. coli* B (here called B) and on *E. coli* K12 (λ) (here called K), but a phage which has lost the function of either gene will not grow on K. Such a phage produces an r plaque on B. Many point mutations of the genes are known which behave in this way. Deletions of part of the region are also found. Other mutations, known as 'leaky', show partial function; that is, they will grow on K but their plaque-type on B is not truly wild. We report here our work on the mutant P 13 (now re-named FC 0) in the $B1$ segment of the B cistron. This mutant was originally produced by the action of proflavin[8].

We[9] have previously argued that acridines such as proflavin act as mutagens because they add or delete a base or bases. The most striking evidence in favour of this is that mutants produced by acridines are seldom 'leaky'; they are almost always completely lacking in the function of the gene. Since our note was published, experimental data from two sources have been added to our previous evidence: (1) we have examined a set of 126 r_{II} mutants made with acridine yellow; of these only 6 are leaky (typically about half the mutants made with base analogues are leaky); (2) Streisinger[10] has found that whereas mutants of the lysozyme of phage $T4$ produced by base-analogues are usually leaky, all lysozyme mutants produced by proflavin are negative, that is, the function is completely lacking.

If an acridine mutant is produced by, say, adding a base, it should revert to 'wild-type' by deleting a base. Our work on revertants of FC 0 shows that it usually

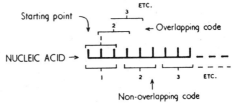

Fig. 1. To show the difference between an overlapping code and a non-overlapping code. The short vertical lines represent the bases of the nucleic acid. The case illustrated is for a triplet code

reverts not by reversing the original mutation but by producing a second mutation at a nearby point on the gene. That is, by a 'suppressor' in the same gene. In one case (or possibly two cases) it may have reverted back to true wild, but in at least 18 other cases the 'wild type' produced was really a double mutant with a 'wild' phenotype. Other workers[11] have found a similar phenomenon with r_{II} mutants, and Jinks[12] has made a detailed analysis of suppressors in the h_{III} gene.

The genetic map of these 18 suppressors of FC 0 is shown in Fig. 2, line a. It will be seen that they all fall in the $B1$ segment of the gene, though not all of them are very close to FC 0. They scatter over a region about, say, one-tenth the size of the B cistron. Not all are at different sites. We have found eight sites in all, but most of them fall into or near two close clusters of sites.

In all cases the suppressor was a non-leaky r. That is, it gave an r plaque on B and would not grow on K. This is the phenotype shown by a complete deletion of the gene, and shows that the function is lacking. The only possible exception was one case where the suppressor appeared to back-mutate so fast that we could not study it.

Each suppressor, as we have said, fails to grow on K. Reversion of each can therefore be studied by the same procedure used for FC 0. In a few cases these mutants apparently revert to the original wild-type, but usually they revert by forming a double mutant. Fig. 2, lines b–g, shows the mutants produced as suppressors of these suppressors. Again all these new suppressors are non-leaky r mutants, and all map within the $B1$ segment for one site in the $B2$ segment.

Once again we have repeated the process on two of the new suppressors, with the same general results, as shown in Fig. 2, lines i and j.

All these mutants, except the original FC 0, occurred spontaneously. We have, however, produced one set (as suppressors of FC 7) using acridine yellow as a mutagen. The spectrum of suppressors we get (see Fig. 2, line h) is crudely similar to the spontaneous spectrum, and all the mutants are non-leaky r's. We have also tested a (small) selection of all our mutants and shown that their reversion-rates are increased by acridine yellow.

Thus in all we have about eighty independent r mutants, all suppressors of FC 0, or suppressors of suppressors, or suppressors of suppressors of suppressors. They all fall within a limited region of the gene and they are all non-leaky r mutants.

The double mutants (which contain a mutation plus its suppressor) which plate on K have a variety of plaque types on B. Some are indistinguishable from wild, some can be distinguished from wild with difficulty, while others are easily distinguishable and produce plaques rather like r.

We have checked in a few cases that the phenomenon is quite distinct from 'complementation', since the two mutants which separately are phenotypically r, and together are wild or pseudo-wild,

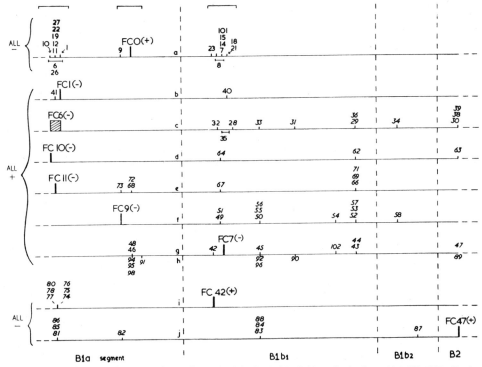

Fig. 2. A tentative map—only very roughly to scale—of the left-hand end of the B cistron, showing the position of the FC family of mutants. The order of sites within the regions covered by brackets (at the top of the figure) is not known. Mutants in italics have only been located approximately. Each line represents the suppressors picked up from one mutant, namely, that marked on the line in bold figures

must be put together in the same piece of genetic material. A simultaneous infection of K by the two mutants in separate viruses will not do.

The Explanation in Outline

Our explanation of all these facts is based on the theory set out at the beginning of this article. Although we have no direct evidence that the B cistron produces a polypeptide chain (probably through an RNA intermediate), in what follows we shall assume this to be so. To fix ideas, we imagine that the string of nucleotide bases is read, triplet by triplet, from a starting point on the left of the B cistron. We now suppose that, for example, the mutant FC 0 was produced by the insertion of an additional base in the wild-type sequence. Then this addition of a base at the FC 0 site will mean that the reading of all the triplets to the right of FC 0 will be shifted along one base, and will therefore be incorrect. Thus the amino-acid sequence of the protein which the B cistron is presumed to produce will be completely altered from that point onwards. This explains why the function of the gene is lacking. To simplify the explanation, we now postulate that a suppressor of FC 0 (for example, FC 1) is formed by deleting a base. Thus when the FC 1 mutation is present by itself, all triplets to the right of FC 1 will be read incorrectly and thus the function will be absent. However, when both mutations are present in the same piece of DNA, as in the pseudo-wild double mutant FC (0 + 1), then although the reading of triplets between FC 0 and FC 1 will be altered, the original reading will be restored to the rest of the gene. This could explain why such double mutants do not always have a true wild phenotype but are often pseudo-wild, since on our theory a small length of their amino-acid sequence is different from that of the wild-type.

For convenience we have designated our original mutant FC 0 by the symbol + (this choice is a pure convention at this stage) which we have so far considered as the addition of a single base. The suppressors of FC 0 have therefore been designated −. The suppressors of these suppressors have in the same way been labelled as +, and the suppressors of these last sets have again been labelled − (see Fig. 2).

Double Mutants

We can now ask: What is the character of any double mutant we like to form by putting together in the same gene any pair of mutants from our set of about eighty ? Obviously, in some cases we already know the answer, since some combinations of a + with a − were formed in order to isolate the mutants. But, by definition, no pair consisting of one + with another + has been obtained in this way, and there are many combinations of + with − not so far tested.

Now our theory clearly predicts that all combinations of the type + with + (or − with −) should give an r phenotype and not plate on K. We have put together 14 such pairs of mutants in the cases listed in Table 1 and found this prediction confirmed.

Table 1. DOUBLE MUTANTS HAVING THE r PHENOTYPE

− With −	+ With +	
FC (1 + 21)	FC (0 + 58)	FC (40 + 57)
FC (23 + 21)	FC (0 + 38)	FC (40 + 58)
FC (1 + 23)	FC (0 + 40)	FC (40 + 55)
FC (1 + 9)	FC (0 + 55)	FC (40 + 54)
	FC (0 + 54)	FC (40 + 38)

At first sight one would expect that all combinations of the type (+ with −) would be wild or pseudo-wild, but the situation is a little more intricate than that, and must be considered more closely. This springs from the obvious fact that if the code is made of triplets, any long sequence of bases can be read correctly in one way, but incorrectly (by starting at the wrong point) in two different ways, depending whether the 'reading frame' is shifted one place to the right or one place to the left.

If we symbolize a shift, by one place, of the reading frame in one direction by → and in the opposite direction by ←, then we can establish the convention that our + is always at the head of the arrow, and our − at the tail. This is illustrated in Fig. 3.

We must now ask: Why do our suppressors not extend over the whole of the gene ? The simplest postulate to make is that the shift of the reading frame produces some triplets the reading of which is 'unacceptable'; for example, they may be 'nonsense', or stand for 'end the chain', or be unacceptable in some other way due to the complications of protein structure. This means that a suppressor of, say, FC 0 must be within a region such that no 'unacceptable' triplet is produced by the shift in the reading frame between FC 0 and its suppressor. But, clearly, since for any sequence there are two possible misreadings, we might expect that the 'unacceptable' triplets produced by a → shift would occur in different places on the map from those produced by a ← shift.

Examination of the spectra of suppressors (in each case putting in the arrows → or ←) suggests that while the → shift is acceptable anywhere within our region (though not outside it) the shift ←, starting from points near FC 0, is acceptable over only a more limited stretch. This is shown in Fig. 4. Somewhere in the left part of our region, between FC 0 or FC 9 and the FC 1 group, there must be one or more unacceptable triplets when a ← shift is made; similarly for

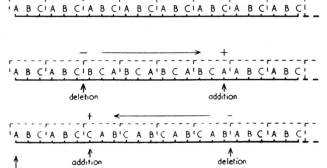

Fig. 3. To show that our convention for arrows is consistent. The letters A, B and C each represent a different base of the nucleic acid. For simplicity a repeating sequence of bases, ABC, is shown. (This would code for a polypeptide for which every amino-acid was the same.) A triplet code is assumed. The dotted lines represent the imaginary 'reading frame' implying that the sequence is read in sets of three starting on the left

the region to the right of the *FC* 21 cluster. Thus we predict that a combination of a + with a − will be wild or pseudo-wild if it involves a → shift, but that such pairs involving a ← shift will be phenotypically *r* if the arrow crosses one or more of the forbidden places, since then an unacceptable triplet will be produced.

Table 2. DOUBLE MUTANTS OF THE TYPE (+ WITH −)

−\+	FC 41	FC 0	FC 40	FC 42	FC 58*	FC 63	FC 38
FC 1	W	W	W		W		W
FC 86		W	W	W	W	W	
FC 9	r	W	W	W	W		W
FC 82	r		W	W	W	W	
FC 21	r	W			W		W
FC 88	r	r			W	W	
FC 87	r	r	r	r			W

W, wild or pseudo-wild phenotype; *W*, wild or pseudo-wild combination used to isolate the suppressor; *r, r* phenotype.
* Double mutants formed with *FC* 58 (or with *FC* 34) give sharp plaques on *K*.

We have tested this prediction in the 28 cases shown in Table 2. We expected 19 of these to be wild, or pseudo-wild, and 9 of them to have the *r* phenotype. In all cases our prediction was correct. We regard this as a striking confirmation of our theory. It may be of interest that the theory was constructed before these particular experimental results were obtained.

Rigorous Statement of the Theory

So far we have spoken as if the evidence supported a triplet code, but this was simply for illustration. Exactly the same results would be obtained if the code operated with groups of, say, 5 bases. Moreover, our symbols + and − must not be taken to mean literally the addition or subtraction of a single base.

It is easy to see that our symbolism is more exactly as follows:

$$+ \text{ represents } +m, \text{ modulo } n$$
$$- \text{ represents } -m, \text{ modulo } n$$

where *n* (a positive integer) is the coding ratio (that is, the number of bases which code one amino-acid) and *m* is any integral number of bases, positive or negative.

It can also be seen that our choice of reading direction is arbitrary, and that the same results (to a first approximation) would be obtained in whichever direction the genetic material was read, that is, whether the starting point is on the right or the left of the gene, as conventionally drawn.

Triple Mutants and the Coding Ratio

The somewhat abstract description given above is necessary for generality, but fortunately we have convincing evidence that the coding ratio is in fact 3 or a multiple of 3.

This we have obtained by constructing triple mutants of the form (+ with + with +) or (− with − with −). One must be careful not to make shifts

Table 3. TRIPLE MUTANTS HAVING A WILD OR PSEUDO-WILD PHENOTYPE

$$FC (0 + 40 + 38)$$
$$FC (0 + 40 + 58)$$
$$FC (0 + 40 + 57)$$
$$FC (0 + 40 + 54)$$
$$FC (0 + 40 + 55)$$
$$FC (1 + 21 + 23)$$

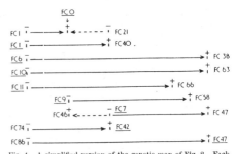

Fig. 4. A simplified version of the genetic map of Fig. 2. Each line corresponds to the suppressor from one mutant, here underlined. The arrows show the range over which suppressors have so far been found, the extreme mutants being named on the map. Arrows to the right are shown solid, arrows to the left dotted

across the 'unacceptable' regions for the ← shifts, but these we can avoid by a proper choice of mutants.

We have so far examined the six cases listed in Table 3 and in all cases the triples are wild or pseudo-wild.

The rather striking nature of this result can be seen by considering one of them, for example, the triple (*FC* 0 with *FC* 40 with *FC* 38). These three mutants are, by themselves, all of like type (+). We can say this not merely from the way in which they were obtained, but because each of them, when combined with our mutant *FC* 9 (−), gives the wild, or pseudo-wild phenotype. However, either singly or together in pairs they have an *r* phenotype, and will not grow on *K*. That is, the function of the gene is absent. Nevertheless, the combination of all three in the same gene partly restores the function and produces a pseudo-wild phage which grows on *K*.

This is exactly what one would expect, in favourable cases, if the coding ratio were 3 or a multiple of 3.

Our ability to find the coding ratio thus depends on the fact that, in at least one of our composite mutants which are 'wild', at least one amino-acid must have been added to or deleted from the polypeptide chain without disturbing the function of the gene-product too greatly.

This is a very fortunate situation. The fact that we can make these changes and can study so large a region probably comes about because this part of the protein is not essential for its function. That this is so has already been suggested by Champe and Benzer[13] in their work on complementation in the *r*_{II} region. By a special test (combined infection on *K*, followed by plating on *B*) it is possible to examine the function of the *A* cistron and the *B* cistron separately. A particular deletion, 1589 (see Fig. 5) covers the right-hand end of the *A* cistron and part of the left-hand end of the *B* cistron. Although 1589 abolishes the *A* function, they showed that it allows the *B* function to be expressed to a considerable extent. The region of the *B* cistron deleted by 1589 is that into which all our *FC* mutants fall.

Joining two Genes Together

We have used this deletion to re-inforce our idea that the sequence is read in groups from a fixed starting point. Normally, an alteration confined to the *A* cistron (be it a deletion, an acridine mutant, or any other mutant) does not prevent the expression of the *B* cistron. Conversely, no alteration within the *B* cistron prevents the function of the *A* cistron. This implies that there may be a region between the two cistrons which separates them and allows their functions to be expressed individually.

We argued that the deletion 1589 will have lost this separating region and that therefore the two (partly damaged) cistrons should have been joined together. Experiments show this to be the case, for now an alteration to the left-hand end of the *A* cistron, if combined with deletion 1589, can prevent the *B* function from appearing. This is shown in Fig. 5. Either the mutant *P*43 or *X*142 (both of which revert strongly with acridines) will prevent the *B* function when the two cistrons are joined, although both of these mutants are in the *A* cistron. This is also true of *X*142 *S*1, a suppressor of *X*142 (Fig. 5, case *b*). However, the double mutant (*X*142 with *X*142 *S*1), of the type (+ with −), which by itself is pseudo-wild, still has the *B* function when combined with 1589 (Fig. 5, case *c*). We have also tested in this way the 10 deletions listed by Benzer[7], which fall wholly to the left of 1589. Of these, three (386, 168 and 221) prevent the *B* function (Fig. 5, case *f*). whereas the other seven show it (Fig. 5, case *e*). We surmise that each of these seven has lost a number of bases which is a multiple of 3. There are theoretical reasons for expecting that deletions may not be random in length, but will more often have lost a number of bases equal to an integral multiple of the coding ratio.

It would not surprise us if it were eventually shown that deletion 1589 produces a protein which consists of part of the protein from the *A* cistron and part of that from the *B* cistron, joined together in the same polypeptide chain, and having to some extent the function of the undamaged *B* protein.

Is the Coding Ratio 3 or 6 ?

It remains to show that the coding ratio is probably 3, rather than a multiple of 3. Previous rather rough extimates[10,14] of the coding ratio (which are admittedly very unreliable) might suggest that the coding ratio is not far from 6. This would imply, on our theory, that the alteration in *FC* 0 was not to one base, but to two bases (or, more correctly, to an even number of bases).

We have some additional evidence which suggests that this is unlikely. First, in our set of 126 mutants produced by acridine yellow (referred to earlier) we have four independent mutants which fall at or

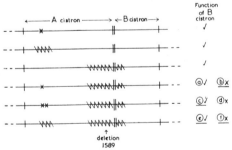

Fig. 5. Summary of the results with deletion 1589. The first two lines show that without 1589 a mutation or a deletion in the *A* cistron does not prevent the *B* cistron from functioning. Deletion 1589 (line 3) also allows the *B* cistron to function. The other cases, in some of which an alteration in the *A* cistron prevents the function of the *B* cistron (when 1589 is also present), are discussed in the text. They have been labelled (*a*), (*b*), etc., for convenience of reference, although cases (*a*) and (*d*) are not discussed in this paper. √ implies function; × implies no function

Fig. 6. Genetic map of *P* 83 and its suppressors, *WT* 1, etc. The region falls within segment *B*9*a* near the right-hand end of the *B* cistron. It is not yet known which way round the map is in relation to the other figures

close to the *FC* 9 site. By a suitable choice of partners, we have been able to show that two are + and two are −. Secondly, we have two mutants (*X*146 and *X*225), produced by hydrazine[15], which fall on or near the site *FC* 30. These we have been able to show are both of type −.

Thus unless both acridines and hydrazine usually delete (or add) an even number of bases, this evidence supports a coding ratio of 3. However, as the action of these mutagens is not understood in detail, we cannot be certain that the coding ratio is not 6, although 3 seems more likely.

We have preliminary results which show that other acridine mutants often revert by means of close suppressors, but it is too sketchy to report here. A tentative map of some suppressors of *P* 83, a mutant at the other end of the *B* cistron, in segment *B* 9*a*, is shown in Fig. 6. They occur within a shorter region than the suppressors of *FC* 0, covering a distance of about one-twentieth of the *B* cistron. The double mutant *WT* (2 + 5) has the *r* phenotype, as expected.

Is the Code Degenerate?

If the code is a triplet code, there are 64 (4 × 4 × 4) possible triplets. Our results suggest that it is unlikely that only 20 of these represent the 20 amino-acids and that the remaining 44 are nonsense. If this were the case, the region over which suppressors of the *FC* 0 family occur (perhaps a quarter of the *B* cistron) should be very much smaller than we observe, since a shift of frame should then, by chance, produce a nonsense reading at a much closer distance. This argument depends on the size of the protein which we have assumed the *B* cistron to produce. We do not know this, but the length of the cistron suggests that the protein may contain about 200 amino-acids. Thus the code is probably 'degenerate', that is, in general more than one triplet codes for each amino-acid. It is well known that if this were so, one could also account for the major dilemma of the coding problem, namely, that while the base composition of the DNA can be very different in different micro-organisms, the amino-acid composition of their proteins only changes by a moderate amount[16]. However, exactly how many triplets code amino-acids and how many have other functions we are unable to say.

Future Developments

Our theory leads to one very clear prediction. Suppose one could examine the amino-acid sequence of the 'pseudo-wild' protein produced by one of our double mutants of the (+ with −) type. Conventional theory suggests that since the gene is only altered in two places, only two amino-acids would be changed. Our theory, on the other hand, predicts that a string of amino-acids would be altered, covering the region of the polypeptide chain corresponding to the region on the gene between the two mutants. A good protein on which to test this hypothesis is the lysozyme of the phage, at present being studied chemically by Dreyer[17] and genetically by Streisinger[10].

At the recent Biochemical Congress at Moscow, the audience of Symposium I was startled by the announcement of Nirenberg that he and Matthaei[18] had produced polyphenylalanine (that is, a polypeptide all the residues of which are phenylalanine) by adding polyuridylic acid (that is, an RNA the bases of which are all uracil) to a cell-free system which can synthesize protein. This implies that a sequence of uracils codes for phenylalanine, and our work suggests that it is probably a triplet of uracils.

It is possible by various devices, either chemical or enzymatic, to synthesize polyribonucleotides with defined or partly defined sequences. If these, too, will produce specific polypeptides, the coding problem is wide open for experimental attack, and in fact many laboratories, including our own, are already working on the problem. If the coding ratio is indeed 3, as our results suggest, and if the code is the same throughout Nature, then the genetic code may well be solved within a year.

We thank Dr. Alice Orgel for certain mutants and for the use of data from her thesis, Dr. Leslie Orgel for many useful discussions, and Dr. Seymour Benzer for supplying us with certain deletions. We

are particularly grateful to Prof. C. F. A. Pantin for allowing us to use a room in the Zoological Museum, Cambridge, in which the bulk of this work was done.

[1] Wittman, H. G., Symp. 1, Fifth Intern. Cong. Biochem., 1961, for refs. (in the press).

[2] Tsugita, A., and Fraenkel-Conrat, H., Proc. U.S. Nat. Acad. Sci., 46, 636 (1960); J. Mol. Biol. (in the press).

[3] Brenner, S., Proc. U.S. Nat. Acad. Sci., 43, 687 (1957).

[4] For refs. see Watson, H. C., and Kendrew, J. C., Nature, 190, 670 (1961).

[5] Crick, F. H. C., Griffith, J. S., and Orgel, L. E., Proc. U.S. Nat. Acad. Sci., 43, 416 (1957).

[6] Benzer, S., Proc. U.S. Nat. Acad. Sci., 45, 1607 (1959), for refs. to earlier papers.

[7] Benzer, S., Proc. U.S. Nat. Acad. Sci., 47, 403 (1961); see his Fig. 3.

[8] Brenner, S., Benzer, S., and Barnett, L., Nature, 182, 983 (1958).

[9] Brenner, S., Barnett, L., Crick, F. H. C., and Orgel, A., J. Mol. Biol., 3, 121 (1961).

[10] Streisinger, G. (personal communication and in the press).

[11] Feynman, R. P.; Benzer, S.; Freese, E. (all personal communications).

[12] Jinks, J. L., Heredity, 16, 153, 241 (1961).

[13] Champe, S., and Benzer, S. (personal communication and in preparation).

[14] Jacob, F., and Wollman, E. L., Sexuality and the Genetics of Bacteria (Academic Press, New York, 1961). Levinthal, C. (personal communication).

[15] Orgel, A., and Brenner, S. (in preparation).

[16] Sueoka, N. Cold Spring Harb. Symp. Quant. Biol. (in the press).

[17] Dreyer, W. J., Symp. 1, Fifth Intern. Cong. Biochem., 1961 (in the press).

[18] Nirenberg, M. W., and Matthaei, J. H., Proc. U.S. Nat. Acad. Sci., 47, 1588 (1961).

11

Reprinted from *Nature*, **201**(4914), 13–17 (1964)

CO-LINEARITY OF THE GENE WITH THE POLYPEPTIDE CHAIN

By Dr. A. S. SARABHAI, Dr. A. O. W. STRETTON and Dr. S. BRENNER

Medical Research Council, Laboratory of Molecular Biology, Cambridge

AND

Dr. A. BOLLE

Institut de Biologie Moleculaire, Université de Genève

THE 'sequence hypothesis' states that the amino-acid sequence of a protein is specified by the nucleotide sequence of the gene determining that protein[1]. It has always been assumed that a simple congruence exists between these two sequences such that the order of the codons in the gene is the same as the order of the corresponding amino-acids in the polypeptide chain. There is no direct evidence for this co-linearity; but it is known that mutations close together in the gene affect the same amino-acid in the protein[2]. In this article we show that a class of suppressible mutations affecting the head protein of bacteriophage *T4D* produce fragments of the polypeptide chain. This property allows us to prove that the gene is co-linear with the polypeptide chain. Dr. C.

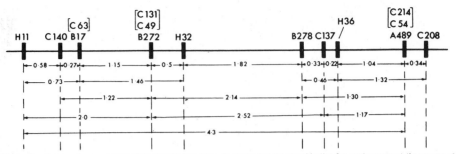

Fig. 1. Genetic map of *amber* mutants of the head protein gene. In many cases the recombination frequencies represent the average of several experiments and may differ slightly from those shown in Table 1

Yanofsky of Stanford University has informed us that he has shown co-linearity in an examination of the tryptophan synthetase of *E. coli*.

Suppressible Mutants

Benzer and Champe[3] have found that certain strains of *Escherichia coli* permit the growth of some phage mutants which do not grow on other strains. Bacteria with this permissive property are said to suppress the effect of the mutation. These suppressible mutants could be divided into different classes depending on the permissive host strain used. Benzer and Champe[4] later showed that one class of these mutants (subset 1), when grown in non-permissive bacteria, appeared to prevent the reading of the genetic message. They suggested that these mutants contained a nonsense codon and that permissive hosts could translate this codon into an amino-acid.

Mutations have been found in other genes in bacteria and bacteriophages which are suppressed by strains suppressing subset 1 mutants[5]. In particular, a large number of suppressible mutants of this class, called *amber* mutants, have been isolated in bacteriophage *T4D* (ref. 6). These are distributed throughout the genetic map of the phage, and it appears that any gene is susceptible to this type of mutation.

In order to characterize the molecular consequences of these suppressible mutations and their mechanism of suppression, we have examined *amber* mutants affecting the head protein of the bacteriophage. About 90 per cent of the protein of the phage particle is head protein[7], and Koch and Hershey[8] have shown that 60–70 per cent of the proteins synthesized late during infection are ultimately incorporated into mature phage particles. Thus more than one-half of the late protein synthesis of the infected

Table 1. ORDERING OF MUTANTS BY 3-FACTOR CROSSES

E. coli CR63, at a concentration of 2×10^8 cells/ml. in buffer, was infected with a mixture of the parental phages, each at a multiplicity of 5. After 8 min at 37° C to allow adsorption, the complexes were diluted 1 : 200 into tryptone broth and lysed with chloroform after 1 h at 37° C. Appropriate dilutions were plated by preadsorption on *E. coli* CR63 to measure total progeny yield and on *E. coli* B to measure wild-type recombinants

Cross	Recombination (per cent)	Order deduced								
B17 × H11	0·80									
B17 × H11 + B278	0·14	H11		B17			B278			
B278 × A489	1·30									
B278 × B17 + A489	0·34			B17			B278		A489	
B272 × B278	2·20									
B272 × H11 + B278	0·32	H11			B272		B278			
B272 × B17	1·30									
B272 × B17 + A489	0·24			B17	B272				A489	
C140 × B17	0·27									
C140 × H11 + B17	0·04									
C140 × B17 + B272	0·20	H11	C140	B17	B272					
H32 × B272	0·5									
H32 × B17 + B272	0·36									
H32 × B272 + B278	0·05			B17	B272	H32	B278			
C137 × B278	0·33									
C137 × B272 + B278	0·28									
C137 × B278 + A489	0·06									
H36 × B278	0·46									
H36 × B272 + B278	0·40									
H36 × B278 + A489	0·06			B272			B278	(C137 | H36)	A489	
H36 × C137	0·22									
H36 × C137 + A489	0·05									
C137 × H36 + A489	0·20							C137	H36	A489	
C208 × A489	0·34									
C208 × B17	4·0									
C208 × B17 + A489	0·33									
C208 × B278 + A489	0·29			B17			B278			A489	C208
Order:		H11	C140	B17	B272	H32	B278	C137	H36	A489	C208

Table 2. DISTRIBUTION OF PEPTIDES IN AMBER-INFECTED CELLS

Segment	H11	C140	B17	B272	H32	B278	C137	H36	A489	C208	Wild	No. of peptides found Chymotryptic	No. of peptides found Tryptic	Example (see Fig. 2)
I	+	+	+	+	+	+	+	+	+	+	+	13	11	—
II	−	+	+	+	+	+	+	+	+	+	+	0	0	—
III	−	−	+	+	+	+	+	+	+	+	+	0	0	—
IV	−	−	−	+	+	+	+	+	+	+	+	8	3	Cys
V	−	−	−	+	+	+	+	+	+	+	+	0	1	His T7c
VI	−	−	−	−	+	+	+	+	+	+	+	10	2	Tyr C12b
VII	−	−	−	−	−	+	+	+	+	+	+	2	2	Try T6
VIII	−	−	−	−	−	−	−	+	+	+	+	5	1	Pro T2a
IX	−	−	−	−	−	−	−	+	+	+	+	8	11	Try T2
X	−	−	−	−	−	−	−	−	−	+	+	1	1	Tyr C2
XI	−	−	−	−	−	−	−	−	−	−	+	7	10	His C6

Note: The number of peptides found in a segment is not a measure of the length of the segment, since in some cases the same peptide sequence may be scored more than once, either because of being labelled by two different amino-acids, or because of partial enzymatic degradation products.

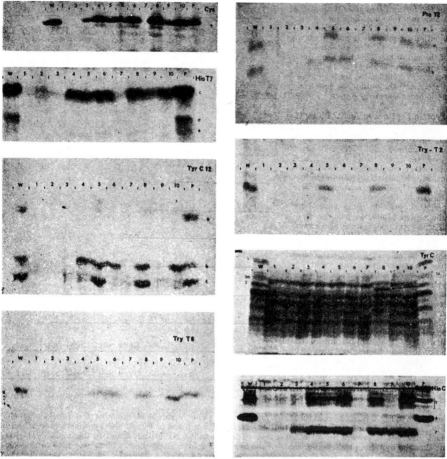

Fig. 2.   Autoradiographs of tryptic (T) and chymotryptic (C) peptides, labelled with radioactive amino-acids. P is wild-type phage, W (and X in one case) is wild-type infected cells, 1–10 are *amber* mutant infected cells, as follows: 1 = *H*11, 2 = *B*17, 3 = *B*272, 4 = *B*278, 5 = *A*489, 6 = *C*137, 7 = *C*140, 8 = *C*208, 9 = *H*32, and 10 = *H*36.
10 ml. cultures of *E. coli* B (5 × 10⁸ cells/ml.) in minimal medium were infected with phage (multiplicity = 5) and, 5 min later, superinfected to produce lysis inhibition.   For labelling with sulphur-35 the minimal medium used contained no sulphur at the time of infection.   10 min after the initial infection, 5 μc. of carbon-14-labelled amino-acid, or 50 μc. of ³⁵SO₄⁻, was added.   After 1–1·5 h protein synthesis was terminated by the addition of chloramphenicol (50 μg/ml.).   The cells were removed by centrifugation, suspended in water, and heated at 100° C for 10 min.   Lysozyme (50 μg/ml.), DNase (10 μg/ml.) and RNase (10 μg/ml.) were added, and, after digestion for 14 h at 37° C, the suspensions were dialysed against 2,500 volumes of distilled water for 18 h.   Each preparation was made 1 per cent in ammonium bicarbonate and divided into two parts, one part was digested with chymotrypsin (10 μg/ml.), the other with trypsin (10 μg/ml.).   After 14 h, digestion was stopped by the addition of acetic acid.   For each radioactive amino-acid, all ten *amber* mutants and a wild-type were treated in parallel as described here.   At the same time, labelled wild-type phage was prepared by allowing 20 ml. cultures of *T4D*-infected cells to assimilate radioactive amino-acid for 4 h.   After lysis with chloroform, the phage was purified by differential centrifugation.   Suspensions were heated, the released DNA digested with DNase, and the denatured ghosts removed by centrifugation.   These were washed, suspended in 1 per cent ammonium bicarbonate, and proteolytic digestion and subsequent treatment carried out as described here.
The same amount of radioactivity of each digest was applied to Whatman 3 *MM* paper, and after high-voltage ionophoresis at *p*H 6·4 the peptide zones located by autoradiography. Selected zones were purified by ionophoresis at *p*H 3·6

cell is devoted to the manufacture of head protein; this has allowed us to detect fragments of the head polypeptide chain without any prior purification.

Isolation and Genetic Mapping of *Amber* Mutants

A previous examination of the *amber* mutants of bacteriophage *T4D* suggested that seven cistrons might be involved in the synthesis of the head protein[6].   We first proved that the cistron characterized by one of these (*am B*17) controlled the head protein.   This was done by a special complementation test between *am B*17 and a known peptide difference[9] in the head protein.

Fourteen mutants, induced in phage *T4D* by base analogue mutagens (5-bromouracil, 2-aminopurine and hydroxylamine) were located in the head protein cistron by their failure to complement *am B*17 on the non-permissive strain (*E. coli* B).   Some of these were shown to be recurrences by spot test crosses.   In all, ten different sites were found, and one representative of each site was chosen.   Double mutants were constructed by crossing two mutants and were tested by back-crossing progeny to the two parents.   In most cases, the double mutants are less well suppressed by the permissive strain (*E. coli CR*63) than are the parents.

Standard crosses were carried out, and the frequency of wild-type recombinants was measured by plating on *E. coli* B.   This was multiplied by 2 to give the recombination frequency.   The results are summarized in

Fig. 1 in the form of a genetic map of the region, showing relevant 2-factor recombination frequencies. The additivity of recombination frequencies is good, but we have not relied on this to establish the order of the mutants. This has been done by 3-factor crosses (Table 1), and although such crosses are complicated by high negative interference[10] they uniquely determine the genetic order shown in Fig. 1.

Head Protein Peptides in *Amber* Mutants

The proteins being synthesized in the phage-infected cells were labelled with radioactive amino-acids. The cells were then lysed, the nucleic acid removed and without further purification the mixture was digested with either trypsin or chymotrypsin. The peptides thus produced were characterized by high-voltage ionophoresis on paper, their position being located by autoradiography. Labelled proteins from purified wild-type phage were also examined in the same way. The technical details are given in the legend to Fig. 2.

The peptides characteristic of the head protein of the phage can easily be recognized in digests of cells infected with the wild-type phage. This shows that the contamination introduced by other proteins synthesized in the infected cells is not serious. We have only scored peptides present in digests of both wild-type phage and the wild-type infected cells. The first general result is that while some of the peptides of the head protein are present in cells infected with a given *amber* mutant, others are absent, and the pattern differs for each *amber* mutant. Hence peptides can be classified as to whether or not they are included in the polypeptide synthesized by an *amber* mutant.

For example, in Fig. 2 it can be seen that the Cys peptide is clearly absent in cells infected with the mutants *H*11, *C*140 and *B*17 but is present in all other cases. This suggests that the mutants produce fragments of the head polypeptide chain such that some of the fragments include the Cys peptide while others do not. The peptide His *T*7c is synthesized in cells infected with *H*32, *B*278, *C*137, *H*36, *A*489 and *C*208 but not in cells infected with *H*11, *C*140, *B*17 and *B*272. This peptide would therefore be absent in the fragments made by the last four mutants. These two peptides distinguish the fragment made by *B*272 from the other fragments, since only in this case is the Cys peptide present and the His *T*7c peptide absent. By extension of the experimental data it should be possible to characterize each fragment uniquely in this way. Some 96 tryptic and chymotryptic peptides, detected by separate labelling experiments with ¹⁴C-tyrosine, phenylalanine, histidine, tryptophan, proline and arginine, and ³⁵SO₄⁻⁻ (methionine + cysteine), have been scored, and examples are given in Fig. 2.

The number of peptides found in any one mutant is a measure of the length of the fragment produced. It is therefore possible to arrange the fragments in order of increasing size: each fragment should include all the peptides present in smaller fragments, together with extra ones. The difference between two fragments defines a segment of the polypeptide chain, and the arrangement of the fragments by increasing size orders these segments. If the gene and the polypeptide chain are co-linear, then the order of the polypeptide segments should be identical with the order of the segments into which the genetic map is divided by the mutants.

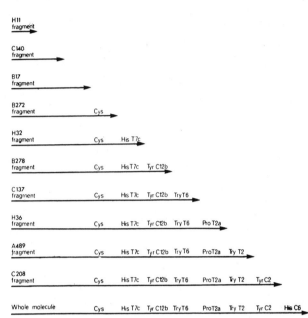

Fig. 3. Diagram showing hierarchical ordering of the fragments and the correlation of the segments with the genetic map. The peptides are those shown in Fig. 2

The hierarchical order of the peptides is shown in Table 2 and in Fig. 3. No peptides scored as head protein peptides have been found to distinguish *H*11, *C*140 and *B*17 from each other. Thus only nine of the eleven segments can be defined. However, we note that since peptides defining segment I are present in all *amber* mutants, there is no internal evidence that they are peptides of the head protein. This difficulty could be resolved by extensive purification, but until this is carried out we have chosen to leave segment I out of account. The order deduced from the fragments is therefore: (*H*11, *C*140, *B*17) *B*272, *H*32, *B*278, *C*137, *H*36, *A*489, *C*208 with eight segments uniquely defined. The order is the same as the genetic order (compare Fig. 1) and hence to a high degree of confidence these experiments show that the two structures are co-linear.

No exception to this order has been found for peptides which are present in the wild-type phage and absent from at least one *amber* mutant. However, peptides are found in some *amber* mutants which are not present in the wild-type phage. We might expect such peptides for two reasons:

(1) The point of termination of a fragment may cut a wild-type peptide, and thus produce a new peptide which occurs in that *amber* mutant only. We appear to have found an example of this in one case.

(2) Some wild-type peptides never show on the paper because they form part of the insoluble 'core'. However, when fragments of the protein are digested such peptides may become soluble, either because they have become accessible to the proteolytic enzyme or because the region of the protein which co-precipitated them is now absent.

Such peptides can also be assigned to segments. We have been able to score ten of these peptides. They agree with the order of segments previously found and in addition there is one peptide which defines segment II; that is, it is present in all *amber* mutants except *H*11.

The remarkable property of the *amber* mutations of the head protein cistron and, by inference, of all other suppressible mutations of this class, is that they lead to the synthesis of a fragment of the polypeptide chain. Qualitatively, fragments appear to be produced in amounts comparable with that of the wild-type protein. Although we do not as yet have any direct evidence, it is likely that the fragments include the N-terminus of the head protein, since proteins are synthesized sequentially from their N-terminal ends[11]. If this is so, then the nonsense codon introduced by the *amber* mutation results in polypeptide chain termination, without permitting re-initiation of synthesis beyond that point. The exact details of this mechanism are unknown, as indeed is the mechanism of suppression itself. However, it is hoped that further analysis of the *amber* mutants should throw more light on these problems.

To summarize: we have shown that suppressible *amber* mutations of the head protein lead to the production of fragments of the polypeptide chain. Analysis of the fragments allows us to define eight segments of the polypeptide chain which are in the same order as the segments defined on a genetic map. We conclude that the gene is co-linear with the polypeptide chain.

We thank Dr. R. H. Epstein for providing some of the mutants used, and Dr. F. H. C. Crick for interesting discussions. We also thank Mrs. R. M. Fishpool, Miss M. I. Wigby and Mr. P. A. Wright for their assistance.

[1] Crick, F. H. C., *Symp. Soc. Exp. Biol.*, **12**, 138 (1958).
[2] Henning, U., and Yanofsky, C., *Proc. U.S. Nat. Acad. Sci.*, **48**, 1497 (1962). See also Rothman, F., *Cold Spring Harbor Symp. Quant. Biol.*, **26**, 23 (1961).
[3] Benzer, S., and Champe, S. P., *Proc. U.S. Nat. Acad. Sci.*, **47**, 1025 (1961).
[4] Benzer, S., and Champe, S. P., *Proc. U.S. Nat. Acad. Sci.*, **48**, 1114 (1962).
[5] Campbell, A., *Virology*, **14**, 22 (1961). Garen, A., and Siddiqi, O., *Proc. U.S. Nat. Acad. Sci.*, **48**, 1121 (1962).
[6] Epstein, R. H., Bolle, A., Steinberg, C. M., Kellenberger, E., Boy de la Tour, E., Chevalley, R., Edgar, R. S., Susman, M., Denhardt, G. H., and Lielausis, A., *Cold Spring Harbor Symp. Quant. Biol.*, **28** (in the press).
[7] Van Vunakis, H., Baker, W. H., and Brown, R. K., *Virology*, **5**, 327 (1958). Brenner, S., Streisinger, G., Horne, R. W., Champe, S. P., Barnett, L., Benzer, S., and Rees, M. W., *J. Mol. Biol.*, **1**, 281 (1959).
[8] Koch, G., and Hershey, A. D., *J. Mol. Biol.*, **1**, 260 (1959).
[9] Brenner, S., and Barnett, L., *Brookhaven Symp. Biol.*, **12**, 86 (1959).
[10] Chase, M., and Doermann, A. H., *Genetics*, **43**, 332 (1958).
[11] Dintzis, H. M., *Proc. U.S. Nat. Acad. Sci.*, **47**, 247 (1961). Goldstein, A., and Brown, B. J., *Biochim. Biophys. Acta*, **53**, 438 (1961).

12

Reprinted from *Cold Spring Harbor Symp. Quant. Biol.*, **28**, 375–392 (1963)

Physiological Studies of Conditional Lethal Mutants of Bacteriophage T4D

R. H. Epstein, A. Bolle, C. M. Steinberg*, E. Kellenberger,
E. Boy de la Tour and R. Chevalley

Biophysis Laboratory, University of Geneva, Geneva, Switzerland

R. S. Edgar, M. Susman†, G. H. Denhardt and A. Lielausis

Division of Biology, California Institute of Technology, Pasadena, California

INTRODUCTION

Following infection of a sensitive bacterium with a phage, a characteristic series of intracellular events occur. In the case of the virulent phage T4, these events include both the cessation of synthesis of many macromolecular constituents characteristic of the growing bacterial cell, and the establishment of a new biosynthetic pattern directed toward the growth and reproduction of the phage. In this new pattern of events, one set of synthetic activities follows another in temporal sequence. For example, a series of enzymes concerned with the synthesis of phage-specific DNA are formed during the first ten minutes following infection while the protein components of the phage particles are synthesized later (see, for example, Kellenberger, 1961). These events are due to the introduction of the phage genome into the bacterial cell and it becomes, therefore, of basic interest to understand how the phage genome is implicated in these processes. This problem which is of importance for our understanding of bacteriophage as a biological entity, has relevance to the general question of the genetic control of growth and development. The special technical advantages of experiments with bacteriophage recommend their use in an attack on this latter problem.

Studies with defective mutants of phage lambda (Jacob et al., 1957; Campbell, 1961) have shown that mutations in the phage genome profoundly affect the pattern of events which occur as a consequence of infection. Yet little is known about what functions are performed by particular phage genes, the distribution of these genes in the phage genome, or the nature of the interactions between these genes. Although for T4 and T2 we have inferential evidence regarding the functions of host range (Streisinger and Franklin, 1956), co-factor (Brenner, 1954), and osmotic shock resistant mutants (Brenner and Barnett, 1959), the function

* Division of Biology, California Institute of Technology, Pasadena, California. Present address: Biology Division, Oak Ridge National Laboratory, Oak Ridge, Tennessee.
† Present address: Genetics Department, University of Wisconsin, Madison, Wisconsin.

of only one gene, the endolysin gene of T4 (Streisinger et al., 1961), is known with complete assurance. In large part, the inability to correlate the genetic structure of the phage with the functions it performs stems from the lack of suitable mutants for such studies.

The establishment of the relationship between a gene and its expression depends upon the isolation of mutants of the gene which affect its activity in a manner that produces an observable change of phenotype. For technical reasons having to do with genetic analysis of the mutants, it is also useful if they are selective. A class of mutations that offer special advantages in these regards are conditional lethal mutations. Recently, two types of conditional lethal mutants in bacteriophage T4D have been described and genetically characterized; temperature sensitive (*ts*) mutants which form plaques at 25°C but not at 42°C (Edgar and Lielausis, in preparation), and amber mutants (*am*) which form plaques on *Escherichia coli* CR63 but not on *E. coli* B (Epstein, R. H., C. M. Steinberg, A. Bolle, in preparation). Neither temperature nor host, under the conditions employed, affect the plaque forming ability of wild type T4. Thus, each type of mutant behaves as a lethal under one set of conditions that we shall refer to as restrictive, while under a second set of conditions the mutant phenotype resembles wild type. Both classes of mutations can occur at only a restricted number of sites within a gene, but may occur in a large number of different genes.

The sites of both types of mutations are widely distributed in the genome of the phage. A comparative genetic study of the two systems (Edgar, R. S., G. H. Denhardt, R. H. Epstein, in preparation) has resulted in a map combining both *ts* and *am* mutants (see Fig. 1). Many *am* and *ts* mutations occur in the same genes. It is probable that the genes already identified represent an appreciable fraction of the total number of genes in the T4 genome. From the genetic analysis of these systems, it appears likely that additional genes will be revealed by further mutant isolations.

Editor's Note: The discussion following this article has been omitted, owing to limitations of space.

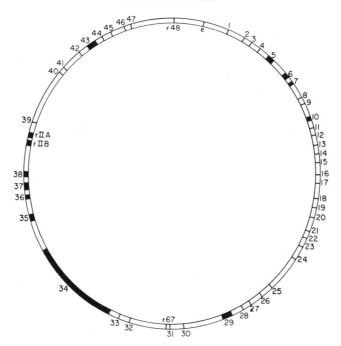

Figure 1. A map showing the distribution of genes in T4D. The extent of some genes is indicated by the filled areas in the circle. The scale is 10° of arc = 15% recombination. The map, taken from Edgar et al. (in preparation) is based upon observed recombination distances, uncorrected for negative interference.

These mutants are especially useful in the study of gene function because, under restrictive conditions, mutant infections are abortive. By comparing a given feature of abortive mutant growth with that of wild type under comparable conditions, it is possible to obtain information relating to the step in development at which the mutation exerts its effect and thus to infer something about the function of the affected gene. Our purpose in the studies reported here is to investigate, in a preliminary way, to what extent various functions initiated by the infecting phage are carried out by different *am* and *ts* mutant phages. We have chosen to study a large number of mutants, thus our study is necessarily a superficial one. It concerns the ability of the various mutants to induce, under restrictive conditions, (1) nuclear breakdown, (2) synthesis of DNA, (3) spontaneous lysis of infected cells, (4) synthesis of tail fiber antigen and (5) the synthesis of phage-related morphological components identifiable in the electron microscope.

As we shall show, in most cases the phenotypes of the various mutants can be described in a qualitative way by the ability or inability of the mutant complex to carry out successfully the above-mentioned processes. The distribution of pheno-types of genes around the genome, when defined in this manner, is strikingly non-random.

MATERIALS AND METHODS

In some cases the materials and methods used in the studies of the amber and temperature sensitive mutants differ. In these cases we have usually described the techniques employed in a section concerning the temperature sensitive mutants (Pasadena) and separately, those features which differ for the amber mutants (Geneva).

Phage strains. (Pasadena) T4D wild type, various *ts* mutants (Edgar and Lielausis, in preparation), and the standard plaque morphology mutants *r*67 and the *r*II mutant *r*73 were used. The preparation of stocks of these strains will be described elsewhere (Edgar and Lielausis, in preparation).

(Geneva) T4D wild-type, various *am* mutants (Epstein et al., in preparation), and the plaque morphology mutants *r*48, and the *r*II mutant *r*61 were used.

Bacterial strains. (Pasadena) *Escherichia coli* strain B, or a one-step mutant B/5,1, were used as host for experiments. *E. coli* strain S/6, or its one-step mutant S/6/5,1, were used as plating indicator.

In some of the serum blocking experiments strain F(λ) (Steinberg and Edgar, 1962) was used as plating indicator. (The method of preparing plating bacteria is given by Edgar and Lielausis, in preparation.) The preparation of host cells for the experiments described here is as follows: A saturated overnight culture of B (or B/5,1) made in SFH broth was diluted 100 times in SFH broth and grown with aeration for $2\frac{1}{2}$ hr at 30°C (to a titer of about 10^9 cells/ml). The cells were centrifuged and resuspended in H broth and adjusted to a concentration of about 2×10^{10} cells/ml.

(Geneva) *E. coli* B was used as host for experiments. Platings were made on the indicator strains *E. coli* S/6 and *E. coli* CR63. Strain K112–12 (λ) of *E. coli* was used in some experiments involving calibration of the serum blocking test. Plating bacteria and host cells for experiments were prepared by diluting an inoculum from a saturated overnight culture 100-fold in prewarmed (37°C) H broth. This culture was then grown with vigorous aeration for 2 hr, centrifuged, and the pellet resuspended in fresh H broth. Bacteria to be used for plating were concentrated ten-fold; the concentration of bacteria which were to serve as hosts in experiments was adjusted during resuspension to give 4×10^8 cells/ml.

Media. (Pasadena and Geneva) H broth, SFH broth (H broth without NaCl) and SPH broth (H broth with one-tenth the amount of NaCl present in H broth) were used as growth media for bacteria and bacteria infected with phage. For plating, EHA top layer agar and EHA bottom agar were used. (The recipes for these media can be found in Steinberg and Edgar, 1962.)

Basic experimental design. (Pasadena) One ml of host bacteria at a concentration of 2×10^{10} cells/ml was equilibrated at 39.5°C. Five ml of phage (also equilibrated at 39.5°C) at a concentration of about 10^{10} phage/ml in H broth was added to the bacteria. After 2 min the total six ml was transferred to 50 ml of SFH broth also equilibrated at 39.5°C. This growth tube now contained bacteria at a concentration of 10^8/ml infected with an average of 2.5 phage. The initial incubation under concentrated conditions and in the presence of NaCl permits the adsorption of at least 99% of the phage in the two min period. Further adsorption is greatly reduced by the low amount of NaCl in the growth tube. The growth tube was aerated by vigorous bubbling and maintained at 39.5°C (± 0.2°C). (In some experiments a temperature of 40.5°C was used.) Samples taken at 10, 20, and 40 min were assayed for infective centers and mature phage. Other samples were taken at various times for measurements described below.

(Geneva) As above except that appropriate volumes of bacteria and phage, each at 30°C were mixed so that the final titer of bacteria was 2×10^8 cells/ml and the multiplicity of infection about five. Samples for the determination of unadsorbed phage and infected bacteria titers were taken at 10 min after infection. All cultures were lysed at 60 min after infection and were assayed for progeny of both *am* and *am*+ genotypes.

Cytological observations. (Pasadena) Samples of infected cells were added to an equal volume of a 40% solution of polyvinylpyrrolidone (PVP) containing 10% formaldehyde. Within ten min after the addition of the formaldehyde to the cells the samples were observed with a phase contrast microscope under oil immersion (mag. $\times$ 1250). As in a previous study using gelatin (Mason and Powelson, 1956), the PVP is used to provide a refractive index in the medium surrounding the cells which permits visualization of cell nuclei. The formaldehyde serves to fix the cells. The cells do not change their optical properties for about 15 min at which time they become opaque. (Even before 15 min, 10 to 50% of the cells are opaque.) Cells infected with wild type phage pass through three clearly distinguishable phases which correspond to (1) uninfected cells with nuclei of normal appearance, (2) nuclear disruption and (3) a stage of general transparency which normally corresponds to the development of the pool of vegetative phage DNA.

(Geneva) As above, except that formaldehyde fixation was omitted and all samples were examined within two min of preparation. In some determinations a gelatin suspension of appropriate concentration was used in place of PVP.

DNA determinations. (Pasadena) The total amount of DNA in samples of the cultures taken at various times was measured by the Keck modification of the Ceriotti method (Keck, 1956). 2.5 ml samples from the infected cultures were added to 2.5 ml of a 0.1 M HCl and 0.001 M indole solution. These samples were heated at 96°C for ten min and then shaken successively with four equivalent volumes of amyl acetate and the aqueous phase retained. The OD of the samples was then measured in a Beckman DU spectrophotometer at 490 mμ. The blank was a control sample of growth medium (SPH broth) treated identically to the growth tube samples.

(Geneva) A number of determinations were made by the Ceriotti method described above. In other measurements of DNA synthesis a slight modification of the method of Burton (1956) was employed. Four ml aliquots containing 2×10^8 infected bacteria/ml were taken from an experimental culture and transferred to centrifuge tubes containing one ml ice cold 70% perchloric acid

(PCA). After two hr in the cold, the mixture of infected bacteria and PCA was centrifuged for 15 min at 6,000 × g, the supernatant discarded, and the pellet resuspended in one ml of 0.5 N PCA. The resuspended samples were hydrolyzed at 75°C for a period of 25 min, cooled, and then centrifuged a second time at 6,000 × g for 15 min. The supernatant was carefully decanted and 0.7 ml added to 0.3 ml of 0.5 N PCA and the resulting one ml sample mixed with two ml of diphenylamine reagent, and incubated at 30°C for 16 hr. The extent of reaction with the reagent was determined as the optical density at 590 mμ. The instrument used for these measurements was a Meunier colorimeter.

Diphenylamine reagent: 1.5 gm of diphenylamine was dissolved in a solution containing 100 ml of concentrated acetic acid and 1.5 ml of concentrated sulfuric acid. This solution was stored in the dark and just before use, 0.25 ml of aqueous acetaldehyde (0.07 M) was added to the solution for each 40 ml of reagent required.

Lysis measurements. (Pasadena) Lysis of the infected cultures was determined from measurements of the optical density of samples taken at different times. The OD was determined at 450 mμ in a Bausch and Lomb Spectronic colorimeter. The OD of the control (r67) cultures drops by at least a factor of two, beginning at about 25 min after infection.

(Geneva) Lysis of most cultures was confirmed by simple visual observation. In a number of cases, lysis was also measured by bacterial counts in a Petroff-Hauser counting chamber; in others, the lysis of a culture was followed by changes in optical density as a function of time. Measurements of optical density at 590 mμ were made with a Meunier colorimeter.

Preparation of lysates. (Pasadena and Geneva) Most mutant-infected cells lyse within 40 min after infection. At 40 min (60 min in Geneva) the lysates were sterilized by the addition of chloroform and debris centrifuged out at low speed (1000 × g for 20 min). However, some mutant infected cultures do not lyse, even after incubation for as long as 2 hr. Lysates of these complexes were obtained by the addition of egg-white lysozyme (10 gamma/ml) and versene (0.1 M). Lysis of the cells of such a treated culture was checked by microscopic observation. These lysates were then sterilized with chloroform and centrifuged to remove large debris.

Serum blocking antigen measurements. (Pasadena) Measurements for the presence of serum blocking antigen were performed essentially by the method of De Mars (1955). The anti-T4 rabbit serum was used at a dilution which gave a k value of 5 × 10⁻²/min. For each experimental series a calibration curve was also made. The serum was

calibrated using an rII mutant (r73) or in some cases ultraviolet inactivated phage. The results of these two calibrations did not differ. Various diluted samples of the ts mutant lysates were mixed with serum and incubated at 48°C for 18 hr. At this time about 5 × 10⁵ r67 "tester phage" were added and the tubes incubated for another two hr. Samples were then plated on F(λ) and the plates incubated at 42°C so that only the tester phage would form plaques. Inactivation of the tester in the absence of serum is negligible. The serum blocking titer of the lysate was calculated on the basis of at least two different dilutions of the mutant lysate which gave tester phage survival on the linear portion of the calibration curves.

(Geneva) Assay and calibrations were the same as in the Pasadena tests except for the following modifications. In most tests the reaction mixtures were incubated at 30°C for 14 hr. Tester phage (wild type) were added at a concentration of 6 × 10⁷ and incubation at 30°C was continued for another 2½ hr. Tester phage survival was measured by plating on S/6. For each series of determinations a few points on the calibration curve were repeated.

Electron microscopy. (Pasadena) Most lysates were examined by two methods. *Method A:* Grids were prepared according to the agar filtration method of Kellenberger and Arber (1957). Polystyrene latex spheres were added to the lysate samples for counting purposes. The concentration of the spheres was determined by calibration against control lysates of wild type phage. Grids were shadowed with gold-platinum-palladium alloy. In many cases the shadow angle was too large (45°) to give good visualization of tail components. Counts were made from photographs of fields chosen on the basis of even dispersion of the spheres. Areas of clumping or streaking of spheres were avoided. Preparations with large holes in the supporting film were also rejected. Only preparations in which the background appeared clean, and in which microsomal particles were clearly visible, were counted. Sufficient counts were made to include at least 50 spheres. This corresponded to 50 phage particles in control lysates. *Method B:* Lysates were treated with DNase and RNase and concentrated 100-fold by centrifugation at 40,000 × g for 30 min. Samples were resuspended in either phosphate buffer or 2% ammonium acetate. Neutralized 1% phosphotungstate acid (PT) with added sucrose (1%) was added to an equal volume of the phage lysate. Samples were placed on grids with parlodion films and the excess liquid removed by the application of filter paper to the edge of the grid. In many cases, more than one preparation was made of a given lysate and for a number of cases, preparations from different lysates of the same mutant (or wild type) were examined. The results of acceptable repeated determinations were similar.

Observations were made with an RCA EMUII or a Philips EM 120 microscope.

(Geneva) The agar filtration method was employed for the preparation of lysates for electron microscopic observation. Latex spheres at known concentration were added to all preparations. Grids carrying the supporting film and sample were shadowed at an angle of 30° with gold-platinum-palladium alloy. For each preparation, grids were examined with an RCA II electron microscope and ten fields of view were photographed at a magnification of 3500×. Films from grids judged to be acceptable with respect to the distribution of latex spheres and of objects were counted. At least 70 but most often 200 or more latex spheres (about 15 to 30 per field) and at least 200 phage-related morphological entities (when present in normal amounts) were counted for each preparation.

From the known latex concentration, the counts of various phage morphological components, and the measurements of infected bacteria used in the preparation of the lysates, the number of equivalents per bacterium for each category of object was calculated. From repeated determinations of the same lysate we estimate that, on the average, two separate determinations will differ from each other by less than a factor of two. The relative amounts of the various components were more reproducible.

In some cases, lysates of mutant-infected cells were also examined in PT preparations; the procedure followed was that described in Brenner and Horne (1959).

RESULTS

With both am and ts mutants, the general experimental design was as follows: Bacteria were infected with a given mutant phage under restrictive conditions. A control culture (wild type or r48) was, in most cases, also included in each experimental set. The vast majority of the mutants produce less than 1% of the viable phage produced by a control culture under comparable conditions. Samples of the infected bacteria were taken at various times during the course of the infectious cycle. On these samples, various determinations were made which are described below.

In general, the results of our studies were unambiguous in the sense that for a given determination, the mutant infected cultures were either comparable to the control or markedly defective. For this reason, we have chosen to present our experimental results in summary form in a table. In Table 1 the mutants studied are grouped by

genes and the genes are listed in the order in which they occur in the linkage structure. The finding in a particular test is reported either as a plus, indicating a result comparable to the control, or as a zero, indicating that the mutant infected culture was defective as compared to the control. Following each description of a particular determination below, we indicate the criteria for the designations plus and zero used in the table.

TESTS AND CRITERIA FOR JUDGING RESULTS

(1) *Nuclear breakdown.* One to three min after infection the nucleus of the bacterium is disrupted. This disruption, termed nuclear breakdown, which has been investigated by electron microscopic studies (Kellenberger, Sechaud, and Ryter, 1959) is also demonstrable under special conditions in phase contrast microscopy (see Materials and Methods). All am and ts mutants studied initiate nuclear breakdown. With some mutants (amN82 in gene 44, amN116 in gene 39, and amN130 in gene 46) the ability to initiate nuclear breakdown has been confirmed by studies of sectioned material in the electron microscope. Since this process occurs in all mutant infected cells, these results have been omitted from the table. In passing, we should note that in all cases, including the mutant complexes in which DNA synthesis is not initiated, the infected bacteria lose the capacity to form colonies.

(2) *DNA synthesis.* Shortly after infection, the total amount of DNA in the cell begins to increase. The increase in DNA is linear with time and reaches, at the end of the latent period, a value 2.5 to 5 times the amount of DNA in the cell at the time of infection. Total DNA was measured in samples taken at early times, before any increase in DNA is observed in controls (one to five min), and at late times, just before or after lysis occurs in the controls (25 to 60 min). Increases in the amount of DNA by a factor of two or more between early and late samples is considered normal and is entered in the table as a plus. Mutant infected cultures which show an increase of less than a factor of two are considered to be defective and the results are scored as zero. Where DNA synthesis was defective, the measurements have been repeated with many samples taken at different times during the course of the infection. In a number of cases, no appreciable DNA increase was detected despite the fact that several phage per bacterium were produced (e.g., ts G25 of gene 42). This is possibly due to conversion of bacterial DNA to phage DNA, resulting in no net synthesis. In some instances, DNA synthesis is detected but is clearly abnormal in kinetics. These cases will be discussed later.

(3) *Serum blocking antigen.* Antibody prepared against phage particles neutralizes phage primarily

TABLE 1. PHENOTYPIC CLASSIFICATION OF *am* AND *ts* MUTANTS

Gene	Mutant	Viable phage/cell	Lysis	DNA	Serum blocking antigen	Normal particles	Contracted particles	Heads	Tails
—	Control Geneva	150	+	+	+	+	0	+	+
	Control Pasadena	200	+	+	+	+	0	0	0
1	*am*B24	0.03	0	0	0	0	0	0	0
	*am*A494	0.005	0	0	0	0	0	0	0
2	*am*N51	2	+	+	+	0	0	+	+
3	*ts*A2	0.01	+	+	+	0	0	+	+
4	*am*N112	2	+	+	+	0	0	+	+
5	*ts*A28	0.4	+	+	+	0	0	+	0
	*ts*B49	0.01	+	+	+	0	0	+	0
	*am*N135	0.008	+	+	+	0	0	+	0
	*am*B256	0.006	+	+	+	0	0	+	0
6	*ts*A25	0.01	+	+	+	0	0	+	0
	*am*N102	0.08	+	+	+	0	0	+	0
	*am*B251	0.002	+	+		0	0	+	0
	*am*B254	0.03	+			0	0	+	0
	*am*B274	0.02	+			0	0	+	0
7	*ts*B98	0.01	+	+	+	0	0	+	0
	*am*B16	0.003	+	+	+	0	0	+	0
	*am*N115	0.009	+			0	0	+	0
	*am*B23	0.002	+			0	0	+	0
8	*ts*B25	0.01	+	+		0	0	+	0
	*am*N132	0.02	+	+	+	0	0	+	0
9	*ts*N11	1.0	+	+	+	0	+	0	0
	*ts*L54	10				+	+	0	0
10	*ts*A10	0.5	+	+	+	0	0	+	0
	*ts*B64	0.02	+	+	+	0	0	+	0
	*ts*B12	0.02	+	+		0	0	+	0
	*am*B255	0.03	+	+	+	0	0	+	0
11	*ts*L140	<0.01	+	+		0	0	+	0
	*am*N93	0.07	+		+	+	+	+	0
	*am*N128	0.02	+	+	+	+	0	+	+
12	*ts*A13	0.02	+	+	+	+	+	+	+
	*ts*B60	0.05	+	+	+	+	+	+	+
	*am*N69	0.003	+	+	+	0	+	+	+
	*am*N104	0.01	+			+	+	+	+
	*am*N108	0.02	+			+	+	+	+
13	*ts*N49	1.0	+	+	+	0	0	+	+
14	*am*N71	0.10	+			0	0	+	+
	*am*B20	0.01	+	+	+	0	0	+	+
	*am*E351	0.1	+	+	+	0	0	+	+
15	*ts*N26	0.02	+	+	+	0	0	+	+
	*am*N133	1.0	+	+	+	0	0	+	+
16	*am*N66	0.01	+	+		0	0	+	+
	*am*N88	0.8	+	+	+	0	0	+	+
17	*ts*L51	2.0	+	+		0	0	+	+
	*am*N56	0.002	+		+	0	0	+	+
18	*ts*A38	0.4	+	+	+	0	0	+	+
19	*ts*N3	0.01	+	+	+	0	0	+	0
	*ts*B31	0.05	+	+		0	0	+	0

131

TABLE 1 (continued)

Gene	Mutant	Viable phage/cell	Lysis	DNA	Serum blocking antigen	Normal particles	Contracted particles	Heads	Tails
20	tsA23	0.01	+	+	+	0	0	0	+
	amB8	0.43	+			0	0	0	+
	amN83	0.002	+	+		0	0	0	+
	amN50	0.007	+	+	+	0	0	0	+
21	tsN8	0.03	+	+		0	0	0	+
	amN80	1.0	+	+		0	0	0	+
	amN121	0.3	+			0	0	0	+
	amN90	0.3	+	+	+	0	0	0	+
22	tsL147	0.5	+	+		0	0	0	+
	amB270	0.001	+	+	+	0	0	+[1]	+
23	tsL65	0.1	+	+		0	0	+	+
	tsN37	1.0	+	+	+	0	0	+	+
	amB17	0.04	+	+	+	0	0	0	+
	amB272	0.006	+	+	+	0	0	0	+
24	tsN29	0.02	+	+	+	0	0	0	+
	amN65	0.01	+	+	+	0	0	0	+
	amB26	0.007	+	+	+	0	0	0	+
25	amN67	1.0	+		+				
	amN61	1.0	+	+	+	0	0	+	0
26	amN131	0.01	+	+	+	0	0	+	0
27	tsN34	0.5	+	+	+	0	0	+	0
	amN120	0.002	+	+	+	0	0	+	0
28	amA452	0.1	+	+	+	0	0	+	0
29[4]	amB7	<0.001	+	+	+	0	0	+	0
	amN85	0.01	+	+	+	0	0	+	0
	tsL103	0.5	+	+		0	0	+	0
30	tsN7	0.5	+	+	+	0	0	0	+
	tsB20	0.01	+	+		0	0	0	+
31	amN54	0.02	+	+	+	0	0	0	+
	amN111	0.005	+	+	+	0	0	0	+
32	amA453	<0.001	0	0	0	0	0	0	0
33	amN134	0.006	0[2]	+	0	0	0	0	0
34	tsA20	0.1	+	+	+	+	0	0	0
	tsN1	0.04	+	+	+	+	0	0	0
	tsB3	0.03	+	+	+	+	0	0	0
	tsB22	0.3	+	+	+	+	0	0	0
	tsB57	0.01	+	+	+	+	0	0	0
	tsA44	0.02	+	+	+	+	0	0	0
	amN58	0.10	+	+	0	+	0	+	+
	amB25	0.04	+	+	0				
	amB258	0.07	+		0				
	amB288	0.03	+		0				
	amB265	0.03	+		0				
35	tsN30	0.1	+	+	+	+	0	0	0
	amB252	0.1	+	+	+	+	0	+	+
36	tsN41	1.0	+	+	+	+	0	0	0
	tsB6	1.0	+	+	+	+	0	0	0
37	tsB78	0.01	+	+	+	+	0	0	0
	tsB32	0.05	+	+	+	+	0	0	0
	tsN5	2.0	+	+	+	+	0	0	0
	amN52	0.03	+	+	0	+	0	+	+
	amN91	0.003	+		0	+	0	+	+
	amB280	0·06	+		0	+	0	+	+
38	amN62	1.2	+		0[1]	+	0	+	+
	amB262	0.1	+	+	0[1]	+	0	+	+

TABLE 1 (continued)

Gene	Mutant	Viable phage/cell	Lysis	DNA	Serum blocking antigen	Normal particles	Contracted particles	Heads	Tails
39	tsA41	~50	+[3]	+[3]					
	tsG41	~50	+[3]	+[3]	+				
	amN116	37	+[3]	+[3]	+[3]	+	0	+	+
40	tsL177	0.01	+[2]	+					
41	tsA14	0.01	0	0	0	0	0	0	0
	amN57	0.07	0	0	0	0	0	0	0
	amB15	0.05	0	0					
42	tsG25	5.0	0	0	0				
	amN122	0.007	0	0	0	0	0	0	0
43	tsL91	5.0	0	0	0				
	tsG37	10.0	0	0	0				
	tsL141	0.5	0	0	0				
	tsL56	0.4	0	0	0				
	tsL107	0.1	0	0	0	0	0	0	0
	amB22	0.08	0	0	0	0	0	0	0
	amN101	16	0	+[3]	+	+	0	+	+
44	amN82	0.001	0	0	0	0	0	0	0
	tsB110	0.01	0	0					
45	tsL159	1.0	0	0					
46	tsL109	1.0	0	+[3]					
	amN130	2.6	0	+[3]	+	+[1]	0	+	+
	amN94	0.9				0	0	+	+
	tsL166	0.01	0	+[3]					
47	tsL86	1.0	0	+[3]					
	tsB10	1.0	0	+[3]					

[1] Present in low but significant amount.
[2] Lysis is incomplete but facilitated by chloroform.
[3] Phenotype abnormal, see text.
[4] Note added in proof: More recent am × ts complementation tests show that gene 29 is actually composed of two separate genes; the sites of amB7 and tsL103 are in one of these genes, the site of amN85 in the other.

by complexing with tail fibers (Franklin, 1961). The presence of tail fiber antigen in lysates can be detected by its ability to combine with neutralizing antibody resulting in a decrease in the neutralizing power of the serum when subsequently tested (De Mars, 1955). We have used this technique to detect tail fiber antigen in lysates of mutant infected cells. Control lysates produce over 50 equivalents of serum blocking power per infected bacterium. Experiments with mutant infected lysates are scored as plus if over 50 equivalents are produced per infected cell. Mutants defective in the production of tail fiber antigen produce less than 15 equivalents per cell and are designated by a zero in the table.

(4) *Lysis.* Although synthesis of the endolysin commences shortly after that of phage DNA, the onset of cell lysis does not occur until a later time. The spontaneous lysis of the infected culture indicates that the endolysin is synthesized and the mechanism triggering the lytic mechanism is operative. Although control cultures infected with r48 or r67 lyse spontaneously, wild type infected cultures do not lyse spontaneously under concen-

trated conditions due to lysis inhibition induced by the phage released from early lysing cells. In most cases mutant infected cultures produce too few progeny phage to induce lysis inhibition. Clearing of these cultures indicates an operative lytic mechanism (+). Lack of clearing in cases where no phage are released indicates a defect in the lytic mechanism (0). Cultures which do not lyse but which produce several infective particles per cell may show no lysis due to lysis inhibition rather than to defects in the lytic mechanism. In these cases, an operative lytic mechanism is indicated if lysis occurs when antiserum is added to the culture to prevent lysis inhibition, or where lysis of the cells can be induced with chloroform. (Chloroform appears to induce lysis of cells only if endolysin is present.)

(5) *Synthesis of morphological components.* Lysates of the mutant infected cells were examined in the electron microscope for the presence of structures identifiable as components of phage particles. We are largely concerned with the presence or absence of: (a) phage head membranes, (b) tail cores with or without surrounding sheaths but with attached endplates, and (c) complete phage

133

particles. Although other phage related structures have been observed in some preparations, they have not been included in the table. In a later section, we shall comment on some of the special observations.

In Geneva, appreciable numbers of free heads and tails as well as complete particles are found in control lysates, while under the conditions employed in Pasadena, the ratio of intact phage to unassembled components was usually more than ten to one. This difference is unexplained. In the *ts* mutant lysates, if the amount of a given component (particles, head membranes, etc.) was comparable to the amount of complete particles in the controls, we conclude that a given component is present in normal quantity and the mutant is scored as plus in the appropriate column. Due to technical difficulties, it was not always possible to get reliable counts on the frequencies of tail cores with attached endplates. For this reason, all lysates which did not have a normal number of intact particles were concentrated by centrifugation and examined in PT. In these preparations, plus means that an abundance of tails were observed in PT preparations, while a zero means that few if any tails were observed.

In Geneva, in almost all shadowed preparations,

both free heads and tails could be readily seen. The number of equivalents per infected cell for each component was calculated. Where the number of equivalents was comparable to that observed in control preparations, the mutant is scored as a plus for that character. If less than 15 equivalents per bacterium of a given component were found in the lysate, the mutant was scored as a zero. In general, as in the case of the other determinations, the distinctions were unambiguous. Mutant lysates classified as defective contain no more than 20% and generally considerably less of a given component than those classified as normal.

In shadowed preparations and, more often, in PT preparations free phage heads which contain electron dense material, presumably DNA, are occasionally seen. The fact that the majority of the free head membranes are empty we interpret as due to the loss of DNA from the head during processing for electron microscopic observation. It is likely that all empty head membranes contain DNA at some stage in their production (see Kellenberger et al., 1959).

GENERAL REMARKS AND OBSERVATIONS

From an examination of the table it is apparent that, with few exceptions, mutations in the same

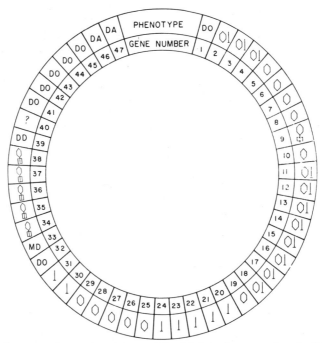

FIGURE 2. Mutant phenotypes of genes containing conditional lethals. Genes are given in their map order. Some mutant phenotypes are symbolized by letters, the meanings of which are described in the text. The symbol designations of the rest of the genes represent the major morphological component present in mutant lysates.

gene give the same general result. This indicates that the phenotype we observe reflects the mutational loss of function of a particular gene under restrictive conditions. We conclude that the differences between a particular mutant and wild type are gene specific and do not represent some property common to the *am* or *ts* mutants as a whole.

In Fig. 2 we have summarized the results of the table, representing each gene by a single symbol indicating its phenotype when mutant. It is clear from the figure that the distribution of the genes throughout the genome is not random with respect to our designated phenotypes. Rather, the genes appear to show a highly clustered arrangement according to function. We will describe below the general phenotypic features of each cluster and additional observations not included in the table.

As a first approximation, the genes appear to fall in two major classes. One class (genes 1, 32, and 39 to 47) exhibits defects in DNA synthesis, the second class (genes 2 to 31 and 33 to 38) exhibits defects in maturation. These two major groups of genes occupy two non-overlapping regions of the genome (although we should point out that the endolysin gene

[Streisinger et al., 1961] lies between genes 47 and 1). The DNA defective genes are classified in the following manner.

DD gene (39). There is a delay in the onset of DNA synthesis after which DNA synthesis proceeds normally. Tail fiber antigen synthesis shows a corresponding delay, active phage are eventually formed, and the complexes finally lyse if serum is present.

DO genes (1, 32, and 41 to 45). No detectable DNA synthesis occurs. (One mutant of gene 43, *am*N101, shows some DNA synthesis and probably is a "leaky" mutant.) In no case was tail fiber antigen, particle components, or lysis of the infected cells detected. The infected cells lose the ability to form colonies.

DA genes (46 and 47). DNA synthesis is initiated normally but ceases after a short interval; few active phage but various phage-related components are synthesized. Cultures clear at least partially if treated with chloroform.

Gene 40 has been incompletely studied. DNA is synthesized but it is not yet known if the kinetics or amount of DNA formed is aberrant. The cells

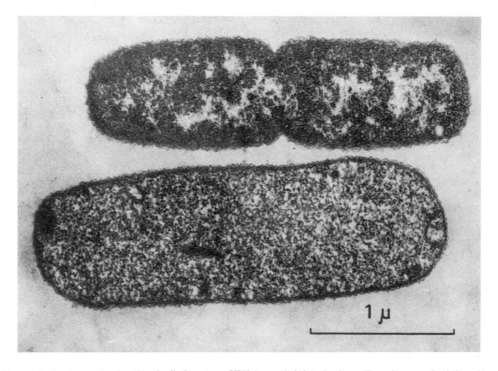

FIGURE 3. A micrograph of sectioned cells from an *am*N122 (gene 42) infected culture. The culture was fixed (formaldehyde, 2% osmium tetroxide, 1%) after 45 min of incubation at 32°C. Note the homogeneous distribution of ribosomes and absence of a DNA pool in the infected cell (lower cell). For purposes of comparison we have chosen a field containing a cell (upper cell) which appears to be uninfected.

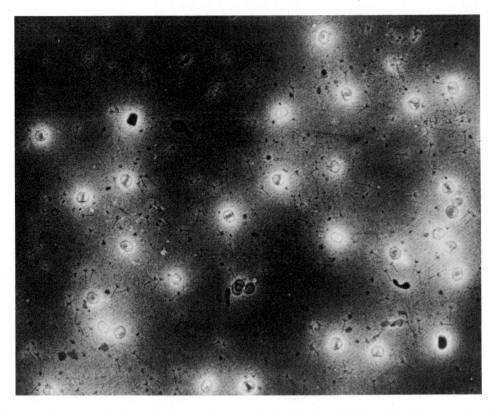

FIGURE 4. A micrograph of a lysate of *ts*A38 (gene 18). As described in Materials and Methods the lysate was concentrated by centrifugation and prepared by the PT method. Note empty heads and free cores with attached endplates.

lyse with chloroform indicating the presence of the endolysin.

In a few cases, observations were made on the intracellular morphology of bacteria infected with mutants defective in DNA synthesis. Samples from mutant infected cultures were taken at various times during the infectious cycle and thin sections of fixed and embedded cells were examined in the electron microscope. The observations which have been made are in accord with the results of the determinations just discussed. Cells infected with a DA mutant, *am*N130, contain condensates (Kellenberger et al., 1959) indicating that maturation is not completely blocked. Sections of cells infected with *am*N116, a DD mutant, contain condensates in samples taken at late times (46 min after infection) but at earlier times no condensates are observed, although they are present in comparable control infected bacteria. For *am*N122 and *am*N81, both DO mutants, no evidence of phage maturation is observed in the sections and further, the characteristic DNA pool is not formed although nuclear

breakdown does occur. The appearance of cells infected with *am*N122 is shown in Fig. 3.

The second class of mutants are those which exhibit defects in maturation. In most respects the pattern of events after infection is normal. DNA synthesis is normal and the infected cultures lyse at the same time as controls. However, almost no viable phage are produced and only certain morphological components are found in the lysates. As a first approximation, the mutants can be divided into several categories based upon the components present in normal amounts in the lysates: (1) normal appearing but inactive particles; (2) particles with contracted sheaths; (3) head membranes and free tails (see Fig. 4); (4) head membranes only; (5) tails only; and (6) few if any components. All of the mutants which lyse normally and synthesize DNA normally fall into one of these 6 categories.

The tail fiber genes (34 to 38). Cells infected with all mutants of these genes produce normal-appearing particles. Free heads and tails are also present

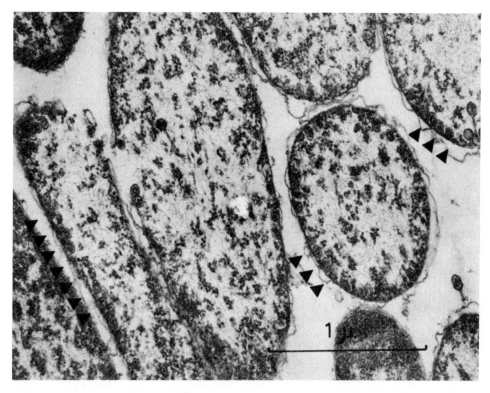

FIGURE 5. *Am*N90 (gene 21) infected *E. coli* B. Cells were fixed after incubation for 40 min at 32°C. Note the abnormal heads arranged along the cytoplasmic membrane (arrows).

in *am* lysates but not in *ts* lysates. However, as pointed out earlier, this difference between the results found in Geneva and Pasadena also occurs for the controls. In the case of certain *am* mutants, it appears likely that the particles do not have tail fibers. In no case are these non-infectious particles capable of killing bacteria and preliminary studies suggest that they cannot attach normally to bacteria. *am* mutants from genes 34 and 37 do not initiate synthesis of tail fiber antigens, while for *am* mutants of gene 38 the level is low. *ts* mutants from all these genes result in normal levels of tail fiber antigen. Mapping experiments have indicated that mutations affecting the co-factor requirement (Brenner, 1957) are located in gene 34 (Edgar et al., in preparation), while the host range character (Streisinger, 1956) is probably located in or near gene 37. These observations, taken as a whole, indicate that these genes are involved in the synthesis and assembly of the tail fibers and that in the absence of their formation inactive particles lacking tail fibers are formed.

The MD (maturation defective) gene (33). While DNA synthesis appears to be normal, the cells show only partial lysis which is facilitated with chloroform. The lysates contain few if any morphological components and tail fiber antigen is absent. While the phenotype suggests a general defect in maturation, more work is required to characterize the nature of the block for this gene.

The head genes (20 to 24 and 30 and 31). These two groups of genes are characterized by the absence of heads in lysates from mutant infected cells and thus would appear to be involved in head synthesis or assembly. In one case (gene 23) heads are formed by the *ts* but not by *am* mutants in the gene. The osmotic shock resistant mutants (Brenner and Barnett, 1959) map in the neighborhood of gene 23.

Sectioned material from bacteria infected with mutants from genes 20–24 provide further information about the defects produced. While lysates of mutants *am*N50 and *am*N90, from genes 20 and 21 respectively, do not contain head membranes, sections of bacteria infected with these mutants contain structures which may be related to head membranes. In the case of amN90, head-like structures are observed against the limiting membrane of the bacterium, but very few of these structures are found in the interior of the cell (see Fig. 5). In bacteria infected with mutant amN50, long cylinders (polyhead) which in cross-section have dimensions similar to phage heads are present in large quantity (see Fig. 6). The fact that these objects are not found in abundance in the lysates is most easily explained by loss during the centrifugation employed in the preparation of lysates for electron microscopic observation. Sections of cells infected with amB17 (gene 23) and amN65 (gene 24) show neither condensates nor aberrant structures similar to those found in the case of amN90 or amN50.

Less can be said about the rest of the maturation defective mutants. We have characterized the phenotypes of these genes in the figure as resulting in the formation of heads only (genes 5–8, 10, 19 and 25–29), unassembled heads and tails (genes 2–4, 11–18) or contracted phage (gene 9). The head and tail classification includes mutants (genes 11

and 12) which give appreciable numbers of complete but inactive particles in the lysates. The proportion of these complete particles varies from mutant to mutant and preparation to preparation. This variability is most easily explained by the formation of particles susceptible to breakdown. The majority of the particles produced by mutants of gene 9 have contracted sheaths although the heads remain filled.

The particular morphological components scored in the lysates were chosen because observation can be made with reasonable confidence. Certain other morphological components have been observed in some mutant lysates but have not been included in our characterization of the mutants because quantitative observations are more difficult. For example, free cores without endplates have been observed in lysates of amB274 (gene 6). It should also be noted that no distinction has been made in reporting our results between the presence or absence of a sheath surrounding the free tail cores. Differences between the various mutants exist in this regard but more work is required to ascertain

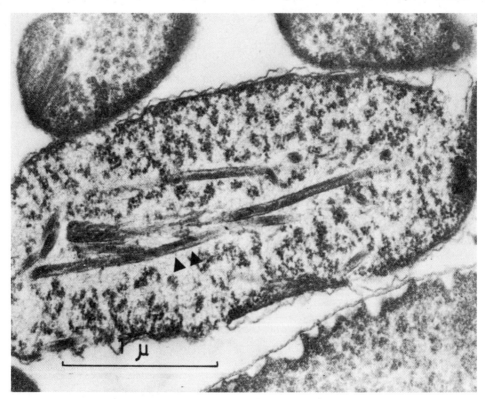

FIGURE 6. AmN50 (gene 20) infected E. coli B. Cells were fixed after incubation for 40 min at 32°C. Note the long cylinders (polyhead) in the interior of the cell (arrows).

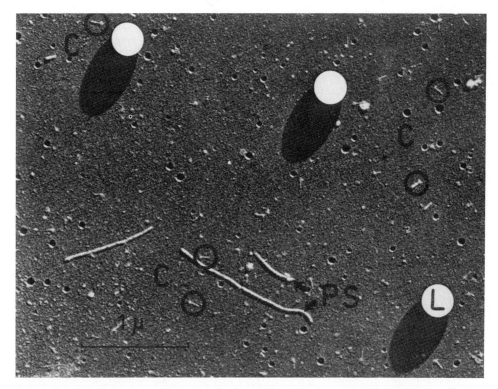

FIGURE 7. A shadowed preparation of a lysate of *am*B17 infected *E. coli* B. This field is taken from a micrograph used for particle counts. Note latex spheres (L), polysheath (PS), and cores with attached endplates (circled).

the distributions of these phenotypes among the mutants.

Of particular interest is the observation that, in most lysates which contain incomplete particles, cylinders of varying lengths and with widths approximately that of contracted sheath are frequently observed (Figs. 7 and 8). Electron microscopic studies of this material (Boy de la Tour and Kellenberger, in preparation) show that these cylinders (which we shall refer to as polysheath) are composed of sheath material.

DISCUSSION

THE NATURE OF *am* AND *ts* MUTATIONS

Although most *am* and *ts* mutants located within the same gene present the same phenotypes, there are instances in which the *am* and *ts* phenotypes are dissimilar. Gene 34 is one example of such dissimilarities. While particle morphology, bacterial killing ability and attachment behavior of the *am* and *ts* mutants of gene 34 are very similar, tail fiber antigen is absent in all *am* lysates and present in all *ts* mutant lysates. The observations hold for *am* and *ts* mutants at a variety of sites within the

gene. The assumption that *am* mutations prevent the formation of protein (or permit only the synthesis of a grossly altered molecule) while the *ts* mutations do not, provides a simple explanation for the differences observed. Thus, we imagine that in *ts* mutant lysates the protein coded by gene 34 is present as an altered and functionally defective molecule which nevertheless has antigenic activity, but that no gene 34 protein is present in the *am* mutant lysates. It is of interest, in this connection, that a preliminary investigation of the particles in the *am* lysates indicates that they lack tail fibers.

The explanation suggested is also supported by a study of a second gene in which dissimilarities exist between the phenotypes of *am* and *ts* mutants. On the basis of the *am* phenotypes we have assumed that gene 23 is involved in the formation of the phage head membrane. Head membranes, however, are present in the *ts* lysates although absent in *am* lysates. Sarabhai and Brenner (personal communication) recently demonstrated that gene 23 is the structural gene for head protein. The study includes the observation that the mutant *am* 17 in this gene prevents the formation of head protein.

Arguments presented elsewhere (Epstein et al., in preparation; Edgar and Lielausis, in preparation; Edgar et al., in preparation) concerning the nature of *am* and *ts* mutants lend additional support to our interpretation of the major difference between these two mutant types.

The Arrangement of Genes with Related Phenotypes

One of the most striking results of our study is that genes with various functions, as inferred from our phenotypic tests, are not randomly distributed in the genome but are grouped into regions within which all genes have similar phenotypes and presumably related functions. Genes which appear to be involved in the early functions of growth (mutations affecting DNA synthesis) form one major region of the genome separate from the second major group of genes which are concerned with late functions (e.g., the maturation of phage). Each of the two major regions can be further sub-divided into sub-regions or clusters of genes with

similar phenotypes. Since our operational definition of a gene (based upon the complementation properties of the mutants,) (Edgar et al., in preparation) is in accord with that used in genetic studies with other organisms, a cluster is, by this criterion, composed of a number of genes and not just one gene. In support of this contention is the finding that mutations in one gene but not in others in the same cluster result in the absence of specific phage-induced proteins. Mutations in gene 42 not only affect the synthesis of dCMP hydroxymethylase but in certain instances modify its structure (Wiberg et al., 1962; Dirsken, this volume; Wiberg, personal communication). As mentioned earlier, there is now evidence that gene 23 specifically controls the structure of the head protein. In both cases these proteins are unaffected by mutations in other genes in the same cluster.

Since a significant proportion, perhaps as much as half, of the genes of the phage have been studied, the clustered arrangement appears to be a general feature of the genome. There are a small number of

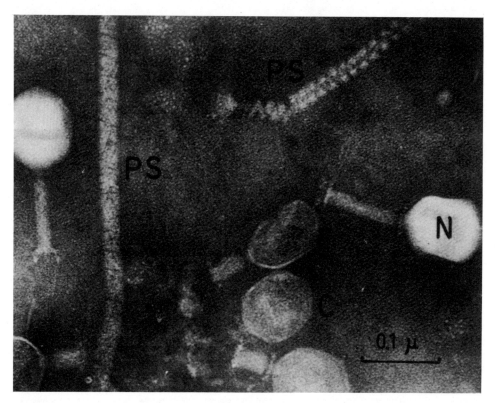

FIGURE 8. Phage stock of *am*N90 (gene 21) made on *E. coli* CR63. The micrograph, a PT preparation, shows polysheath (PS), normal (N) and contracted (C) phage. It should be noted that growth of *am*N90 on CR63 is partially defective.

exceptions, single genes bounded by genes of dissimilar phenotype (e.g., 1, 9, 10, 19, 32, 33, 39, 40). The exceptions may, however, represent single representatives of as yet unidentified clusters, or cases in which the observed phenotype is not diagnostic of the primary gene function.

The cluster pattern bears an obvious resemblance to the operons of bacteria (Jacob and Monod, 1961), and it is tempting to suggest that the phage genome is organized into 10–20 operons. It should be emphasized that the clustering itself is, at present, the only evidence that the operon is a basic unit of organization of phage genes. None of the mutants we have isolated has been demonstrated to have a control function. Although regulator genes analogous to the i gene controlling the β-galactosidase operon of $E. coli$ may be present in our collection of mutants, our methods of analysis, at present, would probably not permit their identification. Operator mutants, because of their cis-dominant phenotypes, should have been revealed by the complementation tests referred to above, yet none were found among more than 400 am and ts mutants examined. The studies of operator mutants in bacteria suggest that the operator does not function through a transcribed gene product, (see Attardi et al., 1963) but rather it is a region of the DNA which is acted upon by a product of a regulator gene. The absence of operator mutants in our set may therefore be due to the fact that the conditional lethal phenotypes of the am and ts mutants are made manifest only through the transcription process, thus restricting these mutants to regions of the DNA which produce a gene product.

GENE INTERACTION

In mutant infected cells in which no detectable DNA synthesis occurs, none of the genes which are involved in the production of components of the mature particle, or the endolysin, are activated. It could be imagined that, in the absence of DNA synthesis, there are not sufficient gene copies in the cell to support the production of late proteins. However, Cohen (1948) and Koch and Hershey (1959) have shown that the rate of protein synthesis in the infected cell is constant over the normal growth cycle; only the type, but not the quantity, of protein differs at different times. For this reason, it is possible that the lack of activity of the genes involved in late function is a direct consequence of the failure of a mutant gene of the DNA negative class to carry out its function. Wiberg et al. (1962) observed that in cells infected with DNA negative mutants, the synthesis of all the early enzymes which were studied continues instead of being arrested shortly after infection as is the case in cells infected with wild type or with am mutants not

exhibiting defective DNA synthesis. These results indicate that mutations blocking DNA synthesis affect the regulation of development in the infected cell such that phage growth cannot proceed past an early stage. The activation of late gene expression may thus require the attainment of a certain cytoplasmic state. The notion that this state may involve the onset of DNA synthesis, or some event associated with it, is supported by experiments with other mutants defective in DNA synthesis. Cells infected with amN116 (gene 39) exhibit a net increase in DNA only after some delay. A comparable delay is observed for the synthesis of tail fiber antigen, the appearance of the first infectious phage progeny, and the onset of cell lysis. In cells infected with amN130 (gene 46) although DNA synthesis is arrested after a normal onset of synthesis, synthesis of tail fiber antigen occurs and the pattern of early enzyme synthesis (Wiberg et al., 1962) is normal.

It is worth noting that this particular type of phenotypic interaction may not be characteristic of all phages. Jacob et al. (1957) found a number of defective mutants of phage lambda which, while blocked in genome replication, and presumably in phage specific DNA synthesis, nevertheless did permit the expression of some of the late functions such as the lytic mechanism and coat protein synthesis. Furthermore, lambda complexes, incapable of DNA synthesis due to a block in thymine synthesis, nevertheless synthesize endolysin and coat protein (Weigle, personal communication; Karamata et al., 1962).

The functional interaction between genes affecting DNA synthesis and those concerned with phage maturation defines two groups of genes which we have called early genes and late genes. These two major groups of gene clusters defined from the point of view of functional interaction have also a physical counterpart in their arrangement in the genetic structure (note that genes 1 and 32 are exceptions). If future investigations support the interpretation of each cluster as a separate operon, the possibility must be considered that operons themselves may sometimes be physically grouped into higher order units of functional significance.

THE MORPHOGENESIS OF PHAGE PARTICLES

The problem of the construction of finished phage particles does not end with an understanding of the synthesis of the individual protein subunits of which the particle is composed. The manner in which these subunits are assembled is of interest and may be considered as a special case of the broader question of the mode of assembly of cell organelles.

Studies on the aggregation properties of the

subunits of the tobacco mosaic virus (TMV) have suggested that the form of the protein shell of the virus is largely determined by the properties of the subunits themselves. The protein subunits can spontaneously aggregate to give the helical form of the virus and complete virus particles may be reconstituted in vitro from these subunits and viral RNA (Schramm et al., 1955; Fraenkel-Conrat and Williams, 1955).

While TMV is a simple rod composed of only a single species of protein, the T4 phage is a more intricate structure composed of a number of different kinds of protein subunits organized in a complex manner (Brenner et al., 1959). Although we have little information concerning the specific mode of assembly of T4 virus particles, it would appear likely that maturation involves the interdependent assembly of many components. This is suggested, for example, by the formation of polysheath by most mutants which give incomplete particle assembly. We might imagine that cores, delimited by endplate and some other element, serve as a matrix for the crystallization of normal sheath. In the absence of this matrix, sheath material might crystallize in a contracted form of indeterminate length.

However, mechanisms only involving interdependent successive crystallization of the various subunits are probably insufficient to account for maturation. We have found 37 genes involved in the morphogenesis of the phage particles, and it is likely that more genes of this type exist. On the other hand, although the phage particle has an intricate structure, it seems unlikely that there are more than 20 species of protein in the particle itself. Thus we might imagine that many of the genes involved in maturation of the phage do so indirectly, playing a role in the assembly rather than in the synthesis of protein subunits which compose the mature particle.

A number of observations suggest that some genes involved in maturation do act in such an indirect manner. The phenotypes of the mutants in genes 20–24 suggest that they are involved in head membrane formation. Sarabhai's observations indicate that head protein is formed by mutants in all of these genes except gene 23. Our preliminary investigations of the intracellular morphology of bacteria infected with mutants from genes 20 and 21 of this region suggest that although head protein is formed, assembly of head structures is aberrant, leading to either polyhead or head-like structures bound to the cell membrane. One interpretation of these observations is that the products of genes 20 and 21 are essential factors in head membrane formation but not in the usual sense of making a material contribution to the finished structure.

If the suggested roles for maturation genes are correct, the phage particle represents a structure whose morphogenesis cannot be conceived of as simple crystallization from subunits, as may be the case with TMV. Thus, the study of those genes with functions ancillary to genes coding for structural proteins of T4 are of great interest as models for "morphogenetic" genes.

ACKNOWLEDGMENTS

The work performed in Geneva was supported by a grant from the U.S. Public Health Service (contract no. E4267). The work performed in Pasadena was aided by a grant from the National Foundation and by a U.S. Public Health Service Grant (no. RG-6965). Part of this work was performed while one of us (R.H.E.) was a U.S. Public Health postdoctoral fellow at the California Institute of Technology. R.H.E. is also greatly indebted to the Department of Bacteriology, University of California at Los Angeles for providing support and facilities for part of this research, and in particular to Mr. F. Eiserling for help with some of the earlier electron microscope studies.

REFERENCES

ATTARDI, G., S. NAONO, J. ROUVIÈRE, F. JACOB, and F. GROS. 1963. Production of Messenger RNA and regulation of protein synthesis. Cold Spring Harbor Symp. Quant. Biol. Vol. 28.

BRENNER, S. 1957. Genetic control and phenotypic mixing of the adsorption cofactor requirement in bacteriophages T2 and T4. Virology 3: 560–574.

BRENNER, S. and L. BARNETT. 1959. Genetic and chemical studies on the head protein of bacteriophages T2 and T4. Brookhaven Symp. Biol. 12: 86–94.

BRENNER, S. and R. W. HORNE. 1959. A negative staining method for high resolution electron microscopy of viruses. Biochim. Biophys. Acta 34: 103–110.

BRENNER, S., G. STREISINGER, R. W. HORNE, S. P. CHAMPE, L. BARNETT, S. BENZER, and M. W. REES. 1959. Structural components of bacteriophage. J. Mol. Biol., 1: 281–292.

BURTON, K. 1956. A study of the conditions and mechanism of the diphenylamine reaction for the colorimetric estimation of deoxyribonucleic acid. Biochem. J., 62: 315–323.

CAMPBELL, A. 1961. Sensitive mutants of bacteriophage λ. Virology, 14: 22–32.

COHEN, S. S. 1948. The synthesis of bacterial viruses. I. The synthesis of nucleic acid and protein in Escherichia coli B infected with T2r+ bacteriophage. J. Biol. Chem. 174: 281–293.

DE MARS, R. I. 1955. The production of phage-related materials when bacteriophage development is interrupted by proflavine. Virology, 1: 83–99.

FRAENKEL-CONRAT, H. and R. C. WILLIAMS. 1955. Reconstitution of active tobacco mosaic virus from its inactive protein and nucleic acid components. Proc. Natl. Acad. Sci., 41: 690–698.

FRANKLIN, N. C. 1961. Serological study of tail structure and function in coliphages T2 and T4. Virology, 14: 417–429.

JACOB, F., C. FUERST, and E. WOLLMAN. 1957. Recherches sur les bactéries lysogènes défectives. Ann. Inst. Pasteur, *93:* 724–753.

JACOB, F. and J. MONOD. 1961. Genetic regulatory mechanisms in the synthesis of proteins. J. Mol. Biol., *3:* 318–356.

KARAMATA, D., E. KELLENBERGER, G. KELLENBERGER, and M. TERZI. 1962. Sur une particule accompagnant le développement du coliphage λ. Path. Microbiol., *25:* 575–585.

KECK, K. 1956. An ultramicro technique for the determination of deoxypentose nucleic acid. Arch. Biochem. Biophys., *63:* 446–451.

KELLENBERGER, E. 1961. Vegetative bacteriophage and the maturation of virus particles. Adv. in Virus Res., *8:* 1–61.

KELLENBERGER, E. and W. ARBER. 1957. Electron microscopical studies of phage multiplication. I. A method for quantitative analysis of particle suspensions. Virology, *3:* 245–255.

KELLENBERGER, E., J. SÉCHAUD, and A. RYTER. 1959. Electron microscopical studies of phage multiplication. IV. The establishment of the DNA pool of vegetative phage and the maturation of phage particles. Virology, *8:* 478–498.

KOCH, G. and A. D. HERSHEY. 1959. Synthesis of phage-precursor protein in bacteria infected with T2. J. Mol. Biol., *1:* 260–276.

MASON, D. J. and D. M. POWELSON. 1956. Nuclear division as observed in live bacteria by a new technique. J. Bact., *71:* 474–479.

SCHRAMM, G., G. SCHUMACHER, and W. ZILLIG. 1955. Über die Struktur des Tabakmosaikvirus. III. Mitt.: Der Zerfall in alkalischer Lösung. Z. Naturforsch., *10b:* 481–492.

STEINBERG, C. M. and R. S. EDGAR. 1962. A critical test of a current theory of genetic recombination in bacteriophage. Genetics, *47:* 187–208.

STREISINGER, G. 1956. The genetic control of host range and serological specificity in bacteriophages T2 and T4. Virology, *2:* 377–387.

STREISINGER, G. and N. FRANKLIN. 1956. Mutation and recombination at the host range genetic region of phage T2. Cold Spring Harbor Symp. Quant. Biol., *21:* 103–109.

STREISINGER, G., F. MUKAI, W. J. DREYER, B. MILLER and S. HORIUCHI. 1961. Mutations affecting the lysozyme of phage T4. Cold Spring Harbor Symp. Quant. Biol., *26:* 25–30.

WIBERG, J. S., M. DIRKSEN, R. H. EPSTEIN, S. E. LURIA, and J. M. BUCHANAN. 1962. Early enzyme synthesis and its control in *E. coli* infected with some amber mutants of bacteriophage T4. Proc. Natl. Acad. Sci. *48:* 293–302.

13

Reprinted from *Proc. Natl. Acad. Sci. (U.S.)*, **51**(5), 775–779 (1964)

CHROMOSOME STRUCTURE IN PHAGE T4, I. CIRCULARITY OF THE LINKAGE MAP*

By George Streisinger, R. S. Edgar, and Georgetta Harrar Denhardt

INSTITUTE OF MOLECULAR BIOLOGY, UNIVERSITY OF OREGON, AND BIOLOGY DIVISION, CALIFORNIA INSTITUTE OF TECHNOLOGY

Communicated by A. D. Hershey, March 26, 1964

On extraction, particles of phage T4 each yield a single DNA molecule;[1-3] and in genetic crosses, all the markers prove to be linked to one another.[4] Crossing also generates short heterozygous regions that are distributed randomly over the genome.[5]

Doermann and Boehner[6] have suggested that at least some of the heterozygous regions involve interruptions of some sort in the DNA chain. Berns and Thomas[7] and Cummings,[8] on the other hand, interpret results of their experiments on the physical properties of T4 DNA to mean that the single strands are continuous.[9]

The paradox could be resolved by making two assumptions about the chromosome of phage T4. First, that it is a linear molecule with a terminal redundancy:

$$a\ b\ c\ d\ e\ f\ldots\ldots w\ x\ y\ z\ a\ b\ c$$

Second, that after several rounds of replication the chromosome becomes circularly permuted as a result of genetic recombination within the region of redundancy.[10] The ends of a population of chromosomes would thus be randomly distributed over the genome:

$$g\ h\ i\ j\ k\ l\ldots\ldots c\ d\ e\ f\ g\ h\ i$$

$$m\ n\ o\ p\ q\ r\ldots\ldots i\ j\ k\ l\ m\ n\ o,\ etc.$$

One of several predictions of the above model is that the linkage map of phage T4 would be circular. A test of this prediction forms the substance of the present communication.

Materials and Methods.—The phage strains used were derivatives of T4D containing the markers $r73$;[4] $tu41$, $tu42b$, $tu44$, $tu45a$, and $r48$;[11] $h42$;[12] $ac41$;[13] rEDb-48;[14] $am85$ and $am54$, obtained from Dr. R. H. Epstein; and $r67$, obtained from Dr. A. H. Doermann. The bacterial strains were *Escherichia coli* B, S/6, K12(λ), CR63, and K12(λ)/4.

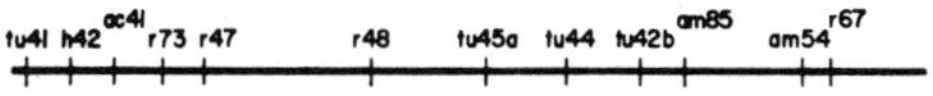

Fig. 1.—A provisional map of T4 showing the relative locations of several markers.

Media: Broth—1 liter H_2O, 10 gm bacto-tryptone, 5 gm NaCl. Tryptone plates—bottom layer, broth with 1.1 per cent bacto-agar; top layer, broth with 0.7 per cent bacto-agar. EHA plates—bottom layer contained 1 liter H_2O, 13 gm bacto-tryptone, 2 gm sodium citrate $\cdot$ 2 H_2O, 1.3 gm glucose, 14 gm bacto-agar, 8 gm NaCl; top layer contained 1 liter H_2O, 13 gm bacto-tryptone, 2 gm sodium citrate $\cdot$ 2 H_2O, 3 gm glucose, 7 gm bacto-agar, 8 gm NaCl. Salt-poor EHA plates— bottom layer was identical to EHA but contained only 2.5 gm NaCl per liter; top layer contained no added NaCl.

Acriflavin-neutral (Nutritional Biochemicals Corp.) was added, when required, to the bottom layer only, at a concentration of 0.25 μg per ml.

Results.—A linkage map of phage T4 is shown in Figure 1. With the exception of *am*85, *am*54, and *r*67, the relative order of the markers indicated in the map was established previously by two- and three-factor crosses.[4,11−13] We now find that *am*54 is closely linked to *r*67 (1.5% recombinants), and *am*85 is closely linked to *tu*42b (3.5% recombinants). The order *tu*44—*am*85—*am*54 is established by means of the three-factor cross *tu*44 *am*54 $\times$ *am*85 (cross 14, Table 1). The order *tu*44—*tu*42b— *r*67 is confirmed by cross 15, Table 1.

The information cited above is compatible with either of two alternatives. The linkage map may have ends, in which case *r*67 lies closer to *ac*41 than to *h*42, as represented in Figure 1. Or the linkage map could be a circle formed by joining the ends shown in Figure 1, in which case *r*67 would be expected to lie closer to *h*42 than to *ac*41.

The cross *r*67 *h*42 *ac*41$^+$ $\times$ *r*67$^+$ *h*42$^+$ *ac*41 (cross 1, Table 1) distinguishes between these alternatives. The progeny of the cross were plated under conditions that permitted the recognition of the *h*42 *ac*41 recombinants. About 65 per cent of these were *r*, and 35 per cent *r*$^+$, showing that the *r* marker is more closely linked to *h* than to *ac*. In other words, the results of cross 1 are inconsistent with previous data summarized in Figure 1 unless the linkages are represented on a circle.

The same linkage was tested in cross 2, Table 1, with a different arrangement of the parental markers. The results are consistent with those of cross 1, showing that the marker frequencies depend on linkages between loci only, and do not reflect a bias associated with particular markers.

Additional three-factor and four-factor crosses are listed in Table 1. They are similar in principle to crosses 1 and 2 and place all the markers in a unique order on a circular map. In some crosses, unequal multiplicities of infection were used to increase the sensitivity of the linkage tests.[4] In all crosses except 3, 14, and 15, the frequency of recombinants was measured among the early phage progeny (1–10 per bacterium), and usually the drift toward genetic equilibrium of unselected alleles was demonstrated by additional sampling at later times during phage growth. Arguments concerning the validity of these types of linkage test have been presented by Streisinger and Bruce.[4] Scoring procedures are described in Table 2. Figure 2 identifies linkages tested in the individual crosses and summarizes the circular map with which all results are consistent.

Discussion.—Our results demonstrate genetic circularity and are thus compatible with the model presented in the introduction to this paper. The circularity of the

145

TABLE 1

No.	Cross Parent 1	Cross Parent 2	Multiplicity Parent 1	Multiplicity Parent 2	Recombinant ratio measured*	Progeny Phage Value of ratio	Progeny Phage Control ratio	Progeny Phage Value of control ratio
1	r67 h42	ac41	9.8	10.3	112/.12	0.65	1../...	0.53
2	h42	r67 ac41	7.5	7.4	112/.12	0.66	1../...	0.54
3	ac41 rEIDb48	h42	4.2	4.7	112/.1·	0.09		
4	r67	r73 r47	5.0	0.5	211/.1·	0.11		
5	r47	r67 r73	3.1	0.7	212/.1·	0.02		
6	r73	r67 r47	5.8	0.6	221/.21	0.58		
7	r67 r73	r47	5.1	0.7	212/.12	0.33		
8	r48	r73 r47	8.6	0.9	221/.21	0.53		
9	r47	r73 r48	7.2	0.8	212/.12	0.29		
10	r47 r73	r48	8.6	0.8	212/21·	0.29		
11	r47	r73 r48	8.4	0.9	122/12·	0.43		
12	tu44	tu45 r48	8.8	9.6	212/21· ; 122/12·	0.27 ; 0.40	1../...	0.60
13	tu44 r48	tu45	7.5	7.4	112/.12	0.74	1../...	0.55
14	am54	am85	6	6	112/.12	0.72		
15	tu44 tu42	am85	6.3	7.8	112/.12 ; 122/12·	0.80 ; 0.74		
16	r48 am54 tu41	am85	7.1	7.6	2121/.12· ; 212·/.12· ; ·121/.12·	0.12 ; 0.46 ; 0.36	2.../.....	0.53
17	am85 tu41	am54	9.7	9.4	1212/.21· ; 121·/.21· ; ·212/.21·	0.13 ; 0.46 ; 0.37	2.../.....	0.54

* 1 or 2 indicates a marker derived from parent 1 or 2, and a dot (·) indicates a marker derived from either parent. The ratio 112/·12 in cross 1, for instance, represents r67 h42 ac41/(r67 h42 ac41 plus r⁺ h42 ac41).

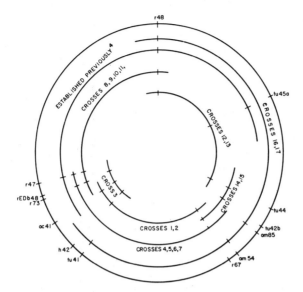

FIG. 2.—The circular map of T4. The locations of markers used for any one cross are connected by an arc.

genetic map of T4 has been confirmed by two-factor crosses made with a large set of amber mutants[15] and temperature-sensitive mutants.[16] Foss[17] has been able to demonstrate genetic circularity in T4 by means of a single, ingeniously devised four-factor cross.

TABLE 2

PROCEDURES FOR SCORING THE PROGENY OF CROSSES

Crosses	Plates	Bacteria	Scoring
1, 2	Salt-poor EHA with acriflavin	Phage preadsorbed to S/6, plated on a mixture of K/4 and S/6	Only $h42$ $ac41$ forms clear plaques, classified as r or r^+ by inspection.
3	EHA with acriflavin	Mixed K/4 and S/6	Only $ac41$ forms plaques, classified as r or r^+ and h or h^+ by inspection.
4–11	Tryptone	K12(λ)	Only $r73^+$ $r47^+$ forms plaques, classified as r or r^+ by inspection.
12, 13	EHA	S/6	$tu45^+$ $tu44^+$ plaques selected and classified as r or r^+ by inspection.
15	EHA	S/6	Genotypes tu^+ r^+ and tu^+ r classified by inspection.
14, 16–17	EHA	S/6	Only am^+ forms plaques, classified as r or r^+ and tu or tu^+ by inspection.

It should be emphasized that our results, while demonstrating genetic circularity, are by no means a critical test of the model we present; a number of other models would account for the results equally well. For instance, Stahl[18] has pointed out that a circular genetic map would be obtained if the chromosome were a nonpermuted rod and if genetic exchanges frequently occurred in pairs. A decision with respect to the molecular basis of circularity calls for other, more critical tests.

Summary.—The linkage map of phage T4 is circular.

The authors wish to acknowledge their great indebtedness to F. W. Stahl, without whose encouragement they would not have taken their rather bizarre notions seriously. Many of the features of the proposed model were developed in the course of conversations and correspondence with F. W. Stahl, M. Meselson, and M. Fox.

* This investigation was supported by research grants from The National Foundation (to R.S.E.), the National Science Foundation (G-14055 to G.S.), and the National Institute of Allergy and Infectious Diseases (E-3892 to G.S.).

[1] Rubenstein, I., C. A. Thomas, Jr., and A. D. Hershey, these Proceedings, **47,** 1113 (1961).

[2] Davison, P. F., D. Freifelder, R. Hede, and C. Levinthal, these Proceedings, **47,** 1123 (1961).

[3] Cairns, J., *J. Mol. Biol.*, **3,** 756 (1961).

[4] Streisinger, G., and V. Bruce, *Genetics*, **45,** 1289 (1960).

[5] Hershey, A. D., and M. Chase, in *Genes and Mutations*, Cold Spring Harbor Symposia on Quantitative Biology, vol. 16 (1951), p. 471.

[6] Doermann, A. H., and L. Boehner, *Virology*, **21,** 551 (1963).

[7] Berns, K. I., and C. A. Thomas, Jr., *J. Mol. Biol.*, **3,** 289 (1961).

[8] Cummings, D. J., *Biochim. Biophys. Acta*, **72,** 475 (1963).

[9] P. F. Davison, D. Freifelder, and B. W. Holloway [*J. Mol. Biol.*, **8,** 1 (1964)] obtained physical evidence suggesting that the strands are not continuous. Thus, the question of strand interruptions is not yet resolved. Although it is true that our experiments were motivated by Berns and Thomas' conclusions and are compatible with them, they do not in fact have any direct bearing on the question of whether strand interruptions do exist.

[10] Particular models for the generation of a population of circularly permuted chromosomes will be discussed in detail in a subsequent communication. The following two alternatives may help to clarify the present discussion: (1) the ends of a particular chromosome, after its injection into a bacterium, could be imagined to pair and to join, forming a circular structure. This circle might then be opened by cuts in the two chains of the DNA molecule, the cuts being staggered in relation to each other. (2) The chromosome could be imagined to replicate and the progeny chromosomes to join end-to-end at the region of terminal redundancy. The polymers thus formed would have to be cut into phage-sized pieces before maturation.

[11] Doermann, A. H., and M. Hill, *Genetics*, **38,** 79 (1953).

[12] Edgar, R. S., *Virology*, **6,** 215 (1958).

[13] Edgar, R. S., and R. H. Epstein, *Science*, **134,** 327 (1961).

[14] Edgar, R. S., R. P. Feynman, S. Klein, I. Lielausis, and C. M. Steinberg, *Genetics*, **47,** 179 (1962).

[15] Epstein, R. H., personal communication.

[16] Edgar, R. S., and I. Lielausis, *Genetics*, in press.

[17] Foss, H. M., and F. W. Stahl, *Genetics*, **48,** 1659 (1963).

[18] Stahl, F. W., personal communication.

14

Reprinted from *Replication and Recombination of Genetic Material*, W. J. Peacock and R. D. Brock, eds., Australian Academy of Science, Canberra, 1968, pp. 206–215

ROLE OF RECOMBINATION IN THE LIFE CYCLE OF BACTERIOPHAGE T4

FRANKLIN W. STAHL
Institute of Molecular Biology and Department of Biology,
University of Oregon, Eugene, Oregon, U.S.A.

SUMMARY OF THE T4 LIFE-CYCLE

Bacteriophage T4 is 0.2 micron long, too small to see in the light microscope. It is composed of one DNA molecule surrounded by a protein shell to which is attached a protein tail. The virus attacks its host cell, which is about ten times as long as itself, by adsorbing to it with its tail fibres and then puncturing its surface with the core of its tail. The DNA molecule, in passing through the tail into the host cell, undergoes a transition from a compactly folded configuration to an extended one. The contour length of the molecule is 60 microns, 30 times the length of the host cell. Its width, on the other hand, is only 20 A. For the first 5 minutes or so after penetration this essentially one dimensional molecule directs the synthesis of a battery of proteins necessary for its own duplication. At the end of this period the DNA undergoes successive duplications until by around 10 minutes there is about 30 times as much DNA as was injected by the infecting particle. The cell at this stage may well be likened to a bag full of spaghetti. During this period of increase in the amount of DNA, the DNA directs the synthesis of those proteins which form the structures of the mature, infectious virus particles; i.e., head and tail proteins. At about 12 minutes, mature particles begin to appear inside the cell; they increase linearly in number until the cell bursts about 22 minutes after the initiation of infection at which time there is an average of about 200 mature particles per cell. Throughout the period of mature phage production both DNA and protein syntheses continue at rates equal to the rates at which those substances are incorporated into the mature particles. The average steady state amount of "naked DNA" remains in the neighbourhood of 30 particle equivalents throughout.

ASSAY METHOD AND DETECTION OF MUTANTS

We enumerate virus particles by the following biological assay. Onto the surface of solidified nutrient agar in a petri plate we distribute about 100 million bacterial cells and a measured volume of appropriately diluted virus suspension containing an estimated hundred or so virus particles. Each one of the 10^8 bacterial cells undergoes successive divisions to form microscopic bacterial colonies each of which abuts neighbouring colonies resulting in a confluency of bacteria except in those areas in which virus particles had been deposited. These particles each attacked a cell, multiplied and were released by lysis of the cell. The particles so released attacked cells in their immediate neighbourhood and repeated the cycle. Successions of such growth result in the clearing of local areas on the surface of the plate otherwise densely populated by intact bacteria. By the time these clear areas (plaques) become visible (about 4 hours after preparation of the plates) there are about 10^8 particles in each plaque. The plates are generally scored after overnight incubation.

Among the many offspring obtained by repeated cycles of growth of a single wild-type phage particle, mutants may be observed which produce plaques of altered morphology. A small variety of mutant classes are distinguishable. Of more general usefulness are mutants which are completely unable to form plaques in some environments which permit the formation of plaques by wild-type phage, but are able to form plaques in other environments. These various hereditary types of T4 represent the starting materials for our genetic analysis of the T4 DNA molecule ("chromosome").

GENETIC CROSSES AND EARLY MAPPING RESULTS

A "cross" in the case of T4 occurs as a consequence of simultaneous infection of a cell by two or more different hereditary types of the same virus. Such mixed multiple infections are usually conducted in the laboratory as follows: Bacteria (at high concentration, say, 10^8 cells per ml., to promote adsorption of phages) are added to a suspension of two phage types, each at a final concentration several times that of the bacterial concentration. For the sake of brevity, let's call the two infecting phage types a^+b and ab^+. Four types of phages are observable among the earliest mature particles — the infecting "parental" types and the recombinant types ab and a^+b^+. Populations of particles maturing in successive intervals contain progressively higher frequencies of recombinants. In no case does a recombinant frequency exceed 50%. The recombinant frequency ascribed to a particular cross is the average frequency among all the mature particles released from a large number of cells after those cells undergo phage-induced lysis and liberate their contents of mature phage particles. For any given pair of mutant types, the observed recombinant frequency is a reproducible characteristic of crosses involving that pair. For different pairs the recombinant frequency ranges from almost 50% to values much less than 1%.

In the case of phage T4, as with other phages and other types of organisms, the data from such crosses can be systematically summarized in the form of a genetic map. To construct a map, a symbol (marker) representing each mutant type is allocated a point in space. Two markers that show low frequencies of recombination with each other are located close together; markers giving large frequencies are put farther away from each other. When a number of markers have been located, the geneticist finds that he can draw one and only one nonbranching line from point to point (until all points lie on the line) such that the distance from one point to another *along the line* is invariably larger for markers that show large recombination values than for those showing smaller values. The data from crosses presented through 1953 for T4 could be summarized as three groups of markers which behaved as follows: within any one group a linear map could be constructed according to the rules; crosses involving markers from *any* two of the three groups gave about 40% recombinants; i.e., the groups themselves couldn't be arranged linearly with respect to each other.[1] In retrospect, three explanations for this relationship seem possible; only the last two received consideration at the time. (1) The three linkage groups are equidistant from each other on an unconventional linkage structure, for instance a branched structure or a circular structure. (2) The three groups are located on separate structures which assort independently from each other in hypothetical individual pairwise matings. The failure to find 50% recombinants between the groups is entirely a consequence of heterogeneity among the progeny particles with respect to mating experience.[2] (3) The three groups are far apart from each other on a conventional linkage structure. Again, the failure to find 50% recombinants between groups is supposed to arise entirely from heterogeneity in mating experiences. A few years later, G. Streisinger and V. Bruce[3] using sensitive three-point linkage tests showed that possibility (2) was wrong. By default, possibility (3) became the accepted picture. This was a nice result since physical studies of the DNA of T4 were soon to demonstrate that each particle carried only one chromosome, not three. At this point we must go back to 1951 to find the beginnings of a line of evidence which was crucial in giving us our current model for the T4 chromosome.

EARLY OBSERVATIONS ON HETEROZYGOSIS

1951-1960: If a bacterium is infected by two hereditary types of phage, say T4r and T4r^+, the cell will produce both types of particles. About 1% of the particles produced are extraordinary. Upon replication these extraordinary particles produce offspring of both r and r^+ genotype and, in addition, an occasional extraordinary particle. Such particles were termed "phage heterozygotes" (HETS) by Hershey and

207

Chase, who reported their studies of such particles in 1951[4]. Hershey and Chase demonstrated that any one of many loci in the T-even phages can become heterozygous as a result of genetically mixed infection; as a rule, however, a particle heterozygous at one locus is also heterozygous at a second only if the second locus is closely linked to the first. It appears as if regions of heterozygosity are not only hereditarily unstable but are also rather short and that the number per phage particle is not very much greater than one. This viewpoint suggested two simple *formal* models for a HET. Model I is the sort of thing which a black-board geneticist would draw if a fragment of DNA were matured along with a whole phage chromosome chosen willy-nilly from the vegetative pool. We'll diagram Model I like this

$$
\text{Formal HET-Model I} \qquad \frac{r}{r^+} \;\; \text{or} \;\; \frac{r^+}{r}
$$

for the instance of heterozygosity for an *r* marker.

In Model II we suppose that heterozygosity results when large parts of chromosomes derived from two different individuals are combined into one mature particle as shown below:

$$
\text{Formal HET-Model II} \qquad \frac{r}{r^+} \;\; \text{or} \;\; \frac{r^+}{r}
$$

The models were distinguished by Levinthal[5] from the results of a three-factor cross of the type—

$$
\frac{A \quad r \quad B}{a \quad r^+ \quad b} \; X
$$

The r/r^+ HETS resulting from the mixed infection were observed to be predominantly of genotypes *Ab* and *aB* arguing that a correct formal representation for most, if not all, HETS is given by Model II.

When Levinthal presented the results of his crosses he suggested two *molecular* models which might correspond to the formal Model II. Both of these models represented minor modifications of the then-recently-announced Watson-Crick structural model for DNA. In Model A heterozygosity is supposed to arise as a result of a region (or regions) of local "diploidy" whose position is variable on the chromosome. An r/r^+ HET chromosome would "look like" this:

Molecular
HET-Model A

i.e., two DNA molecules derived from genetically different phages are matured into the same particle. For reasons which will become clear later, we'll not inquire now into the nature of the forces which unite the two molecules to form a single chromosome.

In Molecular HET-Model B, a heterozygous chromosome has no regions of structural redundancy, and its only genetic redundancy is that implied by the duplex nature of the Watson-Crick DNA model. In Model B, a HET "looks like" this:

208

Molecular
HET-Model B

(The parts of this molecule derived from different parental phage chromosomes are differently shaded.)

Other molecular models for HETS have been proposed, but the two described above are outstanding with respect to both priority and simplicity. In addition, they are sufficiently distinct from each other to encourage experiments designed to choose between them. Experiments by Doermann and Boehner[6] and by Womack[7] succeeded in detecting HETS which have a behaviour which might well be expected of those with a structure like Model A. However, those same experiments demonstrated HETS which behave as one would expect for those with a structure like Model B. The essence of their observations is the following: Two phage stocks differing by 6 or 8 closely linked markers were crossed and the multi-factor HETS resulting were examined for the frequencies with which each of the markers appeared among the progeny of the HET. Some of the HETS gave rise to populations containing approximately equal numbers of each of the markers. Since DNA duplicates semi-conservatively, HETS with the structure of Model B should segregate in that fashion. Others of the HETS, however, showed strikingly different segregation patterns. For one of the outermost heterozygous loci, the frequencies of the two alleles among the offspring were usually grossly unequal. For the outermost one at the other end of the marked sequence of loci a similar inequality was often seen. However, in this case the allele in excess was derived from the parent which was poorly represented at the first heterozygous locus. Loci in between gave inequalities of intermediate magnitude with values varying monotonically from one end to the other. Such segregation behaviour argues for the presence of structural singularities disposed *trans* to each other at each end of the heterozygous region. Model A has exactly that structural feature.

At the same Phage Meeting (Cold Spring Harbor, 1960) at which Doermann's associates presented genetic evidence for the existence of phage chromosomes with structures like that of Model A, K. I. Berns and C. A. Thomas presented physical evidence for their non-existence![8] They testified that their studies of whole T4 chromosomes showed a reduction in molecular weight by just a factor of two upon strand separation brought about by heating the DNA in formaldehyde. If their chromosome population contained many individuals with structures like that of Model A, the reduction in molecular weight could reasonably be expected to be greater than two-fold. At this point G. Streisinger and M. Meselson devised a model for the T4 chromosome to provide a possible solution to this apparent paradox.

THE CHROMOSOME MODEL

(a) *Genetic circularity:* Streisinger and Meselson suggested that the apparent conflict between genetic and physical data regarding the existence of Model A-like structures was resolved if one pictured a phage chromosome like this:

209

Since Model A HET structures can (presumably) occur at any genetic locus, we are led to the following picture of the chromosomes of a (mature) phage population: The genetic sequences from one chromosome to another are circular permutations of each other. Each chromosome has a terminal repetition of its initial loci; i.e.,

abc ...	zab
fgh ...	zabcdefg

etc.

The most obvious prediction of this model was promptly tested by Streisinger, Edgar and Denhardt[9]; upon close scrutiny the linkage map of T4 was indeed found to be circular!

(b) *The mode of origin of HETS:* Emboldened by the initial success of the model, Streisinger resolutely examined its logical consequences.

A "mating" between two chromosomes which are circular permutations of each other may be pictured (on the blackboard) like this:

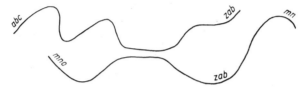

Crossing-over in the synapsed region can lead to the formation of

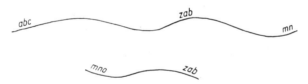

Whatever the precise steps in the crossover process may be, the crossover product arises as if the two parental molecules were each cut on the bias and their parts rejoined

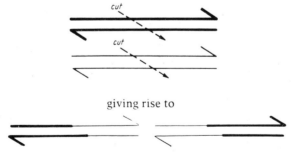

giving rise to

If the two mating chromosomes are genetically distinct in the synapsed region, then the recombinant products may be HETS of type B.

As is implied by the diagram of a "mating" above, reiterated matings lead to the formation of "chromosomes" of indefinite length in which the T4 genome is serially repeated. At some stage T4-sized lengths of DNA are removed from the polymers. It seems most reasonable that this step should occur hand-in-hand with phage matura-

210

153

tion. At maturation, then, and perhaps before, chromosomes with new permutations are created. *A matured chromosome may be a HET of type A if it is cut from the polymer such that it is redundant for a marked locus.*

(c) *Segregation from HETS:* Polarized segregation from type-A HETS was originally explained by Doermann and his collaborators in terms of an "internal copy-choice" scheme. The scheme is most conveniently presented by means of a diagram —

A Type-A structure

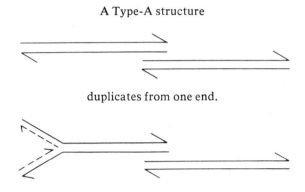

duplicates from one end.

The newly forming chains "switch" in the redundant region, typically before reaching the end of the partial chromosome upon which they started. The two chains need not switch at exactly the same place —

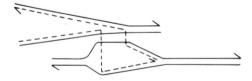

Upon completion of the duplication act, the chains segregate conservatively so that the two daughters are the original type-A HET structure and an all-new chromosome, without chain interruptions, which may be a HET of type B. Subsequent duplication acts of the type-A HET proceed in the same fashion except that the points of switching are variable from one act to another.

This scheme explains the polarized segregation from type-A HETS and the observation[7] that type-B HETS arise rather frequently in cells infected by type-A HETS. While it provides an apparently simple explanation for the behaviour of type-A HETS, the scheme does have two obviously unpleasant features. (1) It proposes a copy-choice mechanism of recombination at a time when other experiments, more critically addressed to mechanism, clearly indicate break-reunion.[10] (2) It proposes a conservative mode of DNA duplication although there is no evidence that DNA duplication ever proceeds in other than a semi-conservative fashion.[11] [10] [12]

In Streisinger's model, polarized segregation from type-A HETS results from "head-to-tail" matings between daughter particles. For instance, a multi-factor HET

211

154

duplicates, then synapses like this

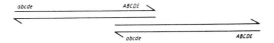

Crossing-over in the synapsed region leads to (for instance)

plus fragments. Duplication of this dimer may be followed by a second head-to-tail mating, etc. The points of crossing-over in these additional rounds will not generally be at precisely the same places as those in the first. Polarized segregation will be the result. Two other features of incestuous head-to-tail matings may be noted here. (1) Type-B HETS arise from type-A HETS by the same cross-over mechanism with which they form in "ordinary" matings. (2) Since polymers are formed, it is possible for monomers with new permutations to arise in singly infected cells. (We may add that crossing-over between the head and the tail of a single chromosome would give rise to a topologically circular chromosome. Such events may happen (it's difficult to see what might prevent them), but they are not a feature of the model in its present form.) At this point we must make a short aside to introduce the genetic material which has been used to test predictions of Streisinger's model.

POINT-MUTANT HETS AND DELETION HETS

Mutants in the *r*II region of T4 can be assigned to either of two categories having the following properties:

Category	*Point-mutants*	*Deletions*
Definitive properties	Reversion rate 10^{-8}. Map at a point.	No detectable reversion. Fail to recombine with each of two or more mutants which are recombinable with each other.

Nomura and Benzer[13] determined the frequency of HETS from crosses of point mutants by wild-type and deletions by wild-type. Crosses of the first type all gave 1.4% mottled plaques (HETS) while crosses of the second type gave 0.4% HETS. This latter value was independent of the extent of the region deleted in the mutant employed. On the basis of this observation, Nomura and Benzer proposed the existence of two classes of HET-structures. One class, which occurs with a frequency of 2%[14] can be heterozygous for wild-type and a point mutation but not for a deletion and wild-type. Both categories of mutants, on the other hand, can form HETS (with wild-type), having structures of the second class, which occurs with a frequency of 0.8%. It occurred to Streisinger that these two classes of HET structures might correspond to the type-B and the type-A structures, respectively. The nucleotide sequences of wild-type and a point mutant differ by only one base pair. The construction of a type-B HET from two such chains represents a minimal, and likely tolerable, violation of the base-pairing rules of Watson and Crick. The construction of such a "heteroduplex" HET from wild-type and a deletion mutant, however, may well represent a violation intolerable to all concerned. On the other hand, there seems to be no reason why both categories of mutants should not *equally* well form HETS of type A ("terminal redundancy HETS").

212

EXPERIMENTAL TESTS OF STREISINGER'S EXPLANATION OF HETS

Hershey and Chase[4] and Levinthal and Visconti[15] reported that the frequency of HETS among phage maturing at different times is essentially constant. Sechaud *et al.*,[16] have observed that this statement is true separately for each of the two classes of HETS; i.e., deletion HETS have a constant frequency of 0.4% while point-mutant HETS occur with a frequency of 1.4% independent of the time after infection at which mature phage are examined. Since at the time of infection neither type of HET exists (although HET *structures* exist in a homozygous state), the values observed among particles sampled from the mating pool may be assumed to result from equilibria between reactions that form and reactions that destroy HETS. Let us examine those reactions which, within the framework of the model, determine the equilibrium values of type-A and B HETS respectively. The frequency of type-A HETS will reach equilibrium when sufficient recombination has ensued that any chromosome-polymer is essentially randomized with respect to the allele present at each representation of the marked locus. Two representatives of a given locus are "maximally unlinked", and the number of "matings" at the time when mature phage first appear is several.[2] Thus, the observed value of 0.4% equals 1/2 times the probability that a matured monomer will be terminally redundant for the marked locus. The equilibrium value for type-B HETS will depend upon the rate at which they are formed as a result of crossing over in their immediate neighbourhood and the rate at which they are destroyed as a result of (semi-conservative) duplication. Thus, the reaction which destroys type-A HETS (recombination) is different from the one which destroys type-B HETS (duplication). A prediction of this notion is that under conditions which permit a normal rate of genetic recombination while depressing the rate of DNA duplication, the frequency of type-B HETS should rise while the frequency of type-A HETS will be unaffected. Deletions can be used to score the frequency of type-A HETS; type-B HETS can be studied using point mutants, which form predominantly (2/3 under normal conditions) HETS of this type.

Fluoro-deoxyuridine (FUDR) inhibits DNA synthesis by interfering with the formation of thymidilic acid. Genetic recombination in T4 is not inhibited by the analogue.[17] In complete agreement with prediction, Sechaud *et al.*,[16] found that they could increase the frequency of point-mutant HETS from 1.4% to about 8% by treatment of infected cells with FUDR. This increase paralleled the increase in recombinant frequency under the same conditions. The frequency of deletion HETS, on the other hand, was not changed by FUDR.

Further substantiation for the differing roles of heteroduplex HETS and terminal redundancy HETS in recombination comes from studies of crosses between closely linked *r*II markers[18]. Wild-type recombinants arising in crosses between *r*II mutants in the same cistron can be selected for by adsorbing the progeny phage to *E. coli* strain K. Among those individuals which grow in K, the "pure wild-type" particles can be distinguished from the "recombinant HETS"[19] by plating the infected K cells on strain B. Recombinant HETS give mottled plaques. In a cross between two point-mutant *r*'s near opposite ends of the B cistron (about 3% recombination), 16% of the wild-type recombinants were HETS. The value rose to 63% when the cross was performed in FUDR. This result is in fine accord with the notion that the primary product of recombination between close markers is usually a recombinant HET which looks like this:

Segregation of this HET by duplication to give "pure wild-type" particles is blocked by FUDR. The same experiment performed with two small deletions near opposite

213

ends of the B cistron gave, as expected, a different result. Only 6% of the wild-type recombinants are HETS, and this value does not change when the cross is performed in FUDR. The value 6%, however, struck us as being immoderately high; it indicated a rather strong correlation between crossing over and terminal redundancy and implied either that ends of T4 chromosomes engage in crossing over at a higher rate than do other regions, or that crossing over determines the location of the end. On second thought, the idea of crossing over near ends is seen to be consistent with the high rate of head-to-tail matings invoked earlier to account for polarized segregation of terminal redundancy HETS.

Streisinger's model proposes that a "machine" (maturation?) which is blind to nucleotide sequence measures monomers of T4 out of the intracellular chromosome polymers. These monomers are of such a size that any one of them contains the complete T4 genome plus a small terminal repetition. Particles which are terminally repetitious for a marked locus can be spotted as deletion HETS. If a matured monomer is somewhere genetically deleted over a length about equal to or larger than the length of a typical terminal redundancy, it should itself carry an abnormally long redundancy. Operationally, phages which carry large deletions in one region of their genome should show a higher-than-ordinary frequency of deletion heterozygosis in another region. To perform an experiment which tested this prediction, Streisinger and his friends apparently needed *two* regions in which deletion mutants were available. The appropriate material was provided by their demonstration that the $h_2{}^+$ and $h_4{}^+$ alleles behave *as if* one of the two were a deletion; i.e., the frequency of $h_2{}^+/h_4{}^+$ HETS is low (somewhat less than 0.5%) and this frequency is not increased by FUDR. The prediction was tested,[20] then, by two crosses of the type *r*-deletion $h_4{}^+$ x r^+ $h_2{}^+$. In one cross the *r*-deletion was very short, in the other, very long. Particles HET for $h_2{}^+$ and $h_1{}^+$ were examined for the allele carried at the *r* locus. In the case of the cross involving the short *r*-deletion, the $h_2{}^+/h_4{}^+$ HETS were equally *r* or r^+. In the case of the long deletion, however, the $h_2{}^+/h_1{}^+$ particles were more frequently *r* than r^+. These results confirmed the results of another set of experiments in which the frequency of $h_2{}^+/h_4{}^+$ HETS arising in mixed infection was compared with that from another. In the first case both phages carried the same short *r*-deletion, in the second, the same long *r*-deletion. The frequency of $h_2{}^+/h_1{}^+$ HETS was observed to be higher in the second case than in the first.

SUMMARY

Streisinger and Meselson proposed a model for the T4 chromosome which resolved apparently contradictory genetic and physical evidence. This model, which has been vigorously tested by Streisinger in Eugene, assumed that the chromosomes in a mature T4 population were circular permutations of each other and that each was terminally redundant. The model successfully predicted a circular map for T4 and, when elaborated in the most obvious manner, successfully predicted several outrageous properties of phage heterozygotes.

REFERENCES

1. Doermann, A. H., and M. B. Hill, Genetics, *38*, 79 (1953).
2. Visconti, N., and M. Delbrück, Genetics, *38*, 5 (1953).
3. Streisinger, G., and V. Bruce, Genetics, *45*, 1289 (1960).
4. Hershey, A. D., and M. Chase, Cold Spring Harbor Symposia Quant. Biol., *16*, 471 (1951).
5. Levinthal, C., Genetics, *39*, 169 (1954).
6. Doermann, A. H., and L. Boehner, Virology, *21*, 551 (1963).
7. Womack, F. C., Virology, *21*, 231 (1963).
8. Berns, K. I., and C. A. Thomas, Jr., J. Mol. Biol., *3*, 289 (1961).
9. Streisinger, G., R. S. Edgar, and G. H. Denhardt, Proc. Nat. Acad. Sci., U.S., *51*, 775 (1964).

214

10. Meselson, M., and J. J. Weigle, Proc. Nat. Acad. Sci. U.S., *47*, 857 (1961).
11. Meselson, M., and F. W. Stahl, Proc. Nat. Acad. Sci. U.S., *44*, 671 (1958).
12. Kozinski, A. W., and P. B. Kozinski, Virology, *20*, 213 (1963).
13. Nomura, M., and S. Benzer, J. Mol. Biol., *3*, 684 (1961).
14. At genetic equilibrium, the frequency of HET-structure is twice the frequency of HETS themselves; i.e., homozygous HET-structures are implied but undetected.
15. Levinthal, C., and N. Visconti, Genetics, *38*, 500 (1953).
16. Sechaud, J., G. Streisinger, J. Emrich, J. Newton, H. Lanford, H. Reinhold, and M. M. Stahl, Proc. Nat. Acad. Sci. U.S., *54*, 1333 (1965).
17. Simon, E., personal communication; and Frey, Sister Celeste, personal communication.
18. Shalitin, C., and F. W. Stahl, Proc. Nat. Acad. Sci. U.S., *54*, 1340 (1965).
19. Edgar, R. S., Genetics, *43*, 235 (1958).
20. Streisinger, G., J. Emrich, and M. M. Stahl, Proc. Nat. Acad. Sci. U.S., *57*, 292 (1967).
21. In these as in some of the other experiments reported here preparative centrifugation in a heavy water gradient was used to remove a third class of "HETS". These "HETS" are simultaneously "heterozygous" for distantly linked markers. The fact that they can be sedimented away from the primary population of particles suggests that they may be clumps or siamese particles.

215

15

Reprinted from *J. Cell. Physiol.*, **70**(2), Suppl. 1, 13–33 (1967)

The Rule of the Ring

C. A. THOMAS, JR.

*Department of Biological Chemistry, Harvard Medical School,
Boston, Massachusetts*

> *In the Land of Mordor where the Shadows lie.*
> *One Ring to rule them all, One Ring to find them,*
> *One Ring to bring them all and in the darkness bind them*
> *In the Land of Mordor where the Shadows lie.*[1]

ABSTRACT   Different species of viruses contain linear or circular DNA molecules. The circular molecules are either single-chained or circular duplexes. The linear molecules from various species are always duplex. However, they may be either unique or circularly permuted collections of sequences. All species of linear duplexes that can be successfully tested can be shown to be terminally repetitious. The two temperate phages that have been studied (λ and P22) are unique and permuted collections, respectively. Shortly after infection both of these molecules form closed helical rings (superhelices). Certain virulent phages show no evidence of super-helix formation. How unique and permuted collections are produced at maturation is a puzzle. In this respect, it is of interest that P22 is a generalized transducing phage, whereas λ is a specialized transducing one.

As Streisinger and Bruce ('60) began putting the pieces of the T4 genetic map together, we began to learn how to isolate T4 DNA molecules without breaking them (Rubenstein *et al.*, '61). Then the genetic map was found to be circular (Streisinger *et al.*, '64). The ingenuity of Stahl and Steinberg ('64) provided three possible explanations, two of them linear and one circular. But the T2 (T4) DNA molecule proved to be *linear*, as first shown by shearing studies (Burgi and Hershey, '61; Rubenstein *et al.*, '61) and as later confirmed by electron microscopy (Thomas and Mac-Hattie, '64). The two remaining models of Stahl were: (1) a unique collection of DNA sequences, but an obligatory two-switch crossing-over scheme, and (2) a circularly permuted collection of sequences with a normal crossing-over scheme (fig. 1). We set about devising experiments to decide between these two models. In the end, we found two different ways of proving that a collection of DNA sequences is unique (nonpermuted) or circularly permuted (Thomas and MacHattie, '64; Thomas and Rubenstein, '64). These methods have both been applied to T2 (T4) DNA, and the results are clear. T2 (T4) DNA is a circularly permuted collection of sequences, and there is apparently no preferred permutation in the collection (they begin at random). This is in exact accord

with model *B* of Stahl and Steinberg and is undoubtedly the basis for the circular genetic map of this phage.

THE PERMUTATION TESTS

Rather than review the published work on T2 (T4) (Thomas and MacHattie, '64; Thomas and Rubenstein, '64; MacHattie *et al.*, '67), I will describe the application of these tests to P22 DNA, which appears to be a circularly permuted collection of sequences (Rhoades, MacHattie, and Thomas, in preparation). These linear molecules are 13.5 ± .7 μ S.D. long, as shown by protein film electron micrographs. The sedimentation rate of this molecule, relative to T7 and λb2b5 DNA, shows it to be slightly longer than these molecules, giving it a molecular weight of close to 27 million.

Test no. 1: Denaturation and annealing

If a solution of this DNA is adjusted to a pH above 12, reneutralized with acidified buffer, and annealed at 65 C for 45 minutes in 2 × SSC, then approximately 50% of the structures seen in the electron microscope are circular. We think that these circles are formed by the mechanism diagrammed in figure 2. P22 is terminally repetitious, as we will show later in the text. These repetitious terminals should be left out of the circular duplex and

---

[1] J. R. R. Tolkien, THE LORD OF THE RINGS.

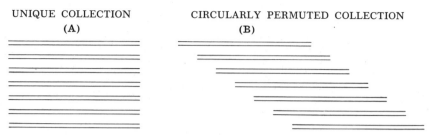

UNIQUE COLLECTION
(A)

CIRCULARLY PERMUTED COLLECTION
(B)

Fig. 1   A nonpermuted (A) and a permuted (B) collection of DNA molecules.

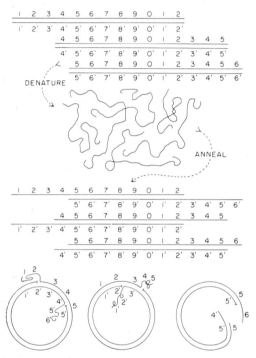

Fig. 2   Circle formation by denaturation and annealing of a permuted collection of duplexes. Notice that each permutation is also terminally repetitious. These repetitious terminals cannot find complementary partners and are left out of the circular duplex. Their separation depends on the relative permutation of the partner chains.

should appear as tangled "bushes," since single chains are not rendered highly visible by our technique. Figure 3 shows a typical circle. An artificial, circular T2 DNA is shown in figure 4 for comparison. The contour length between the two bushes can be measured. In the case of T2, there was no preferred interbush separation, a fact that indicates the presence of all possible permutations. The same "interbush" analysis can be performed with P22, and

again there is no conclusive evidence for preferred spacing, indicating the equal abundance of all permutations.

Marc Rhoades performed two control experiments that should be mentioned. He showed that no circles are formed after a 5-minute exposure to pH 11.8, and that no single polynucleotide chains are seen by zone sedimentation analysis. Upon exposure to pH 11.9, the single chains separate from one another and can be identified by zone sedimentation. Upon annealing this sample, a large fraction of circles are seen. Thus, denaturation to the point at which single chains can be separated is a necessary prerequisite.

Is it necessary, in fact, that they separate as required by the scheme in figure 2? To answer this, crosslinks were introduced by HNO_2 into the native P22 duplexes. The presence of these crosslinks can be recognized by the fact that denatured DNA containing crosslinks will spontaneously renature (or "snap back") when the solution is reneutralized. In short, DNA molecules that contain one or more crosslinks do not form circles. Thus it appears that chain separation is really necessary for circle formation.

Test no. 2: Central deletion experiments

Experiments of this type are based on the ability to mechanically delete a fraction of the middle of a DNA molecule. This can be accomplished by breaking a small fraction of the duplex molecules once and only once. There are two ways in which this has been done: by limited shearing, and by exposure to very small amounts of endonuclease I. In each case multiple breakage is avoided by adjusting conditions to ensure that less than 10% of the molecules are broken. Among the frag-

Fig. 3 An artificial circular P22 DNA molecule formed by denaturation and annealing. Notice the two "bushes" (or faint branches) near the middle and upper portions of the photograph. These are tentatively identified as the repetitious terminal single chains (see fig. 2).

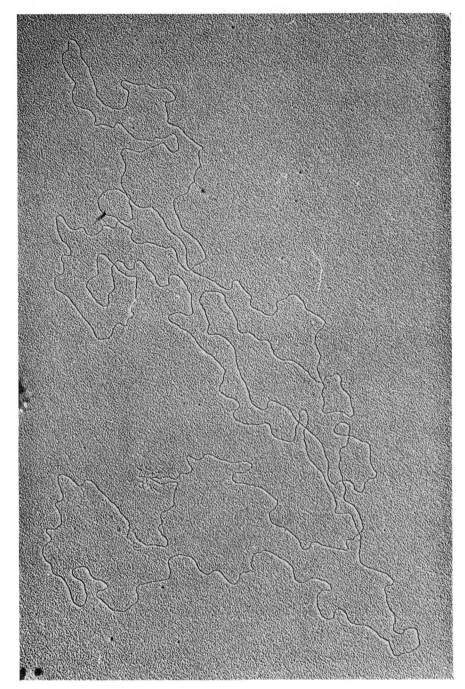

Fig. 4   An artificial circular T2 DNA molecule. See caption for figure 3.

ments of singly broken molecules, the smallest fragments will have originated from the right and left ends of the molecules; therefore, depending upon their length, they are missing a certain fraction of the central region of the DNA molecule. These smallest fragments can be collected by chromatography on MAK columns by the procedure of Mandell and Hershey ('60). If all these molecules have the same (nonpermuted) sequence, then this process removes a certain fraction of the information; the remainder are missing some sequences. If the molecules are a permuted collection of sequences, the process removes a little bit of all sequences, and all sequences remain. These two possibilities can be distinguished by annealing experiments. The actual arrangement is depicted in figure 5. The immobilized, denatured DNA molecule is symbolized by the solid line. The dashes represent ^{14}C-labeled, polynucleotide-chain segments prepared by shearing duplex whole molecules by sonication or by the French Press followed by denaturation. The stars (*) represent ^{32}P-labeled, short polynucleotide-chain segments prepared in the same way from the short fragments. A mixture of these segments is annealed with the immobilzed DNA under conditions of excess segments. This means that there is com-

petition for a limited number of addresses on the immobilized DNA. Clearly, under these conditions more ^{14}C than ^{32}P will complex to the immobilized DNA in the nonpermuted case. In the permuted case, ^{14}C- and ^{32}P-labeled segments are equally abundant for every address, and there should be no fraction of one label predominating. The ratio of ^{14}C/^{32}P should depend on the *fraction of the molecule* that is deleted in the nonpermuted case. In the permuted case, nothing should change the ^{14}C/^{32}P ratio.

These experiments were first done with T5 and T2 DNA, by using the agar column technique (Thomas and Rubenstein, '64). Figure 6 summarizes the results of experiments in which T7 and P22 DNA were analyzed by use of the nitrocellulose filter technique (Rhoades, MacHattie, and Thomas, in preparation). Notice that the value of R increases sharply for T7 DNA as α, the fraction of the central deletion, is increased. With P22 DNA, the value of R

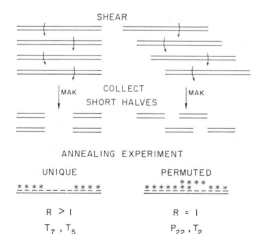

Fig. 5 Central deletion annealing experiment. Single double-chain breaks are made by shear or endonuclease I and by the short fragments collected by MAK chromatography. These provide "centrally deleted" DNA for a double-label annealing experiment.

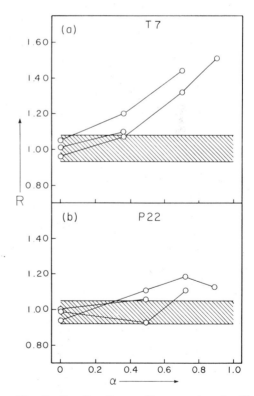

Fig. 6 Results of annealing experiment with centrally deleted T7 and P22 DNA.

does not increase much above 1.0. This is just the behavior one would expect if T7 were a nonpermuted collection and P22 a permuted collection of sequences.

Thus, it appears that both kinds of species of DNA molecules exist: (a) unique or nonpermuted collections of sequences and (b) circularly permuted collections of sequences. It has recently been found that the genetic map of P22, like that of T2 (T4), is circular (Gough and Levine, '67). So far as is known, phage λ has a linear map (Campbell, '61), as does T5 (Fattig and Lanni, '65). To state the obvious conclusion: unique molecules generate linear genetic maps; permuted molecules generate circular ones.

TERMINAL REPETITION

An important feature of the permuted chromosome model for T4 was that each permuted chromosome was pictured to be terminally redundant. There is now genetic and paragenetic evidence to support this idea (Séchaud et al., '65; Shalitin and F. W. Stahl, '65; F. W. Stahl et al., '65). An opportunity to obtain physical evidence for the existence of terminally repetitious DNA molecules was presented by the unique specificity of exonuclease III (exo-III) (Richardson et al., '64). This enzyme removes nucleotides, one at a time, from the 3′ end of a polynucleotide chain, provided it is in the duplex form. The experimental plan is shown in figure 7. As can be seen, circles should be efficiently formed if a sufficient number of nucleotides have been removed from each end. The average number of nucleotides that is removed may be easily determined by measuring the fraction of nucleotides that is rendered acid soluble. Samples are removed at various stages of digestion, annealed briefly, and the frequency of circles, as seen in the electron microscope, determined. An example of a circular P22 molecule is shown in figure 8. An example of an exo-III-produced circle of T*2, the nonglucosylated form of T2, is shown in figure 9. Circular molecules can also be identified by their sedimentation rate. As Hershey, Burgi, and Ingraham demonstrated ('63), circular λ DNA sediments 1.13 times faster than its open linear form. Figure 10 shows the sedimentation profile

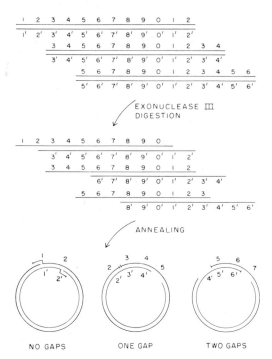

Fig. 7 Terminal repetition experiment — scheme. A collection of T2 DNA molecules is depicted by two parallel lines which correspond to the complementary chains of the duplex. Each molecule is shown having a different circular permutation of a common sequence and a terminal repetition of its first sequence. Exonuclease III exposes the complementary 5′-ended chains at the ends, and circle formation takes place upon annealing. If the degradation proceeds beyond the limits of the terminal repetition, two single chain "gaps" will bracket a duplex segment, the length of which is the length of the terminal repetition.

of a sample of partly degraded P22 that contained about 27% circular structures as seen in the electron microscope. About a third of the material sediments in a zone that moves about 1.14 times faster than that containing linear molecules. If samples that are degraded to different extents are examined by zone sedimentation, no circles are formed until more than 2% of the nucleotides are rendered soluble (fig. 11). The maximum circle frequency is seen between 3–4% degradation. Therefore, we think that the length of the terminal repetition in P22 is near 3%, or 1200 nucleotide pairs.

Thus, permuted species of DNA are terminally repetitious. But what about non-

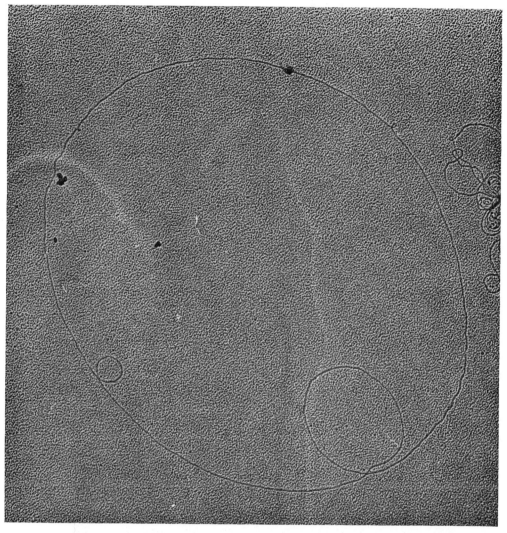

Fig. 8 P22 DNA partly degraded (3.5%) by exo-III and annealed.

permuted species? To answer this, we have performed these experiments with T3 and T7 DNA molecules. Circle formation becomes very efficient when less than 1% of the nucleotides are released by exo-III. If the degradation proceeds beyond the length of the terminal repetition, then the circles should contain two single-chain regions on either side of a duplex segment, or the terminally repetitious region (see fig. 7). So far, this little overlap segment has been seen many times only in T3 and T7 DNA, where it turns out to be 0.10 ± .01 μ long, corresponding to 260 nucleotide pairs (fig. 12). Since these molecules

are nonpermuted, the ends of different molecules are complementary and may anneal with each other. Dimers, trimers, and higher concatemers can be identified by sedimentation or by electron microscopy (fig. 13). Some of these concatemers have a circular form. This cannot happen with a permuted collection because the repetitious terminals are mostly different from each other.

Our first attempts with a vetebrate viral DNA were made on adeno virus in collaboration with Maurice Green (Green *et al.*, '67). The molecules are normally linear (about 12 μ long), but only a low per-

Fig. 9   T*2 DNA partly degraded (3.5%) by exo-III and annealed.

centage (1–2%) can be induced to form circles after treatment with exo-III. An example is shown in figure 14.

CONTROLS

I am sometimes asked whether we have studied any DNA that will not cyclize after partial degradation with exo-III. The an-swer is very clear: DNA fragments will not form circles even though they have suffered partial degradation (Ritchie *et al.*, '67). If someone found certain kinds of DNA fragments that would form circles after partial exo-III degradation, I would be tempted to conclude that these mole-cules were not fragments.

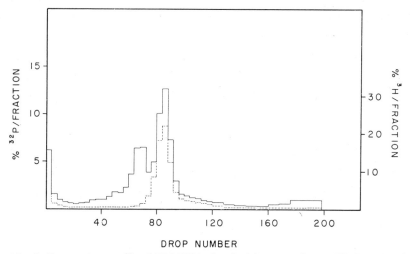

Fig. 10   Sedimentation profile of P22 DNA that had been partly exo-III degraded (3.5%) and annealed. The leading peak containing circular molecules moves 1.4 × faster than the trailing peak containing linear molecules. About 33% of the DNA is in the leading zone (*i.e.*, circularized).

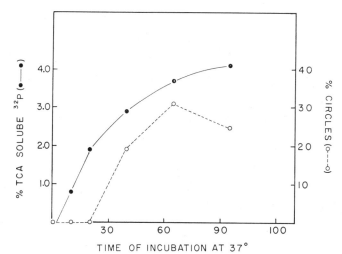

Fig. 11   The frequency of circles as judged by zone sedimentation of P22 DNA that had been partly degraded by exo-III and annealed. Samples were removed from the reaction mixture and assayed for TCA-soluble radioactivity. A portion of the same sample was annealed at 65 C for 45 min in 2 × SSC and sedimented through 5–20% neutral sucrose gradients containing 1.0 M NaCl. Notice that no circular molecules are formed up to 2% degradation. Thus, even though terminal polynucleotide chains are exposed, they do not unite to form circles until further degradation takes place. This observation is a kind of internal control indicating the requirement that complementary polynucleotide chains be exposed.

GENERALIZATION AND SUMMARY

So far we have been considering only three different kinds of experiments. These are shown in table 1. They are all based on annealing being a reliable index of complementarity. Two different kinds of experiments lead to the formation of cir-

cles, yet by quite different routes: one involves denaturation and no exonuclease; the other no denaturation, just exonuclease.

These experiments have been performed on a number of different kinds of viral DNA molecules. They are shown in table 2. As can be seen, terminal repetition oc-

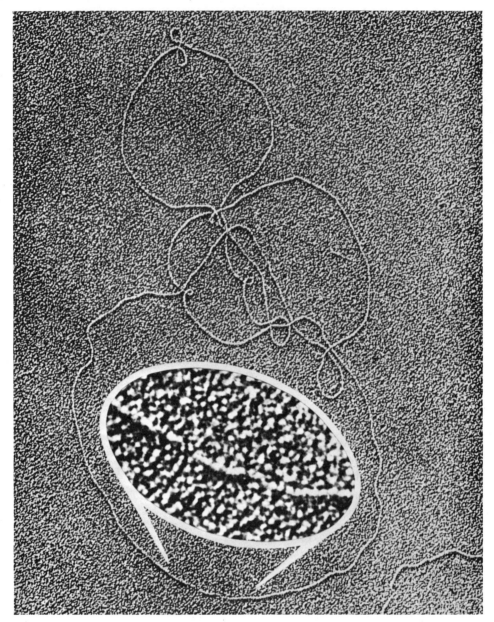

Fig. 12   The duplex between the gaps probably contains the terminally repetitious sequences of T7.

curs in λ, T3, T7, P22, T2 (T4), and possibly in adeno virus DNA. The shortest terminal repetition is about 10 nucleotides and is found in λ, which has its terminal repetition naturally exposed, thereby requiring no exo-III; and the longest, up to 6000 nucleotide pairs, is in T2. Since this list includes both permuted and nonpermuted species, it is tempting to think that all viral DNA molecules are terminally repetitious. If this proves to be true, it leaves one with the following question: Why are viral DNA molecules terminally repetitious?

Fig. 13 A fourfold linear concatemer of T7 DNA molecules formed by partial degradation by exo-III and annealing at higher concentrations (10 μg/ml).

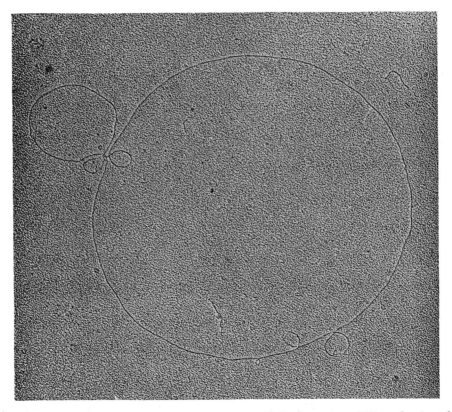

Fig. 14   An example of an exo-III-degraded, annealed adeno virus DNA molecule. The circular molecules were rare (1–2%).

TABLE 1

Tests for terminal repetition and circular permutation

Test for:	Operation	TR	Non-TR	Requirements
Terminal repetition (TR)	Exo-III anneal			Nick-free
		CP	Unique	
Permutation p^1	Center-deletion anneal	R = 1	R > 1	Fractions (with respect to length only)
p^2	Denature-anneal			Nick-free short TR

Most of the species listed contain non-permuted molecules, but two (T2 and P22) contain permuted collections of sequences. What is the origin of these two different forms?

Pleomorphic viral DNA molecules

Although there are no direct answers to these questions, it is informative to review the different forms of viral DNA molecules. DNA molecules from mature virions

TABLE 2

Linear duplex viral DNA molecules

	Terminal repetition	Collection (test: p[1], p[2])	Chain continuity	Type	Intra-cellular circular form	Trans-ducing
λc	10 exposed [a]	nonpermuted (transformation) [b]	2 snicks [c]	temperate	yes	special
λcb2b5	10 exposed	nonpermuted (see above)	2 snicks	(temperate?)	yes	?
T3	260 duplex [e]	nonpermuted p[2] [d]	continuous	virulent	?	—
T7	260 duplex [e]	nonpermuted p[1], p[2]	continuous	virulent	no	—
T1			continuous	virulent	?	—
P22	1200 duplex [e]	permuted p[1], p[2]	continuous	temperate	yes	general
T5		nonpermuted p[1]	4 snicks [f]	virulent	no	—
T2	2000–6000 duplex [e]	permuted p[1], p[2]	continuous	virulent	no?	—
T4	same as T2	permuted p[2]	continuous	virulent	no?	—
SP50		nonpermuted p[2]	1 snick [g]	virulent	?	—
Adeno	some, duplex [i]	permuted [h] p[2]	continuous [i]	?	?	—

[a] Taken from Strack and Kaiser ('65), agrees with estimates made by Wang and Davidson ('66).
[b] The most recent and complete experiments demonstrating this point have been made by Egan and Hogness (see Hogness, '66).
[c] If λ DNA molecules contained two specifically located nicks ("snicks"), ten nucleotides in from each end on the 3'-ended chain, it is doubtful that decamer would be retained in duplex. Thus, λ could be considered similar to other phage DNA molecules, but containing snicks near each end.
[d] p[1], p[2] stand for the two tests for cyclic permutation discussed in the text and listed in table 1.
[e] By exo-III test described in text.
[f] From Abelson and Thomas ('66).
[g] From Reznikoff and Thomas (unpublished experiments).
[h] On the basis of preliminary experiments.
[i] From Green *et al.* ('67). The frequency of exo-III-induced circles was very low.

can be put conveniently into four classes (see table 3). In short, virions contain either linear or circular DNA molecules. However, the term *virus* properly refers to all stages of the viral life cycle. Since a viral DNA molecule spends only a fraction of its existence in the mature virus particle, the intracellular forms must also be considered. What is known about these forms?

Obviously, the prophage is an example of an intracellular viral DNA molecule. Genetic studies have clearly shown (for λ at least) that the prophage is a linear insertion, but the prophage map order is permuted about the prophage attachment site (Franklin *et al.*, '65; Rothman, '65). This could come about by a single reciprocal crossover event, provided that the cohesive ends of the λ DNA molecule had joined to form a ring before the crossover leading to insertion (Kaiser and Inman, '65). Such a crossover should always produce a duplication of the prophage attachment site at the beginning and end of the prophage segment.

The next intracellular forms to consider are transient structures which are involved in replication leading to the vegetative production of more viral genomes, or to the establishment of lysogeny. In view of the success of the Campbell model for prophage insertion, it is to be expected that temperate phage DNA should have an intracellular cyclic form. This proves to be the case in two temperature phages, λ and P22. In fact, the intracellular form shows all the properties of a superhelix, that is, a continuous circular duplex consisting of two topologically linked polynucleotide chains (Vinograd and Lebowitz, '66). This was first demonstrated with λ DNA (Bode and Kaiser, '65) and now with P22 DNA (Rhoades and Thomas, in preparation). Figures 15 A and B are sedimentation profiles made on DNA preparations from λ- and P22-infected cells about 20 minutes after infection. In both cases the parental phage was ^{32}P-labeled. The leading component in both cases has been shown to contain superhelical molecules. The trailing peak almost surely repre-

TABLE 3
Virus particles that contain the following DNA molecules

	Single chain	Duplex
Linear	None known	Large phage, and adeno and pox viruses
Circular	Small phage, MVM	Papova viruses, herpes(?)

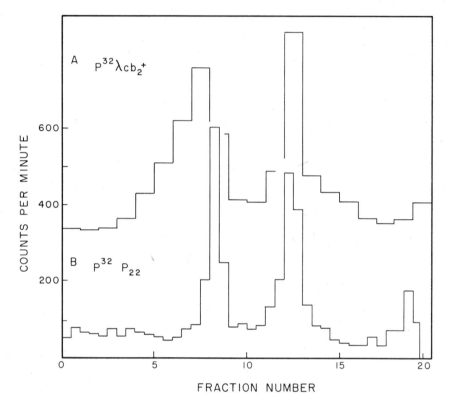

Fig. 15   Sedimentation profiles of DNA extracted from cells infected with labeled λ or P22. A. K12 (λ) infected with ^{32}P-λcb2$^+$: 13 min after infection. B. *S. typhimurium* LT-2 infected with ^{32}P P22: 30 min after infection.

sents linear molecules. Figure 16 (A and B) shows electron micrographs of superhelical λ and P22 that were purified from infected cells. These can be identified with the leading peaks in figure 15. Although there are some differences in the rapidity of the formation of the superhelix, in the sensitivity of its formation to antibiotics, etc. (Rhoades and Thomas, in preparation), I think it is significant that the first two temperate phages that have been examined show an intracellular circular (superhelical) form. This is to be compared with the virulent phage T7. When ^{32}P-labeled T7 phages are allowed to infect growing *coli* and DNA preparations made at various times post infection, the sedimentation profiles reveal no evidence of superhelix formation. A typical experiment (Kelly and Thomas, unpublished experiments, '67) is shown in figure 17. The same general finding also appears to hold for T2/4 and T5 infection. Certainly there is not enough evidence to make a general rule, but these cases just mentioned lead one to pose a question: Are temperate viral DNA molecules the only ones that are formed into a continuous helical ring inside infected cells? Do all virulent DNAs *not* form continuous rings?

Finally, what can be said about the structures that are incorporating labeled

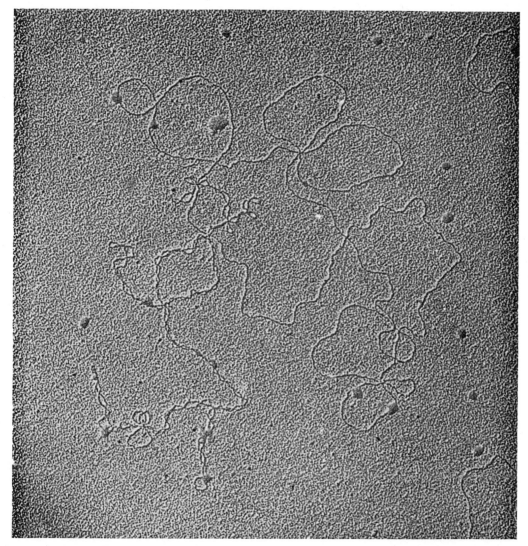

Fig. 16A   Intracellular form of λ DNA; notice the superhelical molecule overlaying an open circle.

nucleotides — the replicating structures? Several groups have examined these forms by zone sedimentation, by CsCl-equilibrium sedimentation, and by preliminary electron micrographs. In the case of T4, it appears that new label is incorporated into very long, shear-fragile structures that may have some of the properties of concatenates or an end-to-end enchainment of genomes by their cohesive terminals. There is some evidence that these structures exist in cells infected with T4, T5, and λ (Frankel, '66; Smith and Skalka, '66).

Thus, making allowances for the fact that the identification of concatenated structures *in vivo* presents some difficulty, there seem to be four basic forms of duplex viral DNA molecules: linear, circular, concatenated, and prophage. These are summarized in figure 18. All of these forms have one thing in common: the viral genetic text is bracketed by two identical sequences at the beginning and at the end. One basic form can be converted into another by genetic recombination

Fig. 16B   Intracellular form of P22 DNA, prepared as described in legend to figure 15.

events taking place which involve these terminal sequences.

A given species of virus may find itself in different basic forms, depending on the stage of its life cycle. We have already seen that different species of virions contain linear or circular duplex molecules. Thus, different basic forms are represented in different species of virus particles. So far no concatenate has been found in a mature virus particle, but according to this view, it would not be unexpected. To use more classical terms: the systematics

of viral DNA molecules appears to reflect their ontogeny.

How are terminally repetitious molecules formed?

Nobody really knows. There are two hypotheses — one for the permuted molecules, and one for those which are non-permuted. The "headfull" hypothesis, generally attributed to Streisinger, supposes that something like a concatemer exists in the vegtative pool and that the geometrical rules which govern the assembly of the

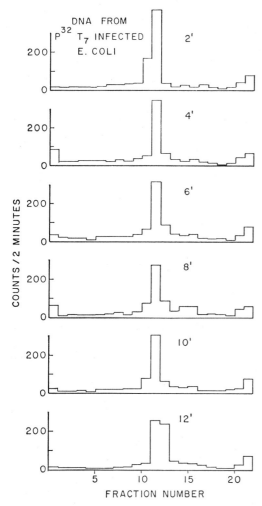

Fig. 17 Sedimentation profile of DNA extracted from cells infected with labeled T7. ^{32}P-labeled T7 were allowed to infect log phase *E. coli*. After 13 min the cells were collected; the DNA was then extracted and sedimented through sucrose.

duplex DNA molecule. The finding that T5 DNA contains 4 specifically located nicks encourages one to think that such nucleases exist (Abelson and Thomas, '66). More generally, the methylating enzymes are known to be sensitive to sequences of the duplex DNA molecule (Gold *et al.*, '66). Therefore, to suspect a specific endonuclease is a reasonable guess. Its mode of action could be as shown in figure 20. This model has some testable aspects, as well as some difficulties. It is included here merely to show how a unique collection of terminally repetitious molecules could be produced in a non-wasteful fashion.

Generalized and special transduction

It is intriguing to notice that P22, whose virions contain permuted molecules, is also a generalized transducing phage. This implies that this particle can enclose a segment of any part of the bacterial chromosome. Phage λ, on the other hand, is only capable of the specialized transduction of nearby *gal* and *bio* genes. This might be expected if the model outlined in figure 20 is correct. It will be interesting to inquire whether all species of phage that are cap-

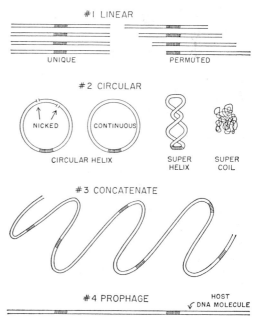

Fig. 18 Summary of basic forms of viral DNA molecules.

phage head define the volume that can be filled with DNA of any sequence. If the volume of the head is slightly greater than a genome, the resulting headfull will be terminally repetitious as well as being a cyclic permutation of its sibling viruses. This is depicted in figure 19.

For viruses containing nonpermuted sequences, some more special process is necessary. We suppose that this is accomplished by an endonuclease that can recognize the local nucleotide sequence of a

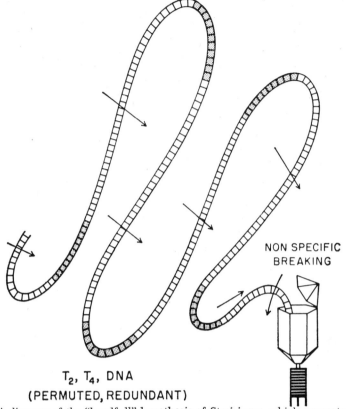

T₂, T₄, DNA
(PERMUTED, REDUNDANT)

NON SPECIFIC
BREAKING

Fig. 19   A diagram of the "headfull" hypothesis of Streisinger, which generates permuted, terminally repetitious molecules.

able of generalized transduction contain permuted molecules in their infective particles; and to the contrary, whether all phage capable of specialized transduction have nonpermuted molecules in their infective particles. This rule holds so far, but only λ and P22 are known.

EPILOGUE

Over the past few years we have learned much about the genetic code and some-

Fig. 20   A diagram of an hypothesis that generates nonpermuted, terminally repetitious molecules. (1) An endonuclease, which can recognize local nucleotide sequence, breaks the polynucleotide chains at specific points. (2) The nick is invaded by DNA polymerase, thereby displacing the 5′-ended chain. (3) A unique collection of terminally repetitious duplex molecules is formed. Notice that this same scheme accounts for the formation of short (10 nucleotides) "exposed" repetitions such as found in λ. Such short overlaps would be substantially less stable, and would be expected to come apart spontaneously.

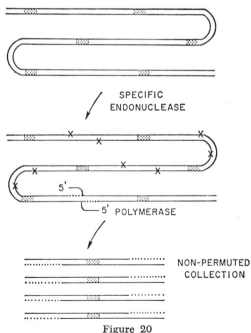

SPECIFIC
ENDONUCLEASE

5′ POLYMERASE

NON-PERMUTED
COLLECTION

Figure 20

thing about the genetic message, together with the punctuation and starting signals that make it interpretable to the translational machinery. The information contained in the entire genome — the genetic text — has been the subject of this paper. It should be reasonable to suppose that the genetic text is obliged to conform to some kind of rules that render it interpretable to the recombinational or replicational machinery of the cell. Indeed, every system of information storage or transmission possesses some sort of grammar in addition to a code. It may be that the basic forms of viral DNA molecules are reflections of some syntactic structure of the genetic text. These forms seem to be unavoidably related to what is emerging as a general feature of viral DNA molecules, namely, terminal repetition. To repeat one's self only when one has reached the end, that may be *The Rule of the Ring*

ACKNOWLEDGMENTS

Much of the factual material included here is the result of the efforts by my collaborators. The contributions of Drs. Donald Ritchie and Lorne MacHattie are now published in primary form. The experiments on P22 DNA are mainly due to Marc Rhoades and will be published shortly. Tom Kelly supplied the preliminary results on intracellular T7. The intracellular λ superhelices are part of a study by Bode and MacHattie (to be published shortly); the sedimentation profile of intracellular λ was provided by Dr. Vladimír Vondrejs. The information on adeno DNA is from a published study with Maurice Green and his associates.

I thank all of these individuals, and we jointly thank our benefactors NIH (E3233), NSF (GB10726), and The Johns Hopkins University.

OPEN DISCUSSION

SIGNER [1]: With regard to your proposal that perhaps generalized transducing phages are permuted, and specialized ones unique, it has been known for some time the P1 is also capable of specialized transduction [S. E. Luria, J. N. Adams, and R. C. Ting, Virology, 8: 348–390, 1960], and there is recent evidence that P22 is as well [P. E. Smith-Keary, Genet. Res., 8: 73–82, 1966; H. O. Smith, personal communication].

THOMAS: Yes, that's true.

SIGNER: This result would indicate that all specialized transducing phages are not unique, although all generalized transducing phages may still be permuted.

THOMAS: But isn't the P22 you refer to a defective variant?

SIGNER: No. It's closer to P1.

THOMAS: I think your previous statement is accurate. Said differently, phages capable of generalized transduction are expected to contain permuted viral DNA molecules; those that can only display specialized transduction are expected to contain nonpermuted DNA molecules. It may be that under certain conditions permuted phage is capable of special transduction.

TRAUTNER [2]: I would like to add another example to your collection of DNA molecules, namely, phase SP-50 DNA, which *obligatorily* has one single-strand break per strand. These breaks are randomly distributed with respect to the ends of the DNA molecules. This was shown by Christoph Spatz, who centrifuged SP-50 DNA in alkali. The sedimentation pattern obtained for denatured SP-50 DNA indicates that there is a continuous distribution of molecular weights, with a calculated average molecular weight equal to one-fourth the molecular weight of the native molecule.

THOMAS: That's very interesting information, and we have been following in Dr. Trautner's footsteps. We have isolated intact strands from the leading edge, and have evidence that they consist of both A and B chains (for they can anneal to form full-length, duplex molecules) and that they are not formed in circles. So breaks can occur in either chain, and they seem, as you say, to appear at random.

COHEN [3]: Is anything known about terminal repetition or circular permutation in RNA viruses?

THOMAS: No. There is no information on this question. We thought at first it might be possible to prove that the double-stranded RNA viruses or certain of the replicating structures were terminally rep-

[1] E. R. Signer, Massachusetts Institute of Technology, Cambridge, Massachusetts.
[2] Thomas A. Trautner, Department of Molecular Biophysics, University of California, Berkeley, California.
[3] J. A. Cohen, Laboratory for Physiological Chemistry, Leiden, The Netherlands.

etitious by the exonuclease-III (exo-III) type experiment. We were discouraged from doing this because I was told that the duplex rA·rU is *not* sensitive to exo-III. It appears that exo-III cannot nibble duplex RNA.

MORA [4]: I would like to say that in the case of one mammalian virus, the polyoma virus, which is one of the circular viruses mentioned by Dr. Thomas, we started to do some homology studies by hybridization, not against the whole mammalian DNA of the transformed cells, but against the mitochondrial DNA of these tumor cells transformed by the virus. As you probably know, the mitochondrial DNA is circular and is roughly the size of the polyoma virus (9×10^6 MW vs. 3×10^6). We have isolated the pure mitochondrial DNA from normal tissues from which these tumors originate. Velocity and sedimentation-equilibrium tests show that it behaves like a circular DNA. We have also isolated the DNA from the tumor mitochondria, but have not done the homology studies as yet.

LERMAN [5]: If we suppose that as an alternative to continuous permutation there is a finite number of points that would serve as initiating sequences, what would be the minimum number of points that would be compatible with your sort of experiments?

THOMAS: We certainly can't rule out ten. But by the millipore filter experiments we can rule out about three or four. It depends on where you want to locate them, of course. Based on measuring the distribution of interbush distances, I think you might say that there must be more than five or six, but certainly there could only be a relatively small number. However, there is no reason to think that there is not a number equivalent to the number of nucleotide pairs in the genome.

LITERATURE CITED

Abelson, J., and C. A. Thomas, Jr. 1966 The anatomy of the T5 bacteriophage DNA molecule. J. Mol. Biol., 18: 262–291.

Bode, V. C., and A. D. Kaiser 1965 Changes in the structure and activity of λ DNA in a superinfected immune bacterium. J. Mol. Biol., 14: 399–417.

Burgi, E., and A. D. Hershey 1961 A relative molecular weight series derived from the nucleic acid of bacteriophage T2. J. Mol. Biol., 3: 458–472.

Campbell, A. 1961 Sensitive mutants of bacteriophage λ. Virology, 14: 22–33.

Fattig, W. D., and F. Lanni 1965 Mapping of temperature-sensitive mutants in bacteriophage T5. Genetics, 51: 157–166.

Frankel, F. R. 1966 Studies on the nature of replicating DNA in T4-infected E. coli. J. Mol. Biol., 18: 127–143.

Franklin, N. C., W. F. Dove, and C. Yanofsky 1965 Linear insertion of a prophage into the chromosome of E. coli shown by deletion mapping. Biochem. Biophys. Res. Commun., 18: 910–923.

Gold, M., M. Gefter, R. Hausmann, and J. Hurwitz 1966 Methylation of DNA. J. Gen. Physiol., 49: 5–28.

Gough, M., and M. Levine 1967 In press.

Green, M., M. Piña, R. Kimes, P. C. Wensink, L. A. MacHattie, and C. A. Thomas, Jr. 1967 Adenovirus DNA. I. Molecular weight and conformation. Proc. Natl. Acad. Sci. U.S., 57: 1302–1309.

Hershey, A. D., E. Burgi and I. Ingraham 1963 Cohesion of DNA molecules isolated from phage λ. Proc. Natl. Acad. Sci. U.S., 49: 748–755.

Hogness, D. S. 1966 The structure and function of the DNA from bacteriophage lambda. J. Gen. Physiol., 49: 29–57.

Kaiser, A. D., and R. B. Inman 1965 Cohesion and the biological activity of bacteriophage λ DNA. J. Mol. Biol., 13: 78–91.

MacHattie, L. A., D. A. Ritchie, C. A. Thomas, Jr., and C. C. Richardson 1967 Terminal repetition in permuted T2 bacteriophage DNA molecules. J. Mol. Biol., 23: 355–364.

Mandell, J., and A. D. Hershey 1960 Chromatography of intact and fragmented DNA molecules on methylated albumin-Kieselghur columns. Anal. Biochem., 1: 66–73.

Richardson, C. C., R. B. Inman, and A. Kornberg 1964 Enzymic synthesis of DNA XVIII. The repair of partially single-stranded DNA templates by DNA polymerase. J. Mol. Biol., 9: 46–69.

Ritchie, D. A., C. A. Thomas, Jr., L. A. MacHattie, and P. C. Wensink 1967 Terminal repetition in non-permuted T3 and T7 bacteriophage DNA molecules. J. Mol. Biol., 23: 365–376.

Rothman, J. L. 1965 Transduction studies on the relation between prophage and host chromosome. J. Mol. Biol., 12: 892–912.

Rubenstein, I., C. A. Thomas, Jr., and A. D. Hershey 1961 The molecular weights of T2 bacteriophage DNA and its first and second breakage products. Proc. Natl. Acad. Sci. U.S., 47: 1113–1122.

Séchaud, J., G. Streisinger, J. Emrich, J. Newton, H. Lanford, H. Reinhold, and M. M. Stahl 1965 Chromosome structure in phage T4 II. Terminal redundancy and heterozygosis. Proc. Natl. Acad. Sci. U.S., 54: 1333–1339.

---

[4] Peter T. Mora, National Cancer Institute, National Institutes of Health, Bethesda, Maryland.
[5] Leonard S. Lerman, Vanderbilt University, Nashville, Tennessee.

Shalitin, C., and F. W. Stahl 1965 Additional evidence for two kinds of heterozygotes in phage T4. Proc. Natl. Acad. Sci. U.S., 54: 1340–1341.

Smith, M. G., and A. Skalka 1966 Some properties of DNA from phage-infected bacteria. J. Gen. Physiol., 49: 127–142.

Stahl, F. W., H. Modersohn, B. Terzaghi, and J. Crasemann 1965 The genetic structure of complementation heterozygotes. Proc. Natl. Acad. Sci. U.S., 54: 1342–1345.

Stahl, F. W., and C. M. Steinberg 1964 The theory of formal phage genetics for circular maps. Genetics, 50: 531–538.

Strack, H. B., and A. D. Kaiser 1965 On the structure of the ends of λ DNA. J. Mol. Biol., 12: 36–49.

Streisinger, G., and V. Bruce 1960 Linkage of genetic markers in phages T2 and T4. Genetics, 45: 1289–1296.

Streisinger, G., R. S. Edgar, and G. H. Denhardt 1964 Chromosome structure in phage T4. I. Circularity of the linkage map. Proc. Natl. Acad. Sci. U.S., 51: 775–779.

Thomas, C. A., Jr., L. A. MacHattie 1964 Circular T2 DNA molecules. Proc. Natl. Acad. Sci. U.S., 52: 1297–1301.

Thomas, C. A., Jr., and I. Rubenstein 1964 The arrangement of nucleotide sequences in T2 and T5 bacteriophage DNA molecules. Biophys. J., 4: 93–106.

Vinograd, J., and J. Lebowitz 1966 Physical and topological properties of circular DNA. J. Gen. Physiol., 49: 103–125.

Wang, J. C., and N. Davidson 1966 Thermodynamic and kinetic studies on the interconversion between the linear and circular forms of phage λ DNA. J. Mol. Biol., 15: 111–123.

IV
Getting It All Together

Editor's Comments on Papers 16 Through 19

Anyone who has tried to assemble a complex mechanism such as a clock will have discovered that certain steps must be performed in a particular sequence. There are two obvious reasons for this. First, some of the parts are located internal with respect to other parts, making it often impossible to insert an interior part after an exterior part has been put in place. It may be possible to assemble the remaining pieces, but the clock may not work. Second, some of the parts act as a framework to hold other parts in their proper position. If a framework part is omitted, further assembly may be impossible.

With the elucidation of the complex structure of T4 it became apparent that assembly of the phage particle from its DNA and protein subunits must likewise involve some orderly sequence of events, although it was not possible merely from knowledge of the final structure to deduce the sequence. For example, one of the substructures, the core, is totally surrounded by another, the sheath. This geometry poses the question of whether the sheath and the core are assembled independently, followed by insertion of the latter into the former, or whether they are assembled dependently with either one or both components directing the assembly of the other. An obvious experimental approach would be to look for partially assembled intermediate structures, but for the wild-type phage these appear to be too short-lived to be observed.

As discussed earlier, the discovery of conditional lethal mutants of T4, described in Paper 12, made it possible to study this kind of question, since many of the mutants were found to be defective in assembly. For some mutants, lysates prepared in the restrictive host were found to be missing one or another of the phage components. For other mutants, all the components were present but unassembled.

By merely observing which components are absent for different mutants it was possible to deduce or at least to guess something about the sequence of events. For example, mutants whose lysates lack either heads, tails, or fibers were found to lack only one of these components at a time, indicating that these three major components are assembled independently of one another. Assembly of the tail subcomponents (sheath, core, and baseplate), however, appears to take place in sequentially dependent steps. Sheath assemby depends on prior core assembly and core assembly depends on prior baseplate assembly. Thus any mutant blocked in baseplate formation is missing all three tail components.

Assembly of the tail was studied in more detail by King *(111)*, who concluded that the core acts as a template on which the sheath subunits are polymerized. He found, moreover, that the sheath rapidly depolymerizes unless two phage gene products act to stabilize it. One of these products appears to form a small cap at the top of the core. King's studies *(113, 114)* have identified the structural genes for many of the proteins that compose the parts of the tail and suggest that other gene products may act catalytically in the assembly process.

Many questions concerning assembly could not be answered by merely observing which components are present in mutant lysates. What was needed was a way to study the process in vitro. The geometrically simple virus TMV had been assembled from its dissociated protein and RNA as early as 1955 *(67)*, but for the more complex T4, prospects for achieving a similar reconstitution were not good. Edgar and Wood reasoned that, although complete in vitro assembly would be unlikely, some of the late steps of the process might nevertheless occur outside the cell. As described in Paper 16, they mixed together two cell-free lysates, one of a mutant that produces free fibers and another of a mutant that produces fiberless phage particles. Neither of these lysates alone contained any plaque-forming particles (other than rare revertants) but when mixed together, plaque-forming particles increased by over 1000-fold in 20 minutes. They had thus achieved in vitro attachment of the free fibers to the fiberless phage. By using a mutant blocked in head assembly and another mutant blocked in tail assembly, Edgar and Wood also demonstrated the attachment of heads to tails in vitro. For some mutants that make morphologically normal but unassembled heads and tails, this in vitro assembly assay enabled them to determine whether the head or the tail was defective. For certain mutants it was found that the normal-appearing heads could nevertheless not join with a complete tail. By supplying the missing gene product, in certain cases, the defective heads could be converted in vitro to complete functional heads *(52)*.

In Paper 17 Wood et al. describe the general methods used to establish the sequence of the assembly steps (Table 1 and Fig. 2) and present the overall picture of the assembly process indicated by their studies (Fig. 3). The head, tail, and fiber pathways require, respectively, 18, 21, and 5 genes to produce the finished components. Complete heads then join with complete (but fiberless) tails and the fibers attach to the particle only after the head–tail union has occurred. The head–tail union occurs spontaneously, that is, without the need for an additional assembly factor. In contrast, the fiber–tail union was found to require the product of T4 gene 63 as a catalyst *(207)*.

Using an improved method of gel electrophoresis, Laemmli detected 28 different proteins in purified T4 particles. In the course of this study, described in Paper 18, Laemmli discovered that certain of the phage proteins undergo a reduction in molecular weight upon being assembled into the phage particle. The most striking case is that of the major head protein, the product of gene 23, which before assembly has a molecular weight of 55,000. This protein as extracted from the phage particle, however, has a molecular weight of 45,000. Two other head proteins were found to be similarly cleaved, and a fourth protein, the product of gene 22, appears to be degraded to such an extent that its fragments cannot be identified on gels. These cleavage reactions are dependent on head assembly, since all four proteins remain uncleaved if head assembly is blocked by a mutation in any one of the seven genes which function in the early steps of head formation. The role of proteolytic cleavage in the assembly of macromolecular structures is not clear, but the phenomenon is ubiquitous, having been observed for other phages *(71, 217)* as well as for certain animal viruses *(94, 102, 176)*.

In the last paper of this section (Paper 19) Kellenberger discusses some general features of virus assembly. He distinguishes between self-assembly, in which the properties of the repeating unit or protomer fully determine the form of the final structure, and aided assembly, in which the products of helper genes function in specifying one particular assembly over other possible alternatives. The concept of aided assembly derives from the fact that mutations in some of the T4 assembly genes result not in the accumulation of intermediate precursor structures but in the formation of aberrant structures. The most striking are the tube-like structures called polyhead, which form, for example, when gene 20 is defective *(60, 122)*. These tubes, which are composed of the uncleaved major head protein P23, can attain lengths many times the normal head length, the elongation apparently being limited only by the dimensions of the cell. When core formation is blocked by mutation, the sheath subunit can also polymerize aberrantly to form polysheath *(109)*. These structures are of indeterminant length, often much longer than the normal sheath, and appear to have a subunit configuration similar to that of a contracted sheath. These aberrant forms thus appear to be alternate modes of assembly which occur when the normal mode is blocked either because a shape-specifying component is absent, as in the case of polyhead, or because the foundation or template structure is missing, as in the case of polysheath.

Kellenberger points out that sequential assembly as observed for T4 may often depend upon conformational changes of the protein subunits induced by their interaction. For instance, two different kinds of subunit, say of the sheath and the core, do not interact when present as free monomers. Only after the core subunits have polymerized do they have an affinity for the sheath subunits. In this example of aided assembly the core not only induces polymerization of the sheath but determines its length as well. How the length of the core is determined is still unanswered (but see the discussion following Paper 19). A major insight gained from studies on T4 assembly is that the sequence of assembly is not determined by the time of synthesis of the component proteins but by the specificity of their interaction. This principle may apply as well to the morphogenesis of other supramolecular structures such as cellular organelles.

The mechanism of head assembly and DNA packaging is the least understood aspect of phage assembly. The earliest careful kinetic studies of head formation by Koch and Hershey *(116)* revealed that the phage head structure is formed from amino acids in about 1 minute. For the next 5 minutes, the head exists in an unstable state which breaks down to fragments if the cell lyses. This unstable structure is then converted within a period of about 1 minute to the complete infective phage particle. A long-standing question has been whether the head is made first and then filled with DNA, or whether the DNA first condenses into a packet which is then coated with protein subunits. Luftig et al. *(129)* have presented evidence that preformed head structures can be filled with DNA. The cleavage of certain proteins associated with head assembly suggests that this may be a mechanism that regulates the sequence of interaction of the head proteins with each other and with the DNA, and may represent, at least in part, the events that convert the unstable head to a stable form. More specifically, Laemmli and Favre *(119)* have proposed that the peptide fragments produced by the cleavage of the gene 22 protein are directly involved in the DNA packaging process.

A tentative general model of T4 head assembly is illustrated in Fig. 9 of Paper 19. Several current lines of investigation suggest that a complex of several phage proteins acts as a head assembly initiator *(158)*, that assembly is initiated at sites on the host cell membrane *(159)*, and that a phage gene product is responsible for cutting the replicating phage DNA to phage-size pieces *(66, 128)*, but many questions remain to be answered.

A major difficulty in studying head assembly has been the extreme fragility of the intermediate structures. This is at least in part responsible for the failure to effect the early steps of T4 head assembly in in vitro systems, although success in this direction has recently been reported for phage λ *(107)*.

Various aspects of phage assembly and virus assembly in general have recently been reviewed in detail *(54, 110, 126)*.

16

Reprinted from *Proc. Natl. Acad. Sci. (U.S.)*, **55**(3), 498–505 (1966)

MORPHOGENESIS OF BACTERIOPHAGE T4 IN EXTRACTS OF MUTANT-INFECTED CELLS*

By R. S. Edgar and W. B. Wood

DIVISION OF BIOLOGY, CALIFORNIA INSTITUTE OF TECHNOLOGY, PASADENA

Communicated by Max Delbrück, January 27, 1966

The complex structure of bacteriophage T4 includes a variety of proteins[1] which become assembled into mature particles during intracellular development of the virus. Some insight into the genetic control of this process has been provided by physiological studies with conditional lethal mutants, which show that over 40 phage genes are involved in T4 morphogenesis[2] (Fig. 1). However, the mechanisms by which components are assembled have remained obscure, due in part to the lack of a suitable system for their study. In the experiments reported below, conditional lethal mutants of strain T4D have been exploited to develop an *in vitro* system in which several of the steps in phage morphogenesis can be demonstrated.

Methods and Materials.—Incubations and growth of liquid cultures were carried out at 30°C unless otherwise indicated. Previously described procedures were employed for preparation and assay of phage stocks, and for complementation tests between amber (*am*) mutants.[2]

Phage strains: All strains are derivatives of T4D unless otherwise indicated. Wild type, the *r*I mutant *r*48, and most of the *am* mutants employed have been described previously.[2] *Am* mutants of phage T2L were isolated and characterized by R. L. Russell. Mutations carried by the T2 *am* strains were assigned to homologous T4 genes on the basis of T2 *am* × T4 *am* complementation tests. Additional information on the phenotypic defects of the various mutants is given in Figure 1 and the tables.

Escherichia coli host strains were employed as follows: CR 63 (permissive for *am* mutants) for all phage assays and preparation of stocks; B/5 (nonpermissive) for preparation of infected cell extracts; S/6 (nonpermissive) as a selective plating indicator for *am*⁺ phage; B/2, S/4 (resistant to adsorption of T2 and T4, respectively) for adsorption experiments (see Table 1).

Media and reagents: H broth, used for growth of bacteria, and EHA top and bottom agar, used for plating assays, were prepared as described previously.[3] Buffer contained Na_2HPO_4 (0.039 M), KH_2PO_4 (0.022 M), NaCl (0.07 M), and $MgSO_4$ (0.01 M) at pH 7.4. Crystalline bovine pancreatic DNase was obtained from Sigma Chemical Co.

Tail-fiberless particles were prepared using a multiple *am* mutant (X4E) defective in the tail-fiber genes 34, 35, 37, and 38 (*am* mutations: B25, A455, B252, N52, B280, and B262). A culture of *coli* B/5 was grown to 4×10^8 cells/ml, infected with X4E phage at a multiplicity (m.o.i.) of 4, aerated for 3 hr, and then treated with $CHCl_3$ to lyse the infected cells. The defective particles were purified by two cycles of low- and high-speed centrifugation and suspended in buffer. The particle concentration was estimated from the optical density of the suspension at 265 mμ, assuming $OD_{265} = 1.0$ for a suspension of 1.2×10^{11} particles/ml.[4] Fewer than 0.01% of the particles in such preparations formed plaques when plated on CR 63 indicator bacteria.

Infected-cell extracts: A culture of B/5 was grown in H broth at 37°C to 4×10^8 cells/ml, cooled to 30°C, infected (0 min) at m.o.i $\sim$4 with phage of the desired genotype, and aerated vigorously at 30°C. Assays of surviving bacteria were generally made to verify that most of the cells were infected. At 30 min the culture was rapidly chilled by pouring into large iced Erlenmeyer flasks. (Following infection with wild-type T4D at 30°C, intracellular phage first appear at 23 min and spontaneous cell lysis does not begin until about 40 min.) The chilled culture was concentrated about 200 times by centrifugation at $5000 \times g$ for 8 min and resuspension of the viscous pellet in buffer containing DNase (10 μg/ml). Microscope counts indicated 10–20% recovery of intact cells. (The low recovery is probably due to the fragility of the cells at this stage of the latent period.) The resuspended pellet was frozen at -70°C in a dry ice–ethanol

498

bath, thawed at 30°C, and either used immediately or refrozen at −70°C and stored at −20°C. (Extracts retain considerable activity upon storage of up to 1 month.) As determined by microscope count, this procedure disrupted 99% of the cells. Protein content of the extracts, estimated colorimetrically,[5] varied between 20 and 30 mg/ml. Infectious phage were present at levels of 10^8 to 3×10^{10} per ml, due to unadsorbed phage and the low but finite transmission of *am* mutants.

Extracts are referred to in the text by the number of the defective gene of the *am* mutant used to infect the cells.

Results.—Activation of tail-fiberless particles: Since gene 23 controls the synthesis of the major structural component of the head membrane,[6] a 23-extract

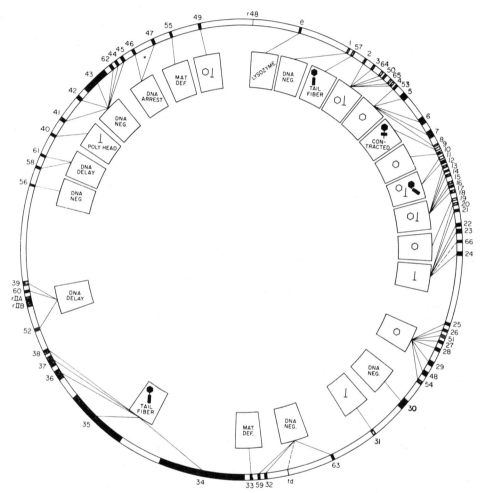

Fig. 1.—Defective phenotypes of conditional lethal mutants of T4D under restrictive conditions. Characterized genes are represented by shaded areas illustrating relative locations and, if known, approximate map lengths.

The enclosed symbols indicate defective phenotypes as follows: *DNA NEG.*, no DNA synthesis; *DNA ARREST*, DNA synthesis arrested after a short time; *DNA DELAY*, DNA synthesis commences after some delay; *MAT DEF.*, maturation defective, DNA synthesis is normal but late functions are not expressed; a hexagon indicates that free heads are produced, an inverted T, that free tails are produced; *TAIL FIBER*, fiberless particles produced; gene 9 mutants produce inactive particles with contracted sheaths; gene 11 and 12 mutants produce fragile particles which dissociate to free heads and free tails. Based on previously published[2] and unpublished experiments.[2]

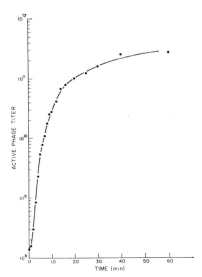

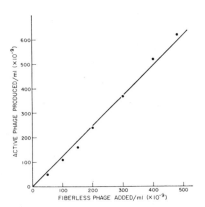

Fig. 3.—Proportionality between fiberless particles added and final yield of active phage in the presence of 23-extract. Buffer, fiberless particles, and 50 μl of 23-extract were mixed as indicated in total volumes of 100 μl. After 200 min of incubation, aliquots of each reaction mixture were assayed for plaque-forming phage. The 20% excess of active phage recovered over fiberless particles added probably reflects the uncertainty in determining titer of the particle suspension by optical density (see *Methods and Materials*).

Fig. 2.—Kinetics of activation of fiberless particles in the presence of 23-extract. Buffer containing 8×10^{11} purified fiberless particles/ml was mixed with an equal volume of 23-extract and incubated at 30°C. At the times indicated, samples were removed for plaque assay on CR 63 indicator bacteria. Titers shown represent active phage/ml of reaction mixture.

(prepared from nonpermissive cells infected with *am* B17, a T4D mutant defective in gene 23) should contain all of the components required for tail-fiber assembly but no phage heads or complete particles. When purified tail-fiberless particles (see *Methods and Materials*) are incubated with 23-extract, the titer of active phage in the mixture rapidly increases over 3 orders of magnitude to a level approaching that of the fiberless particles initially added (Fig. 2). The kinetics of active phage production are not linear. No increase in active titer is observed upon incubation of either the extract or the fiberless particles alone.

As shown in Figure 3, the number of active phage produced is proportional to the number of fiberless particles added to the mixture. Within the error of measure-

TABLE 1

Adsorption and Serological Specificities of Activated Phage

Phages	Fraction Unadsorbed to:			Fraction surviving anti-T4 serum
	B/5	B/2	S/4	
T4r48	0.002	0.01	0.8	0.01
T2L	0.001	0.9	0.00	0.43
X4E-4	0.004	0.003	1.00	0.01
X4E-4P	0.06	0.1	0.7	0.01
X4E-2	0.001	0.9	0.2	0.13
X4E-2P	0.1	0.2	1.2	0.01

Phages: T4r48: T4 control; T2L: T2 control; X4E-4: fiberless T4 particles activated by 23-extract; X4E-4P: progeny of X4E-4; X4E-2: fiberless T4 particles activated by T2 *am* 108 (gene 20) extract; X4E-2P: progeny of X4E-2.

Adsorption experiments: Phage were mixed with suspensions of the indicated bacteria (4×10^8/ml) in H broth containing 0.004 *M* KCN. After 15 min, incubation samples were shaken with CHCl₃-saturated broth and assayed to determine the fraction of input phage remaining unadsorbed.

Inactivation by antiserum: Hyperimmune anti-T4 rabbit serum was used at a dilution giving 1% survival of T4D phage in H broth after 5 min at 30°C.

ment, activation is quantitative when the incubation period is extended to 200 min, and no further increase in active titer is observed beyond this point. Lowering the concentration of 23-extract by a factor of 2 decreases the rate of the reaction, but not the final yield of active phage. No reaction is observed at 0°C.

The following experiments provide evidence that activation reflects the attachment of tail-fiber components present in the extract to the fiberless phage particles.

(1) Complementation spot tests were used to determine the genotype of 40 of the phage particles activated in two separate experiments. All 40 were of genotype X4E, that of the fiberless particles added.

(2) An extract of cells infected with a T2L mutant (*am* 108) defective in gene 20 (required for head membrane formation) was found to activate fiberless T4D particles as described above. The progeny of particles activated in the presence of this extract and of particles activated with a T4D 23-extract were obtained by growth for one cycle on strain CR 63. Samples of the T2 extract-activated and T4D extract-activated phage and their progenies were then compared with T2L wild type and T4D *r*48 for adsorption to B, B/2, and S/4 bacteria, and for neutralization by anti-T4 serum (Table 1). Phage activated with the T2 extract showed the adsorption characteristics, and, to some extent at least, the serological properties of T2L. Their progeny, however, behaved like the T4D extract-activated phage and the T4D control samples. These results support the view that activated phage are T4D particles with either T2 or T4D tail fibers, depending upon the extract used in the activation reaction.

(3) As a more direct test for attachment of fibers, samples of phage from the reaction mixture of Figure 2 were removed after 0, 12, and 200 min of incubation and photographed in the electron microscope by Dr. M. Moody, after purification by three cycles of high- and low-speed centrifugation and negative staining with uranyl acetate. The fractions of active particles in the three samples were, respectively, 0.1, 10, and 100 per cent. Counts of the number of fibers per particle in the three samples (Fig. 4) indicate that the phage acquire tail fibers during incubation with the extract. The results suggest that more than one fiber is necessary for activity, since over 70 per cent of the particles in the 12-min sample show one or more fibers, whereas only 10 per cent are able to form plaques no CR 63.

Extract complementation among tail-fiber mutants: Mutants defective in one or more of the genes 34, 35, 36, 37, 38, and 57 fail to produce active progeny under restrictive conditions *in vivo*. They do, however, produce a normal yield of noninfectious phage particles

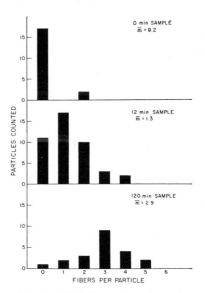

FIG. 4.—Tail-fiber counts from electron micrographs of extract-activated phage (see text). Counts were made on two or more fields from each sample, scoring only particles with clearly visible end plates on an unobstructed background. For each time point the average number of visible fibers per phage ($\overline{m}$) is given. It should be noted that six fibers per phage are seldom visible even in preparations of wild-type T4D.

which lack tail fibers.[2] Antigenic studies with *am* mutants indicate that genes 34, 36, and the gene pair 37–38 control the synthesis of three antigens, designated A, B, and C, respectively, which are found in wild-type T4 lysates associated with the tail fibers of active phage.[7] Under restrictive conditions, gene 57 mutants produce low levels, and gene 35 mutants give normal levels of all three antigens. However, the inactive particles produced by all of the above tail-fiber mutants are devoid of associated antigens (J. King, unpublished).

The preceding experiments indicate that fiber attachment to fiberless particles can proceed *in vitro*. By mixing extracts made with mutants defective in different tail-fiber genes, some of the earlier steps in fiber assembly can be shown to occur as well. We shall refer to the production of active phage under these conditions as "extract complementation."

No increase in active phage titer is observed upon incubation of 34, 35, 36, 37, or 38 extracts alone or with added fiberless particles. However, when pairwise combinations of extracts are allowed to react, the fiberless particles present in the extracts become activated in some, but not all, of the mixtures (Table 2). Of particular interest are combinations such as 34 + 36 and 34 + 37. Since each of the extracts lacks one of the three antigens, the observed activity suggests that at least two steps occur in the mixture: association of the antigens and their attachment to the phage. The partial or complete inactivity of all pairwise combinations among 36, 37, and 38 indicates that not all of the reactions in fiber assembly proceed efficiently in the *in vitro* system as presently constituted. (The results of tests with

TABLE 2

EXTRACT COMPLEMENTATION AMONG PAIRS OF TAIL FIBER MUTANTS

Tail-fiber antigens present:	A, B, C	B, C	A, B, C	A, C	A	A
Defective gene no.	23*	34	35	36	37	38
38	1000	370	187	9.4	2.6	2.6
37	269	38	65	2.6	2.2	
36	268	238	70	0.4		
35	1050	55	3.3			
34	835	2.8				
23*	0.8					

Equal volumes of two extracts were mixed, incubated 90 min, and then assayed for plaque-forming phage. Results are expressed as titer (active phage/ml) $\times 10^{-9}$ in the reaction mixture. Numbers at the bottom of each column represent active titers observed after 90 min incubation of each extract alone. *Am* mutants employed for extract preparation were as follows: 38, *am* C290; 37, X2j (double mutant: *am* N52 and *am* B280); 36, *am* E1; 35, *am* B252; 34, X2d (double mutant: *am* B25 and *am* A455); 23, *am* B17. The first column heading gives the antigenic components found in defective lysates of the mutants listed in the second column heading.
* Control; defective in head formation.

TABLE 3

EXTRACT COMPLEMENTATION BETWEEN HEAD AND TAIL MUTANTS

Defective gene number	5	6	10	27	29	2	23	31
31	—	91	20	—	21	1.0	2.3	0.8
23	283	180	176	445	278	3.7	3.8	
2	189	218	137	352	132	5.3		
29	1.5	1.8	0.9	9.9	3.5			
27	1.4	1.5	0.7	0.7				
10	1.6	2.2	0.4					
6	2.6	3.4						
5	2.7							

Pairwise mixtures of extracts were allowed to react, and the results are expressed as in Table 2, except that incubation time was increased to 120 min. The *am* mutants employed are given in Table 5.

TABLE 4

FUNCTIONAL PROPERTIES OF MORPHOGENETIC-MUTANT EXTRACTS

Extract			Incubated with		Mutant Phenotype	
Defective gene	Am mutant	Self	23-Extract (tail donor)	6-Extract (head donor)	Functional components	Components visible in EM
2	N51	5.3	3.7	218	T	HT
64	E1102	1.8	8.9	205	T	HT
50	A458	20	23	229	T	HT
65	E348	5.2	9.5	133	T	HT
4	N112	36	80	407	T	HT
53	H28	1.4	181	1.6	H	H
5	N135	2.7	283	2.6	H	H
6	N102	3.4	180	—	H	H
7	B16	1.2	263	1.6	H	H
8	N132	1.3	518	3.5	H	H
9	E17	7.8	76	458	HT	Cφ
10	B255	0.38	176	2.2	H	H
11	N93	3.8	15	14	HT	φHT
12	N69	0.9	104	261	HT	φHT
13	E609	0.5	636	216	HT	HT
14	B20	1.1	337	426	HT	HT
15	N133	5.6	936	923	HT	HT
16	N66	1.7	1.8	149	T	HT
17	N56	0.5	0.5	96	T	HT
18	E18	0.3	447	775	HT	HT
20	N50	0.4	1.6	57	T	T
21	N90	20	8.9	165	T	T
22	B270	2.6	1.1	141	T	T
23	B17	0.83	—	180	T	T
24	N65	0.7	0.8	135	T	T
25	N67	29	608	16	H	H
26	N131	4.1	144	3.6	H	H
51	S29	1.6	67	1.8	H	H
27	N120	0.7	1670	1.5	H	H
28	A452	3.0	91	2.3	H	H
29	B7	3.5	278	1.8	H	H
48	N85	0.7	107	1.6	H	H
54	H21	0.9	137	6.4	H	H
31	N54	0.8	0.9	350	T	T
49	E727	1.3	1.4	125	T	HT

Extracts are listed in the map order of the corresponding defective genes (see Fig. 1). Incubations were carried out and the results are expressed as in Table 3. Interpretation of the results is given under the heading "Functional components." Extracts complementing 6-extract (reference head donor) but not 23-extract (reference tail donor) to produce active phage are designated as tail donors (T); those complementing 23 but not 6 as head donors (H), and those complementing both reference extracts as both head and tail donors (HT). The last column lists for comparison the previously determined defective phenotypes of the various mutants.[2] The symbols indicate the presence of free heads (H), free tails (T), complete particles (φ), and particles with contracted sheaths (Cφ) in electron micrographs of defective lysates.

57-extracts, not shown, were ambiguous due to the high leakage of gene 57 mutants.)

Genotype tests of phage produced in the active combinations show that endogenous particles from *either* extract can be activated by tail-fiber attachment. Assays of the activated phage on the restrictive host S/6 revealed no am^+ recombinants, a further indication that the increase in active titer is due to an extracellular process. If the observed complementation were due to mixed infection by the two mutant phages of uninfected cells present in the extracts, then the resulting progeny would be expected to include 5–35 per cent am^+ recombinants, depending upon the map interval between the two defective genes employed.

Extract complementation among other morphogenetic mutants: The experiments described above indicate that the attachment of tail fibers to fiberless particles and at least some of the steps in fiber assembly can proceed in extracts with high efficiency. To test the *in vitro* efficiency of other steps in the maturation process, extracts were prepared from cells infected with a number of mutants defective in various morphogenetic genes. These were examined in pairs for extract complementation as above (Table 3). Platings on the restrictive host S/6 indicated that $< 10^{-3}$ am^+ recombinants were generated in the many mixtures tested. In most cases the results of the test were unambiguous, either showing no increase over the controls (no complementation) or more than a tenfold increase (complementation).

A striking correlation is found between the extract complementation behavior of mutants and their defective phenotypes.[2] Mutants which produce heads but no tails as determined in the electron microscope (EM), and thus presumably are blocked in tail assembly, do not complement among themselves (genes 5, 6, 10, 27, 29) but do complement with mutants which produce tails, but no heads (genes 23, 31). Many other tests not shown in Tables 3 or 4 support this generalization. These results suggest that the attachment of heads to tails can proceed *in vitro*, and that mutant extracts can be classed as "head donors" or "tail donors," extracts from one class being active only with extracts from the other.

To test this generalization further, infected-cell extracts were prepared using *am* mutants defective in each of 35 genes known to be involved in morphogenetic steps other than tail-fiber assembly.[2] (Genes identified only with temperature-sensitive mutants were not tested.) Each extract was tested for activity against a reference head donor (6-extract) and a reference tail donor (23-extract). As shown in Table 4, every mutant showing the EM phenotype of heads but no tails behaves as a head donor only (14 tested), while every mutant showing tails but no heads behaves as a tail donor only (6 tested). Most of the class of mutants producing both heads and tails (unattached; see Fig. 1) also behave as tail donors only (8/12). Apparently the heads observed by EM are either incomplete or nonfunctional by-products of defective maturation which cannot be efficiently activated in extract mixtures. However, extracts made with the remaining four mutants of this class (genes 13, 14, 15, and 18) behave as both head and tail donors, suggesting that they contain functionally competent heads and tails but lack components required for their union. These preparations, in contrast to those of head donors only and tail donors only, also complement extracts made with double mutants which produce neither heads nor tails (one defective in genes 10 and 31, another in genes 27 and 23).

Mutants defective in genes 9, 11, or 12 also show extract complementation with both the reference head donor and the reference tail donor, as well as with the double mutants. Under restrictive conditions *in vivo*, 9, 11, and 12 mutants produce apparently complete but inactive phage particles characterized in the electron microscope by contracted sheaths (gene 9 mutants) or a tendency to dissociate into free

TABLE 5

GENOTYPES OF EXTRACT-ACTIVATED PHAGE; HEAD AND TAIL MUTANTS

Extract A		Extract B		Number of Phage	
Gene defect	Functional components	Gene defect	Functional components	Genotype A	Genotype B
29	H	23	T	20	0
14	HT	23	T	20	0
14	HT	6	H	11	9
6	H	2	T	20	0
14	HT	2	T	20	0
14	HT	16	T	20	0
6	H	16	T	20	0

Genotypes of the phage particles activated during 120-min incubation of the indicated extract mixtures were determined by spot testing for complementation with appropriate known mutant strains.

192

heads and tails upon storage (gene 11 and 12 mutants). The extract complementation results suggest that these defective particles can be activated when a missing component is provided by another extract.

Complementation spot tests of phage produced in the active combinations of Tables 3 and 4 show that their genotype is exclusively that of the head donor (Table 5). When 14-extract, a head and tail donor, complements a tail donor extract, the phage produced are of genotype 14; when it complements a head donor extract, comparable numbers of the two genotypes are found among the active particles.

Discussion and Summary.—The results reported here show that several of the steps in the morphogenesis of bacteriophage T4D can take place in extracts of infected cells. These steps include the assembly of tail fibers and their attachment to the virus particle, as well as the union of the head and the tail. Apparently many of the larger components—fiberless particles, fibers, heads, and tails—which accumulate in mutant-infected cells under restrictive conditions are not aberrant byproducts of defective synthesis, but intermediates in the assembly process which may be utilized *in vitro* for the morphogenesis of active virus.

The limited success of our attempts at complementation in extracts indicates that many of the steps in the maturation process do not proceed efficiently under the conditions presently employed. It remains to be seen whether some of these steps, such as the formation of the head or of the tail, can be demonstrated under altered conditions, for example, at higher concentration of reactants.

As an extension of the previously reported EM characterization of defective phenotypes,[2] the extract complementation studies provide some further insight into the functions controlled by the various mutationally defined morphogenetic genes. An example is the group of 12 genes whose mutants produce unattached heads and tails recognizable in electron micrographs. While four of these are apparently involved in the union of heads and tails, the remaining eight may be required to complete or alter the head in some manner which activates it for tail attachment.

Perhaps most significantly, our results show that at least portions of the morphogenetic pathway are open to direct attack by biochemical methods. It may be hoped that purification of components and further study of individual reactions will lead to some understanding of the interactions and specificities involved in the assembly of complex supramolecular structures such as bacteriophage T4.

We wish to thank Mrs. Ilga Lielausis for excellent technical assistance and Dr. M. Moody for the preparation of electron micrographs.

* This research was supported by grants from the National Science Foundation (GB3930) and the U.S. Public Health Service (GM06965).

[1] Brenner, S., G. Streisinger, R. W. Horne, S. P. Champe, L. Barnett, S. Benzer, and M. W. Rees, *J. Mol. Biol.*, 1, 281 (1959).

[2] Epstein, R. H., A. Bolle, C. M. Steinberg, E. Kellenberger, E. Boy de la Tour, R. Chevalley, R. S. Edgar, M. Susman, G. H. Denhardt, and A. Lielausis, in *Cold Spring Harbor Symposia on Quantitative Biology*, vol. 28 (1963), p. 375, and Edgar, R. S., and R. H. Epstein, unpublished.

[3] Steinberg, C. M., and R. S. Edgar, *Genetics*, 47, 187 (1962).

[4] Winkler, U., H. E. Johns, and E. Kellenberger, *Virology*, 18, 343 (1962).

[5] Lowry, O., N. J. Rosebrough, A. L. Farr, and R. J. Randall, *J. Biol. Chem.*, 193, 265 (1951).

[6] Sarabhai, A. S., A. O. W. Stretton, S. Brenner, and A. Bolle, *Nature*, 201, 13 (1964).

[7] Edgar, R. S., and I. Lielausis, *Genetics*, 52, 1187 (1965).

$$17$$

Reprinted from *Fed. Proc.*, **27**(5), 1160–1166 (1968)

Bacteriophage assembly[1]

WILLIAM B. WOOD, ROBERT S. EDGAR, JONATHAN
KING, ILGA LIELAUSIS AND MELVA HENNINGER

Division of Biology, California Institute of Technology,
Pasadena, California

Extensive research during the past ten years has given us a fairly complete understanding of how genes direct the formation of individual protein molecules. Considerably less is known about the assembly and genetic control of supramolecular protein complexes. Several such structures composed of few components, such as the tobacco mosaic virus (1), the hemoglobin molecule (14), and the α-keto acid dehydrogenase complexes of *Escherichia coli* (17, 18), have now been disassembled by mild chemical procedures and successfully reconstituted. In these simple systems specification of the subunit protein structures is sufficient to direct their aggregation, through noncovalent interactions, into complexes of unique size and shape. Still unclear is the extent to which such a self-assembly mechanism can explain the morphogenesis of more complex structures composed of many different gene products.

For the exploration of this question some of the larger bacteriophages offer several attractive features. Their complex capsid structures include many different proteins (2). They are well suited to genetic and physiological studies, which have already provided considerable insight into the process of intracellular phage development (5, 10, 12). Perhaps most important, recent experiments have shown that the assembly of bacteriophages T4 (8), lambda (22), and P22 (13), can be studied in vitro using viral components derived from infected cells.

In this paper we shall review the recent developments and current status of our work on T4 morphogenesis. Experimental details not included here have been (7, 8, 16) or will be published elsewhere. We shall consider the results in three categories: *1*) genetic and physio-

logical evidence, *2*) in vitro complementation experiments in cell extracts, and *3*) ordering and characterzation of individual assembly steps.

EVIDENCE FROM GENETIC AND
PHYSIOLOGICAL STUDIES

The experiments of Epstein et al. (10) showed that many conditionally lethal mutations in T4 lead to accumulation of recognizable phage components under restrictive conditions.[2] These and subsequent similar observations (Edgar and Epstein, unpublished) on the components missing from electron micrographs of mutant-infected cell lysates (defective lysates) have provided information regarding the function of 46 genes affecting viral morphogenesis (Fig. 1). Two features of the results should be noted. *1*) They are consistent with a morphogenetic pathway of three major branches. These lead independently to the formation of head, tail, and tail fibers, respectively, and are followed by steps in which the completed components are assembled into infectious viral particles. *2*) They indicate that a surprisingly large number of gene products is required for virus assembly. Since it is unlikely that all of the morphogenetic genes have been mutationally identified, this number is probably greater than 50. This has two implications. The number of different proteins in the phage particle may be considerably larger than the 15–20 for which there is now evidence (2, 3, 12, 19). Alternatively or in addition, some gene products may play directive or catalytic roles in the assembly process, rather than contributing materially to the phage structure. New approaches to evaluating these possibilities have become available with the demonstration that several steps in T4 assembly can be carried out in vitro.

From the Biochemistry Symposium on "Biosynthesis and Function of Supramolecular Structures" presented at the 52nd Annual Meeting of the Federation of American Societies for Experimental Biology, Atlantic City, N.J., April 16, 1968.

[1] This study was supported by National Science Foundation Grant GB3930 and the Public Health Service Grant GM06965.

[2] All of the T4 mutants used in this work are of the suppressor-sensitive class known as amber mutants (5). Restrictive conditions refer to growth in the *E. coli* host strain B/5(su^-). Strain CR63 (su^+) was used for growth under permissive conditions.

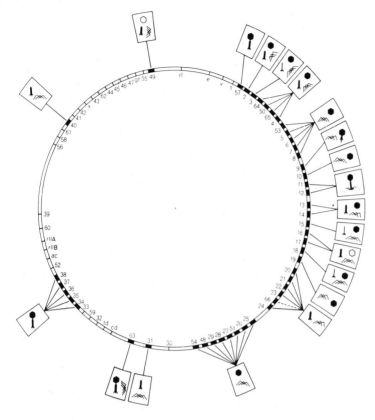

FIG. 1. Defective phenotypes of conditionally lethal mutants of T4. Mutationally identified genes affecting phage morphogenesis are represented by broad black segments indicating relative locations on the circular genetic map. Narrow lines represent genes controlling phage "early functions" (3, 10, 12) not involved in morphogenesis. Defective phenotypes are indicated by the boxed symbols, which represent the phage structural components observed in electron micrographs of defective lysates prepared under restrictive conditions with mutants representative of the various genes (10). Gene 9 mutants produce fiberless particles whose sheaths sometimes appear contracted in electron micrographs; gene 11 and 12 mutants produce fragile noninfectious particles with tail fibers (10, 16).

COMPLEMENTATION EXPERIMENTS IN EXTRACTS OF INFECTED CELLS

To determine the number and nature of the assembly reactions accessible to in vitro study, we have performed a large number of extract complementation tests (7, 8, and unpublished observations). When two extracts made from mutant-infected cells defective in different gene products are incubated together, the extracts can be said to complement if infectious phage are produced. For this to occur one must supply a functional viral component which is lacking or nonfunctional in the other.

Almost all such pairwise tests have given unambiguous positive or negative results: either no detectable increase or more than a 10-fold increase in the infective titer. On this basis we can arrange the defective extracts into 14 complementation groups, defined such that pairs derived from any two different groups will complement, whereas pairs from within the same group will not (Table 1). Since each group defines a functional component for assembly, we have 14 components whose interaction to produce complete virus can be studied in extracts under present experimental conditions.

The above interpretation is consistent with the physiological studies, in that mutants comprising a complementation group exhibit the same defective phenotype (except in group I; see footnote to Table 1) as determined by electron microscopy. From the latter evidence, we can infer the nature of the missing components corresponding to most groups (Table 1), and therefore the kinds of assembly reactions which can occur in vitro. Groups I and II, representing the head and the tail baseplate, respectively, each include a large number of defective extracts. This indicates that these components cannot be built in vitro, but must be supplied in nearly finished form to be functional in subsequent reactions.[4] Among the steps which can take place in extracts, as seen from the properties of the remaining groups, are the union of heads and tails, the assembly of the tail

---

[4] In recent experiments (Edgar and Lielausis, unpublished data) significant complementation has been observed in some pairwise mixtures of group II extracts.

TABLE 1. *Extract complementation groups*

Group	Mutant Genes		Components Present		Inferred Defect
I [a]	20, 21, 22, 23, 24, 31	—	Tail	Fiber	Head (formation)
I [b]	49, 2, 64, 50, 65, 4, 16, 17	Head	Tail	Fiber	Head (completion)
II	53, 5, 6, 7, 8, 10, 25, 26, 51 27, 28, 29	Head	—	Fiber	Tail (baseplate)
III	48, 54	Head	Baseplate	Fiber	Tail (core, sheath)
IV	13, 14				
V	15	Head	Tail	Fiber	?
VI	18				
VII	9	Contracted particle (fiberless)		Fiber	?
VIII	11	Defective phage particle (fibers attached)			?
IX	12				
X	37, 38				
XI	36	Fiberless particle		—	Fiber assembly
XII	35				
XIII	34				
XIV	63	Fiberless particle		Fiber	Fiber attachment

See text for definition of complementation groups. Each entry under the heading "Mutant Genes" represents an infected-cell extract, prepared under restrictive conditions as described previously (8), using an amber mutant defective in the numbered gene. The conditions for carrying out complementation tests and the criteria for scoring them have been previously described (8).

Groups I and II are not rigorously defined in that not all of the possible pairwise tests have been performed. Each extract in both groups was tested against a representative extract of I (23-defective extract)[3] and II (27-defective extract). Assignments were made on the basis of these results and the previously observed defective phenotypes, which were mutually consistent. Extracts of group Ia do not complement those of Ib; i.e., the observed Ib heads are nonfunctional for in vitro assembly. Extracts of groups XI, XII, and XIII contain structures corresponding in dimensions to one arm of the complete tail fiber. Not tested were extracts representing genes 3, 19, and 40, defined by temperature sensitive mutants only at the time of these experiments, and genes 57 and 66, defined by leaky amber mutants which yield too high a background level of infectious phage to permit meaningful extract complementation experiments.

Heads with attached tails are designated "particles." All structural components listed as present are unattached to each other unless otherwise indicated. Description of these as heads, tails, etc. does not imply that they are complete, but merely that they are identifiable in electron micrographs.

fibers, and the attachment of tail fibers to fiberless particles (8).

ORDERING OF STEPS IN PHAGE ASSEMBLY

The foregoing analysis provides only limited information concerning the functional roles of the 14 components, and the order in which they normally interact to form complete virus. To determine the sequence of assembly reactions, we have isolated precursor structures from the various defective extracts and characterized them by several methods, including electron microscopy, serological analysis, and complementation behavior in appropriate extracts.

A first step in this phase of the work was to confirm the apparent independence of the head and tail branches of the pathway, by obtaining additional evidence that most of the heads present in group II extracts and the tails in group I extracts represent complete structures. We found that 27-defective heads and 23-defective tails,[3]

---

[3] Extracts of phage-infected cells (8) are referred to by the number of the defective gene in the mutant used for infection. Phage precursor structures isolated from such extracts (7, 8, 16) are referred to similarly; e.g., heads isolated from a 27-defective extract are designated 27-defective heads.

isolated from the corresponding extracts by sucrose gradient sedimentation, appear complete in electron micrographs, and unite when mixed together, forming tail fiberless particles (7).

These isolated heads and tails have been used in complementation tests to characterize the ambiguous defects of extracts in groups IV, V, and VI (Table 1). The above result, indicating that the head-tail reaction does not require other gene products in addition to the structures themselves, suggests that in these extracts it is defects in either head or tail completion which prevent head-tail union. This is confirmed by the following results. Extracts lacking the group IV component are complemented by 27-defective heads but not 23-defective tails. Since a complete head can supply the missing function, component IV (genes 13 and 14) must be involved in head completion. Conversely, extracts defective in the group V or group VI functional components are complemented by tails but not heads, indicating that genes 15 and 18 control steps in tail completion (7). The nature of the latter two steps has been clarified by electron microscopy (16) as described below.

The general method used in sequencing the remain-

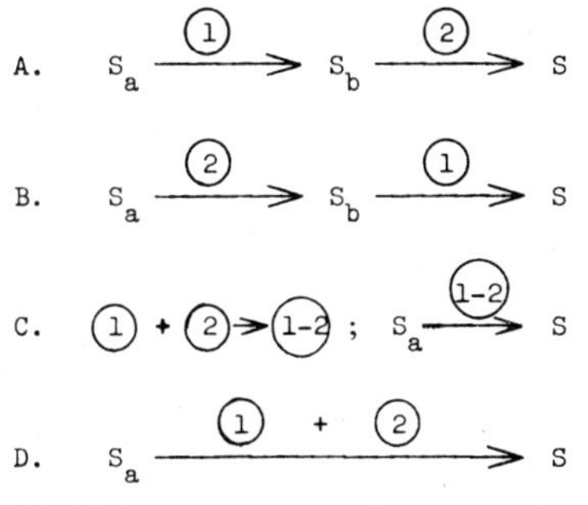

A. $S_a \xrightarrow{\enspace ① \enspace} S_b \xrightarrow{\enspace ② \enspace} S$

B. $S_a \xrightarrow{\enspace ② \enspace} S_b \xrightarrow{\enspace ① \enspace} S$

C. $① + ② \rightarrow \boxed{1\text{-}2}\; ;\quad S_a \xrightarrow{\enspace \overset{\text{1-2}}{} \enspace} S$

D. $S_a \xrightarrow{\enspace ① + ② \enspace} S$

FIG. 2. Possible interaction sequences of two gene products with a structural intermediate. See text for explanation.

ing assembly reactions in each of the three branches may be explained with the aid of the schematic diagram shown in Fig. 2. Assume that S and its precursors S_a and S_b are large components, for example, head structures, which can be isolated easily from extracts by sucrose gradient sedimentation. The mutationally defined gene products 1 and 2 are known to be involved in the completion of the S structure; neither a 1-defective extract nor a 2-defective extract contains complete S, but when incubated together they complement to produce it. This complementation could occur by any one of the four paths shown. Possibilities A, B, and C represent possible sequential interactions, and D represents the possibility of no sequence—that is, where no two of the components will interact in the absence of the third. The precursor structures are isolated from each extract and tested to ascertain which intermediate form is present, S_a or S_b. Examination of the possibilities diagramed in the figure will show that the results allow a distinction between possibilities A, B, and C or D.

Experimentally of course the problem lies in distinguishing one intermediate form from the other; that is, in determining whether or not a gene product has acted. If S_a and S_b can be distinguished by sedimentation behavior in sucrose gradients, antigenic properties, or appearance in electron micrographs, this can be determined directly. Several examples of such cases have been encountered (7, 16, and Wood and King, ms in preparation). If these methods fail, the intermediates can always be distinguished by testing the isolated structure from one extract for its ability to complement the other. For example, if isolated precursor from 2-defective extract complements 1-defective extract, whereas the reciprocal test gives no complementation, it is clear that the precursor in 2-defective extract has been acted upon by gene product 1, and that the correct pathway is possibility A. If neither of the reciprocal tests give complementation, then the correct pathway is probably C or D.

GENE CONTROL OF THE ASSEMBLY PROCESS

Evidence obtained by all of the methods described above is summarized in Fig. 3, which illustrates the pathway of T4 morphogenesis as presently understood. Dashed arrows are used to indicate steps which have not yet been demonstrated in vitro; solid arrows indicate those which have. Numbers refer to the gene control of the various steps. The structural intermediates are drawn according to electron micrographs of components found in or isolated from the appropriate defective extracts (10, 16).

Head morphogenesis can be divided into four stages, only the last of which is so far accessible to in vitro study. Since gene defects in group Ia of Table 1 prevent the formation of any recognizable head precursor, we assume that the earliest steps are controlled by these

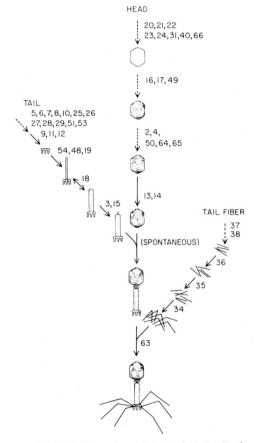

FIG. 3. Pathway of T4 morphogenesis. See text for explanation of symbols.

genes, as well as genes 40 and 66 (10, 15). Mutants defective in genes 16, 17, or 49 accumulate head membranes which appear in electron micrographs to be normal in size and shape but empty of DNA (16). Recent "pulse-chase" experiments on the flow of radioactive amino acids through these structures during phage maturation in vivo suggest that the empty membranes are precursors of complete heads (Luftig and Wood, ms in preparation). These findings bear on the perplexing problem of how DNA becomes packaged within the viral capsid. While they do not rule out the proposal that head membranes are laid down around DNA "condensates" (15), they do indicate we must also consider the alternative view that the DNA is somehow introduced into preformed head membranes by a process requiring the function of genes 16, 17, and 49. The remaining head structures in the pathway are so far indistinguishable in electron micrographs, and the nature of the head completion steps is still unclear. We can conclude, however, that these steps are prerequisites for head-tail union, since heads with tails attached have not been observed in any of the group I*b* or group IV extracts (Table 1).

The earliest structure so far identifiable in tail morphogenesis is the baseplate, whose formation is controlled by the genes of group II (Table 1), as well as genes 9, 11, and 12 (see below). The tail core, which has never been observed in mutants lacking the baseplate, is built onto it under the control of genes 54, 48, and 19 (7, 16, and Edgar and Lielausis, unpublished observations). The gene 18 product appears to be the major structural protein of the sheath, whose assembly around the tail core is reversible by dilution until the tail is completed under the control of genes 3 and 15 (16). As in the preceding branch of the pathway, evidence from electron microscopy indicates that tail completion must precede union with the head, possibly because gene products 3 and 15 make up the connector at the top of the sheath (16) through which attachment occurs (Fig. 3). There are, however, three exceptions to the requirement for tail completion; the steps controlled by genes 9, 11, and 12 can be bypassed to produce particles with defective tails, as discussed below.

As noted earlier, the completed head and tail combine in the absence of additional factors. This appears to be a simple second-order reaction whose rate is independent of incubation temperature between 0 and 30 C (7). The resulting particle is the substrate for tail fiber attachment. Evidence from electron micrographs and from serological analysis of precursor tail structures (see below) indicates that fibers cannot attach to the baseplate until the union of head and tail has taken place (16).

Seven mutationally identified gene products are required for the formation and attachment of tail fibers (see Fig. 1). Preliminary evidence regarding their sequence of action was provided by Edgar and Lielausis (6) who used serum blocking techniques (4) to study the gene control of three tail fiber antigens which they designated "A," "B," and "C." We have extended their

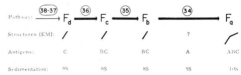

FIG. 4. The tentative pathway of T4 tail fiber morphogenesis. Heavy arrows indicate steps which have been demonstrated in vitro. Sedimentation constants of fiber structures were estimated by centrifugation of the appropriate defective extracts through sucrose gradients, followed by localization of antigenic components in the collected fractions using adsorbed sera as previously described (6). Structures of intermediates observed in the electron microscope by Eiserling et al. (9) have been confirmed in our laboratory.

experiments using the methods outlined above to give the tentative pathway of tail fiber morphogenesis shown in Fig. 4 (King and Wood, ms in preparation), which is also supported by recent experiments of Eiserling, Bolle and Epstein (9). In the first step, not yet demonstrated in extracts, a structure corresponding in dimensions to one arm of the finished tail fiber is formed under the control of genes 38 and 37. This component interacts sequentially with the products of genes 36, 35, and 34 in that order, to produce the complete structure. The nature of these interactions remains unclear, as does the presumed role of gene 57 in the process (6). It is of interest that the sequence of gene product action so far elucidated parallels the map order of the corresponding genes. As in the preceding pathways completion of the fiber appears to be a prerequisite for attachment to the baseplate; individual antigens or fiber precursor structures have never been detected on the particles produced by mutants blocked in the various steps of tail fiber formation (King and Wood, ms in preparation).

CHARACTERIZATION OF THE TAIL FIBER ATTACHMENT REACTION

The terminal step in T4 morphogenesis appears to be the attachment of completed tail fibers to the otherwise finished phage particle. We have studied this process in some detail (Wood and Henninger, ms in preparation) in reaction mixtures containing purified fiberless particles (8) and the high-speed supernatant fraction from a 23-defective extract as a source of tail fibers. The initial kinetics of phage production are exponential, and in good agreement with predictions based on the assumption that fibers are attached one at a time and randomly among the population, and that a particle must acquire 3–4 fibers to become infectious. The reaction shows an absolute but nonspecific ionic requirement, which can be satisfied by any of several divalent cations at low (0.02 M) or monovalent cations at high (2.0 M) concentration. No other dialyzable factors appear to be required. The rate of fiber attachment is temperature dependent, with a Q_{10} of approximately 2 between 20 and 30 C. No reaction is detectable at 0 C.

Fractionation of the 23-defective extract has led to

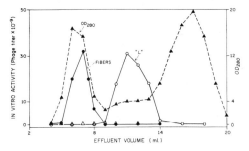

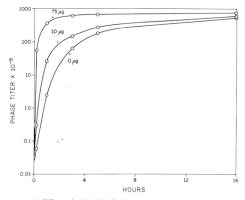

FIG. 5. Fractionation of 23-defective extract by gel filtration. A 1-ml sample of the extract, prepared as previously described (8) was clarified by ultracentrifugation for 1 hr at 49,000 rpm, run onto a 1 x 25 cm column of Bio-Gel P300 (Calbiochem), and eluted with buffer (0.01 M Tris pH 7.4 containing 0.005 M $MgSO_4$) at a flow rate of approximately 1 ml/hr. Fractions of 1 ml were collected and their optical density at 280 mμ determined (solid triangles). Assays for activity were carried out as indicated below at 30 C for 10 min in reaction mixtures of 100 μliters containing 0.01 M Tris pH 7.4, 0.02 M $MgSO_4$, and 3×10^{10} tail fiberless particles prepared as previously described (8). Where indicated, an extract prepared as above using a multiple mutant defective in genes 23, 34, 36, 37, and 38 was used as a source of labile factor ("L") free of tail fibers and particles. A sample of 23-defective extract heated 10 min at 50 C was used as a source of tail fibers free of "L." Reaction mixtures contained either 70 μliters of the indicated column fraction alone (open triangles— assay for fibers plus "L"), or 70 μliters of column fraction plus 20 μliters of labile factor (solid circles—assay for fibers), or 30 μliters of column fraction plus 60 μliters of heated 23-defective extract (open circles—assay for "L"). Infectious phage titers were determined by plaque assay of an appropriate dilution of the reaction mixture on permissive *E. coli* bacteria as described earlier (8).

the discovery of an additional factor required for tail fiber attachment (Fig. 5). It is inactivated by heating (10 min, 50 C), and its elution profile on a calibrated gel filtration column is consistent with a molecular weight of approximately 10^5 daltons. When added in increasing amounts to reaction mixtures containing purified fibers and fiberless particles, it increases the rate but not the final level of infectious virus production (Fig. 6). This provides a method of assay for the factor, and suggests that it acts catalytically rather than stoichiometrically in the fiber attachment reaction.

Labile factor activity has not been detected in extracts of uninfected *E. coli* cells. Following infection at 30 C, activity appears at 4 min and continues to increase up to 30 min (unlike previously described early functions (3, 12), whose synthesis stops at about 15 min). The same pattern has been found following infection of restrictive host cells with amber mutants defective in any gene except 63. The mutant defining this gene produces normal levels of fiberless particles and unattached fibers, but less than 1 % of normal labile factor activity.[5] When the

FIG. 6. Effect of added labile factor on the rate and extent of fiberless particle activation. Tail fibers relatively free of labile factor were prepared from 23-defective extract using the column fractionation described under Fig. 5. Three reaction mixtures were made up as in Fig. 5 except that 2×10^{11} fiberless particles and 40 μliters of the tail fiber preparation were present in each. Labile factor was added in the amounts indicated, and the kinetics of infectious phage production were followed for 16 hr at 30 C.

mutant is grown in a permissive strain carrying the single amber suppressor *su-3* (11, 21), it produces labile factor activity which is abnormally temperature sensitive, showing a first-order inactivation rate constant at 37 C nine times greater than the activity induced by wild-type phage in the same host. This finding is strongly suggestive of a suppressor-specific structural alteration, which would indicate that the labile factor is a protein whose structure is determined by phage gene 63.

SOME FEATURES OF THE MORPHOGENETIC PATHWAY

The assembly steps so far observed in vitro (Fig. 3) represent only perhaps one-fourth to one-fifth of the total number. It is already apparent, however, that there is a stringent sequential order in the morphogenetic process. As a consequence mutational blocks cannot in general be bypassed, so that characteristic structural intermediates accumulate. This feature of the pathway has made possible the experiments described above. So far we have found only three exceptions: the steps controlled by genes 9, 11, and 12. The phenotypes of mutants defective in these genes (Fig. 1, Table 1) suggest that the corresponding gene products act near the end of the pathway, since nearly completed defective particles accumulate. Moreover, these particles[6] can be converted to infectious phage

[5] The amber mutant defining gene 63 was previously classified as defective in DNA synthesis (Epstein and Bolle, unpublished data, and (8)). However, genetic analysis showed that in addition to the gene 63 defect this strain carried a second mutation affecting

DNA synthesis. The second mutation was removed by backcrossing to the wild type, yielding the gene 63 mutant described in the text (Wood and Henninger, ms in preparation).

[6] Only a fraction of the particles found in 9-defective extracts show contracted sheaths in electron micrographs; the majority appear uncontracted (16). It is presumably the latter which are converted to infectious phage when incubated with an extract supplying the gene 9 product and tail fibers (Flatgaard, unpublished observations).

by incubation with extracts containing the missing gene products. However, experiments using tail precursors to complement 9-, 11-, and 12-defective extracts show that these gene products can also act early in the tail pathway, during formation of the baseplate.[7] Unlike the remaining steps, therefore, these three can apparently occur at more than one point in the morphogenetic sequence (7, 16, and Flatgaard, unpublished observations).

It appears likely that the sequential order in the pathway is imposed entirely by structural features of the intermediates, at the level of gene product interaction rather than gene action. The kinetics of intracellular phage production (3), the nearly identical times of appearance of the T4 late proteins studied by Hosoda and Levinthal (12), and the finding of several extract complementation groups corresponding to single gene products (Table 1) all argue against control by sequential gene induction, and support the view that all morphogenetic precursor components are synthesized simultaneously and independently[8] throughout the latter half of the latent period.

The information obtained so far is primarily descriptive. Although we can write a partial sequence of assembly steps, we do not yet understand, except in a few cases, the mechanisms which dictate their sequential order, or the manner in which the various gene products contribute to the process. To obtain this information, more detailed biochemical characterization of individual steps will be required. There is evidence that the six genes 10 (12), 15 (16, and Edgar, unpublished observations), 18 (16), 22 (12), 23 (12, 19), and 34 (6, 12) direct the synthesis of proteins which are incorporated into the virus structure. By contrast, gene 63 codes for a protein which appears to act catalytically in tail fiber attachment. Proof for this suggestion, currently being sought through purification of the labile factor and study of its mode of action, would establish that T4 morphogenesis cannot be explained on the basis of self-assembly alone. However, suggestions as to the possible nature and importance of alternative mechanisms must await the outcome of this and similar studies of the remaining assembly reactions.

REFERENCES

1. ANDERER, F. A. *Z. Naturforsch.* 146: 642, 1959.
2. BRENNER, S., G. STREISINGER, R. W. HORNE, S. P. CHAMPE, L. BARNETT, S. BENZER AND M. W. REES. *J. Mol. Biol.* 1: 281, 1959.
3. CHAMPE, S. P. *Ann. Rev. Microbiol.* 17: 87, 1963.
4. DEMARS, R. I. *Virology* 1: 83, 1955.
5. EDGAR, R. S., AND R. H. EPSTEIN. *Genetics Today.* New York: Pergamon, 1964, p. 1.
6. EDGAR, R. S., AND I. LIELAUSIS. *Genetics* 52: 1187, 1965.
7. EDGAR, R. S., AND I. LIELAUSIS. *J. Mol. Biol.* 32: 263, 1968.
8. EDGAR, R. S., AND W. B. WOOD. *Proc. Natl. Acad. Sci., U.S.* 55: 498, 1966.
9. EISERLING, F. A., A. BOLLE AND R. H. EPSTEIN. *Virology* 33: 405, 1967.
10. EPSTEIN, R. H., A. BOLLE, C. M. STEINBERG, E. KELLENBER-

GER, E. BOY DE LA TOUR, R. CHEVALLEY, R. S. EDGAR, M. SUSMAN, G. H. DENHARDT AND A. LIELAUSIS. *Cold Spring Harbor Symp. Quant. Biol.* 28: 375, 1963.
11. GAREN, A., S. GAREN AND R. C. WILHELM. *J. Mol. Biol.* 14: 167, 1965.
12. HOSODA, J., AND C. LEVINTHAL. *Virology* 34: 709, 1968.
13. ISRAEL, J. V., T. F. ANDERSON AND M. LEVINE. *Proc. Natl. Acad. Sci., U.S.* 57: 284, 1967.
14. ITANO, H. A., E. A. ROBINSON AND A. J. GOTTLIEB. *Brookhaven Symp. Biol.* 1964, p. 194.
15. KELLENBERGER, E. In: *Ciba Found. Symp. Principles Biomolecular Organization.* 1966, p. 192.
16. KING, J. *J. Mol. Biol.* 32: 231, 1968.
17. KOIKE, M., L. J. REED AND W. R. CARROLL. *J. Biol. Chem.* 238: 30, 1963.
18. MUKHERJEE, B. B., J. MATTHEWS, D. L. HORNEY AND L. J. REED. *J. Biol. Chem.* 240: PC2268, 1965.
19. SARABHAI, A. S., A. O. W. STRETTON, S. BRENNER AND A. BOLLE. *Nature* 201: 13, 1964.
20. STAHL, F. W., N. E. MURRAY, A. NAKATA AND J. M. CRASEMANN. *Genetics* 54: 223, 1966.
21. WEIGERT, M. G., E. LANKA AND A. GAREN. *J. Mol. Biol.* 14: 522, 1965.
22. WEIGLE, J. *Proc. Natl. Acad. Sci., U.S.* 55: 1462, 1966.

[7] Based on the defective phenotype of gene 9 mutants, previous versions of the pathway (7, 16) have shown the step controlled by this gene as occurring after head-tail union. The revision of this conclusion (Fig. 3) was necessitated by the results described here, which were obtained in more recent experiments (Flatgaard, unpublished observations).

[8] Except for polarity effects seen with a few amber mutants (20).

18

Reprinted from *Nature*, **227**(5259), 680–685 (1970)

Cleavage of Structural Proteins during the Assembly of the Head of Bacteriophage T4

by

U. K. LAEMMLI

MRC Laboratory of Molecular Biology,
Hills Road, Cambridge

Using an improved method of gel electrophoresis, many hitherto unknown proteins have been found in bacteriophage T4 and some of these have been identified with specific gene products. Four major components of the head are cleaved during the process of assembly, apparently after the precursor proteins have assembled into some large intermediate structure.

BACTERIOPHAGES of the T-even type are complex structures containing many different proteins and specified by many genes. Using an improved technique of electrophoretic separation I have found that the phage particle contains at least twenty-eight components, eleven of which are in the head. In the course of identifying the genes specifying these proteins I discovered that four major components of the head, the product of gene 22, 23, 24 and a protein called IP of unknown genetic origin are cleaved during the process of assembly. The head of bacteriophage T4 is therefore no longer a self-assembly system in the narrow sense, because the bonding properties of the various components become altered during the assembly process.

The product of gene 23 is the principal protein component of the head of bacteriophage T4 (refs. 1–3). Two minor components of unknown genetic origin have also been found in capsids[2,3]. Besides the product of gene 23, the products of genes 20, 21, 22, 24, 31, 40 and 66 are required to determine the size and shape of the head-shell (refs. 4 and 5 and unpublished work of F. A. Eiserling,

E. P. Geiduschek, R. H. Epstein and E. J. Metter). To several of these genes shape-specifying functions have been tentatively assigned[5]. Gene 22 is associated with the diameter selecting (initiation) process of head formation, gene 66 with the elongation of the particle and genes 20 and 40 with the formation of the hemispherical cap. Gene 31 somehow modifies or activates the major subunit for ordered assembly[6]. Ten more proteins, the products of genes 2, 4, 13, 14, 16, 17, 49, 50, 64 and 65, are thought to control later steps in head formation[7].

Structural Components of the Phage

Many phage proteins can be separated with our improved method of disk-electrophoresis in sodium dodecyl sulphate (SDS). This system, to be described in detail elsewhere (U. K. L. and J. V. Maizel), combines the high resolution power of disk-electrophoresis[8] with the capability of SDS to break down proteins into their individual polypeptide chains[9]. The proteins are also separated according to their molecular weight as was first reported for a continuous system[10]. All the proteins the genetic

Cleavage of Structural Proteins During the Assembly of the Head of Bacteriophage T4

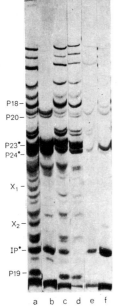

P18—
P20—
P23*—
P24*—
X₁—
X₂—
IP*—
P19—

a b c d e f

Fig. 1. Autoradiogram of ^{14}C-labelled T4 phage proteins separated in acrylamide gels. ^{14}C-amino-acid-labelled preparations were analysed in 10 per cent acrylamide gels containing SDS. a, Wild type lysate; b, purified heads; c, purified phage particles; d, purified "ghosted" phage particles; e, supernatant of "ghosted" phage particles; f, "early labelled" phage particles. The prefix P is used to designate the protein of a particular gene: for example, P20 stands for the product of gene 20; the asterisk indicates that this protein is derived from a large precursor protein and has become modified during head assembly.

[14]C-labelled lysates. Ten ml. cultures of Escherichia coli Bᵇ (the restrictive host for phage carrying amber mutations) in 'M9' medium[13] grown at 37° C to 2 × 10⁸ cells/ml. were infected with the various phages at a multiplicity of five and superinfected at 8 min with the same phage and the same multiplicity to ensure lysis inhibition. Two μCi of ^{14}C-amino-acid mixture ('CFB 104', Radiochemical Centre, Amersham) with a specific activity of 45 mCi/m atom was added to each culture 13 min after the first infection. Three ml. of per cent 'casamino-acids' mixture (Difco) was added to each sample at 30 min, and the infected cells were concentrated by low speed centrifugation 35 min following infection. The pellets were drained and directly resuspended in "final sample" buffer (see gel electrophoresis).

[14]C-labelled phage and head particles. Ten ml. cultures were grown and infected as described above (^{14}C-labelled lysates). The double mutant (B255–E18) in genes 10 and 18 was used for production of tailless heads. Ten μCi of the ^{14}C-amino-acid mixture was added 13 min after the first infection to each culture. The infected cells were concentrated by a low speed centrifugation 35 min after infection and the pellet was resuspended in 1 ml. neutral phosphate buffer containing 10⁻³ M MgSO₄, 20 μg deoxyribonuclease and a drop of chloroform. The pellet was resuspended by repeated pipetting, and incubated for 15–30 min at room temperature before layering on a CsCl step gradient[2]. The latter was prepared in tubes for a Spinco 'SW50' rotor, with 0·8 ml. layers, and the following densities starting at the bottom of the tube: 1·55, 1·46, 1·38 and 1·29 g/cm³. Furthermore, a 10 per cent sucrose solution (in neutral phosphate buffer and 10⁻³ M MgSO₄) was layered on the last CsCl step to prevent precipitation of soluble proteins at the CsCl interface. Centrifugation was for 1 h at 40,000 r.p.m. The phage and heads, which form a sharp visible band two-thirds down the tube, were collected through the bottom of the tube and dialysed against water. The band containing the heads was always viscous, indicating that the heads lost their DNA in CsCl although the DNA was still confined within the band. Occasionally the heads lost their DNA before centrifugation particularly in more concentrated lysates, which had to be treated with deoxyribonuclease for a much longer time. "Early labelled" phages were prepared identically, but the label was added 1 min and chased 6 min after infection by the addition of 3 ml. of 3 per cent 'casamino-acids'.

"Ghosted" phage particles and supernatant of "ghosted" phage particles. Crystalline NaCl was added to ^{14}C-labelled purified phage particles in water to a final concentration of 5 M. The preparation was then repeatedly frozen in a solid CO₂-acetone bath and thawed in warm water. This procedure releases the DNA and its internal proteins from the phage head. The sample was dialysed against neutral phosphate buffer containing 10⁻³ M MgSO₄ and the DNA digested by the addition of a small amount of crystalline deoxyribonuclease. The ghost was finally separated from the soluble proteins (internal proteins) by centrifugation and layered on a step gradient identical to that already described. Centrifugation was at 35,000 r.p.m. for 1 h. The top of the gradient containing the internal proteins was collected with a pipette and the ghosts, which band at an approximate density of 1·3 g/cm³, were collected through the bottom of the tube. SDS at a final concentration of 2 per cent was added to both samples and the samples were dialysed against water containing 2 per cent SDS.

Gel electrophoresis. Gels containing 3 per cent (stacking gel), 8·0 per cent or 10 per cent acrylamide were prepared from a stock solution of 30 per cent by weight of acrylamide and 0·8 per cent by weight of N,N'-bis-methylene acrylamide. The final concentrations in the separation gel were as follows: 0·375 M Tris-HCl (pH 8·8) and 0·1 per cent SDS. The gels were polymerized chemically by the addition of 0·025 per cent by volume of tetramethylethylenediamine (TEMED) and ammonium persulphate. Ten cm gels were prepared in glass tubes of a total length of 15 cm and with an inside diameter of 6 mm. The stacking gels of 3 per cent acrylamide and a length of 1 cm contained 0·125 M Tris–HCl (pH 6·8) and 0·1 per cent SDS and were polymerized chemically in the same way as for the separating gel. The samples (0·2–0·3 ml.) contained the final concentrations ("final sample buffer"): 0·0625 M Tris–HCl (pH 6·8), 2 per cent SDS, 10 per cent glycerol, 5 per cent 2-mercaptoethanol and 0·001 per cent bromophenol blue as the dye. The proteins were completely dissociated by immersing the samples for 1·5 min in boiling water[5]. Electrophoresis was carried out with a current of 3 mA per gel until the bromophenol blue marker reached the bottom of the gel (about 7 h). The proteins were fixed in the gel with 50 per cent trichloroacetic acid (TCA) overnight, stained for 1 h at 37° C with a 0·1 per cent Coomassie brilliant blue solution made up freshly in 50 per cent TCA. The gels were diffusion-destained by repeated washing in 7 per cent acetic acid. Autoradiograms of gels were prepared by a modified version (U. K. L. and J. V. Maizel, unpublished) of Fairbanks et al.[20] (autoradiograms are shown in Figs. 1–7).

origin of which was determined are labelled in Fig. 1, and their molecular weights are listed in Table 1. At least 28 bands can be distinguished in the autoradiogram of radioactively labelled, purified phage particles (Fig. 1c). The 28 proteins found in dissociated phage particles do not include proteins with molecular weights less than about 15,000. Those proteins are not sieved in gels of 10 per cent acrylamide (unpublished results of U. K. L. and J. V. Maizel) and migrate with the marker dye. In gels of higher acrylamide concentration another three low molecular weight proteins have been separated (results not shown). The largest protein in the phage has an approximate molecular weight of at least 120,000 and no label stays at the top of the gel, indicating complete dissociation of the particles by the method used.

Eleven of these 28 proteins are found in the purified head preparation (Fig. 1b). The remaining 17 proteins absent from the head but present in the whole phage pattern are presumably structural proteins of the tail and tail fibres. The complete absence of these proteins from the head gel pattern indicates the high degree of purity of the preparation. The classification of the proteins into tail and head components may not be valid for proteins making up the head to tail junction.

Only two minor proteins besides the major components P23 were found by others[2,3] in phage capsids purified in the same way but fractionated in urea gels. The larger number of proteins found in SDS gels is to be expected, for at least 46 genes are known to affect T4 morphogenesis[4,7] although the proteins of these genes may not all be incorporated into the particle.

The gel pattern of a total lysate of wild type infected cells radioactively labelled at late times is presented in Fig. 1a. Note that most of the resolved proteins that

are synthesized late in infection are structural phage components. A few rather intense bands (X₁ and X₂), however, are completely missing in the phage particles, also demonstrating that the separation of the phage particles from the soluble proteins is complete.

When phages are subjected to osmotic shock a number of proteins are released[11]. The principal one is IP*, an internal protein (Fig. 1). It is present in complete phage particles, is extracted from the phage particles by freezing and thawing in high salt concentrations and is quantitatively recovered in the supernatant. The supernatant fraction also contains many minor components which are found both in phages as well as phage ghosts, indicating that the

Table 1. MOLECULAR WEIGHTS OF PHAGE PROTEINS DETERMINED BY COMPARING THEIR MOBILITY IN SDS GELS WITH THOSE OF MARKER PROTEINS WITH KNOWN MOLECULAR WEIGHTS[10,12]

Gene product	Observed value in SDS gels	Published value
P18	69,000	50,000[9]
P20	63,000	
P23	56,000	
P23*	46,500	46,000[11]
P24	45,000	
P24*	43,500	
P22	31,000	
IP	23,500	
IP*	21,000	
P19	18,000	

The following marker proteins were used: serum albumin (68,000), γ-globulin, heavy chain (50,000), ovalbumin (43,000), γ-globulin, light chain (23,500) and TMV (17,000), to calibrate 10 per cent acrylamide gels. In this improved gel system the distances of migration of the various marker proteins relative to the distance of migration of bromophenol blue are also a linear function of the logarithm of the molecular weight of the marker proteins, as has been described[10,12]. Radioactively labelled phage proteins were mixed with unlabelled marker proteins before electrophoresis; the distance of migration of the phage proteins was determined from the autoradiogram and those of the marker proteins from the stained gel. It is assumed that the phage proteins also separate in SDS gels solely according to their molecular weight.

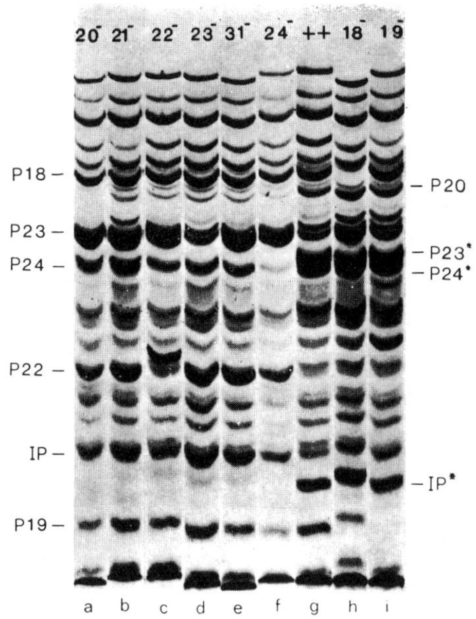
20⁻ 21⁻ 22⁻ 23⁻ 31⁻ 24⁻ ++ 18⁻ 19⁻
P18 –
P23 –
P24 –
P22 –
IP –
P19 –
– P20
– P23*
– P24*
– IP*
a b c d e f g h i

Fig. 2. Identification of gene products on 10 per cent acrylamide gels. The lysates were prepared as described in Fig. 1 and analysed on 10 per cent acrylamide gels. *a*, Gene 20-defective lysate, mutant N50; *b*, gene 21-defective lysate, mutant N90; *c*, gene 22-defective lysate mutant B270; *d*, gene 23-defective lysate, mutant H11; *e*, gene 31-defective lysate, mutant N54; *f*, gene 24-defective lysate, mutant N65; *g*, lysate from wild type infected cell; *h*, gene 18-defective lysate, mutant E18; *i*, gene 19-defective lysate, mutant E1137. Mutants N90 (gene 21), E18 (gene 18) and E1137 (gene 19) carried second mutations in gene 10 (mutant B255). The following amber fragments may be detected. A rather intense band just below P23 is consistently seen in 18-defective lysates, and is presumed to be the amber fragment of mutant E18 in gene 18. Furthermore, the gel pattern of defective lysates of all double mutants in genes 18–10, 21–10, and 19–10 possess a band just above P23 which is probably the amber fragment of mutant B255 in gene 10. It is striking that the amber fragment of mutant B270 gene 22 behaves anomalously in the gel: it migrates more slowly than the wild type product, although SDS gels are known to separate on the basis of molecular weights. This anomalous behaviour of certain proteins will be discussed elsewhere (U. K. L. and J. V. Maizel).

separation of the released proteins and the ghosted particles by centrifugation was not complete. Some of these proteins, however, are extracted into the supernatant quantitatively by the freezing and thawing procedure.

Only the protein IP*, and perhaps some low molecular weight proteins which migrate with the marker dye, are structural phage components synthesized at early times (early proteins). This can be seen in the gel pattern of purified phage labelled at early times only (Fig. 1*f*). IP* is also labelled at late times. All the other structural phage components are synthesized only late in infection. Some of the principal late proteins show up on the autoradiogram, probably because of residual incorporation of radioactive amino-acids following the chase with unlabelled amino-acids.

Identification of Gene Products

The products of genes 18, 19, 20, 22, 23 and 24 were identified by comparing the gel pattern of extracts of cells infected with wild type phage with those infected with amber mutants in various genes. The identification of the tail and tail fibre proteins will be described later (J. King and U. K. L.). Amber mutations produce only fragments of the protein chain of the mutant gene on infection of restrictive bacteria[1]. These fragments migrate differently in the gel from the complete proteins. The molecular weights of the proteins identified are listed in Table 1. Fig. 2 is the autoradiogram from dried and sliced gels for various mutants. The product of gene 20 is identified by its absence in the gel pattern of a 20-defective* lysate (Figs. 2 and 3*a*), and the product of gene 22 by its absence in the gel pattern of a 22-defective lysate (Figs. 2 and 3*c*). Note the amber fragment of mutant B270 in gene 22. This fragment was also identified by Hosoda and Levinthal[12] in urea gels (see legend to Fig. 2).

The product of gene 23 is easily identified by its absence in the gel pattern of a 23-defective lysate (Fig. 2*d*). It can also be seen that P23 overlaps with two minor tail components. If the 23-defective lysate is analysed on gels of lower acrylamide concentration another important observation is made. A band, P23*, which is detected in variable amounts in the other head defective lysates, is completely missing in the 23-defective lysate (Fig. 3*d*). This band, P23*, overlaps with P24 in Fig. 2, but is better separated from P24 in less concentrated gels (Fig. 3). As with the product P23, the product of P24 was identified by its absence in the gel pattern of a 24-defective lysate (Figs. 2*f* and 3*e*).

So far, no missing bands have been found in the gel patterns of 31 or 21-defective lysates (Fig. 2*b* and *e*), but analysis of the 21-defective lysate on gels of lower acrylamide concentration (Fig. 3*b*), which resolves higher molecular weight proteins better, clearly shows that a band is missing. This protein, however, is the product of gene 10, a baseplate gene. The mutant N90 in gene 21 in fact carries a second mutation in gene 10 (mutant B255).

In comparing the gel pattern of the head-defective lysates (Fig. 2*a*–*f*) with that of wild type (Fig. 2*g*), further important differences are observed, which shed light on the precursor–product relationship of the head components.

The principal fraction of the gene 23 product has a molecular weight of 56,000 in all the head-defective lysates, but in wild type or tail-defective lysates it appears at the position of P23*, with a molecular weight of 46,500. Small but significant amounts of P23* are also observed in head-defective lysates (Fig. 3*a*–*f*). In lysates prepared identically about 20 per cent of the total P23 is converted to P23* in the 20-defective lysate, 10 per cent in 21 and 2–3 per cent in 22, 24 and 31-defective lysates, as determined from densitometer tracing of the autoradiographs.

The bands P22 and IP are both absent or considerably less intense in the wild type gel pattern (each overlaps with two other proteins). A new band, IP*, is seen at the bottom of the gel, which is completely missing in the head defective lysates (Fig. 2*a*–*f*).

The band P24, which is found in all head-defective lysates is missing in the wild type pattern, but a new band, P24*, which migrates slightly faster is observed. This is difficult to visualize in Fig. 2, but will become evident in Fig. 6.

Also included in Fig. 2 is the gel pattern of two tail defective lysates. The product of gene 18 (mol. wt 69,000), the principal protein of the tail sheath[13], is identified by its absence in an 18-defective lysate (Fig. 2*h*) and the product of gene 19 (mol. wt 18,000) by its absence in a 19-defective lysate (Fig. 2*i*). P19 is thought to be the chief component of the tail tube[14]. This demonstrates that the differences in the head proteins are not related to tail attachment, for the gel patterns of the tail-defective lysates are identical with that of wild type.

Evidence will be presented that the proteins P23, P22, P24 and IP are cleaved in wild type infected cells and are precursors to proteins P23*, P24* and IP* found in the final head structure. (The cleavage product of P22 was not detected.) This precursor–product conversion is strongly inhibited by mutations in genes 20, 21, 22, 23, 24 and 31. Two important conclusions can be drawn: (*a*) the aberrant head-related structures single and

* Henceforth, lysates of cells infected at restrictive conditions with amber mutants in various genes will be referred to as, for instance, a "21-defective lysate", where the amber mutant used was in gene 21.

multi-layered polyheads[4,5,15], τ-particles[4,5,16] and lumps[6] known to be produced in these mutant infected cells—chiefly consist of the precursor protein P23; (*b*) because P22 is not cleaved in these mutant infected cells, but is required for polyhead and τ-particle formation, as has been established by genetic means[5], it is strongly suggested that P22 is incorporated into these structures as such.

Kinetics of the Cleavage Reactions

The following experiments were designed to study the precursor relationship of the proteins P23, P24 and IP with P23*, P24* and IP*, respectively. In this experiment the infected cells were pulse-labelled with radioactive amino-acids for a short period (1 min) and the modification of the various proteins was then followed by analysis of the samples taken at intervals in SDS gels. The results of autoradiography of the dried gels are presented in Fig. 4.

Cleavage of P23. It is readily seen in Fig. 4 that most of the 23 protein is at the position of P23 immediately following the pulse of the radioactive amino-acids. It then rapidly disappears, and a new band, P23*, appears simultaneously. That P23 is cleaved and gives rise to P23* is suggested by the fact that both are principal components and this is reinforced by the absence of P23 and P23* from the gel patterns of a 23 defective lysate (Fig. 3*d*). The kinetics of the cleavage reaction P23→P23* are plotted in Fig. 5*a*. Cleavage is very rapid: about 50 per cent of the precursor is cleaved within the first 2 min following chase of the label. P23* appears at about the same rate, in a satisfactorily correlated way. The total labelled protein in P23 and P23* is also plotted in Fig. 5*a*. The total label increases during the first minute, which reflects the completion time of the chase of the labelled amino-acid, but finally the total falls off by 25–30 per cent. This final decrease of the total labelled protein can be nicely explained, for a cleavage from about 56,000 to 46,500 corresponds to a loss of about 20 per cent by weight of protein.

Cleavage of P22 and IP. In the pulse-chase experiment of Fig. 4, two other protein bands, P22 and IP, disappear with time. (The disappearance of P22 in wild type infected cells

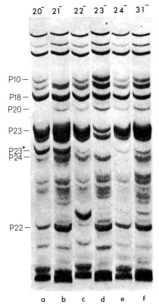

Fig. 3. Identification of gene products on 8 per cent acrylamide gels. ^{14}C-labelled lysates were prepared as described (Fig. 1) and analysed on 8 per cent acrylamide gels. *a*, Gene 20-defective lysate, mutant N50; *b*, gene 21-defective lysate, mutant N90; *c*, gene 22-defective lysate, mutant B270; *d*, gene 23-defective lysate, mutant H11; *e*, gene 24-defective lysate, mutant N65; *f*, gene 31-defective lysate, mutant N54.

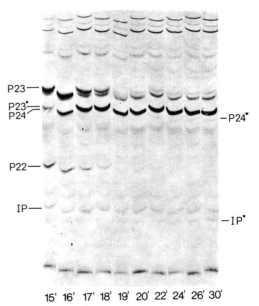

Fig. 4. Cleavage of products of genes 22, 23, 24 and protein IP (10 per cent acrylamide gels). A 10 ml. culture grown at 37° C was infected with a double mutant defective in genes 10 and 18 (mutant B255 and E18) as described (Fig. 1). The radioactive amino-acid mixture (10 μCi) was added 14 min after the first infection, and chased 1 min later with an excess of unlabelled amino-acids (final concentration 1 per cent). The chase of the label was verified by measuring the counts in the total TCA precipitable proteins. One ml. samples were prepared at intervals after the chase and immediately frozen in a solid CO₂-acetone bath. SDS was added after thawing to a final concentration of 2 per cent and the samples were carefully dialysed into 2 per cent SDS in water. The samples were finally mixed with an equal volume of twice concentrated "final sample buffer" and boiled for 1 min before electrophoresis. The sampling time is indicated at the bottom of the gels. All the preliminary experiments were done with wild type phage, but this experiment was performed with the double mutant in genes 10 and 18 in view of plans for future experiments.

has also been observed by M. Showe, personal communication.) A new band, IP*, appears at the bottom of the gel pattern. The kinetics of these cleavage reactions are plotted in Fig. 5*b*. P22 disappears with approximately the same initial rate as P23: about 50 per cent is cleaved 2–3 min following chase of the labelled amino-acids. I have not found a band in the gel pattern which may be derived from P22. Hosoda and Levinthal[12] reported indirect evidence that P22 is a structural phage component, but they considered the possibility that P22 might become altered during head formation.

Evidence for the precursor–product conversion IP→IP* is provided by the observation that the disappearance of IP and the appearance of IP* is coordinated in time (Fig. 5*b*). Furthermore, the total label lost at the IP position is recovered in the IP* band. The reaction is slower than that of P23 and P22. Only about 50 per cent of the total label at the IP position disappears. This could be explained either by another protein band overlapping with the IP band or synthesis in excess. IP* cannot arise from P22. IP* must be derived from a precursor which is synthesized early, for I have shown (Fig. 1) that IP* is strongly labelled in the "early labelled" phage preparation. P22, however, is reported to be synthesized at late times only[12], which I have confirmed. The precursor conversion of IP→IP* has also been observed in a pulse-chase experiment in which the label was added between 4 and 5 min following infection, thus labelling only early proteins (results not shown). The band IP was easily recognized in these gels and the disappearance of IP and the appearance of IP* were again correlated in time. This quantitative agreement and the absence of other unaccounted changing bands in the gel pattern support the argument for the IP→IP* relationship. Of course, a final proof awaits chemical analysis.

It was also observed that, although the pulse was performed between 4 and 5 min after infection, cleavage of IP starts only

4

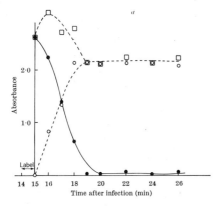

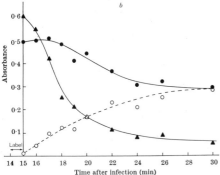

Fig. 5. Kinetics of cleavage of P23, P22 and IP. The kinetics of cleavage were measured using a microdensitometer (double beam recording microdensitometer, Joyce-Loebl) to record the autoradiogram. The exposure of the autoradiogram was chosen so that the absorbance of the band to be measured did not exceed 1 unit. The abscissa represents the integrated absorbance over the relevant peaks. *a*: ●—●, integrated absorbance over the P23 peak; ○- - -○, integrated absorbance over the P23* peak; □- - -□, total absorbance in P23 and P23*. *b*: ▲—▲, integrated absorbance over peak P22; ●—●, integrated absorbance over peak IP; ○- - -○, integrated absorbance over peak IP*.

at late times (after 17 min). Phage assembly starts at about this time and it is therefore thought that the cleavage of IP is linked to phage assembly.

Cleavage of P24. P24* (mol. wt 43,500) appears coordinately with P23* and P22* (Fig. 4) (the kinetics are not plotted). The precursor product relationship P24→P24* is more difficult to demonstrate, because P24 migrates only slightly faster than P23*, but the following experiment proves that P24 is missing in a pulse-chase wild type lysate. P24 separates somewhat better from P23* in a gel of lower concentration. The samples of the pulse-chase experiment were analysed on 8 per cent acrylamide gels and four time points are presented in Fig. 6. P24 is easily distinguished from the small amount of P23* existing immediately after the chase of the radioactive label (Fig. 6, 15 min). P24 disappears at the same time as P24* appears while P23* increases. One might argue that P24 is obscured by the heavy P23* band. This possibility was excluded by adding a lysate (23-defective lysate) containing P24 to the final samples of the pulse-chase experiment. P24 was then detected and it can be concluded that measurements of P24 are reliable, thus showing that P24 most likely gives rise to P24*. The integrated absorbance values over these two bands are about equal, but the small loss of protein weight by the P24→P24* reaction is not likely to be detected by the densitometric measurements. P24 could not give rise to IP*, because the total label in IP* is two or three times larger than that of P24 and P24 is synthesized late. Moreover, the precursor relationship of P24→P24* is considerably strengthened by recent observations on head maturation genes (my unpublished results). P24 does not seem to be cleaved at all in 50-

defective cells and, indeed, no P24* is found. Cleavage of P23, P22 and IP does occur in 50-defective cells, although at a reduced rate.

Fate of the Small Fragments

Where are the small fragments of these cleavage reactions? The expected molecular weights for the small fragments stemming from P23, IP and P24 would be about 10,000, 2,500 and 1,500 respectively. Peptides of this size are not sieved on 10 per cent acrylamide gels and migrate with the marker dye (unpublished results of U. K. L. and J. V. Maizel). Attempts to find at least the 10,000 molecular weight fragment from P23 on gels of higher acrylamide concentration have failed. Possibly the fragments are further broken down to undetectable sizes. Fragments of P22 also have not been detected. Acid-soluble components which are derived from an acid-insoluble precursor are known to exist in T4 infected cells[17]. Two are associated with the phage particle and they are released with the DNA from the head upon osmotic shock[17]. The genetic determinant of one of these internal peptides has been mapped recently and lies in the neighbourhood of genes 20 and 21 (ref. 18). My results definitely rule out gene 20, which is incorporated unmodified into normal phage. Unfortunately, I have not discovered the product of gene 21 in the gel pattern. These results do not rule out the possibility that one of the internal proteins is derived from the small cleavage fragment of P22, P23 or P24. The appearance of the internal peptides seems indeed to be coordinated with the cleavage reactions of P23, P22 and P24. Genes 20, 21, 22, 23, 24 and 31, which affect the cleavage of P23, P22 and IP, are known also to affect the appearance of the internal peptides[17].

It has been pointed out to me by S. Brenner and A. Stretton that the cleavage point must occur at the N-terminal end of the P23 protein. In establishing the co-linearity of gene 23 and its polypeptide[1], they observed peptides in 23-defective lysates which contain the amber fragments, but these are absent in wild type lysates or in purified phage particles. These peptides may be derived from the N-terminal end of the amber fragment, which is cleaved off in the protein P23*. These observations also suggest that the small cleavage fragment, with an expected molecular weight of about 10,000, is fragmented to even smaller pieces.

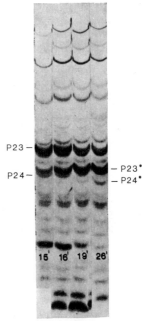

Fig. 6. Cleavage of the product of gene 24 (8 per cent acrylamide gels). Some samples (15, 16, 19 and 26 min) of the pulse chase experiment of Fig. 4 were analysed on 8 per cent acrylamide gels. Only the relevant part of the gel pattern is shown.

Cleavage occurs in a Large Structure

The observation that all the proteins P20, P21, P22, P23, P24 and P31 are required for efficient cleavage of P23, P22, P24 and IP suggests that these precursor proteins aggregate first to form an oligomeric structure, and are cleaved subsequently, rather than being cleaved first, and then assembled. The following experiment supports this view. The experiment is based on the observation that most of the precursor proteins are soluble and monomeric in SDS at room temperature, but that phage particles are not totally disrupted, as is also true in urea[12]. The gel pattern of purified phage treated with SDS at room temperature is compared with completely degraded phage, boiled for 1 min in SDS (Fig. 7a and b). Only a small fraction of P23* is extracted from phage treated with SDS at room temperature. Most of the proteins stay at the top of the gel or enter the gel as high molecular weight aggregates. Some proteins are not extracted at all. Note, however, that a few proteins are almost quantitatively extracted from the phage particles. The samples from the pulse-chase experiment treated with SDS at room temperature only are shown in Fig. 7. P23 disappears with time but no P23* appears, suggesting that P23 enters an SDS-resistant structure before being cleaved. Of course, these experiments cannot rule out the possibility that P23* is converted to an SDS-resistant structure so rapidly that it is not detected. A high molecular weight protein appeared at the top of the gel, which could be an aggregate of P23 but only accounts for part of the label which disappears from P23. Most of the label stays at the top of the gel. The molecular weight of this structure must therefore be greater than 300,000, for such a molecular weight is excluded from these gels (unpublished results of U. K. L. and J. V. Maizel). It is possible that this structure has a capsid-like shape. IP* and P24* are not resolved in these gels. This is because of the high salt concentration in these samples ('M9' growth medium) which impairs the resolution of the gel in the low molecular weight region.

Maturation of the Head

My experiments demonstrate that the assembly of the head of bacteriophage T4 is not a simple, straightforward self-assembly, because several structural proteins are chemically altered at some stage of assembly. The uncleaved precursor protein P23 can, however, be polymerized into single and multilayered polyheads and τ-particles, if its cleavage is blocked as a result of mutation.

Investigations of genes 2, 4, 13, 14, 16, 17, 49, 50, 64 and 65, which supposedly control late steps in head formation, are in progress. It is interesting that the cleavage of P23, P22, P24 and IP seems to be normal in cells infected with phage carrying mutations in these genes with the exception of genes 2, 50 and 64 (my unpublished results).

Why are these structural head components cleaved ? The finding that IP gives rise to an internal IP* sheds light on a possible consequence of the cleavage reactions. Internal proteins are thought to bind to DNA and the cleavage reactions may possibly trigger the necessary DNA–protein interactions, which result in orderly packing of the DNA within the shell of proteins. The following model may be proposed. The proteins P20, P21, P23, P24, P31 and IP form an intermediate structure (SDS-resistant) which combines with an end of a DNA strand. Cleavage of P23, P22, P24 and IP proceeds from the DNA attachment site. During this process more and more DNA binding sites may be formed at the inside of this structure, perhaps by the formation of IP*, thus winding up the DNA strand successively. The small acid-soluble peptides formed during this process may also interact with the packed DNA[17] to neutralize charges. My results do not decide whether DNA packing proceeds simultaneously with the polymerization of the head membrane or follows thereafter.

I thank Dr A. Klug for encouragement and facilities, my colleagues J. King, J. Maizel and S. Altman for many valuable suggestions (J. Maizel in particular for advice on gel techniques and J. King for help in preparing the manuscript), and S. Brenner (among others) for critically reading the manuscript. U. K. L. holds an EMBO fellowship.

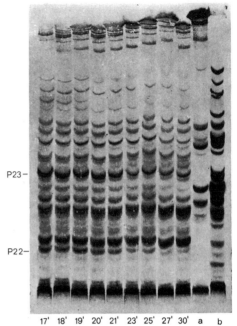

Fig. 7. Chase of P23 into a product stable to SDS at room temperature. A culture was infected and pulse labelled as described in Figs. 1 and 3 with the following changes. The culture was grown at 30° C, infected with wild phage, and labelled for 2 min from 15–17 min after infection. The samples were mixed at room temperature with an equal volume of twice concentrated "final sample buffer" without previous dialysis. The samples were then directly applied to the gels without being boiled. The sampling time is indicated at the bottom of the gels. As a control a purified phage preparation is analysed. *a*, Phage treated in SDS at room temperature only; *b*, phage boiled in SDS for 1·5 min.

Note added in proof. During the preparation of this manuscript I was informed that the alteration of P23 has also been observed by other workers: E. Kellenberger and C. Kellenberger-van der Kamp, *FEBS Lett.*, 8, 3, 140 (1970); R. C. Dickson, S. L. Barnes and F. A. Eiserling, *J. Mol. Biol.* (in the press); and J. Hosoda and R. Cone, *Proc. US Nat. Acad. Sci.* (in the press).

Received May 7, 1970.

[1] Sarabhai, A. S., Stretton, A. O. W., Brenner, S., and Bolle, A., *Nature*, 201, 13 (1964).
[2] Kellenberger, E., *Virology*, 34, 549 (1968).
[3] Baylor, M. B., and Roslansky, P. F., *Virology*, 40, 251 (1970).
[4] Epstein, R. H., Bolle, A., Steinberg, C. H., Kellenberg, E., Boy de la Tour, E., Chevalley, R., Edgar, R. S., Susman, M., Denhardt, G. H., and Lielausis, J., *Cold Spring Harbor Symp. Quant. Biol.*, 28, 375 (1963).
[5] Laemmli, U. K., Molbert, E., Showe, M., and Kellenberger, E., *J. Mol. Biol.*, 49, 99 (1970).
[6] Laemmli, U. K., Beguin, F., and Gujer-Kellenberger, G., *J. Mol. Biol.*, 47, 69 (1970).
[7] Edgar, R. S., and Wood, W. B., *Proc. US Nat. Acad. Sci.*, 55, 498 (1966).
[8] Davis, B. J., *Ann. NY Acad. Sci.*, 121, 404 (1964).
[9] Maizel, J. V., *Fundamental Techniques of Virology* (edit. by Habel, K., and Salzman, N. P.), chap. 32, 334 (Academic Press, New York, 1969).
[10] Shapiro, A. L., Viñuela, E., and Maizel, J. V., *Biophys. Biochem. Res. Commun.*, 28, 815 (1967).
[11] Minigawa, T., *Virology*, 13, 515 (1961).
[12] Hosoda, J., and Levinthal, C., *Virology*, 34, 709 (1968).
[13] King, J., *J. Mol. Biol.*, 32, 231 (1968).
[14] King, J., *FEBS Symp.* (edit. by Ochoa, S., Nachmansohn, D., Asensio, C., and Heredia, C. F.), 21, 156 (Academic Press, London, 1970).
[15] Favre, R., Boy de la Tour, E., Segre, N., and Kellenberger, E., *J. Ultrastruct. Res.*, 13, 318 (1965).
[16] Kellenberger, E., Eiserling, F. A., and Boy de la Tour, E., *J. Ultrastruct. Res.*, 21, 335 (1968).
[17] Eddleman, H. L., and Champe, S. P., *Virology*, 30, 471 (1966).
[18] Sternberg, N., and Champe, S. P., *J. Mol. Biol.*, 46, 337 (1970).
[19] Laemmli, U. K., and Eiserling, F. A., *Molec. Gen. Genet.*, 101, 333 (1968).
[20] Fairbanks, G., Levinthal, C., and Reeder, R. H., *Biochem. Biophys. Res. Commun.*, 20, 393 (1965).
[21] Larcom, L., Bendet, I., and Mumma, S., *Virology*, 41, 1 (1970).
[22] Weber, K., and Osborn, M., *J. Biol. Chem.*, 244, 4406 (1969).

Printed in Great Britain by Fisher, Knight & Co., Ltd., St. Albans.

Assembly in biological systems

E. KELLENBERGER

Abteilung für Mikrobiologie, Biozentrum der Universität Basel, Basel

Abstract The characteristic features of protein assemblies are discussed, that is: (*a*) the type and strength of interactions; (*b*) the regular crystalline arrangement in spatially folded and deformed two-dimensional crystal sheets; and (*c*) the characteristic determination of form (size and shape).

To obtain a better understanding of this determination of form two polymorphic systems are further discussed:

(1) The protein subunit of tobacco-mosaic virus is form-determining and needs no further protein for its assembly into rods. As Durham, Klug, and collaborators have shown, at least two physical variants of the rods occur—the helix and the stacked disc—depending on the ionic strength and pH. Both are relatively stable products which, at the end, are rather resistant towards dissociation.

(2) More complex systems (bacteriophages) provide examples of genetically determined polymorphisms. In phage T4, several aided assemblies are observed. The one on head-related particles is particularly interesting. According to what the active 'helping' genes are, tubular forms and tau particles (including their variants) are formed, made essentially of the 'uncut' gene product P23 (molecular weight 61 000). The capsids of isometric, normal (prolate) and giant heads are made of P23^c, the 'cut' form of the same gene product (mol. wt. 49 000).

A preliminary report is given of experiments by R. Bijlenga, D. L. Anderson and E. Lüscher which suggest that a species of tau particles is a precursor to phage heads and that the cutting of P23 to P23^c occurs *in situ* on the particle in formation. Some involvement of phage DNA in this process is suspected and experimentally suggested.

Many enzymes, virus shells, bacterial flagella, bacterial pili, microtubules and other biological structures are assemblies of protein subunits. Each differs in the number of different protein species involved and in the number of identical subunits used.

Oligomeric enzymes are assemblies of from two to six identical protomers, while multifunctional enzymes are composed of as many non-identical subunits as there are functions, each subunit appearing mostly once, but sometimes twice or more. Larger structures may also be made up of a large number of identical protomers (bacterial flagella, pili, or some viruses such as tobacco mosaic virus [TMV]), or—in more complex systems—contain different numbers of various protein species. In such complex cases as bacteriophages T4 and lambda we can also distinguish morphologically distinct structural parts, most of which consist mainly of one merid, that is of an ordered assembly of a relatively large number of identical protomers.

GENERAL FEATURES OF PROTEIN ASSEMBLIES, PARTICULARLY MERIDS

In view of the previous contributions to this symposium, it may be interesting to outline some of the characteristic and unique features of protein assemblies. In contrast to what we have heard about biopolymers, protein assemblies are not covalently bonded, although the rare occurrence of disulphide bonds has not yet been ruled out. It is generally believed that weak interactions such as hydrophobic, electrostatic and hydrogen bonding are prevalent. These are the same types of interactions as occur in the tertiary structures of proteins.

Within or between merids the bonding of protomers is of variable strength In some cases intramerid bonds are so strong that they can be opened only by denaturants at concentrations which also fully denature proteins. For example the following merids, enumerated in order of decreasing bond strength, are found in bacteriophage T4 (To et al. 1969): contracted tail sheath, tail fibre, head capsid, tail tube. The known intermerid bonds are much weaker and—in the above series—follow after the tail tubes. The order in the above series is independent of the chemical nature of the denaturant, in other words of organic solvent concentration (replacement of water: hydrophobic bonds), and of the nature and concentrations of ions (electrostatic interactions). The action of high ion concentrations and that of organic solvents are generally combined (acetic acid, guanidine hydrochloride, urea pH). Ionic strength and pH alone are frequently sufficient to disrupt bonds, as well as urea alone if at adequate concentration. Ionic detergents at room temperature were found to be active only in disrupting intermerid bonds of T4, inducing for example the sheath contraction. For rupture of the stronger interactions within merids, sodium dodecyl sulphate (SDS) has to be used in conjunction with high temperature— conditions in which proteins are likely to be fully denatured.

These and other observations strongly suggest that intramerid bonds are usually mixtures of several different types of weak interactions: the pattern of their arrangement within the interacting area would account for the very high specificity.

The high flexibility of most large protein assemblies, as deduced from electron microscopy, therefore cannot be accounted for by single bonds with a flexible angle. Since the nature and the strength of bonding within the interacting area are similar to those of intramolecular bonding, one must necessarily conclude that the deformation associated with this flexibility is not confined to the interaction area between protomers but extends into the tertiary structure of the protomers themselves. For the same reason protomers in quasi-equivalent positions in virus shells (Caspar & Klug 1962) must have slightly modified conformations. In the cases of polymorphism, one species of protomer can be assembled into several distinct morphological variants. In these variants, the bonding between protomers is frequently of different strengths and thus it can be predicted that these protomers will also have different conformations.

The following model is therefore postulated for such protein assemblies: on interaction, the protomers adjust their tertiary structure so as to form a continuous hydrophobic core which is not interrupted between protomers.

Many studies, based mainly on the work and influence of Klug, now make it certain that the merids are regular arrays which may be considered as two-dimensional 'biocrystals' folded and bent into three dimensions. I am deliberately introducing the term 'biocrystals' here because of the unique features which distinguish them from 'normal' crystals:

(1) These two-dimensional biocrystals have the property that in at least one dimension, and more frequently in both, they grow only to a precisely determined specific size.

(2) With very few exceptions (which in reality are abnormal non-physiological cases), finished biocrystals are not in dynamic equilibrium with the solution of their protomers. In a physiological environment the equilibrium shifts to a unidirectional reaction, apparently without the aid of a catalytic enzyme.

These two unique and specific features are certainly linked to high cooperativity within the biocrystals. Again, it is very difficult to explain such high cooperativity without introducing conformational changes in each of the component protomers.

In the discussion that follows, I shall select observations concerning polymorphism at different levels and introduce the concepts of sequential regulation through successively induced conformational changes.

Protomer assembly ranges from the simplest case of self-assembly, where all the information for the form of the final assembly product is contained in the geometry and bonding properties of the protomer itself, to the more complex cases, where the final form is attained only through the previous or concomitant help of other proteins (morphopoietic gene products: Kellenberger 1966). A protomer might be composed of several different proteins, but this case is obviously trivial and can be brought back to that of self-assembly. The situation becomes fundamentally different when proteins are involved that are *not* part of the protomer. Whether a protein found in an assembly is a helper protein or a constituent part of a protomer therefore has to be decided by experiments. If the numbers of molecules of two or more species of proteins involved in a complex structure are equal or integer multiples of one another, then at least one must suspect that together they might form a single protomer species. A real helper protein is certainly involved, for example, in all those cases where it can be demonstrated that it has only a transient function and is no longer present in the end-product. (For discussion of these cases see Kellenberger & Edgar 1971.)

ASSEMBLIES OF FULLY FORM-DETERMINING SUBUNITS

Among the representatives of this case, TMV has been studied most adequately. Purified virus protein can be reassembled into rods. This process has been investigated in thermodynamic terms and it was found that the bonding

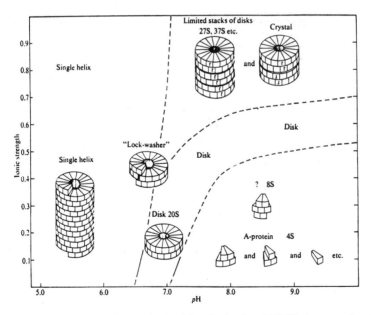

FIG. 1. Preponderant forms of assembly of subunits of TMV virus as a function of pH and ionic strength (Durham & Klug 1971)

is preponderantly of the hydrophobic type; the energies involved are related to organized water which, before aggregation, has to be 'melted' away from the interacting sites. The energy involved is indeed comparable to that of melting ice (Stevens & Lauffer 1965). Consequently, aggregation occurs at high temperature and dissociation at a lower temperature.

Reassembled rods are in chemical equilibrium with their subassemblies (Lauffer & Stevens 1968). A complete study of the equilibrium between different subassemblies and rods, as a function of pH and ionic strength, has been published recently (Durham & Klug 1971; Durham *et al.* 1971), and Dr Klug will discuss these results here (Klug 1972). The situation is summarized by the phase diagram of Fig. 1. Most important is the finding that discs are the subassembly through which either stacked discs or helical rods are assembled. The two-layer discs are formed by two rings of the same polarity, each consisting of 17 subunits, while in the helical form the rings are open and slightly changed, to contain only 16.34 subunits per turn. This *physical polymorphism* could well be related to the two carboxyl groups whose abnormal behaviour in titration Caspar (1963) has already pointed out. It is likely that these carboxyl groups interact differently, depending on the environmental concentration of protons and thus control the conformation of each protein subunit. Two pathways lead to structures of which the equilibrium becomes 'irreversible': in certain conditions, not yet clearly defined, the stacked discs assemble into long rods of stacked discs (Fig. 2) in which the protomers are now bound very strongly (A. Klug, personal communication).

The other pathway, which involves the viral RNA as well as discs, leads to the infectious viral particles. These particles suddenly become very resistant to dissociation and only strong denaturants such as 60% acetic acid are able to solubilize them again into subunits.

Among other interesting systems of self-assembly we might mention bacterial flagella (Asakura 1968) and the shells of some spherical viruses (reviewed by Hohn & Hohn 1970). The catalytic effect of nucleic acids or nucleic acid analogues on the assembly of the capsids of small spherical viruses is worth mentioning because here—unlike TMV—other nucleic acids, and even analogues, seem to be nearly as efficient as the one proper to the species (Hohn 1969).

AIDED ASSEMBLY OF MERIDS IN COMPLEX SYSTEMS

Aided assembly probably also occurs in relatively simple systems. Aid is however easier to recognize in those systems where genetics can be used as an experimental tool. Well-developed genetics is available for rather complex structures like bacteriophages T4 and λ. Fortunately, the complexity does not seem to be fundamental but is due rather to a high number of structural parts. Many of these structural parts considered individually seem to follow simple pathways of assembly. In these phages mutants are available for most genes, so that each gene action can be studied (see Epstein *et al.* 1963 for T4; and Kellenberger & Edgar 1971 for a review of λ and a summary of the experimental procedures).

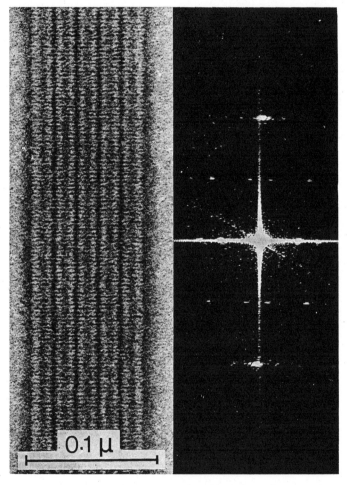

Fɪɢ. 2. A negatively stained preparation of stacked disc rods of TMV protein with the corresponding optical diffraction, showing the arrangement of rings in discs and the disc as repeat units. (Preparation kindly provided by A. C. H. Durham, Cambridge, micrograph and diffraction by J. Dubochet, Basel.)

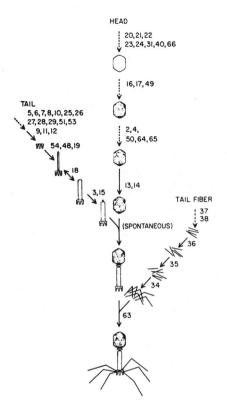

FIG. 3. Pathways of assembly of bacteriophage T4 according to Wood *et al.* (1968; reproduced by permission of *Federation Proceedings*). The numbers indicated for each step are those of the genes known to be involved at this step. A more recent path for head maturation is given in Fig. 9.

The pathways of assembly of phage T4 have been elucidated most elegantly by Edgar, Wood and collaborators (Wood *et al.* 1968); they are shown schematically in Fig. 3.

The assembly of these complex viruses occurs under most interesting conditions: the proteins involved all belong to the same class of the so-called 'late proteins'. In other words they all start being made simultaneously. Within a cell, the assembly of phage particles is far from being synchronized. The pathways in Fig. 3 are therefore not the result of a chronologically timed production of each gene product; the correct sequence of interactions is locally achieved through another mechanism. Such a mechanism has to account for the fact that the two partners of a specific protein-protein interaction do not react appreciably with each other when both are free in solution, but only when one of the partners is sitting in or on the maturing particle. There are three pos-

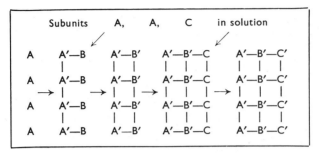

FIG. 4. Schematic mechanism of sequential regulation through induced conformational changes. Subunits A, B, C, etc. are free in solution (or on a membrane) but do not interact. B interacts with A' only; A' is induced through polymerization of A. The same is true for the pair B, C which free in solution does not interact except when B is modified into B' after interaction with A' (and possibly also polymerization).

sibilities which have to be considered, alone or in combination: (1) The maturing precursor particle may act as a nucleating centre and help to initiate the first steps of assembly of a merid. It is indeed known that the initial phases of assembly from subunits in solution are more difficult and more energy-consuming than later, when at least three subunits are bound together (Durham & Klug 1971). The immobilization of the first two subunits on a 'substrate' (precursor particle, cell membrane) would help to overcome this initial difficulty. (2) New specific sites of interaction may be made by the juxtaposition of proteins. (3) New specific sites may appear through induced conformational changes (Fig. 4).

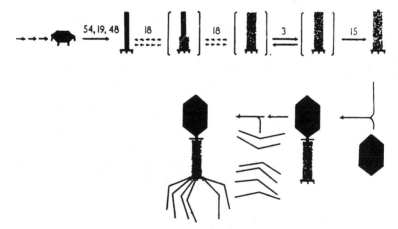

FIG. 5. Detailed scheme of the tail assembly according to King (1968; reproduced by permission of Academic Press). The action of genes 3 and 15 is required to stabilize the finished sheath. Until this takes place, sheaths can be dissolved again.

A good example for this sequentially induced specificity is given by the tail assembly, as shown in Fig. 5 (King 1968): according to recent studies P19 is the sole component of finished tubes (F. A. Eiserling, personal communication). Tail tubes without base-plates are never found *in vivo*. Hence the assembly of P19 into tail tubes is induced on the already assembled base-plates. When the tail tube is finished, or possibly even before, P18 is added to form the tail sheath in its extended form. This assembly is not stable and is still reversible. Only when further gene products are added does it become an irreversibly stable

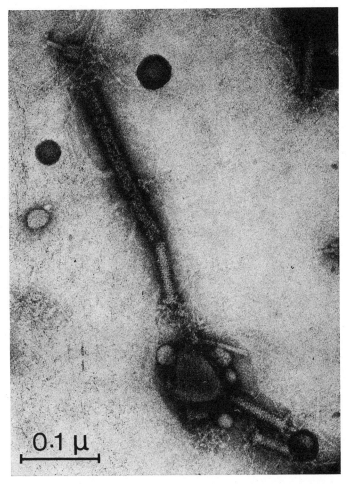

Fig. 6. A negatively stained preparation showing different forms of assembly of phage T4 tail sheath subunits: from upper left to lower right, a contracted sheath still sitting on the tail tube, polysheath showing 'crystal defects', a tail with extended sheath, a tail tube and a complete phage with artificially distorted head.

tail which can now, and only now, interact with a finished head. When no tail
tubes are available P18 can assemble, after a certain interval, into polysheaths
of undetermined length (Kellenberger & Boy de la Tour 1964). This polysheath
has the same structure as a contracted sheath (Moody 1967). Contraction of the
sheath occurs, for example, *in vivo* after absorption of phage to bacteria. The
different variant forms of sheath are shown in Fig. 6.

This potentially very interesting system shows that the normal extended
sheath has an aided assembly, through the tail tube acting as a morphopoietic
core. The length of sheath is predetermined by the length of the tube. But how
is the length of the tube determined? When measured in a fresh lysate, tail

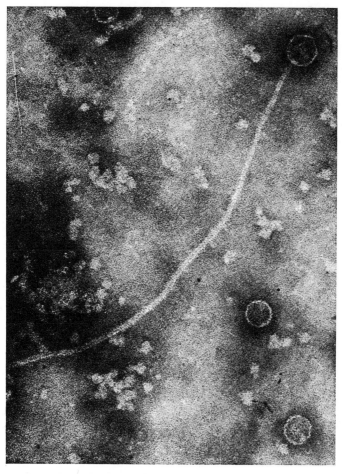

FIG. 7. A giant tail of bacteriophage lambda observed in a lysate of a mutant in gene *U*
(Mount *et al.* 1968; reproduced by permission of the editors, *Virology*).

tubes produced *in vivo* show an extremely narrow length distribution, suggesting a constant number of rows of subunits (22 rows in tubes 95 nm in length, according to Eiserling [personal communication]). Only in old lysates (Epstein *et al.* 1963) and in old suspensions of separated tail tubes (To *et al.* 1969) is the distribution of lengths found to be much broader; these observations show that some conditions exist in which the length determination is not functioning. Unlike the situation in λ, no gene has yet been found that controls length in T4. Lambda tails are made of gene product *V*; gene *U*, however, determines the length without being part of the final tail (Murialdo & Siminovitch 1971). If gene *U* is not acting, tails grow to undetermined lengths (Mount *et al.* 1968) (Fig. 7).

The T4 tail tube may provide a good experimental system for testing the hypothesis that length is determined through cumulated strain (Kellenberger 1969; see also this volume, p. 296).

Another example of genetic polymorphism is given by the T4 capsid and capsid-related structures (= capsoids). The product of gene 23, in its native or modified form, is the major component of seven different head-related variants; their capsids or capsoids are shown schematically in Fig. 8, where they are

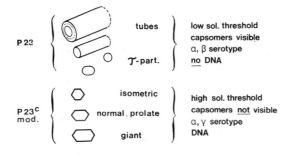

FIG. 8. T4 head polymorphism. The different variants are shown: multilayered tubular forms, single-layered tubular forms and τ particles (prolate and spherical) all made of P23, and phage heads made of P23^C in the three variant forms: isometric (Eiserling *et al.* 1970), normal prolate, and giant (A. H. Doermann & F. A. Eiserling, unpublished).

enumerated in the hierarchy of an increasing number of morphopoietic genes needed for contributing supplementary information (Laemmli *et al.* 1970). Through the study of lysates in SDS-acrylamide gels, Dickson *et al.* (1970), Hosoda & Cone (1970) and Laemmli (1970) have found that P23 undergoes a modification of electrophoretic mobility during phage development. A similar modification occurs also for some other gene products (Laemmli 1970). Simultaneously we showed that capsoids of the tubular form as well as lumps are made of the uncut native gene product (61 000 apparent mol. wt.) while the normal capsids are made of the cut (or cleaved) P23^c (49 000 apparent mol. wt.) (Kellenberger & Kellenberger 1970). Since then, A. H. Doermann and F. A. Eiserling in our laboratory (unpublished) have also shown that the capsids of both the small (short) and the giant (long) variants of T4 heads (containing DNA) are made of P23^c. According to whether they are constituted of P23 or P23^c, the seven variants therefore belong to one of two classes, as shown in Fig. 8. The protomers in the capsoids made of P23 are bonded much less strongly than in capsids made of P23^c.

The molecular weight differences were deduced from SDS–acrylamide gels. It is likely, but not yet proven, that the molecular weight modification is real and due to a reduction of about 20% in the length of the polypeptide chain. Such a cutting obviously provides a most efficient way to change the conformation of the protomer.

Recent experiments in our laboratory (to be published) suggest very strongly that the modification P23 to P23^c occurs *in situ*, on the maturing particles. The tentative pathway of head maturation that is emerging is summarized in Fig. 9:

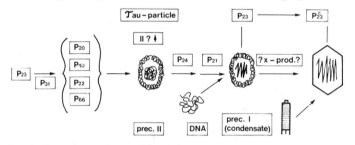

Fig. 9. Tentative pathway of head formation, as discussed in more detail in the text. The identity of precursor particle II with the τ particle observed in the electron microscope is not yet definitely established. Accordingly the aspect of this precursor II is assumed to be that of τ particles. The exact time of DNA packing has not been definitely established. (From unpublished data of R. Bijlenga, D. L. Anderson and E. Lüscher.)

a first precursor ('head precursor II') is made of P23, probably independently of the presence or absence of phage DNA. It is later filled with DNA, increases in size and becomes 'head precursor I', a very fragile structure still made with uncut P23. This precursor is likely to be identical with the so-called condensates (Kellenberger *et al.* 1959, 1968) and the particles produced by mutants in gene 50 (U. K. Laemmli, personal communication). Only during the last steps of head maturation would the transformation P23 to P23^c occur and lead to the stable strongly bonded capsid.

Most of these results were obtained by using a temperature-sensitive reversible mutant in gene 24 and by following the fate of radioactively labelled P23 (unpublished work with R. Bijlenga). P23 is pulse-labelled and chased before a shift is made from a non-permissive (40.5 °C) to a permissive (30 °C) temperature. After the shift, most P23 is transformed into P23^c and is found in mature phage particles. Simultaneously, the τ particles produced at the non-permissive temperature decrease in number, as evidenced by electron microscopy counts. This suggests rather strongly that the P23 containing 'precursor II' is identical with the τ particles. We are now doing experiments to prove this identity.

According to an older concept, DNA condenses first and capsids form around these condensates (Kellenberger *et al.* 1959). At that time it had already been clearly demonstrated that protein ('condensing principle') is necessary for this condensation to occur. It was also postulated that one or several proteins involved in condensation might act independently of DNA and form particles (Kellenberger 1963, 1966, 1969). Later it was shown that the condensates possess a very fragile membrane (Kellenberger *et al.* 1968). Luftig *et al.* (1971) then demonstrated experimentally, by using a temperature-sensitive (*ts*) mutant in gene 49, that organized protein particles are made first and then filled with DNA. However, in the interpretation of their very elegant experiments they assumed that these protein particles are final capsids, an assumption which now seems to be only a very crude approximation. We think that most of the 'empty capsids' observed in the lysate of mutant 49 are breakdown products of initially DNA-containing but extremely fragile particles; this is suggested by our two observations: (1) in thin sections most of the particles indeed contain DNA in variable amounts (Granboulan *et al.* 1971), and (2) the capsids isolated from lysates are made of P23^c (unpublished). The observed small proportion of pre-shift protein which is filled with post-shift DNA could be due to some precursor II which is also present. Fortunately the model of Fig. 9 reconciles most of the experimental data. Experimental proof for each point of the model is complicated by the great fragility of all the postulated precursors. They are indeed very difficult to observe in unaltered form by electron microscopy; DNA is particularly easily lost (Granboulan *et al.* 1971), τ particles cannot be purified without losing their supposedly proteinaceous core, and condensates lose their DNA on lysis and become so distorted that they can no longer be recognized on micrographs.

It was previously shown that the τ particles are already under the control of the form-determining genes, as for example those responsible for the short-headedness (Eiserling *et al.* 1970). All previous experiments demonstrated τ particles as abortive products (work done with U. K. Laemmli, unpublished; D. Scraba *et al.*, unpublished). With our new experimental data obtained on a reversible *ts*-mutant, these particles now become eligible candidates for head precursors. Hence, one has to search for the mechanisms of form determination at the level of τ particles.

References

ASAKURA, S. (1968) A kinetic study of in vitro polymerization of flagellin. *J. Mol. Biol.* **35**, 237–239

CASPAR, D. L. D. (1963) Assembly and stability of the TMV particle. *Adv. Protein Chem.* **18**, 37–118

CASPAR, D. L. D. & KLUG, A. (1962) Physical principles in the construction of regular viruses. *Cold Spring Harbor Symp. Quant. Biol.* **27**, 1–24

DICKSON, R. C., BARNES, S. L. & EISERLING, F. A. (1970) Structural proteins of bacteriophage T4. *J. Mol. Biol.* **53**, 461–474

DURHAM, A. C. H. & KLUG, A. (1971) Polymerization of tobacco mosaic virus protein and its control. *Nat. New Biol.* **229**, 42–46

DURHAM, A. C. H., FINCH, J. T. & KLUG, A. (1971) States of aggregation of tobacco mosaic virus protein. *Nat. New Biol.* **229**, 37–42

EISERLING, F. A., GEIDUSCHEK, E. P., EPSTEIN, R. H. & METTER, E. J. (1970) Capsid size and DNA length: the petit variant of bacteriophage T4. *J. Virol.* **6**, 865–876

EPSTEIN, R. H., BOLLE, A., STEINBERG, E. M., KELLENBERGER, E., BOY DE LA TOUR, E., CHEVALLEY, R., EDGAR, R. S., SUSMAN, M., DENHARDT, G. H. & LIELAUSIS, A. (1963) Physiological studies of conditional lethal mutants of bacteriophage T4D. *Cold Spring Harbor Symp. Quant. Biol.* **28**, 375–394

GRANBOULAN, P. H., SÉCHAUD, J. & KELLENBERGER, E. (1971) On the fragility of phage T4 related particles. *Virology* **46**, 407–425

HOHN, T. (1969) Role of RNA in the assembly process of bacteriophage fr. *J. Mol. Biol.* **43**, 191–200

HOHN, T. & HOHN, B. (1970) Structure and assembly of simple RNA bacteriophages. *Adv. Virus Res.* **16**, 43–98

HOSODA, T. & CONE, R. (1970) Analysis of T4 phage proteins. I. Conversion of precursor proteins into lower molecular weight peptides during normal capsid formation. *Proc. Natl Acad. Sci. U.S.A.* **66**, 1275–1281

KELLENBERGER, E. (1963) Morphogenesis of phage and its genetic determinants. In *New Perspectives in Biology* (SELA, M., ed.), pp. 234–245, Elsevier, Amsterdam

KELLENBERGER, E. (1966) Control mechanisms in bacteriophage morphopoiesis. In *Principles of Biomolecular Organization* (WOLSTENHOLME, G. E. W. & O'CONNOR, M., eds.), pp. 192–226, Churchill, London

KELLENBERGER, E. (1969) Studies on the morphopoiesis of the head of phage T4. VII. Polymorphic assemblies of the same major virus protein subunit. In *Symmetry and Function of Biological Systems at the Macromolecular Level* (ENGSTROM A. & STRANDBERG, B., eds.), pp. 349–366, Almquist & Wicksell, Stockholm

KELLENBERGER, E. & BOY DE LA TOUR, E. (1964) On the fine structure of normal and 'polymerized' tail sheath of phage T4. *J. Ultrastruct. Res.* **11**, 545–563

KELLENBERGER, E. & EDGAR, R. S. (1971) Structure and assembly of phage particles. In *The Bacteriophage Lambda* (HERSHEY, A. D., ed.), pp. 271–295, Cold Spring Harbor Laboratory, Cold Spring Harbor, N.Y.

KELLENBERGER, E. & KELLENBERGER-VAN DER KAMP, C. (1970) On a modification of the gene product P23 according to its use as subunit of either normal capsids of phage T4 or of polyheads. *FEBS (Fed. Eur. Biochem. Soc.) Lett.* **8**, 140–144

KELLENBERGER, E., SÉCHAUD, J. & RYTER, A. (1959) Electron microscopical studies of phage multiplication. VI. The establishment of the DNA-pool of vegetative phage and the maturation of phage particles. *Virology* **8**, 478–498

KELLENBERGER, E., EISERLING, F. A. & BOY DE LA TOUR, E. (1968) Studies on the morphopoiesis of the head of phage T-even. III. The cores of head-related structures. *J. Ultrastruct. Res.* **21**, 335–360

KING, J. (1968) Assembly of the tail of bacteriophage T4. *J. Mol. Biol.* **32**, 231–262

KLUG, A. (1972) The polymorphism of tobacco mosaic virus protein and its significance for the assembly of the virus. *This volume*, pp. 207–214.

LAEMMLI, U. K. (1970) Cleavage of structural proteins during the assembly of the head of bacteriophage T4. *Nat. New Biol.* **227**, 680–685

LAEMMLI, U. K., MÖLBERT, E., SHOWE, M. & KELLENBERGER, E. (1970) Form-determining function of the genes required for the assembly of the head of bacteriophage T4. *J. Mol. Biol.* **49**, 99–113

LAUFFER, M. A. & STEVENS, C. L. (1968) Structure of the tobacco mosaic virus particle; polymerization of tobacco mosaic virus protein. *Adv. Virus Res.* **13**, 1–63

LUFTIG, R. B., WOOD, W. B. & OKINAKA, R. (1971) Bacteriophage T4-head morphogenesis. On the nature of gene 49 defective heads and their role as intermediates. *J. Mol. Biol.* **57**, 555–573

MOODY, M. F. (1967) Structure of the sheath of bacteriophage T4. I. Structure of the contracted sheath and polysheath. *J. Mol. Biol.* **25**, 167–200

MOUNT, D. W. A., HARRIS, A. W., FÜRST, C. R. & SIMINOVITCH, L. (1968) Mutation in bacteriophage lambda affecting particle morphogenesis. *Virology* **35**, 134–149

MURIALDO, H. & SIMINOVITCH, L. (1971) The morphogenesis of bacteriophage lambda III. Identification of genes specifying morphogenetic proteins. In *The Bacteriophage Lambda* (HERSHEY, A. D., ed.), pp. 711–723, Cold Spring Harbor Laboratory, Cold Spring Harbor, N.Y.

STEVENS C. L. & LAUFFER, M. A. (1965) Polymerization—depolymerization of TMV protein. IV. The role of water. *Biochemistry* **4**, 31–37

TO, C. M., KELLENBERGER, E. & EISENSTARK, A. (1969) Disassembly of T-even bacteriophage into structural parts and subunits. *J. Mol. Biol.* **46**, 493–511

WOOD, W. B., EDGAR, R. S., KING, J., LIELAUSIS, A. & HENNINGER, M. (1968) Bacteriophage assembly. *Fed. Proc.* **27**, 1160–1166

Discussion

Levitt: Is there a convincing mechanism for length determination that does not involve a template? You mentioned the core of T4 tails. Here two gene products are involved; one could imagine that one gene made the subunit and the other a long α-helical molecule that determined the length of the core.

Kellenberger: The tail core or tube seems to be composed of one type of subunit only (product of gene 19). The two other gene products seem to be necessary to initiate the tube on the base plate. If they or other genes were involved in the length determination through an α-helix, for example, one should find amber mutants which influence the length. But even if the final tail tube is composed of only one type of subunit this does not exclude a transient action of one or several other proteins during assembly; after completion they would disappear from the structure.

Levitt: It is rather difficult to imagine length determination that does not involve the use of a template.

Kellenberger: We have proposed the 'cumulated strain' model (Kellenberger 1969).

Levitt: That mechanism would give a Gaussian-like distribution of lengths; the width of the Gaussian curve (i.e. the range of lengths) would depend on the increase in strain energy per added subunit. Cumulative strain may give a single length with a few subunits, but here you are dealing with hundreds of subunits.

Kellenberger: Some calculations are probably needed. It will certainly not work with hundreds of subunits! But the tail tube seems to have 22 rows of rings. It is very difficult to make predictions about the energies involved in the deformations. We need the help of protein chemists. Personally I do not think that templates are involved. Do you think of RNA or DNA as templates?

Levitt: No; in this case it would be a long α-helical protein. How long is the core of the phage tail?

Kellenberger: About 90 nm. It should be possible to investigate the length-determining mechanism experimentally on the T4 phage tail system. On base plates as initiators the *in vitro* assembly of gene product P19 to tail tubes should be possible, and also it would provide an excellent system for all sorts of studies.

Katchalski: Dr Kellenberger, can you give some more details concerning the proteolytic enzyme and the assembly mechanism that you have mentioned?

Kellenberger: Experiments are now being done in several laboratories. Nobody has yet been able to identify the proteolytic protein which is responsible for cutting the major head protein, P23. Mutants in about ten different genes result in not cutting the protein P23. The cutting is therefore an obligatory corollary to the right assembly process. Indeed several aberrant assemblies can prevent the cleavage from taking place. We have observed an interesting phenomenon with a temperature-sensitive (*ts*) mutation in the 23 gene itself. Under non-permissive conditions so-called 'crummy' heads are produced (Favre *et al.* 1965). We have now found that in these heads the protein subunit is not cleaved. This does not necessarily mean that the specific site for cleavage on P23 is altered; it is more likely that the wrong type of assembly is abortive and cannot proceed to those later steps of head maturation which involve the cleavage. In summary, we cannot yet say exactly when and how the cleavage occurs in the scheme shown in my Fig. 9 (p. 200). When we know more about precursors I and II and possible steps in between, we can then try to investigate the process of cleavage more thoroughly.

Klug: Although the details are not known, it seems clear from Laemmli's work (1970) that there is a large assembled structure as an intermediate. In other words assembly precedes cleavage, so that the cleavage is not simply a matter of preparing a protein for assembly.

Kellenberger: From recent work in our laboratory (unpublished, with D. L. Anderson and E. Lüscher) with a triple mutant of T4 phage in which the production of late proteins is no longer linked with DNA synthesis (Riva *et al.* 1970), it appears that cleavage needs the physical presence of DNA. It has to be definitively established, however, that the small proportion of cleaved $P23^c$ is dependent on the amount of DNA present in the cell. Organized capsid-like particles are found in very small numbers, barely accounting for the $P23^c$ found. Most of the P23 made is uncut and aggregated in (unorganized) 'lumps'. We have combined the triple mutant with a mutation in gene 21 which normally leads to τ particles. The quadruple mutant still produces τ particles but with only about 30% of the normal yield. This reduction corresponds roughly to the general reduction of phage proteins synthesized. The yield of τ particles in the quadruple mutant is, however, about 100 times higher than that of the capsid-like particles in the triple mutant. Several inconsistencies will have to be settled by experiment before a final model emerges.

Tanford: Do all your numbers from 1 to 60 represent distinct polypeptide chains, or are some of them mutations in different places on the same polypeptide chain?

Kellenberger: Each of these numbers represents one gene, and therefore—by definition—corresponds to one peptide chain. In many genes several mutations in different places are available. They have numbers and letters—e.g. ambermutation N90 in gene 21, abbreviated as 21 (am N90).

Tanford: So this virus has something of the order of 60 structural pieces?

Kellenberger: More than 80 genes are known, but we estimate T4 to have 100 to 200 genes. About half the known genes are involved in the replication of phage DNA, the others in assembly.

Tanford: Are many of the proteins not incorporated in the virus, but just used to synthesize it and then thrown away?

Kellenberger: It is not yet really known which proteins go into the final structure or, if they are found on the final structure, whether they are structurally relevant there or not. We can easily imagine that a protein, after it has done its job, still stays adsorbed, but no longer has any function. One can test this by seeing whether the functions and activities of the virion are unimpaired by partial extraction of some proteins. Furthermore, the interpretation of electrophoretic lines follows two 'philosophies': in one, the more lines there are, the better it is; in the other, one tries to get a minimum number. Critical work has still to be done.

References

FAVRE, R., BOY DE LA TOUR, E., SEGRÉ, N. & KELLENBERGER, E. (1965) Studies on the morpho-poiesis of the head of phage T-even. I. Morphological, immunological and genetic characterization of polyheads. *J. Ultrastruct. Res.* **13**, 318–342

KELLENBERGER, E. (1969) Studies on the morphopoiesis of the head of phage T4. VII. Poly-morphic assemblies of the same major virus protein subunit. In *Symmetry and Function of Biological Systems at the Macromolecular Level* (ENGSTROM, A. & STRANDBERG, B., eds.), pp. 349–366, Almquist & Wicksell, Stockholm

LAEMMLI, U. K. (1970) Cleavage of structural proteins during the assembly of the head of bacteriophage T4. *Nat. New Biol.* **227**, 680–685

RIVA, S., CASCINO, A. & GEIDUSCHEK, E. P. (1970) Uncoupling of late transcription from DNA replication in bacteriophage T4 development. *J Mol. Biol.* **54**, 103–119

V
The Virtues of Temperance

Editor's Comments on Papers 20 Through 27

Infection of sensitive bacteria by any one of the T-even phage results in the inevitable production of progeny phage particles and death of the cell. In the early decades of phage research, evidence accumulated which suggested that some phage do not behave in this manner but rather enter into some kind of symbiotic relationship with the host. Bacterial strains isolated from nature were often found to contain phage in the culture medium even after repetitive re-cloning, as if the cells could generate phage without prior infection (see *171*). However, the concept of *de novo* generation of phage, called lysogeny by its advocates, was open to the objection that bacteria might carry the phage on their surface without, in general, replicating them. The phage might occasionally replicate by infecting rare phage-sensitive bacterial mutants arising in the population. Bacterial strains that carry phage in this manner had, in fact, been demonstrated, and it was assumed by many that a similar mechanism could explain all such cases of phage-host symbiosis.

Not until 1950 did Lwoff and Gutmann (*131*) demonstrate conclusively that the property of lysogeny could be passed on from parent to daughter cell in the absence of preformed phage particles. Using a micromanipulator, they followed the line of descent of one lysogenic cell of *B. megaterium*, isolating the individual daughter cells into microdrops at each generation. At each generation the fluid of the microdrop was assayed for phage. In one experiment they followed the line of descent for 19 generations with no appearance of free phage in the microdrops. Nevertheless, the culture originating from the nineteenth-generation cell contained phage in the medium. They argued that if the original cell had phage adsorbed to it which were then distributed uniformly without replication to its descendants, the number of phage would have had to be at least 2^{19}, or roughly 500,000. Since the volume of this number of phage is an order of magnitude greater than the volume of the cell, Lwoff and Gutmann concluded that the phage must be propagated as an intracellular *anlage* (the prophage) along with the bacterium.

In these experiments it occurred occasionally that a single cell isolated in a micro-drop would vanish, leaving several hundred phage in the droplet. The trigger that induced the cell to make phage and lyse was not understood, but the frequency of its occurrence seemed to be influenced by external factors. Shortly thereafter, Lwoff, Siminovitch, and Kjeldgaard *(132)* found that exposure of a bacterial culture to an appropriate dose of ultraviolet light would induce the entire population to lyse and yield phage.

Phage of the type studied by Lwoff were called *temperate*. Temperate phage can multiply lytically, as do the T-even virulent phage, but they also have the option of becoming prophage upon infecting a sensitive host and thus transforming the host into a lysogen.

The temperate phage λ, which has been extensively studied and which is the subject of an excellent recent compendium *(88)*, resided in laboratories many years before its presence was detected. It was hiding as a prophage in the strain of *E. coli* called K12, which Lederberg and Tatum *(124, 125)* had used in their initial genetic studies of bacteria. The reason that λ was elusive has to do with a fundamental property of lysogens: They are immune to infection by the phage for which they are lysogenic. Thus, although cultures of K12 contained λ particles, these particles would not form plaques on K12. It was not until E. Lederberg *(123)* accidentally isolated a variant of K12 which had lost its λ prophage that the plaque-forming particles present in the original K12 culture were revealed. Infection of the λ-sensitive strain gave rise to surviving cells, which were again lysogenic for and immune to λ. Immunity of a lysogen to the phage it harbors is almost a logical corollary of stable lysogeny, for if lysogens were not immune to the phage which are occasionally released spontaneously, it would be impossible to grow cultures of lysogenic cells.

The phenomenon of lysogeny posed two questions that were to occupy researchers for the next two decades, the solutions of which turned out to have far-reaching implications. The first was the nature of the association of the prophage with the host. The second was the mechanism of immunity.

The Lederbergs quickly supplied part of the answer to the first question. As they describe in Paper 20, no evidence could be found for the transmission of λ prophage as a cytoplasmic factor. Moreover, they identified a genetic locus for lysogeny at a specific site on the *coli* chromosome in close linkage to the *gal* locus. Further studies *(100)* showed that this locus is the prophage itself and not a gene that merely affects prophage maintenance. This indeed was a novel genetic situation: A group of alien and potentially lethal genes enters a cell, finds its way to a specific chromosomal site, and thereafter behaves as a genetic element of the cell. Subsequently, other phages related to λ were discovered *(99)*, such as φ80 and 434 (mentioned below). These phages can recombine with λ in mixed infection, but each has its own specific chromosomal locus.

Two general ways were suggested by which the prophage might associate with the host chromosome *(100)*: Either the prophage somehow adheres to the chromosome or it is inserted into the continuity of the chromosome. Both models had conceptual difficulties. The kind of permanent synapse implied by the "sticking" model could at best be defined only vaguely, and the kind of specific recombination events required for the total insertion of one linear structure into another were unprecedented. Here the problem simmered for several years, with opinion, and some inconclusive evidence, favoring a sticking model.

In 1962 Campbell suggested that if the λ genome were circular, it could be inserted into or excised from the bacterial chromosome by a single conventional reciprocal recombination event. That part of Campbell's review article in which he puts forward this model is reproduced here (Paper 21). It is one of those rare instances in science in which a simple idea gives an entirely new perspective to a variety of questions. As Campbell describes, his model could explain a curious apparent inconsistency that had arisen in genetic mapping experiments with λ. For a temperate phage, gene order can be determined either by conventional phage crosses, yielding a map of the *vegetative phage*, or by crosses between lysogens each of which carries genetically marked prophage, yielding a map of the *prophage*. Calef and Licciardello *(27)* had found that for three phage genetic markers the order in the vegetative map was *h-cl-mi*, whereas the order in the prophage map was *h-mi-cl*. If the phage chromosome is assumed to be circular, at least at some stage of its replicative cycle, the cutting of this circle between *h* and *cl* when it inserts into the host chromosome, as illustrated in Fig. 2, would produce the observed reversal of marker order. Based, as it was, on only three markers, the argument was hardly compelling, but as more markers became available, the prediction that the prophage and vegetative maps are cyclic permutations of one another was confirmed *(32)*.

In another section of the same article from which Paper 21 was excerpted Campbell proposed that circularization could explain more generally how all extra-chromosomal genetic elements (episomes) such as fertility factors become integrated into and excised from chromosomes. These ideas, the status of which has been recently reviewed by Campbell in an excellent monograph *(33)*, have since proved to be largely correct.

The studies of Franklin, Dove, and Yanofsky described in Paper 22 offer convincing evidence in favor of prophage insertion. They found that mutations which delete host genes often extend for various distances into the prophage from one end, a result that would be difficult to understand unless the prophage were inserted into the bacterial chromosome. The genetic manipulations employed in this study illustrate the great versatility provided by the λ group of phage. The phage they used was not ordinary λ but a hybrid between λ and its close relative, φ80. This hybrid included about half of the λ genome, in which many genetic markers were available. Unlike λ, this hybrid prophage was located very close to the tryptophan operon of *coli*, a region that also includes the gene responsible for resistance to phage T1. It was this particular location that made their analysis possible because deletions of this region could be isolated by selecting for mutants that concomitantly were resistant to T1 and required tryptophan for growth.

Another prediction of the insertion hypothesis is that an inserted prophage, in contrast to an attached prophage, should increase the genetic distance between bacterial markers that flank the prophage site. Such an effect on linkage by λ prophage was demonstrated by Rothman *(154)*.

If a stock of λ is prepared by induction of a lysogen, rare phage particles in the stock (about 1:10[6]) have the ability to transfer certain bacterial genes to recipient cells *(139)*, a phenomenon called transduction. The transducing ability of λ is restricted to those genes closely linked to the chromosomal locus of the prophage and transducing particles are not found in stocks prepared by infection of a sensitive host. These features of λ-mediated transduction are in contrast to generalized transduction observed with other phages in which any host gene can be transduced and in which the transducing particles are formed in a lytic infection cycle as well as upon induction *(216)*. In the case of generalized transduction it is now known that the transducing particles contain no phage DNA but are merely phage shells filled with a piece of host DNA *(96)*. The transducing particles of λ are, however, quite different, containing a DNA molecule derived in part from the prophage and in part from the host.

Campbell *(29)* and Arber et al. *(5)* found that the λ transducing particles, although they can be propagated as prophage, cannot multiply unless ordinary λ is present in the same cell. The name given these extraordinary particles was λdg—*d* for defective and *g* for galactose, since genes of the galactose operon were the only ones known at the time to be transduced. More detailed genetic analysis revealed a remarkable regularity: Each independent isolate of a λdg was found to be missing a contiguous block of genes extending from the terminus of the prophage *distal* to the *gal* region to an internal point of the prophage, the internal point differing for each isolate *(30)*. This observation, that the deleted segments were in each cast terminal, hinted that a λdg is generated not by a random substitution of host genes for phage genes, but in some regular prescribed way.

An important question relating to the origin of transducing particles was whether or not these particles contain the same amount of DNA as ordinary λ, that is, whether or not the loss of phage DNA is equal to the gain of host DNA. In Paper 23, Weigle, Meselson, and Paigen describe experiments that measure the ratio of DNA to protein of a number of independent isolates of λdg. Assuming that the protein content is constant, they conclude that the DNA content of the λdg's differs among themselves and from ordinary λ. Moreover, they found that the transducing particles can have either more or less DNA than λ.

These facts, difficult to reconcile if the prophage were stuck to the host chromosome, were neatly explained by Campbell's model of prophage insertion and excision. Referring again to Paper 21, the chromosome of λdg can be imagined as arising by a rare illegitimate recombination event occurring at sites indicated by the arrows in Fig. 2—one site in the prophage region and the other in the *gal* region—the two sites being juxtaposed by looping of the chromosome. Just as in the case of normal prophage excision, the recombination event would excise a circular segment of DNA, but in this case the segment would include a piece of bacterial DNA and would lack a region of the prophage distal to *gal*. The amount of phage DNA lost and host DNA gained would depend on the locations of the particular (but variable) points at which the synapse occurred. The transducing phage would thus have either more or less DNA than ordinary λ, depending on the points of synapse.

Campbell's model was thus a great success in explaining a variety of phenomena in a unified way, but at the time it was proposed its basic premise, that λ DNA is or can become circular, had not been demonstrated. The first systematic studies of the physical properties of λ DNA were undertaken by Hershey, Burgi, and Ingraham in the same year that Campbell put forth his model. As they describe in Paper 24, measurement of the centrifugal sedimentation rate of DNA extracted from phage particles revealed that the DNA can exist in several physically distinct forms, depending on the DNA concentration and other factors. One of these forms was concluded to be a linear monomer and was the *predominant* form if the DNA was extracted from the phage at low concentration. At high DNA concentrations the linear monomers formed aggregates composed of two and three monomers, but even at low DNA concentrations the linear monomers could undergo a change of state, yielding what were called "folded" molecules. Of special significance was the finding that aggregation competed with folding as if the same cohesive sites were responsible for both processes. Their conclusions are best summarized in their own words: "Since folded molecules exist in only one stable configuration, and since molecular folding and aggregation are mutually exclusive processes, we postulate that each molecule carries two cohesive sites in prescribed locations, and that these are responsible for both processes. To account for the considerable effect of molecular folding on sedimentation rate, the sites must lie rather far apart along the molecular length. To account for the moderate effect of dimerization on sedimentation rate, the cohesive sites must be small compared to the total molecular length."

It was soon determined that these cohesive sites are located at the two ends of the linear DNA molecule and that the "folded" molecules are in fact circular molecules formed by joining the two cohesive ends *(89, 151)*. In Paper 25, Kaiser and Wu report the results of their investigations on the structure of the cohesive ends. They found them to be 5′ terminal single-stranded regions, less than 20 nucleotides in length, which are complementary to one another in sequence, thus allowing the two ends to join by hydrogen bonding. Later refinements of the sequence *(208)* showed that the single-stranded region is exactly 12 nucleotides in length and that complementarity is exact. This feature of the λ DNA molecule could thus provide the circular structure postulated by Campbell's insertion model. In fact, it has been found that upon injection of the linear phage DNA into the host cell, the formation of hydrogen-bonded circles is followed by covalent joining of the ends *(213)*.

The site of insertion of λ prophage is now known to be determined by an attachment region of the phage genome (see Paper 27 and the λ genetic map in the Appendix). Insertion itself is mediated by a phage gene *int* (for integration) whose protein product promotes site-specific recombination between the circularized λ DNA and the host DNA. Gottesman and Weisberg have recently reviewed the insertion process in detail *(75)*.

It should be noted that insertion of a phage genome at a specific chromosomal site is not characteristic of all temperate phage. Phage P1 can apparently lysogenize without insertion, and the prophage is not localizable on the chromosome *(100)*. Another phage, Mu-1, can become inserted at many (perhaps any) sites in the chromosome, even in the middle of a host gene, resulting in inactivation of that gene *(178)*. Mu-1 is thus called a mutator phage.

As an object for genetic studies, a temperate phage has the advantage over a virulent phage that mutants incapable of vegetative replication can be isolated and propagated in the prophage state. Such mutants are recognized by the fact that the lysogen retains its immunity for the homologous phage but fails to produce infectious phage particles upon induction. A number of different mutants, defective lysogens as they were called, were charcterized in detail in 1957 by Jacob et al., who found that different mutants were blocked at different stages of replication *(101)*. Some appeared to be defective in a late step involving maturation of the phage particle; others could not initiate vegetative replication and thus appeared to be blocked in an early step. This was the first solid evidence that phage replication is a complex process that involves a considerable number of functions controlled by phage genes. Later, Campbell *(31)* discovered mutants of λ that could replicate on one host but not on another. These host-dependent *(hd)*, later renamed supressor-sensitive *(sus)*, mutants were also blocked in a variety of functions in the nonpermissive host. The *sus* mutations of λ and the *amber* mutations of T4 in fact generate the same class of nonsense codon, which, as discussed previously, is translatable only in a cell carrying the appropriate suppressor gene. The present extent of knowledge of the λ genome, as well as that of T4, derives largely from studies of these and other kinds of conditional lethal mutants.

Although exploitation of conditional lethals unveiled many of the steps of vegetative replication, understanding the most distinctive feature of a temperate phage—the property of immunity—was aided by another kind of mutant called *c* (for clear plaque). The plaques formed by a temperate phage generally have a turbid appearance, owing to the outgrowth of microcolonies derived from cells that have become lysogenic and thus immune. Occasional plaques can be found in which this turbidity is absent, and the phage isolated from them retain this *c* phenotype.

Kaiser *(105)* isolated a large number of *c* mutants and found that they defined three cistrons, *c*I, *c*II, and *c*III, the *c*I mutants being the weakest in ability to lysogenize and thus the *c*I gene being the most stringently required. It was later determined that *c*I is required for maintenance of the lysogenic state, while *c*II and *c*III function together in its initial establishment (see Paper 27).

In further experiments, Kaiser and Jacob *(106)* studied the immunity specificity of the group of lambdoid phages (phages 434, 82, and 21) which can recombine with λ during vegetative growth but whose prophages are located at different chromosomal sites. They found that a cell lysogenic for one of these phages is immune only to the homologous phage; i.e., each of the phages has a different immunity specificity. Since these phages could recombine and thus must be largely genetically homologous, it was possible to ask which region of the phage genome determines the immunity specificity.

A hint was obtained first from crosses between one of the lambdoid phages (phage 434) and a suitably marked λ. Recombination was found to occur in all regions examined *except* in the region containing the *c*I gene, suggesting that this is a unique region of nonhomology. This was pursued by constructing a hybrid phage for which the *c*I region of λ was replaced by the *c*I region of 434, the remainder of the genome being derived from λ. This hybrid was found to have the immunity specificity not of λ by that of 434.

The insight derived from these studies is that the cI gene of the prophage is required both for the maintenance of the lysogenic state and for the immunity of the lysogen to infection by the homologous phage, thus implying that the two phenomena have a common basis. This could be understood if the product of the cI gene is a cytoplasmic factor that inhibits release from the prophage state and, in addition, inhibits replication of an infecting homologous phage DNA. Strong evidence for the existence of such a cytoplasmic factor was obtained from conjugation experiments between a lysogenic donor and a nonlysogenic recipient. Jacob and Wollman *(99)* found that entrance of the prophage by conjugation into the nonimmune recipient results in immediate vegetative replication of the phage and lysis of the recipient cell, a phenomenon called zygotic induction.

In their classic 1960 paper on genetic regulatory mechanisms, Jacob and Monod *(97)* point out the analogy between the lysogenic state and the repression of inducible enzymes and suggest that maintenance of the prophage state may be a special case of a more general mechanism whereby negative control elements, which they termed repressors, act to inhibit the expression of groups of genes, probably by preventing transcription of messenger RNA. At the time there was some indirect evidence that repressors were composed of RNA, but in the several cases to date where the repressor has been isolated, it has turned out to be a protein.

The merging of biochemistry and genetics finally enabled the isolation of the λ repressor in 1966, and a subsequent in vitro demonstration of its mode of action. As Ptashne points out in Paper 26, the repressor is probably present at a level of only about 1 part in 10^4 among the proteins of a lysogenic cell. To find this needle in the haystack, Ptashne sought to raise the level of repressor synthesis to a significant fraction of the total protein synthesis of the cell. He achieved this by employing tricks to depress host protein synthesis and phage protein synthesis (other than repressor) while maximizing the number of functioning repressor genes. Column chromatographic fractionation of the proteins synthesized by infected cells under these conditions revealed a component that had attributes expected of the protein product of the cI gene, the most significant being that this component is modified by a cI temperature-sensitive mutation.

In further studies, Ptashne and Hopkins *(146, 148)* showed that this protein binds specifically to λ DNA at two sites, O_L and O_R (see the Appendix), which flank the cI gene. These sites had previously been identified as genetic loci which when mutationally altered confer virulence to λ, that is, which render the phage insensitive to repression *(98, 143)*. The interpretation given these findings, supported by genetic and in vitro transcription studies *(147)*, is that the two binding sites are operators of two transcription units, one to the left and one to the right of cI, whose functions are required for the expression of all other λ genes. The binding of the cI gene product to these operators prevents transcription of the genes to the immediate right and left of cI and thus indirectly of all other phage genes.

Studies with λ have been particularly valuable in elucidating not only negative control mechanisms, as mediated by the repressor, but also positive control mechanisms whereby the products of certain genes are required for the expression of other genes or groups of genes. The last paper of this section (Paper 27) is a current review by Echols, summarizing what is known about these control mechanisms and how they function in the phage's two vital decisions: between lysis and lysogeny when it infects a cell, and between the maintenance of lysogeny and induction when it is in the prophage state. Echols points out that one obvious advantage of lysogeny from the phage's point of view is that it allows the phage to propagate itself when the supply of host cells becomes scarce. Induction, on the other hand, is a device that allows the phage to escape when the host's ability to synthesize DNA is threatened, a behavior not unlike that of rats in a doomed ship.

20

Reprinted from *Genetics*, **38**, 51–64 (1953)

GENETIC STUDIES OF LYSOGENICITY IN ESCHERICHIA COLI [1]

ESTHER M. LEDERBERG AND JOSHUA LEDERBERG

Department of Genetics, University of Wisconsin, Madison, Wisconsin

Received May 8, 1952

R ECENT research on *Escherichia coli* phages has outlined the biology of viruses that promptly lyse their bacterial hosts (DELBRÜCK 1950). In addition to the progressive parasitic relationship that these studies have analyzed, many phage-bacterium complexes persist in a more enduring symbiosis, lysogenicity. The experiments to be described in this paper were designed to probe two related questions: how is the virus of a lysogenic bacterium transmitted in vegetative and sexual reproduction? and how is a symbiotic complex established following infection by the virus, as an alternative to the parasitization and lysis of the host bacterium? Complementary problems, especially concerning the growth and release of virus in lysogenic bacteria have received more emphasis from other workers (BERTANI 1951; LWOFF and GUTMANN 1950; WEIGLE and DELBRÜCK 1951).

Our interest in lysogenicity was provoked by the discovery that *E. coli* strain K-12 was lysogenic. On two occasions, mixtures of certain mutant stocks appeared to be contaminated with bacteriophage. The plaques were unusual in showing turbid centers, suggesting those figured by BURNET and LUSH (1936). It soon became apparent that practically all K-12 cultures carried this latent phage. The novelty consisted of two exceptional mutant substrains, W-435 and W-518 which were sensitive to the phage, now referred to as λ. These two strains had been maintained in our stocks as nonfermenting mutants for lactose (Lac_3^-) and galactose ($Gal_4^- Lac_1^-$), respectively, isolated from ultraviolet-treated suspensions. Both cultures are derived from 58-161, a methionine-requiring auxotroph previously used in many recombination experiments (TATUM 1945; TATUM and LEDERBERG 1947). The lysogenicity of strain K-12 had remained unsuspected despite its maintenance for over 25 years and close study as the subject of mutation and recombination experiments since 1944. However, the only objective criterion of a lysogenic symbiosis is the lysis of another sensitive strain that functions as an indicator. Thus, in the absence of an appropriate conjunction of strains the virus carried by the K-12 subline would remain undetected. Because of the low frequency of sensitive strains, such opportunities are rare. The development of crossing techniques in strain K-12 has allowed the virus to be studied as a genetic factor. Intercrosses among strains differing with respect to λ and the development of lysogenic from sensitive strains are the main subjects of this report.

[1] Paper No. 490 of the Department of Genetics, University of Wisconsin. The investigation was supported in part by a research grant (E72-C3) from the National Microbiological Institute of the National Institutes of Health, Public Health Service.

GENETICS 38: 51 January 1953.

Although the adoption of a fixed terminology would be premature, for convenience, a few terms will be defined for use in this account. Lysogenicity will be understood as the regular and persistent transmission of virus potentiality during the multiplication of a bacterium, without overt lysis. When tested directly with the phage, a bacterial culture is *sensitive* (lysed) or *resistant* (not lysed). When tested with a sensitive indicator strain, the bacteria are *lysogenic* (carriers of λ) if the indicator is lysed, or *nonlysogenic* if not. Bacteria that are resistant to λ but nonlysogenic are termed *immune*. The virus as transmitted in lysogenic bacteria will be referred to as *latent virus*.

MATERIALS AND METHODS
Preparation of free phage

Suspensions of λ were first obtained from filtrates of 6–8-hour bacterial cultures developed from mixed inocula of λ-sensitive and λ-lysogenic strains in nutrient broth. Thereafter, further batches were prepared by growing λ-sensitive cells with virus according to the usual methods (ADAMS 1950). Lysis in broth is indicated by decreased turbidity rather than marked clearing. A convenient method for obtaining high titer λ directly from lysogenic bacteria has been developed by WEIGLE and DELBRÜCK (1951) from the methods described by LWOFF *et al.* (1950). A lysogenic strain grown in a yeast-extract broth is subjected to a dose of ultraviolet irradiation which kills only a small fraction of a genetically comparable λ-free strain. After 40–50 minutes incubation in yeast-extract broth the majority of lysogenic cells lyse with a burst of about 100 phage particles each. Virus titrations were made by established methods (ADAMS 1950; DELBRÜCK 1950). All lysates were sterilized by filtration through nine- or fourteen-pound test Mandler candles.

Some pertinent physical and morphological characteristics of λ have been described by other investigators (WEIGLE and DELBRÜCK 1951).

Media

The media recommended for observing phage-bacterium interactions are less useful with the λ system because of the presence of bacterial survivors (which prove to be either resistant or sensitive) in the plaques. The lysed areas are, however, accentuated by their discoloration on an eosin–methylene blue agar medium without the fermentable sugar customarily added (EMB base, LEDERBERG 1950). Plaques from free λ suspensions were counted on TSA (tryptone saline agar, WEIGLE and DELBRÜCK 1951). It was sometimes supplemented with ten percent citrated bovine blood to test the release of hemolysins during bacterial lysis by phage (SCHIFF and BORNSTEIN 1940). A positive reaction is the clearing of the blood at the zone where sensitive bacteria are exposed to λ or to lysogenic bacteria. It must be cautioned, however, that occasional cultures are normally hemolytic, perhaps owing to a high rate of spontaneous lysis.

Scoring for sensitivity and lysogenicity

Susceptibility to λ is tested by streaking a phage suspension across a dry EMB agar plate with a broad loop. A small loopful of the cells to be tested is then streaked at right angles to the phage. To test for lysogenicity, the bacteria are similarly streaked against a sensitive indicator. As a precautionary measure, the tested cells are also deposited at a control spot. As shown in figure 1, positive tests consist of the interruption in the continuity of growth of the indicator, or plaques and discoloration at the conjunction of phage with sensitive bacteria. The technique of replica plating (LEDERBERG and LEDERBERG 1952) facilitates large-scale tests of lysogenicity. Instead of individual tests on bacterial colonies from a plate, these are transferred en masse by means of velveteen fabric to a TSA plate previously layered with 10 ml of TSA seeded with about 10^8 indicator cells. On the replica, each lysogenic colony is surrounded by a zone of lysis.

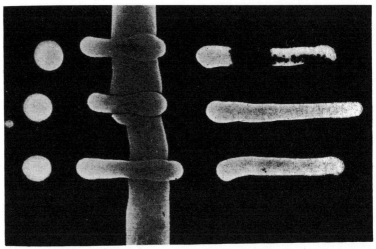

FIGURE 1.—Reactions of sensitive, lysogenic and immune. Extreme left: control spots. Left center: cross-streaks against sensitive indicator bacteria. Right: cross-streaks against λ. From top to bottom Lp_1^s, Lp_1^+, and Lp_1^r respectively.

Crosses

Crosses are carried out by plating washed cultures differing in nutritional characters on minimal agar (TATUM and LEDERBERG 1947; LEDERBERG et al. 1951) or with added streptomycin where streptomycin sensitive (S^s) prototrophs are crossed with resistant (S^r) diauxotrophs (LEDERBERG 1951a). The resulting progeny are picked and purified by streaking on a complete medium, and from this, single colonies are isolated for further characterization of segregating markers, including lysogenicity.

Selection of lysogenic and nonlysogenic cultures

Lysogenic bacteria may be routinely isolated from turbid plaques on sensitive bacteria plated with λ or from the residual growth after mixed inoculation of sensitive bacteria and phage in broth on agar. Successive single colony purifications result in stably lysogenic isolates free of extraneous λ and sensitive bacteria. By isolating one lysogenic derivative from a series of single plaques, it was demonstrated that the transfer of λ from well-marked lysogenic to previously sensitive stocks occurs without any alteration of the known genetic markers of the new host other than its reaction to λ.

The isolation of nonlysogenic (immune or sensitive) types from lysogenic bacteria is less predictable but they have been obtained by the following procedure. EMB or blood agar plates were spread with 10^8 cells and exposed to ultraviolet light so that about 100 to 200 colonies survived. New types have been sporadically detected either by testing large numbers of normal-appearing colonies or by the partially lysed appearance of phage-contaminated sensitive colonies. These consist of lysogenic and sensitive sectors, and free λ. Immune mutants have arisen from sensitive bacteria after selection with phage. The various occurrences of nonlysogenic derivatives are listed in table 1.

RESULTS

Intercrosses of various phenotypes

Crosses among the sensitive, lysogenic, and immune strains are all fully fertile. They have been repeated many times with the following qualitative results, based on 200 or more tests for each cross.

1. Lysogenic × lysogenic: all progeny lysogenic.
2. Sensitive × sensitive: all progeny sensitive.
3. Sensitive × lysogenic: the progeny segregate into sensitive and lysogenic, with ratios depending on the nutritional markers of the parents.

The total number of tests of crosses 1 and 2 is actually much larger, for exceptional progeny would have been apparent upon inspection of similar crosses conducted for other purposes.

Since only the parental types are found in cross 3, it might be inferred that lysogenic differs from sensitive only by one factor, the presence of the λ. However, the consideration of λ as a cytoplasmic factor leads to a possible paradox: when λ is contributed by just one parent, it segregates among the progeny, but it always appears when contributed by both parents. It should be emphasized that the same segregation ratios for other markers have been obtained regardless of the presence or absence of λ in the parents. No evidence has been found to date for the functioning of λ as a gamete or other sexual form (cf. HAYES 1952; LEDERBERG, CAVALLI and LEDERBERG 1952).

Further crosses involving two immune parents gave the following results:

4. Immune-1 × sensitive: parental only.
5. Immune-2 × sensitive: parental only.
6. Immune-1 × lysogenic: parental only.
7. Immune-2 × lysogenic: parental and sensitive.
8. Immune-1 × immune-2: parental and sensitive.

TABLE 1

Principal stocks used in lysogenicity studies.

Strain number	Source strain	History	Genotype[1]
		Sensitive (Lp_1^s)	
W-435	58-161	UV (ultraviolet)	$M^-Lac_3^-$
W-518	Y-87	UV	$M^-Lac_1^-Gal_4^-$
W-1267		W-518 × W-588, f-1 segregant	$T^-L^-Lac_1^-Gal_4^-$
W-1485	K-12	UV; blood agar	wild type sensitive
W-1486	W-811	plating with streptomycin	$M^-Lac_1^-Gal_4^-S^r$
W-1487	W-1405	plating with streptomycin	$T^-L^-Lac_1^-Gal_4^-S^r$
W-1502	W-1245	spontaneous variation	M^-
W-1503	W-1296	spontaneous variation	T^-L^-
W-1655	58-161	UV	M^-
W-1872	K-12	UV	wild type sensitive
		Immune-1 and -2 (Lp_1^r and Lp_2^r)	
W-1027	Y-70	UV	$T^-L^-\ Lac_1^-Lp_1^r\ Lp_2^s$
W-1924	W-518	selection with λ	$M^-Lac_1^-Gal_4^-Lp_1^r\ Lp_2^s$
W-1248	W-518	selection with λ	$M^-Lac_1^-Gal_4^-Lp_1^s\ Lp_2^r$
W-1603	W-1177	UV	T^-L^- etc., $Lp_1^s\ Lp_2^r$
W-1245	W-478	UV	M^-; unstable immune
W-1296	W-588	UV	T^-L^-; unstable immune
		Lysogenic (Lp_1^+ : Lp_2^r or Lp_2^s)	
58-161		standard parent	$M^-\ Lp_2^s$
W-1177		multiple marker parent	$T^-L^-\ Lac_1^-Mal_1^-Xyl^-Gal_5^-S^rLp_2^r$
W-811	W-518	infection with λ	$M^-\ Lac_1^-Gal_4^-\ Lp_2^s$
W-1439	W-811	selection with λ-2	$M^-\ Lac_1^-Gal_4^-\ Lp_2^r$

[1] The significance of the genotypic symbols, and further details of ancestry of many stocks are given in LEDERBERG *et al*. (1951) and LEDERBERG (1952).

The appearance of a sensitive recombination class in cross 8 implicates two loci in resistance to λ. Sensitive can be described as $Lp_1^s\ Lp_2^s$, immune-1 as $Lp_1^r\ Lp_2^s$ and immune-2 as $Lp_1^s\ Lp_2^r$. From the result of cross 6, in contrast to cross 7, lysogenicity is also determined at the Lp_1 (latent phage) locus. Evidence for two kinds of lysogenic, Lp_2^s (those so far discussed) and Lp_2^r, respectively, will be presented in another section.

Occasional sensitive progeny would have been anticipated in cross 6 on the hypothesis that lysogenic is genotypically equivalent to sensitive, and differs only by the presence of cytoplasmic λ, but were not found. The independent segregation of λ (cross 3) and of the genetic factor Lp_1 (cross 4) would have resulted in some λ-free recombinants sensitive to the virus. The results of all these crosses hinted at a primarily " chromosomal " determination of lysogenicity.

Linkage behavior of lysogenicity

The concept of an Lp_1 locus was strengthened by the outcome of linkage tests in which various markers were segregating. A loose linkage of Lp_1 to Xyl and to S was indicated in preliminary crosses with a multiple marker stock. However, Lp_2 was also segregating, thus doubling the number of genotypic classes, and perhaps confusing the issue. The closest linkage of Lp_1 thus far found has been to Gal_4, as shown in table 2. As it happens, this is the distinctive marker of W-518, in which λ-sensitivity was first noticed.

The linkage of Lp_1 with Gal_4 has been verified by crosses with various combinations of lysogenic stocks resynthesized from sensitive auxotrophs. Some of the latter were newly developed from W-1485, a λ-sensitive directly derived from strain K-12. There can be little doubt, therefore, that the segregating

TABLE 2

Linked segregation of Gal_4 and Lp_1 among prototrophic recombinants.

Parents		Prototroph recombinants: $M^+T^+L^\dagger$...			
A $M^-T^+L^+$	$\times M^+T^-L^-$	Gal^+Lp^+	Gal^+Lp^s	Gal^-Lp^+	Gal^-Lp^s
1 Gal^+Lp^s	$\times Gal^-Lp^+$	1	83	90	2
2 Gal^+Lp^+	$\times Gal^-Lp^s$	33	1	3	41
3 Gal^-Lp^s	$\times Gal^+Lp^+$	55	0	5	53
4 Gal^-Lp^+	$\times Gal^+Lp^s$	1	42	44	1
B $M^-H^+L_2^+$	$\times M^+H^-L_2^-$				
1 Gal^-Lp^+	$\times Gal^+Lp^s$	0	34	40	1
C $M^-M_2^+G^+$	$\times M^+M_2^-G^-$				
1 $Gal^-Lp_1^+$	$\times Gal^+Lp^s$	0	40	39	1
2 Gal^-Lp^s	$\times Gal^+Lp^+$	64	2	1	67

The crosses were conducted on EMS galactose medium, from which approximately equal numbers of Gal^+ and Gal^- prototrophs were picked for further test. Similar results were obtained when the proportion of Gal^+ and Gal^- was not thus fixed, as on non-indicator glucose minimal agar, but the preponderance of one parental type among the prototrophs limited the usefulness of unselected isolates for linkage tests. The $H^+L_2^-$ and $M_2^-G^-$ parents indicated in B and C are histidine-leucine and methionine-glycine auxotrophs, respectively, recently derived from W-1485. All parents in these crosses were Lp_2^s, but V_1, Lac_1, and S were segregating in their usual patterns.

factor is directly associated with lysogenicity. The linked segregations justify the assignment of a new allele, Lp_1^+, characteristic of lysogenicity. The result indicated for cross 6 points to this as a third allele at the same locus as the contrasting Lp_1^r (immune-1) and Lp_1^s (sensitive).

Segregation of λ from diploids

Heterozygotes selected as Lac^+/Lac^- or Gal^+/Gal^- were obtained and shown to be segregating for a number of other factors (LEDERBERG 1949), but these selections were either λ-sensitive or λ-lysogenic. Similar results were obtained in immune, Het crosses. It was thought, however, that the λ-determinant might be hemizygous in these diploids, like the Mal and S factors previously studied (LEDERBERG et al. 1951). This difficulty has been circumvented by the use of diploid $\times$ haploid crosses, in which the segmental elimination (of *Mal* and *S*) apparently does not occur. A lysogenic diploid parent $(T^- L^- Gal^+ Lac^+ Mal^+/Lac^- Mal^-)$ was crossed with a sensitive, haploid auxotroph $(M^- Gal_4^- Lac^- Mal^+)$ on minimal agar. The resulting prototrophs were almost all diploid, and several were identified as lysogenic, but segregating Gal^+/Gal^- as well as other factors. As shown in table 3, presence vs. absence of λ segregated in the same coupling as shown by the parents: Gal^+ lysogenic/Gal^- nonlysogenic. Unfortunately, this diploid is also segregating Lp_2, so that the nonlysogenic segregants include immune-2 as well as λ-sensitive.

The linkage and segregation evidence shows that a chromosomal factor is altered when a cell becomes lysogenic. In addition, a cytoplasmic factor (λ itself) may be postulated, but genetic evidence for it is entirely inconclusive. Two possible interpretations may be considered: 1) The virus or provirus occupies a definite niche on the chromosome, near Gal_4. Lysogenicity results from the cellular or even chromosomal fixation of the latent virus. 2) The chromosomal factor is a gene, Lp_1^s, which mutates spontaneously to an allele Lp_1^+ that potentiates a symbiotic relationship of λ in the bacterial cytoplasm.

TABLE 3

Segregation of Gal_4 and Lp_1 from heterozygous diploids.

	Gal^+Lp^+	Gal^+Lp^s	Gal^-Lp^+	Gal^-Lp^s
H-295	36	1	1	39 (19 Lp_2^s)
H-297	29	0	0	11 (3 Lp_2^s)

Segregant (pure) Gal^+ and Gal^- colonies were picked from EMB galactose agar at random, and tested for susceptibility to λ and λ-2, and for lysogenicity. The phenotypically λ-sensitive (Lp_1^s) moiety of the Lp_1^s segregants is shown in parentheses. Almost all of the Lp_1^+ were Lp_2^r.

On this hypothesis, the role of λ in the induction of lysogenicity is confined to the selection of the pre-adaptive mutation, Lp_1^+. A similar dilemma in the determination of the killer trait in *Paramecium aurelia* has been resolved in terms similar to the second interpretation (SONNEBORN 1950), although the first was originally favored (SONNEBORN 1945). Its substantiation for lysogenicity would require the recognition of the possible genotypes: Lp_1^s no-λ (presumably the sensitives); Lp_1^s infected with λ (presumably lethal); Lp_1^+ with lambda (the lysogenic); and a new combination, Lp_1^+ no-λ. This last type, genetically pre-conditioned for lysogenicity, would presumably be recognized as an apparently immune form that would promptly absorb λ to become lysogenic. It has not yet been identified among immune stocks of K-12, or immune progeny collected from a variety of strain intercrosses.

Mechanism of infection

When λ-sensitive bacteria are plated with λ, survival ratios in the range of ten to fifty percent are usually encountered. Many of the survivors are apparently lysogenic. The hypothesis of spontaneous variation at the Lp_1 locus

would be untenable if, as these facts appear to show *prima facie,* several percent of sensitive bacteria became lysogenic under the direct influence of the
virus. Only preliminary experiments have been done on this aspect of the
problem, with results that are not yet conclusive. A striking feature of platings
of diluted bacteria-virus mixtures of varying relative multiplicity has been
the development of contaminated colonies, similar to those figured by BOYD
(1951). These colonies displayed a very characteristic appearance on EMB
agar. They were often delayed in their development, lagging a few hours
behind their neighbors, and later show either a central " necrosis " or plaquing,
or often single or multiple pericentric plaques. When the contaminated colonies
were restreaked, they typically gave rise to a mixture of contaminated, sensitive and lysogenic colonies.

Many of the latter are only apparently lysogenic, for they include sensitive
bacteria as shown by serial restreaking of single colonies. It is not unlikely
(though not yet proven) that contaminated colonies may arise from single
infected cells. If this is the case, the determination of lysis versus lysogenicity
is effected during the development of a contaminated clone, and there would
be greater opportunity for genetic variation and natural selection. On the other
hand, if a fair proportion of infected cells are actually converted directly into
lysogenics, it would be concluded that λ itself induces or fixes the mutation
from Lp_1^s to Lp_1^+.

Virus and host mutations

Following irradiation of a type lysogenic, a self-lysed colony was noted from
which a distinctive virus was isolated. This virus, λ-2, differs from λ in its
ability to destroy Lp_1^+ bacteria. Attempts to develop a symbiosis of λ-2 with
each of a variety of bacterial stocks have been unsuccessful. Its relationship
to λ as a " host range mutant " is supported by the concurrent development
of resistance to λ with mutations from sensitivity to resistance to λ-2. Several
recurrences of λ-2 have been detected in lysed colonies after ultraviolet irradiation, and in λ stocks grown on sensitive cells. It has not, however, been observed in routine bacterial cultures, although it would presumably have been
conspicuous. This is in contrast to the rapid accumulation of comparable virus
mutants in cultures of the lysogenic staphylococci studied by BURNET and
LUSH (1936).

Immune bacterial mutants have been observed among survivors of both
irradiated lysogenic cultures and sensitive cultures exposed to the viruses.
Immune-1 has occurred very infrequently, and is resistant to (and nonlysogenic for) λ, but sensitive to λ-2. Immune-2 is resistant to both phages, showing
neither lysis nor the development of lysogenicity. As already mentioned, different loci, Lp_1 and Lp_2, appear to be involved. Although immune-2, $Lp_1^s Lp_2^r$,
does not respond to free λ, selection for resistance to λ-2 in a lysogenic stock
gives the genotype $Lp_1^+ Lp_2^r$ which remains lysogenic for λ. Crosses of such
lysogenics with sensitive ($Lp_1^+ Lp_2^r \times Lp_1^s Lp_2^s$) gave all four of the expected
types: immune-2 ($Lp_1^s Lp_2^r$) and type lysogenic ($Lp_1^+ Lp_2^s$), in addition to
the parents. Current stocks of K-12 are mixed populations with respect to Lp_2.

It is not surprising, therefore, that several mutant derivatives, notably W-1177 extensively used in crossing experiments, are already $Lp_2{}^r$. Two λ-immune selections have been found, both sensitive to λ-2, which were unstable and frequently engendered λ-sensitive colonies. Tests for allelism with $Lp_1{}^r$ were inconclusive owing to this instability.

Other mutants of the virus have been sought, but only plaque variants not readily scored were observed. Resistance to λ and λ-2 is concomitant with resistance to p-14, a phage isolated from sewage. Morphologically, the plaques of p-14 are intermediate between those of λ and λ-2, with turbid centers associated with a spurious or unstable lysogenicity which persisted in slow-growing isolates at 30° and was rapidly lost at 37°. Despite its initial promise as a selective agent for other bacterial mutations related to λ, p-14 did not elicit any otherwise unrecognized types.

A " weakly lysogenic " bacterium was recovered after ultraviolet irradiation of a typical lysogenic form. When inoculated with the indicator strain, the variant induced very few plaques, so that it was not readily distinguished from immune nonlysogenic forms. When the virus was transferred from the weakly lysogenic form to sensitives normal lysogenicity ensued. This suggests that reduced lysogenicity was a property of the host rather than of the phage. It was conceivable, however, that the plaques of free virus represent reverse-mutants from a virus population that otherwise remains entirely latent within the infected variant bacterium. To eliminate this possibility, sensitive recombinants from crosses of the weak lysogenic with sensitive were infected individually with type λ. Both types of lysogenicity were expected on the hypothesis of bacterial mutation, and this was actually observed. A modifier locus is thus revealed, but its relationships with other factors have not been explored.

Another intermediate reaction type was isolated from plates spread with 10^8 bacteria and λ-2. Most of the survivors were fully resistant to both λ and λ-2, but some exhibited a partial resistance to λ and λ-2, which was reflected in overgrowth of cross-streaks and reduced efficiency of plating and plaque size for both viruses, similar to the expression of $V_1{}^p$ (partial resistance to T_1, LEDERBERG 1951b; WAHL and BLUM-EMERIQUE 1952). λ-lysogenic derivatives were prepared which were still semi-resistant to λ-2. The mutation thus involved either a third allele, $Lp_2{}^p$, at the Lp_2 locus or mutation at another locus.

In view of speculation concerning the dispersion of lytic phages into genetic subunits during intracellular growth, the possibility that fragments of λ might persist in apparently nonlysogenic cells was considered. The reconstitution of lytically active λ from components carried in different nonlysogenic recombinants or variants would be relevant evidence. However, such a recurrence of phage from appropriate mixtures and crosses has hitherto not been demonstrated.

Disinfection

Two lysogenic streptomycin-sensitive (S^s) cultures plated on streptomycin agar have been observed to yield large numbers of resistant (S^r) mutant colonies which showed the characteristically mottled margins of phage attack. These colonies gave rise to S^r λ-sensitive isolates. Reconstruction experiments with these mutants or their re-infected derivatives failed to establish any foundation for either a selective advantage or a specific inductive effect of streptomycin to explain the accumulation of λ-sensitive. By indirect selection (LEDERBERG and LEDERBERG 1952), it was possible to extract the S^r components, and show their λ-sensitive character without exposing them to streptomycin. The λ-sensitive and S^r characters were not distinguishable from mutations previously isolated in single steps. No explanation for this remarkable association can be offered.

Systematic attempts were made to remove λ from lysogenic bacteria by a number of other methods. As none were successful, details will be omitted. The treatments that were tried included cultivation at limiting temperatures and pH ranges (as originally suggested by D'HERELLE 1926), and exposure to antibiotics and antiviral chemicals, including streptomycin, aureomycin, chloromycetin, Phosphine GNR, 2-nitro-5-aminoacridine, citrate ion, cobaltous ion, and desoxypyridoxine. A serious limitation to this type of investigation is the inadequacy of earlier methods of detecting disinfected variants, if they occur infrequently. Replica plating should help to surmount this problem, but was not available at the time of these experiments.

Almost all of our original λ-sensitive stocks in strain K-12 have been noticed following exposure to treatment with ultraviolet light. Inasmuch as this agent, under certain conditions, preferentially kills lysogenic cells by inducing lysis (WEIGLE and DELBRÜCK 1951), it cannot be concluded whether a selective or inductive (disinfective) action is involved.

Lysogenicity and other E. coli strains

The λ reaction of about 2000 strains under investigation for intercrossability has been routinely tested. No recurrence of λ itself has been identified, but five new strains are sensitive to λ and λ-2. One apparently unstable immune strain gave rise to sensitive subtypes, which, however, could not be made lysogenic on K-12 line indicators for either virus. All of the new sensitive lines, including NTCC 123 (CAVALLI and HESLOT 1949) are fertile with K-12, suggesting a statistical correlation of λ receptors with compatibility. Most of the 50 or so interfertile strains that have been screened are, however, immune to λ.

Although a large proportion of the strains tested produced an antibiotic or colicin (FREDERICQ 1948) active on K-12, less than one percent were lysogenic. The lysogenic cultures (which include, for example, the Waksman strain used in biochemical genetic studies, DAVIS 1950) carry what appear to be quite distinctive phages, judging from plaque type and resistance patterns. Two of the new latent phages have been successfully transferred to the K-12 line. Triply lysogenic K-12 strains were maintained without any overt effects on the λ system or other characters of the bacteria. The genetic determination of lysogenicity for other phages may differ from that of λ, however, in so far as clear-cut segregation for them was not observed in crosses or from diploids also segregating λ.

This work was initiated in the expectation that λ would behave as an extra-nuclear factor, and might indeed provide a favorable model system for studies of cytoplasmic heredity. Phenotypic changes associated with the transfer of λ have, so far as known, been confined to the direct consequences of virus infection. For example, lysogenic bacteria are more susceptible to ultraviolet light, owing to the "induction" of the latent phage and lysis of the bacterium (WEIGLE and DELBRÜCK 1951). In other systems, latent viruses have been shown to determine the pattern of susceptibility to other viruses, the "lyso-type" (NICOLLE and HAMON 1951; WILLIAMS-SMITH 1948; ANDERSON 1951), by a mutual exclusion effect. With one dubious exception, no phages that would differentiate λ-sensitive from λ-lysogenic were found in tests of some thousands of coliphage plaques from sewage. In principle, however, a virus-symbiosis might be detected in terms of the intercellular transfer of a genetically active agent not readily recognizable as a lytic phage (LOMINSKI 1938).

This view of λ may have to be qualified in view of the genetic tests discussed in this paper. No genetic evidence of λ as a cytoplasmic agent was found. In the most critical tests, segregation from heterozygote diploids, lysogenicity behaved precisely as if it were controlled by a nuclear factor, linked to other segregating factors. This result provides strong support for the "provirus" concept of the symbiosis. The segregation of uninfected, virus-sensitive haploids from a lysogenic diploid is not readily compatible with the presence of free, mature virus in the latter. It is not, however, conclusive against a cyto-plasmic provirus. The segregation of lysogenicity/sensitivity may reflect the overriding control by a segregating nuclear factor which is concerned with the maintenance of the pro-λ. The mutational origin of this segregating factor is, however, still in question.

It should not be assumed that these results can be generalized to other lyso-genic symbioses. In *Salmonella typhimurium*, BOYD (1951) has shown that the multiplicity of infection is an important element in the determination of lyso-genicity. This would leave little room for bacterial variability, but a closer analysis of the incidents immediately related to the development of lysogenic cells might reveal a situation more comparable to that in *E. coli*. In preliminary studies of the transmission of other viruses, transferred to K-12 from other lysogenic strains, diploids lysogenic for two phages showed segregation for λ but not for the second phage. The apparent difference with respect to nuclear determination may be a consequence of the antiquity of the association of K-12 with λ in contrast to the newly introduced phages.

It may be noteworthy that λ has not recurred in extensive samplings of other *E. coli* strains and of sewage. The occurrence of λ-sensitive isolates has already been mentioned. It is rather striking that all five of these isolates should be cross-fertile, compared to the four to five percent of the whole population. Whether this speaks for a close genetic relationship or for the closer attention given these lines cannot be said. It should be emphasized that all of the evidence argues against any functional relationship between lysogenicity and sexual fertility. The most decisive point, perhaps, is that nonlysogenic crosses are as fertile as crosses involving one or both lysogenic parents, both within strain K-12, and as between strains.

The biological significance of the lysogenic symbiosis is attested to not only by the behavior of individual examples, but by its prevalence in many groups of bacteria. BURNET (1945) and others have emphasized the biological advantages to the parasite as well as the host of symbiotic adaptation. In addition, the virus genotype represents an additional reservoir of genetic material subject to adaptive variation. This adaptation will often lead to an amelioration of the pathogenic effects of the virus. One can imagine a situation in which a virus remains trapped within a host that it never lyses. A bacterial mutation for weak lysogenicity illustrates this trend, and it has perhaps been realized in LOMINSKI's (1938) experiments. The extreme case would however restrict the migration of the virus to other genotypes, as well as our ability to recognize it as a virus. It is conceivable that the immune-1 $(Lp_1{}^r)$ mutation represents such a bound virus, although free λ has not recurred even in its crosses to λ-sensitive.

The most prominent mutation of λ has, according to this picture, only short-term evolutionary advantages. The virulent mutant, λ-2 will rapidly destroy lysogenic bacteria, and thus displace λ from viral populations. The exhaustion of sensitive hosts will, however, limit its long-term survival. The early literature on bacteriophage contains many references to the so-called spontaneous generation of bacteriophage in bacterial cultures. While some of these reports are possibly founded on technical faults, probably most of them represent instances of the mutation of virus in lysogenic bacteria not recognized as such. If it were not for the availability of an indicator strain for λ, the occurrence of lysis due to λ-2 in platings of ultraviolet irradiated K-12 would have passed either for a contamination or such " spontaneous generation."

For technical reasons, the phages of the T series acting on *E. coli* B have received considerable attention. These phages have been used for such work precisely because they are atypical in their prompt destruction of sensitive bacteria, high efficiency of plating, the limited number of secondary resistants, and clear plaques. A plating of sewage with indicator bacteria shows at a glance that phages of this kind are relatively infrequent. Although the analysis of phage-bacterium relationships on a logically sound, particulate basis has demanded systems with these technical properties, it would be a fallacy to generalize too hastily on virus biology from the study of a restricted set of materials.

SUMMARY

Escherichia coli strain K-12 carries a symbiotic phage, λ. This phage was discovered only by the occurrence of " mutant " substrains sensitive to λ, and serving as indicators for it. In addition to the lysogenic (carrier) and sensitive bacteria, two immune types (" 1 " and " 2 ") were found. These are defined as resistant but nonlysogenic.

The various types have been intercrossed to elucidate the genetic basis of lysogenicity. The crosses lysogenic × lysogenic; immune-1 × immune-1 and sensitive × sensitive have yielded only the parental class. Similarly, only the two parental classes were found in lysogenic × sensitive; lysogenic × immune-1; and sensitive × immune-1. The segregation of lysogenicity has been confirmed by the synthesis of diploid stocks heterozygous for lysogenicity, which behaves as a factor linked to Gal_4 (galactose fermentation). Genetic evidence of the transmission of λ as a cytoplasmic factor was not found. A locus for latent phage, Lp_1, which controls the maintenance of λ, or to which λ is bound is postulated. The detailed role of λ in the alteration of the Lp_1 locus that is associated with the resynthesis of lysogenic from sensitive has not been clarified.

Mutation of λ to a more virulent mutant λ-2 has been observed. λ-2 lyses λ-lysogenic as well as λ-sensitive bacteria. Immune-2 confers resistance both to λ and to λ-2. It does not, however, interfere with the maintenance of λ in bacteria already lysogenic. It is genetically separable from immune-1. A few additional E. coli stocks sensitive to λ, or lysogenic for other phages, have been found. In an extensive survey, λ itself has not recurred.

LITERATURE CITED

ADAMS, M. H., 1950 Methods of study of bacterial viruses. Methods in Medical Research 2: 1–73.

ANDERSON, E. S., 1951 The significance of Vi-phage types F1 and F2 of Salmonella typhi. J. Hyg. 49: 458–470.

BERTANI, G., 1951 Studies on lysogenesis. I. The mode of phage liberation by lysogenic Escherichia coli. J. Bact. 62: 293–300.

BOYD, J. S. K., 1951 Observations on the relationship of symbiotic and lytic bacteriophage. J. Path. Bact. 63: 445–457.

BURNET, F. M., 1945 Virus as Organism. 134 pp. Cambridge, Mass.: Harvard University Press.

BURNET, F. M., and D. LUSH, 1936 Induced lysogenicity and mutation of bacteriophage within lysogenic bacteria. Austr. J. Exp. Biol. Med. Sci. 14: 27–38.

CAVALLI, L. L., and H. HESLOT, 1949 Recombination in bacteria: outcrossing Escherichia coli K-12. Nature 164: 1057–1058.

DAVIS, B. D., 1950 Studies on nutritionally deficient bacterial mutants isolated by means of penicillin. Experientia 6: 41–50.

DELBRÜCK, M. (ed.), 1950 Viruses 1950. Pasadena: California Institute of Technology.

D'HERELLE, F. D. 1926 The Bacteriophage and Its Behavior. Transl. G. H. Smith. Baltimore: Williams and Wilkins. See p. 237.

FREDERICQ, P., 1948 Actions antibiotiques réciproques chez les Enterobactériaceae. Rév. belge path. med. exp. 19 (Supplement 4): 1–107.

HAYES, W., 1952 Recombination in Bact. coli K 12: unidirectional transfer of genetic material. Nature 169: 118.

LEDERBERG, E. M., 1952 Allelic relationships and reverse mutation in Escherichia coli. Genetics 37: 469–483.

LEDERBERG, J., 1949 Aberrant heterozygotes in Escherichia coli. Proc. Nat. Acad. Sci. 35: 178–184.
1950 Isolation and characterization of biochemical mutants of bacteria. Methods in Medical Research 3: 5–22.
1951a Prevalence of Escherichia coli strains exhibiting genetic recombination. Science 114: 68–69.

1951b Genetic studies with bacteria. Genetics in the 20th Century. viii + 634 pp. New York: Macmillan.

LEDERBERG, J., L. L. CAVALLI, and E. M. LEDERBERG, 1952 Sex compatibility in *Escherichia coli*. Genetics **37**: 720–730.

LEDERBERG, J., and E. M. LEDERBERG, 1952 Replica plating and indirect selection of bacterial mutants. J. Bact. **63**: 399–406.

LEDERBERG, J., E. M. LEDERBERG, N. D. ZINDER, and E. R. LIVELY, 1951 Recombination analysis of bacterial heredity. Cold Spring Harbor Symp. Quant. Biol. **16**: 414–443.

LOMINSKI, I., 1938 Souches lysogènes nonactives (cryptolysogènes). Compt. Rend. Soc. Biol. **129**: 151–153.

LWOFF, A., and A. GUTMAN, 1950 Recherches sur un *Bacillus megaterium* lysogène. Ann. Inst. Pasteur **78**: 711–739.

LWOFF, A., L. SIMINOVITCH, and N. KJELGAARD, 1950 Induction de production de phage dans une bactérie lysogène. Ann. Inst. Pasteur **79**: 815–858.

NICOLLE, P., and Y. HAMON, 1951 Recherches sur les facteurs qui conditionnent l'appertenance des bacilles paratyphiques B aux différents types bacteriophagiques de Felix et Callow. III. Nouvelles études sur les transformations de types par l'action des bactériophages extraits des bacilles lysogènes. Ann. Inst. Pasteur **81**: 614–630.

SCHIFF, F., and S. BORNSTEIN, 1940 Hemolytic effect of typhoid cultures in combination with pure lines of bacteriophage. J. Immunol. **39**: 361–367.

SONNEBORN, T. M., 1945 The dependence of the physiological action of a gene on a primer and the relation of primer to gene. Amer. Nat. **79**: 318–339.

1950 The cytoplasm in heredity. Heredity **4**: 11–36.

TATUM, E. L., 1945 X-ray induced mutant strains of *Escherichia coli*. Proc. Nat. Acad. Sci. **31**: 215–219.

TATUM, E. L., and J. LEDERBERG, 1947 Gene recombination in the bacterium *Escherichia coli*. J. Bact. **53**: 673–684.

WEIGLE, J. J., and M. DELBRÜCK, 1951 Mutual exclusion between an infecting phage and a carried phage. J. Bact. **62**: 301–318.

WILLIAMS-SMITH, H., 1948 Investigations on the typing of staphylococci by means of bacteriophage. J. Hyg. **46**: 74–81, 82–89.

WAHL, R., and L. BLUM-EMERIQUE, 1952 Les bactéries semi-resistantes au bactériophage. Ann. Inst. Pasteur **82**: 29–43.

21

Reprinted from *Advan. Genetics,* **11**, 109–114 (1962)

Episomes

ALLAN M. CAMPBELL

III. The Integrated State

A. CHROMOSOMAL LOCALIZATION OF PHAGE AND OTHER EPISOMES

A lysogenic culture is defined as one in which the capacity to form phage is perpetuated intracellularly and transmitted from one generation to the next. Since the phage has a specific set of genetic information, each cell of a lysogenic culture must contain one or more copies of this information, and to these copies the name *prophage* is given. It seemed of fundamental importance to know whether the prophage was present in a few copies, distributed regularly between the daughter cells at each division, or in many copies, distributed at random at division. Studies on superinfection of *Shigella* lysogenic for phage P2 (Bertani, 1953a) and of *E. coli* lysogenic for lambda (Jacob and Wollman, 1953) indicated that the former hypothesis was correct, and this was soon confirmed by the results of bacterial crosses of *Escherichia coli* (E. Lederberg and J. Lederberg, 1953; Appleyard, 1954a), which showed that, in a lysogenic cell, the genome of a temperate phage behaves as though it were located at a specific point on the bacterial chromosome. Crosses between lysogenic parents carrying genetically different lambda prophages revealed (1) that the attachment sites of the two prophages were allelic and (2) that there was linkage between the prophage and other genes of the bacterium. This observation has been generalized to other systems, although the facts do not completely justify such a generalization.

On the basis of those lysogenic systems for which crosses have been made, three types can be distinguished. (1) The lambda type, in which the phage has a unique point of attachment to the bacterial chromosome. Many different phages, some of them related to lambda, have been shown each to have its own specific attachment site (Jacob and Wollman, 1957b). (2) The P2 type, in which the phage preferentially occupies a particular site, but where it is possible to form, under appropriate conditions, quite stable lysogens in which the preferred site is vacant and one of the several possible secondary sites is instead occupied (Bertani and Six, 1958). (3) The P1 type, for which the segregation in crosses does not permit the assignment of any chromosomal location to the prophage (Jacob and Wollman, 1957b). A separate transfer mechanism for P1 is possible (Boice and Luria, 1961).

The difference between lambda and P2 may be technical rather than fundamental. These two phages differ in the behavior of homopolylysogens, which are bacterial lines perpetuating simultaneously more than one related prophage. In P2 double lysogens the two prophages are located at different places on the bacterial chromosome, whereas with lambda they occupy the same site (or extremely closely linked sites). Double lysogens of lambda are also less stable than those of P2. It seems possible that P2 can also form double lysogens of the lambda type, but that they are even more unstable and are never recovered, thereby rendering easier the isolation of double lysogens in which distinct locations are occupied.

In the case of prophage, one can infer a chromosomal location of the phage genome itself rather than of some other gene affecting prophage maintenance because each parent of a bacterial cross can be marked with a genetically different prophage. It has not been possible to do the analogous experiment with other episomes. One cannot therefore exclude totally the possibility that the $F+ \rightarrow$ Hfr change, for example, might involve, rather than the fixation of F itself at a particular site, a mutation at that site which affects the behavior of F during conjugation. Richter's (1957, 1961) finding of $F-$ recombinants which become Hfr upon infection with F fits this hypothesis. It can be explained alternatively by assuming that in these strains a *part* of F has adhered to the chromosome. Wollman and Jacob (1958a) have presented other evidence which can also be interpreted in this manner.

B. Mode of Attachment of the Prophage to the Bacterial Chromosome

The manner of attachment of the prophage to the bacterial chromosome has been much discussed in the past. At one time, the problem was how a small linear structure (the phage genome) could act as part of a large linear structure (the bacterial chromosome). The most appealing solution was that the small structure was inserted into the continuity of the large one so that one linear genome would result. In recent years, many data have appeared which seem to contradict such a co-linear

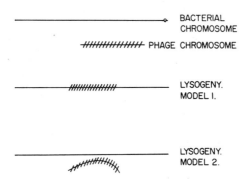

Fig. 1. Possible modes of attachment of prophage and bacterial chromosome. Model 1: Insertion. Model 2: Partial synapsis of central region of phage to bacterial chromosome.

structure, and popular taste has tended to favor some sort of branched model, in which the prophage is not really attached to the bacterial chromosome at all but rather portions of the two are permanently synapsed together. These two extreme cases are illustrated in Fig. 1 as Model 1 and Model 2, respectively. Bertani (1958) has categorized more completely the various possible modes of juxtaposition.

To evaluate much of the recent evidence on this subject, one must understand first that it is primarily evidence *against* the insertion hypothesis rather than *for* any particular kind of branched structure. One is thus really setting a model which makes quite specific predictions against *all* other possibilities, and whenever one prediction fails, the specific model is discarded. This seems a somewhat unfair procedure to the reviewer, and we will react to it by discussing primarily insertion hypotheses. To explain some of the facts in this way requires additional *ad hoc* assumptions, but insertion, with complications, is not inherently less desirable than an undefined model of branching or sticking together.

Besides such data which bear on the answer, there is also information which changes the question. Our large linear structure (the bacterial chromosome) behaves, on formal genetic analysis, not as a line but as a circle (Jacob and Wollman, 1957a). This really does not matter very much, but what if the small structure is also a circle? Detailed linkage studies lead to the conclusion that the genome of one phage (T4) is indeed circular Streisinger, Edgar, and Harrar, quoted by Stahl (1961). If circularity is a property of phages in general, the equivalent of the insertion hypothesis is to make one circle out of two, and we will discuss later the simplest model for accomplishing this.

The most obvious approach to distinguishing the various models is to examine the results of crosses between two lysogenic parents in which both the prophages and the bacterial chromosomes are well marked. The most complete study of this type thus far published is that of Calef and Licciardello (1960) on phage lambda. The assortment of prophage markers among bacteria recombinant for a pair of bacterial genes on either side of the prophage indicates a linear order with the prophage genes lying between those of the bacterium. However, the order of prophage markers on this map is different from that of the vegetative lambda phage. Whereas crosses of vegetative phage give the order "*h-cl-mi*," the order in the lysogenic cell is "*try-h-mi-cl-gal*."

This surprising result is supported by some other data suggesting a singularity in the region between *h* and *cl*. For example, Whitfield and Appleyard (1958) found with doubly lysogenic strains marked at these two loci that one recombinant type was liberated in excess of either parental type. The type preferred depended on the order of lysogenization and on the parental couplings, not on any selective advantages of the markers employed. This seems quite possible if the phage genome splits into two or more pieces at the time of lysogenization and is reassembled later. The results of Calef and Licciardello also allow one to contemplate mechanisms for the transduction of the galactose genes by phage lambda which otherwise would be unthinkable. This point will be amplified in a later section.

If the phage genome is circular rather than linear, the lambda chromosome need not be split into parts, but rather could be cut at a specific point on the circle when it lysogenizes. It is actually very simple (on paper) to insert a circular phage chromosome into a linear bacterial chromosome by reciprocal crossing-over (Fig. 2).

Figure 3 shows the genetic constitutions predicted for the chromosome of doubly lysogenic bacteria and for those carrying the defective, galactose-transducing lambda (Section VIII,A) on the model described here. If induction is imagined to entail a reversal of the process shown in Fig. 2, it is easily seen that one can make many different complete loops from a double lysogen, which would explain the result of Whitfield and Apple-yard. The instability of double lysogeny in lambda (Appleyard, 1954b; Arber, 1960), the unstable lysogeny of transducing lambda (Campbell, 1957), and the apparent correlation between loss of transducing lambda and recombination in the *gal* region (Arber, 1958) could all be explained as consequences of the presence of a longer region of duplication.

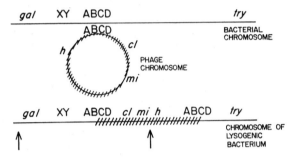

FIG. 2. Possible mechanism of lysogenization by reciprocal crossing-over between a circular phage chromosome and a linear bacterial chromosome. Arrows indicate possible rare points of breaking and joining in the formation of transducing lambda. (See Section VIII.) The genes ABCD are hypothetical and indicate a small region of homology between host and phage. X and Y are unspecified bacterial genes.

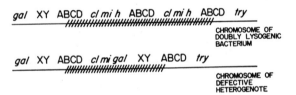

FIG. 3. Genetic constitution of doubly lysogenic bacterium and of defective heterogenote, assuming the mechanism of Fig. 2. The origin of the transducing lambda is described in the text (Section VIII).

On the other hand, this model would predict that it should be easier to lysogenize an already lysogenic strain than a non-lysogenic one, since the former will present a larger region of homology to an entering phage. This prediction is not fulfilled. Lysogeny creates a strong steric hindrance against lysogenization by another phage at the same site (Six, 1961a).

This illustrates well what we pointed out to begin with—that *ad hoc* hypotheses are required to fit most insertion models to the facts. We would assume that the pairing between phage and bacterial chromosomes may have other requirements besides the homology of the regions which pair. This is almost required by the model anyway, because the specificity of the attachment is determined by the immunity region (Kaiser and

Jacob, 1957), which does not coincide with the region of lambda where we wish pairing to take place in this case.

Evidence indicating that the prophage is not inserted into the bacterial chromosome has been presented by Jacob and Wollman (1959b). The most important concerns the non-inducible prophage 18, which is located close to two methionine markers (M1 and M2). Two Hfr strains by which this region was injected early but in opposite directions were isolated, and it was shown that the prophage appears to enter the bacterium after the marker M1 in both cases. They therefore suggest that the prophage may be synapsed parallel with the bacterial chromosome, overlapping the M1 gene, inasmuch as "prophage entrance" would require the entrance of the entire prophage genome. They also observed that the distance between M1 and M2 is not altered by lysogenization of both parents in a cross.

These facts should not be disregarded. They do not, in the reviewer's opinion, necessitate the abandonment of the insertion hypothesis, although it requires some ingenuity to circumvent them. In this review we will discuss mostly insertion hypotheses, not because we strongly favor them, but because we believe alternative ideas have received rather more than their share of attention at the hands of others.

Any model for prophage attachment must ultimately be extended to explain the behavior of homopolylysogenic strains of the lambda type. Strong asymmetries in the patterns of phage liberation and of segregation are seen with such strains, and there are reproducible differences between strains in the type of asymmetry shown (Arber, 1960a). Our picture in Fig. 2 is therefore at best a naïve first approximation to the true situation. It seems permissible at present because no simple explanation of the facts has been provided by alternative hypotheses.

22

Reprinted from *Biochem. Biophys. Res. Commun.*, **18**(5–6), 910–923 (1965)

THE LINEAR INSERTION OF A PROPHAGE INTO THE CHROMOSOME

OF E. COLI SHOWN BY DELETION MAPPING[1,2]

Naomi C. Franklin, William F. Dove[3] and Charles Yanofsky

Departments of Biological Sciences and Biochemistry,

Stanford University, Stanford, California

Received January 25, 1965

Evidence bearing on the relationship of a prophage to the chromosome of its host bacterium has been lucidly reviewed by Hayes (1964),
who considered that a clear decision between attachment and insertion
of prophage could not then be made. Just prior to Hayes' summation,
Campbell (1962) had formulated a largely theoretical model which would
allow for the linear insertion of phage genetic material into a bacterial chromosome, and would account for the permuted gene sequence of
the prophage as compared to vegetative phage. The permuted gene
sequence had been indicated by comparisons of the gene order in phage
recombination experiments and in crosses between bacteria lysogenic for
genetically marked λ prophages. (Calef and Licciardello, 1960). Evidence
in support of the model was subsequently supplied by Campbell (1963) in
studies of the segregation patterns of lysogenic heterogenotes, and by
Rothman (1965) in studies of P1 cotransduction of galactose and λ markers.

We describe in this paper observations which substantiate Campbell's
model by showing that deletions in particular lysogenic bacteria

---

[1] Supported by grants from the National Science Foundation and the
United States Public Health Service.

[2] W. F. D. would like to acknowledge an independent proposal of similar
experiments by Mr. R. Freedman.

[3] Public Health Service Postdoctoral Fellow. Present address: McArdle
Memorial Laboratory, University of Wisconsin.

910

simultaneously eliminate bacterial genes and segments of a prophage
genome. The missing phage segments are overlapping, allowing sequencing of the prophage genes.

The prophage used is a hybrid with the immunity of $\emptyset$80 and the host
range of λ. Like the $\emptyset$80 prophage (Matsushiro, 1963), it resides in
E. coli K12 close to the genes which determine the structures of several
enzymes essential for the synthesis of tryptophan (Yanofsky, 1960).
Mutations of E. coli K12 to T1-resistance (T1^r) not infrequently result
from deletions which extend into the tryptophan (tryp) operon for varying distances (Yanofsky and Lennox, 1959). T1^r mutants isolated from
lysogenic bacteria are also either tryp$^+$ or tryp^{deletion}, and in addition may lack different amounts of prophage genetic material.

MATERIALS AND METHODS

Bacteria: E. coli K12 #1485, a non-lysogenic prototroph; a T5-
resistant mutant, 1485T5^r (T5 resistance also confers resistance to T1,
$\emptyset$80 and $\emptyset$80h, but not to λ); a T1-resistant mutant, 1485T1^r (T1 resistance also confers resistance to $\emptyset$80 but not to $\emptyset$80h nor to λ); lysogenic derivatives of these bacteria.

E. coli K12 #W3101, non-permissive for the λ sus mutants
used (Campbell, 1961).

Phages: Virulent phages T1 and T5.

Transducing phage Plkc = Pl (Lennox, 1956).

Temperate phage λ wild type, obtained from Dr. D. K. Fraser,
and several of its suppressor-sensitive (sus) mutants (Campbell, 1961):
sus A (= sus 11), sus F (= sus 96b), sus I (= sus 2), sus J (= sus 6),
sus K (= sus 24), sus L (= sus 63), sus N (= sus 7) and sus R (= sus 5).

Temperate phage $\emptyset$80 (Matsushiro, 1963). From a sample
kindly sent by Dr. Matsushiro, a single plaque was isolated and high
titer stocks were prepared by Dr. R. L. Somerville in this laboratory.
A spontaneous mutant $\emptyset$80h was selected from $\emptyset$80 by its ability to plate
on 1485T1^r.

Temperate phages λ and $\emptyset80$ differ in immunity specificity (i) and in ability to adsorb to various bacteria (h), as well as in plaque type, serological properties, density and other characteristics. Nevertheless, recombination between $\emptyset30$ and λ occurs (R. L. Somerville, personal communication; Signer, 1964). From 1485 doubly lysogenic for $\emptyset30$ and λ a lysate was obtained after UV induction which contained infective phages with different bouyant densities and recombinant characteristics of λ and $\emptyset80$. By CsCl density gradient centrifugation of the lysate, a phage type was recovered with a density of 1.510 g/ml (slightly denser than λ, considerably denser than $\emptyset80$), which was able to form plaques on $1485(\lambda)Tl^r$. This phage was therefore recombinant for the immunity determinant of $\emptyset80$ and the host range determinant of λ, and is designated $i^{\emptyset80}h^{+\lambda}$. A clear mutant of this hybrid, $i_c^{\emptyset80}h^{+\lambda}$, was used to test immunity to $\emptyset80$ in bacteria which are Tl^r. Bacterium 1485 lysogenized with $i^{\emptyset80}h^{+\lambda}$ was the parental stock for the present studies.

The extent of genetic homology between λ and $i^{\emptyset80}h^{+\lambda}$ was scanned by performing crosses of $i^{\emptyset80}h^{+\lambda}$ with various λ sus mutants. Hybrid $i^{\emptyset80}h^{+\lambda}$ gave evidence of containing sus A^+, B^+, C^+, E^+, F^+, N^+, P^+ and R^+. That the whole left hand arm (sus A through h, Fig. 1) of the hybrid stemmed from λ is suggested by the fact that $\emptyset80$ itself cannot contribute functions of sus A^+, B^+ or C^+ in complementation tests with λ sus$^-$ for these alleles, whereas $i^{\emptyset80}h^{+\lambda}$ can. Since $\emptyset80$ can provide sus N^+, P^+ and R^+, as shown in complementation and recombination tests, it is likely that the right-hand arm of $i^{\emptyset80}h^{+\lambda}$ stems from $\emptyset80$.

Media: Tryptone broth: 10 g Bacto tryptone, 5 g NaCl, 1 liter H_2O; 12.5 g Bacto agar added for plates; 7 g Bacto agar added for soft agar top layer.

Minimal medium (Vogel and Bonner, 1956) with 0.2% glucose and 0.2% acid hydrolyzed casein (= MCA), supplemented when desired with 20 γ/ml L-tryptophan, solidified with 1.5% Bacto agar.

912

<u>Induction</u> <u>and</u> <u>superinfection</u> <u>of</u> <u>lysogenic</u> <u>bacteria</u>: Bacteria were
grown in MCA + tryptophan or in broth to a concentration of 3×10^8
cells/ml, UV-irradiated (broth cultures are transferred to saline before
irradiation), diluted 2X into broth with superinfecting phage, diluted
further 0-100X after adsorption of the superinfecting phage, and allowed
to lyse during 2 hours at 37°C with aeration. The lysates were sterilized with chloroform. For superinfection with λ, the bacteria must be
starved by aerating in 0.01 M $MgSO_4$ at 37°C for 40 minutes before UV
irradiation.

RESULTS

Initially it was observed in $1485(\underline{i}^{\emptyset 80}\underline{h}^{+\lambda})$ that Tl^r mutants which
simultaneously suffered deletion of <u>tryp</u> genes of the bacterium also
occasionally became defective lysogens: the bacteria were still immune
to $\emptyset 80$; however, few if any infective phage particles were released. A
similar observation has been made in <u>E</u>. <u>coli</u> $B(\underline{i}^{\lambda}\underline{h}^{+\emptyset 80})$ (Signer, in
preparation). This situation was investigated further.

Tl^r mutations occur spontaneously in <u>E</u>. <u>coli</u> at a frequency of 10^{-6}
to 10^{-7}. A large number of Tl^r mutants were isolated, some of which
required tryptophan. To assure that each mutant was the result of an
independent mutational event, the following procedure was employed. A
series of broth culture tubes was inoculated with $1485(\underline{i}^{\emptyset 80}\underline{h}^{+\lambda})$ to a concentration of 10^3 cells/ml. These were grown with shaking at 37°C to
4×10^8 cells/ml. All these initial sub-cultures were shown to produce
infective phage in quantity. From each of twenty individual cultures,
about 10^9 bacteria were spread on broth-agar with 10^9 Tl phages. Of the
Tl^r survivors, those mutated at the Tl-receptor gene adjacent to <u>tryp</u>
(as opposed to the Tl-T5-receptor gene unlinked to <u>tryp</u>) were recognized
by their small colony size. These were purified by streaking twice, and
were tested for their ability to produce infective phage. They were also
tested for immunity to $\emptyset 80$ by spot testing with $\underline{i}_c^{\emptyset 80}\underline{h}^{+\lambda}$. The Tl^r <u>tryp</u>⁻

913

colonies were located by their ability to grow when replicated to MCA agar (tryptophan-free); they constitute about 5% of the $T1^r$ mutants. $T1^r$ mutants are considered to represent independent mutational events if they arise in different subclones, or if they differ from the other $T1^r$'s from the same tube in any of the characteristics tested (Table I).

TABLE I

Number of independent isolates from $1485(\underline{i}^{\emptyset 30}\underline{h}^{+\lambda})$ which were $T1^r$ and distinguishable by the characteristics given.

	$\underline{tryp}^{+\cdot}$	$\underline{tryp}^-$
active lysogenic	15	3
immune defective	15	19
not immune	0	3

From the independent $T1^r$ mutants of $1485(\underline{i}^{\emptyset 30}\underline{h}^{+\lambda})$, 33 isolates, immune-defective or non-immune, were selected for further study of the amount of prophage material that was retained. The following properties were examined.

1) <u>Rare production of infective phage</u>. The parental strain and each mutant clone was grown in MCA + tryptophan to 2×10^8 cells/ml, UV-irradiated, mixed 1:1 with broth, shaken at 37°C for 3 hours, and sterilized with chloroform. Although no lysis was evident, some $T1^r$ clones did produce infective phages in small amounts (about 10^{-3} X normal yield of $\underline{i}^{\emptyset 30}\underline{h}^{+\lambda}$). The plaques formed on 1485 were weakly centered, indistinguishable from those of $\underline{i}^{\emptyset 30}\underline{h}^{+\lambda}$, indicating normal phage growth. Nevertheless, lysogenic clones could not be derived from these plaques. A search for more defective phage types by plating on UV-induced $1485(\lambda)$ was not successful. Lysates containing the non-lysogenizing phages were able to transduce the <u>tryp</u> A gene (10^{-3} transductants per plaque forming phage); the transductants, though initially immune to $\underline{i}_c^{\emptyset 80}\underline{h}^{+\lambda}$, became sensitive after restreaking. These sensitive <u>tryp</u>$^+$ clones did not yield high-frequency transducing lysates, even if superinfected with $\emptyset 30$, indi-

914

cating that transduction by lysogenization cannot occur with these lysates.

2) <u>Ability</u> <u>to</u> <u>contribute</u> <u>the</u> $\underline{h}^{+\lambda}$ <u>gene</u>. The parental culture and the Tl^r mutants were UV-irradiated and superinfected with $\emptyset$80$\underline{h}$ at a multiplicity of infection (moi) = 5. Lysates were tested for presence of the $\underline{h}^{+\lambda}$ allele from $\underline{i}^{\emptyset 80}\underline{h}^{+\lambda}$ prophage by spot testing or plating on 1485T5^r, which can adsorb phage with $\underline{h}^{+\lambda}$ phenotype, but not $\underline{h}^{\emptyset 80}$.

3) <u>Ability</u> <u>to</u> <u>contribute</u> $\underline{i}^{\emptyset 80}$. The parental strain and its Tl^r mutants were UV-induced and superinfected with λ at moi = 3, or with <u>sus</u>$^-$ mutants of λ (see 4). Lysates were tested for the presence of infective phage with the $\underline{i}^{\emptyset 80}$ gene by plating on 1485(λ).

4) <u>Ability</u> <u>to</u> <u>contribute</u> <u>sus</u>$^+$ <u>alleles</u> <u>of</u> λ <u>sus</u> $\underline{A}^-$, $\underline{F}^-$, $\underline{I}^-$, $\underline{J}^-$, $\underline{K}^-$, $\underline{L}^-$, $\underline{N}^-$, or $\underline{R}^-$. Strain 1485($\underline{i}^{\emptyset 80}\underline{h}^{+\lambda}$) and its Tl^r mutants were UV-irradiated and superinfected at moi = 1 with each of several <u>sus</u>$^-$ mutants of λ. Lysates were assayed for the presence of <u>sus</u>$^+$ infective phage by plating on a non-permissive bacterium, W3101. These lysates were also tested for the presence of infective phage with $\underline{i}^{\emptyset 80}$ by plating on 1485(λ).

The presence of <u>sus</u>$^+$ alleles in defective prophages can be screened easily by a quick spot test procedure (Campbell, 1964). Non-permissive W3101 is poured in soft agar onto broth-agar plates. The defective lysogenic bacteria to be tested are spotted over the surface. A series of λ <u>sus</u>$^-$ mutants (10^8 phage/ml) is then spotted over the defective bacteria, and the plates are UV-irradiated with 1/8 inducing dose. Any infective <u>sus</u>$^+$ recombinant phages will be detected by lysis of W3101. Control spots are necessary on W3101 of each defective lysogenic bacterium alone, and of each λ <u>sus</u>$^-$, alone or with 1485; the last test is needed because growth of λ <u>sus</u>$^-$ on permissive bacteria permits a certain accumulation of <u>sus</u>$^+$ revertants.

The characteristics of the defective Tl^r mutants of 1485($\underline{i}^{\emptyset 80}\underline{h}^{+\lambda}$) as shown by the above tests are given in Table II. On the basis of deletion endpoints, an unambiguous linear map can be derived for most of the

915

TABLE II

Examination of 1485($i^{\emptyset 80} h^{+\lambda}$) and its Tl^r mutants for phage and bacterial markers

	Infective phage average burst size	Markers detected or recovered									
		tryp $h^{+\lambda}$	sus J^+	sus I^+	sus K^+,L^+	sus F^+	sus A^+	sus R^+	sus N^+	by immunity	$i^{\emptyset 80}$ by recombination
1485($i^{\emptyset 80} h^{+\lambda}$)	4-20	+	+	+	+	+	+		+	+	
Tl^r mutant # 1	10^{-2}-10^{-3}	++	+	+	+	+	+		+	+	+
3		+	+	+	+	+	+		+	+	
4, 5, 14, 31		+	+	+	+	+	+		+	+	
10, 22		++	+	+	+	+	+		+	+	
11, 23		++	+	+	+	+	+		+	+	
12, 16		+	+	+	+	+	+		+	+	
20, 28		++	+	+	+	+	+		+	+	
32		+	+	+	+	+	+			+	
13, 29, 33	$<10^{-5}$	-	+	+	+	+	+	+	+	+	
25		+	+	+	+	+	+	+	+	+	
24		-	+	+	+	+	+	+	+	+	
27		-	-	-	-	+	+	+	+	+	
7, 8		-	-	-	-	-	+	+	+	+	
6, 17, 30		+	-	-	-	-	+	+	+	+	+
21		+	-	-	-	-	-	(+)	(+)	+	(+)
2		-	-	-	-	-	-	-	-	+	-
9		- B^+A^-	-	-	-	-	-	-	-	+	-
18		+	-	-	-	-	-	-	-	+	-
15		-	-	-	-	-	-	-	-	+	-
19, 26		-	-	-	-	-	-	-	-	-	-

916

260

prophage genes that were scored (Fig. 1).

That a defective prophage resulting from Tl^r mutation in $1485(\underline{i}^{\emptyset 80}\underline{h}^{+\lambda})$ is actually located on the bacterial chromosome close to Tl^r and residual

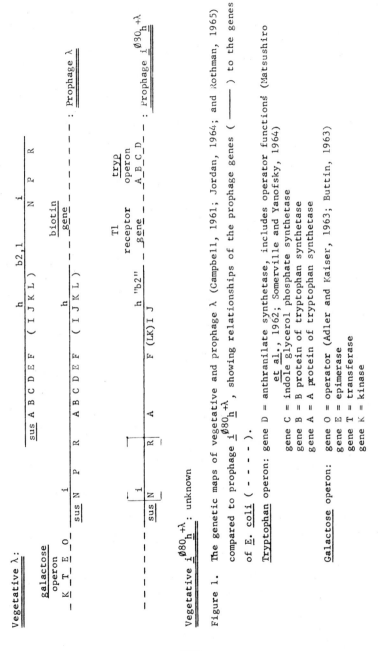

Figure 1. The genetic maps of vegetative and prophage λ (Campbell, 1961; Jordan, 1964; and Rothman, 1965) compared to prophage $\underline{i}^{\emptyset 80}\underline{h}^{+\lambda}$, showing relationships of the prophage genes (————) to the genes of E. coli (- - - -).

Tryptophan operon: gene D = anthranilate synthetase, includes operator functions (Matsushiro et al., 1962; Somerville and Yanofsky, 1964)
gene C = indole glycerol phosphate synthetase
gene B = B protein of tryptophan synthetase
gene A = A protein of tryptophan synthetase

Galactose operon: gene O = operator (Adler and Kaiser, 1963; Buttin, 1963)
gene E = epimerase
gene T = transferase
gene K = kinase

917

tryp genes could be confirmed by Pl transduction. Phage Pl grown on Tl^r defective #9, tryp $B^+A\underline{\underline{del}}$, was used to transduce a tryp B^- recipient (point mutant). The B^+ transductants, selected by their ability to utilize indole in place of tryptophan, were of two types only: 10% $B^+A^+Tl^S$-$\emptyset 30^S$, 90% $B^+A^-Tl^r\emptyset 30\underline{\underline{immune}}$. Thus immunity to $\emptyset 30$ was cotransduced with B^+, in close linkage with tryp $A\underline{\underline{del}}$ and Tl^r. The absence of recombination between tryp $A\underline{\underline{del}}$, Tl^r and immunity to $\emptyset 80$ is one criterion for considering that these three were lost simultaneously by deletion.

The information in Table II does not permit us to distinguish between several of the more defective prophages: defective lysogenic Tl^r mutants #17, 21 and 30, which all contribute genes sus $A^+R^+N^+$ and $i^{\emptyset 30}$ upon superinfection with λ sus$^-$, and mutants #9, 13 and 15 which do not contribute any of the above mentioned genes, although they show immunity to $\emptyset 80$. Mutant #2 is intermediate between these two groups in that, unable to contribute sus A^+ and immune to $\emptyset 30$, it provides sus R^+ and N^+ poorly. When Tl^r mutants #17, 21 and 30 are superinfected with λ sus N^- or R^-, substantially more i^λ sus R^+ are recovered than i^λ sus N^+, showing the closer linkage of sus N to i than sus R to i, as in phage λ itself. The order of sus R, sus N and i, however, is not given by these experiments.

DISCUSSION

A large proportion of the Tl^r mutants of 1435($i^{\emptyset 30}h^{+\lambda}$) become defective lysogens; i.e., immune to $\emptyset 30$, but unable to yield infective phages in normal amount, if at all. When these defective prophages are tested for their ability to contribute $i^{\emptyset 80}$, $h^{+\lambda}$ and various sus$^+$ alleles by recombination, it is found that the defective prophages lack varying amounts of their parental genome. [A partially deleted λ prophage has recently been described by Fischer-Fantuzzi and Calef (1964).] The deletions of $i^{\emptyset 30}h^{+\lambda}$ constitute a coherent series, allowing the interpretation that deletions of the Tl-receptor bacterial gene can extend for

918

varying distances into a prophage whose chromosomal location is adjacent
to this gene. The deletions can terminate at various points within the
prophage. Therefore, the prophage must be linearly inserted into the
bacterial chromosome. This finding of linear prophage insertion is supported by recent work showing increased recombination frequencies between
two bacterial markers after lysogenization with a prophage whose chromosome location lies between the two markers studied (Rothman, 1965;
Signer, in preparation).

It was known previously that mutation to T1 resistance in E. coli
occasionally results in the simultaneous loss of tryptophan-synthesizing
ability (Anderson, 1946; Yanofsky and Lennox, 1959). Physiological and
genetic studies of these $T1^r$ $\underline{tryp}^-$ mutants indicated that deletions of
the bacterial genes had occurred. In the present study it is seen that
$T1^r$ mutants may simultaneously lose prophage genes as well as $\underline{tryp}$ functions, supporting the interpretation that the prophage genes are also
deleted. On the other hand, loss of $\underline{tryp}$ functions and prophage genes
may occur independently in conjunction with mutation to T1 resistance,
showing that $\underline{tryp}$ and prophage must lie on opposite sides of the T1-
receptor gene. This gene order has been independently deduced (for prophage $\emptyset 80$) by transduction mapping with phage P1. The deletions show
great variability in their points of termination on either side of the
T1-receptor gene, though in K12 strains they rarely extend beyond the
$\underline{tryp}$ operon on the one side, or beyond the prophage immunity gene on the
other. Deletions extending into the prophage are about ten times more
frequent than those affecting tryptophan prototrophy.

The coherent set of $T1^r$ deletions entering the $\underline{i}^{\emptyset 80}\underline{h}^{+\lambda}$ prophage
allows mapping of that prophage in a manner analogous to the deletion
mapping described by Benzer (1961) and Campbell (1961). The gene order
found (Fig. 1)/clearly the permutation of the vegetative λ gene order
which has already been described for prophage λ by Campbell (1963) and
by Rothman (1965). Although we do not know the order of genes in vege-

919

tative $\underline{i}^{\phi 30}\underline{h}^{+\lambda}$, we think it likely that it is the same as the order in vegetative λ. Certainly homology between λ and $\underline{i}^{\phi 30}\underline{h}^{+\lambda}$ is revealed by recombination between several of their genes.

Deletion mapping of prophage $\underline{i}^{\phi 80}\underline{h}^{+\lambda}$ can be used to order some of the λ genes which could not previously be ordered. The $\underline{sus}^{-}$ alleles I, J, K and L are all closely linked to $\underline{h}$, all five being regularly replaced by $\underline{gal}$ in λ $\underline{dg}$ (Arber, Kellenberger and Weigle, 1957; Campbell, 1961). Results here show that the order of loci is (KL) I J $\underline{h}$. This location for $\underline{sus}$ J is consonant with the finding that $\underline{sus}$ J mutants are deficient in antiserum-sensitive antigens of λ (W. F. Dove, unpublished).

Among the Tl^{r} mutants, a class of defective lysogens is found in which the prophage retains the $\underline{h}^{+\lambda}$ gene and all $\underline{sus}$ genes tested, yet loses almost all ability to produce infective phage particles. A deletion of prophage genetic material at the $\underline{h}$ terminus is therefore indicated. This class of defective lysogen produces a few infective particles with an average burst size of 0.01, showing that an event of low efficiency is able to restore infectious phage. But the few phages produced, though apparently able to replicate normally, are unable to lysogenize. This behavior is ostensibly similar to that of the λ $\underline{b}2$ mutant and the λ $\underline{l}$ mutant, both of which are associated with deletions of the λ chromosome (Kellenberger, Zichichi and Weigle, 1961). Furthermore, the terminal position of our deletions in apparent proximity to $\underline{h}$ is in accord with the known position of $\underline{b}2$, linked to $\underline{h}$ (Kellenberger et al., 1961), lying between $\underline{i}$ and $\underline{h}$ on the vegetative map of λ (Jordan, 1964). It appears, therefore, that a terminal prophage deletion may be the source of $\underline{b}2$-type mutants.

The fact that prophages with terminal deletions are able to reconstitute replicating though defective structures is of importance to Campbell's (1962) model for lysogenization and induction of temperate prophages. This model predicts that in a lysogenic bacterium the prophage is bounded by duplicated regions which arose during an integra-

920

264

tional recombination between homologous regions of the phage and the
parental bacterium. Recombination between such duplicated regions would
allow escape of the prophage from the bacterial chromosome. A deletion
of part of the member of the duplication proximal to the T1-receptor
gene might account for reduced frequency of phage release. But it would
not account for the release of an incomplete phage genome. Thus the
incomplete phages would arise as exceptions to the above model, just as
would transducing phage particles, which also are found in lysates of
these terminally deleted strains.

The present experiments indicate that prophage fragments remaining
after deletions even more drastic than the terminal deletion type can
replicate after induction. Prophages which have suffered deletions
extending beyond gene h are still, upon superinfection, able to contribute sus K^+ or sus A^+ with an efficiency near to that of a non-deleted
prophage. Thus prophages deleted even beyond the terminus may be able
to replicate normally.

When the deletion extends into sus A, however, none of the distal
markers can be recovered, although the immunity function may still be
in evidence. This may simply reflect limited homology between the distal prophage half, stemming from $\emptyset$80, and that of the superinfecting λ
test phage. But it is tempting to speculate that the boundary observed
beyond prophage sus A has greater significance, since sus A marks an end
of the vegetative phage.

Finally there is interest in the relationships of the prophage genes
to the bacterial genes. In comparing λ and $i^{\emptyset 80}h^{+\lambda}$ prophage maps (Fig. 1),
it is seen that the biotin gene of E. coli bears the same relationship
to the λ prophage as does the tryp operon to $i^{\emptyset 80}h^{+\lambda}$. It is noteworthy
that respective non-defective transducing phages can be recovered carry-
ing these bacterial markers (Wollman, 1963; Matsushiro et al., 1964),
but not the galactose genes. Furthermore, the tryp and gal operons read
in the same direction (right to left as written) relative to the order

921

of the adjacent prophage markers. If the prophages are considered as having a polarity, reading from early functions at the <u>sus</u> N end to late functions at the <u>h</u> end, then the prophage polarity is opposite, in both cases, to the operon polarity of the bacterial genes. The many implications of these relationships remain to be examined.

ACKNOWLEDGMENT

The work reported above is superficially far removed from the gene-enzyme relationships in <u>Neurospora</u> which first occupied me as a graduate student under Dave Bonner. His lack of inhibition fostered mobility in terms of experimental systems and approach. Yet the interest that he inspired in the examination of genetic phenomena is evident. A veteran of much excellent guidance through subsequent years of study, I still count myself most fortunate to have started in research with Dave, to have witnessed over many years his extraordinary quality and endearing nature. N. C. F.

REFERENCES

Adler, J. and Kaiser, A. D., Virology <u>19</u>, 117 (1963).
Anderson, E. H., Proc. Natl. Acad. Sci. U.S. <u>32</u>, 120 (1946).
Arber, W., Kellenberger, G., Weigle, J., Schweiz. Z. path. Bakt. <u>20</u>, 659 (1957).
Benzer, S., Proc. Natl. Acad. Sci. U.S. <u>47</u>, 403 (1961).
Buttin, G., J. Mol. Biol. <u>7</u>, 164 (1963).
Calef, E. and Licciardello, G., Virology <u>12</u>, 81 (1960).
Campbell, A., Virology <u>14</u>, 22 (1961).
Campbell, A., Adv. Genetics <u>11</u>, 101 (1962).
Campbell, A., Virology <u>20</u>, 344 (1963).
Campbell, A., in "The Bacteria", eds. I. C. Gunsalis and R. Y. Stanier, Academic Press, New York, p. 49 (1964).
Fischer-Fantuzzi, L. and Calef, E., Virology <u>23</u>, 209 (1964).
Hayes, W., "The Genetics of Bacteria and Their Viruses", Blackwell Scientific Publications, Oxford (1964).
Jordan, E., J. Mol. Biol. <u>10</u>, 341 (1964).
Kellenberger, G., Zichichi, M. L. and Weigle, J., J. Mol. Biol. <u>3</u>, 399 (1961).
Lennox, E. S., Virology <u>1</u>, 190 (1956).
Matsushiro, A., Virology <u>19</u>, 475 (1963).
Matsushiro, A., Kida, S., Ito, J., Sato, K. and Imamoto, F., Biochem. Biophys. Res. Comm. <u>9</u>, 204 (1962).

922

Matsushiro, A., Sato, K. and Kida, S., Virology 23, 299 (1964).

Rothman, J. L., Ph.D. Dissertation, Massachusetts Institute of Technology (1965).

Signer, E. R., Virology 22, 659 (1964).

Somerville, R. L. and Yanofsky, C., Bact. Proc., p. 88 (1964).

Vogel, H. and Bonner, D. M., Microb. Genetics Bull. No. 13, 43 (1956).

Wollman, E. L., Compt. Rend. Acad. Sci. (Paris) 257, 4225 (1963).

Yanofsky, C., Bact. Revs. 24, 221 (1960).

Yanofsky, C. and Lennox, E. S., Virology 8, 425 (1959).

923

Reprinted from *J. Mol. Biol.*, 1(3), 379–386 (1959)

Density Alterations Associated with Transducing Ability in the Bacteriophage Lambda†

J. Weigle and Matthew Meselson

Division of Biology, Norman W. Church Laboratory for Chemical Biology and Gates and Crellin‡ Laboratories of Chemistry, California Institute of Technology, Pasadena, Cal., U.S.A.

AND

Kenneth Paigen

Roswell Park Memorial Institute, Buffalo, N.Y., U.S.A.

(*Received 2 October 1959*)

The transducing variant of the bacteriophage λ is thought to arise by a double crossover in which a section of the normal chromosome is exchanged for a region of the bacterial chromosome containing galactose fermentation markers. If the exchange were unequal, the resulting transducing phages should have an altered DNA content which might be detected as a modification of the density of the phage particle. An examination of ten independently arising populations of transducing phages showed that each had a different but essentially uniform density. In contrast, the non transducing phage contained in these lysates possessed a constant density. The density of a transducing phage was stable to a cycle of lysogenization, prophage replication and multiplication following induction, as well as to genetic interactions with normal λ and with the bacterial chromosome.

Assuming the observed density changes to be due to changes in DNA content, it was calculated that the maximum change corresponded to 10^7 mol wt units, or 1.5×10^4 nucleotide pairs, and the smallest to 800 nucleotide pairs.

It is proposed that the unequal exchanges giving rise to the density alterations are the result of poor homology and incomplete pairing between the *Gal* region of the bacterial chromosome and the region of λ which is deleted.

1. Introduction

The bacterial strain *E. coli* K12, lysogenic for the bacteriophage λ, liberates upon either induced or spontaneous lysis a population of phages which contain one exceptional phage among approximately 10^4 "ordinary" λ. The exceptional quality of this phage is its transducing ability: it carries genes for galactose fermentation (*Gal*) from the original host bacterium into the bacterium it subsequently lysogenizes. Only *Gal* markers, which are closely linked to the prophage attachment site of λ, are known to be transduced by λ. Lysates from cells lysogenic for ordinary λ and containing this low proportion of transducing particles are called LFT (low frequency transducing). (Morse, Lederberg & Lederberg, 1956.)

Transducing λ is defective in the sense that it cannot multiply in the absence of ordinary λ and it is thus called λ*dg* (λ defective, galactose) (see Arber, Kellenberger & Weigle, 1957; Campbell, 1957; Arber, 1958). It can, however, establish itself as a prophage and be replicated as such when the bacterium multiplies. If ordinary λ is

†Aided by grants from the National Foundation and the National Institutes of Health.
‡Contribution No. 2463.

present in a cell together with λdg, as in the case of cells lysogenic for both λ and λdg, induction yields a lysate containing an equal proportion of both phages. Such lysates are called HFT (high frequency transducing).

The original event which gives rise to λdg may be thought of as a double crossover in which a piece of the λ chromosome has been exchanged for a section of the bacterial chromosome containing the *Gal* genes. In accord with this view the transducing phage is known from genetic experiments (Arber, 1958) not only to carry the *Gal* genes of the bacterium, but also to lack approximately one-fourth of the mapped length of the ordinary λ chromosome.

One might ask whether this exchange is exactly reciprocal, and thus whether λdg contains the same amount of genetic material as λ. A change in the quantity of genetic material might result in an altered ratio of DNA to protein and might thereby alter the density of the phage. To study this question we have investigated the density of phages from HFT lysates, assuming them to be the same as those from the LFT from which they were derived. The method of equilibrium density-gradient centrifugation (Meselson, Stahl & Vinograd, 1957) was used in these studies. It was found that the density of ordinary λ is constant whereas, *mirabile dictu*, the density of the λdg of each independent HFT has a characteristic and unique value different from that of λ. These densities are genetically stable. It is possible to explain the density changes as a consequence of the original event giving rise to λdg.

2. Material and Methods

(a) *Bacterial strains*

The bacterial strains were lysogenic 112–12 (Wollman, 1953) and a series of strains derived from Lederberg's 3110 (F^-, Gal^+, prototroph) either by lysogenization or by transduction of the *Gal* markers 1 and 2 (Morse *et al.*, 1956). The latter strains were used for most of the experiments described and are designated by their galactose markers and any prophage they carry. For example, $Gal\ 1^+\ 2^+$ (λ) is the original strain 3110 lysogenic for λ, while $Gal\ 1^-\ 2^-$ (λ) (λdg $Gal\ 1^+\ 2^+$) is a strain having the 1^- and 2^- *Gal* markers, and is lysogenic for λ and for λdg carrying the *Gal* markers 1^+ and 2^+.

(b) *Bacteriophage*

Phage λ*ref* is the reference type of Kaiser (1955). Phage λ*h* was derived from λ*ref* and has extended host range. Phage λ*vir* is the virulent mutant described by Lederberg & Lederberg (1953). Phage λ*vir h* was obtained from a cross of λ*h* by λ*vir*.

The techniques for assaying plaque-forming particles have been described by Adams (1950) and those for the assay of transducing phages and for u.v. induction have been given by Arber (1958).

(c) *Media*

Growth medium was either tryptone broth (10 g Difco tryptone, 5 g NaCl, 1 l. H_2O) or K medium (100 ml. of double strength M9, 100 ml. of 3% casamino acids treated with decolorizing charcoal to remove u.v. absorbing material, 0·1 g NaCl, 0·06 g $MgSO_4$ and 0·4 g glucose). Suspension medium contained 6 ml. M-tris (2-amino-2-hydroxymethyl-propane-1:3-diol) at pH 7·5, 0·12 g $MgSO_4$, 4 g NaCl, 0·05 g gelatin, and 1 l. H_2O. To distinguish Gal^+ from Gal^- colonies, tetrazolium triphenyl chloride was used as a fermentation indicator in a plating medium of the following composition: 15 g agar, 10 g tryptone, 5 g NaCl, 0·5 g tetrazolium triphenyl chloride, 4 g galactose, and 1 l. H_2O.

Cesium chloride obtained from the American Potash Company was recrystallized from water and treated in solution with decolorizing charcoal to remove u.v. absorbing material.

(d) *Preparation of purified phage*

Lysogenic bacteria were grown in K medium to a titer of about 4×10^8 per ml. The cultures were induced in samples 2 mm deep and were allowed to lyse at 37°C with aeration. Several drops of chloroform were added and the suspensions were centrifuged at low speed before being filtered through a No. 9 candle. The filtrates contained 2 to 4×10^{10} plaque-forming particles per ml. They were centrifuged for 90 min at 30,000 rev/min in a No. 30 Spinco preparative ultracentrifuge rotor and the pellet was resuspended by standing overnight in suspension medium. Another low speed centrifugation brought down debris which had passed the filter.

(e) *Analytical density-gradient centrifugation*

A portion of purified phage was added to a 57% (w/w) stock solution of CsCl buffered at pH 8·0 with 0·015 M-tris buffer to give a final concentration of 1 to 2×10^{10} plaque-forming particles per ml. after the density of the sample had been adjusted to $1·50 \pm 0·01$ g cm^{-3} by the addition of water. The density at 20°C was calculated according to the relation

$$\sigma^{20} = 10·860\, n_D^{25} - 13·500$$

where n_D^{25} is the index of refraction of the solution for sodium light at 25°C determined in an Abbé refractometer. A 0·70 ml. portion of the adjusted phage suspension was then placed in a cell equipped with a plastic (Kel-F) centerpiece and was centrifuged in a Spinco model E analytical ultracentrifuge at 27,690 rev/min at 20°C.

The various phages move through the CsCl concentration and density gradient, which forms during the first few hours of centrifugation, to the region at which their buoyant density equals that of the solution, thereby forming bands. Equilibrium is closely approached in 10 hr, whereupon ultraviolet absorption photographs are taken.

The determination of the relative buoyant densities of banded phage was greatly facilitated by the addition to each sample of a density reference material. This was a dense λ variant obtained from cells of the lysogenized K12 strain 112–12 whose density was increased by growth in a medium containing ^{15}N. This phage has a buoyant density in CsCl solution of 1·525 g cm^{-3}, considerably greater than that of any other λ phage we have studied.

(f) *Preparative density-gradient centrifugation*

To recover the phages from each band separately, CsCl phage suspensions prepared as described above were centrifuged in "lusteroid" tubes in a Spinco SW39 swinging bucket rotor at 22,000 rev/min. After 16 hr, equilibrium is closely approached and the rotor was allowed to come to rest without braking. The plastic tube was withdrawn and its bottom was pierced with a fine needle. Successive emergent drops were collected for subsequent examination. A 3 ml. sample yielded approximately 120 drops, each drop comprising a successive density layer in the gradient. This technique permits even a single virus particle to be isolated and characterized biologically.

3. Results

(a) *Centrifugation of ordinary lambda*

The phage λ formed a single sharp band at equilibrium in a CsCl density gradient as shown in Plate I. The buoyant density of λ in CsCl solution at 20°C calculated from the position of the band, the value of the density gradient and the initial density of the suspension is 1·508 g cm^{-3}. This value is close to the average of the buoyant density of DNA (1·7 g cm^{-3}) and of protein (1·3 g cm^{-3}), suggesting that the virus contains approximately equal volumes of each.

With one exception, the densities of mutants of non-transducing λ were indistinguishable even when the viruses were grown under widely different conditions.

The various preparations of phages examined were obtained:

1. by the induction of different strains of K12 lysogenic for λref.
2. by the induction of Gal^- doubly lysogenic for λref and λh.
3. by infection of Gal^+ with $\lambda vir\ h$.
4. by infection of strain C, showing host controlled variation, with $\lambda vir\ h$.
5. by induction of lysogenic bacteria such as $Gal\ 1^-\ 2^+\ (\lambda)\ (\lambda dg\ Gal\ 1^+\ 2^-)$ or $Gal\ 1^-2^-\ (\lambda)\ (\lambda dg\ Gal\ 1^+2^+)$.
6. by induction of the defective lysogenic $Gal\ 1^-\ 2^+\ (\lambda dg\ Gal\ 1^-\ 2^+)$ and super-infection with λref.
7. by infection of Gal^+ with four independent spontaneous clear mutants of λref.

The exceptional phages were obtained by induction of the lysogenic strain 112–12. These phages have a uniform density greater than that of any other λ. Grown in ^{15}N

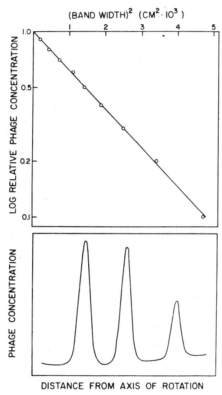

Fig. 1. Photometer tracing of Plate I. The Gaussian character of the center band is shown in the upper portion in which the logarithm of the concentration is plotted as ordinate against the square of the band width for ordinary λ.

medium to increase their density, they have been used, as mentioned above, as density reference material.†

The concentration distribution of λ and λdg in a band is Gaussian within experimental error as may be seen in Fig. 1. If all the phages had the same density the width of the band, due to thermal motion, would indicate a value of $3\cdot8\times10^7$ for their molecular weight. This value is in disagreement with the approximate value of $1\cdot4\times10^8$ which may be calculated from the inactivation cross section or phosphorus content per

†Subsequent experiments done in collaboration with Mrs. G. Kellenberger and Mrs. M. Zichich have shown that the λ mutants $c\ mi$, $c\ m6$, $m5\ co$ of Kaiser (1955) are much less dense than λref.

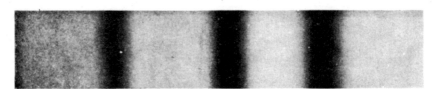

PLATE I. Photograph by u.v. absorption of the centrifuge cell after sedimentation equilibrium has been closely approached (21,740 rev/min at 20°C for 48 hr). The initial density of the suspension was 1·522 g cm^{-3} and it contained ordinary λ (left band), λdg (center band) and λ dense grown in ^{15}N (right band).

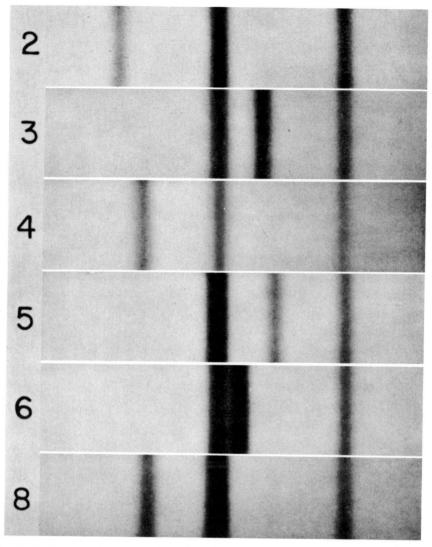

PLATE II. Photographs by u.v. absorption of the centrifuge cell after the phage distribution has reached approximate equilibrium (27,690 rev/min at 20°C for 12 hr). The cell contained different independent HFT lysates and some λ dense *vir h* grown in ^{15}N. The photographs have been aligned using this phage as density marker; it forms the band on the extreme right in each photograph. It can be seen that each independent HFT lysate contains two bands. One of these bands has the density of ordinary λ and the other band (the transducing phage) has a density which is either greater or less than that of ordinary λ. The numbers appearing on the left of each photograph refer to the HFT's listed in Fig. 2.

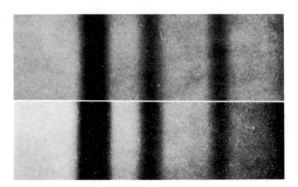

PLATE III. Photographs by u.v. absorption of the density-gradient distribution of the HFT lysate No. 5 and of the HFT lysate of an induced strain transduced by the λdg of lysate No. 5. It is seen that the density of λdg has not been altered. Upper photograph: original lysate No. 5; lower photograph: lysate of strain transduced by λdg of No. 5.

plaque-forming unit (Stent & Fuerst, 1955) when taken together with the percentage of DNA in the phage. It would thus appear that λ phages are not uniform with respect to density in cesium chloride solution. A population of particles uniform with respect to molecular weight but heterogeneous with respect to density forms a Gaussian band if the density distribution is itself Gaussian (Baldwin, 1959). One may calculate that a standard deviation of only 0.0003 g cm^{-3}, that is 0.02% of the density of λ, would account for the low apparent molecular weight obtained from the observed band width.

(b) Centrifugation of transducing λdg

To insure that each HFT lysate studied was the product of an independent genetic event the following procedure was employed:

A culture of Gal 1^+2^+(λ) derived from a single cell was streaked out; ten single colonies were picked and grown into ten cultures. Each culture was induced and the resulting LFT lysate was used to infect sensitive Gal 1^-2^- bacteria at a multiplicity of approximately three phages per bacterium. Each of the ten infected suspensions was plated on indicator plates and from each plate a single colony of transduced cells was selected. These Gal 1^-2^-(λ) (λdg Gal 1^+2^+) colonies were purified twice by streaking and then were grown to yield ten separate cultures. Upon induction each gave an HFT lysate. The phages of these HFT lysates were purified and examined by density-gradient centrifugation.

The results of some of the ten independent density determinations are presented in Plate II. The density differences between ordinary λ and the ten independent λdg's are shown in Fig. 2. The separation of the bands of λ and λdg from another HFT

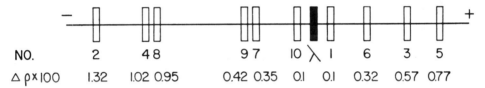

FIG. 2. Densities of ten independent λdg's. The number given identifies each of the ten independent isolates of λdg. For each of them the difference in density from that of ordinary λ is shown multiplied by 10^2.

lysate was carried out by centrifuging a suspension of phage in CsCl solution in the swinging bucket rotor as described in the section on methods. Fig. 3 shows the titer of plaque-forming and transducing phages as a function of distance along the centrifuge tube as expressed by the drop number. The difference in density between λ and λdg obtained in this experiment was 0.0052 g cm^{-3}, equal, within the experimental error, to that found from centrifugation in the analytical machine (0.0060 g cm^{-3}).

Similar separations of the bands of λ and λdg were performed for a fairly large number HFT lysates.

(c) The association of altered density with transducing ability

From Fig. 3 it may be seen that no λdg with normal density was found in an HFT lysate. It thus appears that the characteristics of density and transducing ability have not reassorted during multiplication of λdg in the presence of ordinary λ, even though genetic interactions have been shown to occur (Arber, 1958).

To test the stability of the density of λdg upon relysogenization followed by prophage replication and then induction with subsequent phage multiplication, the following experiment was performed. The HFT number 5 was used to infect a sensitive Gal 1⁻2⁻ culture at multiplicity of 4 plaque-forming phages per bacterium. At this multiplicity all the transduced cells carry both λ and λdg prophages. A single colony of transduced cells was selected and from this a pure clone of Gal⁺ cells was obtained. A culture of these cells was grown, induced, allowed to lyse, and the phage was purified and examined in a CsCl density gradient. Plate III shows the density gradient photograph obtained during centrifugation of this lysate together with the photograph of the banded phage of the parent number 5 lysate. It can be seen that the density of the phages in the new lysate is the same as in the original HFT No. 5.

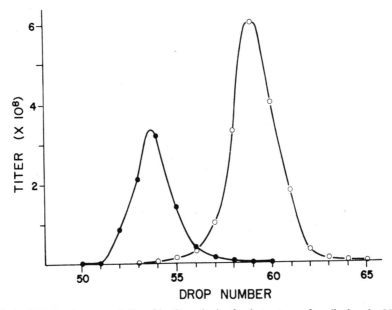

Fig. 3. An HFT lysate was centrifuged in the swinging bucket rotor as described under Methods, and the droplets were collected in separate tubes containing 1 ml. of suspension medium. Each tube was assayed for plaque-forming phages and for transducing phages. Increasing tube number represents decreasing density. Filled circles: transducing phages; empty circles: plaque-forming phages.

The density of λdg appears also to remain unchanged after genetic recombination with the bacterial chromosome in the Gal region, as shown by the following experiment. Sensitive Gal 1⁻2⁺ bacteria were infected with an HFT containing λdg Gal 1⁺2⁺ phages of known density. A colony from a cell Gal 1⁻2⁺(λ) (λdg Gal 1⁺2⁺) was picked, purified and grown into a culture. From this culture a strain Gal 1⁻2⁺(λ) (λdg Gal 1⁻2⁺) was selected (Morse et al., 1956). Density-gradient centrifugation of the HFT lysate of this strain showed that its λdg had a density not detectably different from that of the λdg Gal 1⁺2⁺ which had been used for the original transduction.

4. Discussion

Let us summarize the main features of the alterations in density of transducing λ. Induction of a bacterial culture doubly lysogenic for normal λ and any one of ten transducing λ of independent origin yields a population of phage particles which are genetically of two types, plaque-forming and transducing λ; and physically of two types, those having the density of normal λ and those having a new characteristic density. Plaque-forming λ is identified with the particles of normal density, and transducing λ with those particles showing the new density. No reassortment of density and transducing ability is observed. Each of ten independent λdg has its own characteristic density which is genetically stable and is not altered by a cycle of growth involving transduction and multiplication. Finally, an exchange of galactose markers between a transducing phage and its host cell had no apparent effect on the characteristic density of the transducing phage.

These facts may be accounted for simply by the following model in which no distinction is made between the structure of the original transducing phage from an LFT and that of the λdg in the HFT derived from it. The prophage λ has a region, extending on both sides of the h marker, which is sufficiently similar to the region of the bacterial chromosome surrounding the Gal markers to permit occasional pairing between the two chromosomes. Upon rare occasions a double crossover occurs and the Gal markers are integrated into the prophage chromosome where they replace that part of the prophage genome surrounding the h region.

This event renders the prophage defective through its loss of genes essential for independent multiplication, but able to transduce through its acquisition of the Gal markers. Because the transducing phage chromosome carries a segment strictly homologous to a part of the bacterial chromosome, genetic exchange at normal crossover frequencies is now possible in the Gal region between transducing phage and bacterium, and outside the Gal region between transducing and normal phages.

The region transferred from the bacterial chromosome includes genetic determinants for the density of λdg. The altered density of transducing λ could result from changes in the intrinsic densities or relative amounts of DNA, protein, or other possible viral constituents or from changes in packing or solvation. None of these possibilities is exclusive of the others and direct experimentation alone can ultimately decide among them.

A simple and appealing hypothesis is that the genetic determinant of density and the physical factor responsible for the determination of density are in fact identical, namely the quantity of DNA per phage. This assumption fits the proposed model if the chromosomal segments lost by the phage and acquired from the bacterium are of unequal size. The λdg chromosome would then differ in size from the original λ chromosome, and if replicated as such would give rise to a genetically stable population of phages of a new density. Since the genetic exchange in which λdg arises might involve a bacterial segment either greater or smaller than that deleted from the phage, the transducing phage may be either more or less dense than ordinary λ.

The absence of a double occurrence of a given density among the various independently arisen transducing phages implies that the possible number of unique exchanges between phage and bacterium must be much larger than the number of cases studied.

As this model suggests, the determinants of altered density are not affected when the transducing prophage exchanges homologous Gal markers by normal recombination with the bacterium nor are they ever a part of a plaque-forming non-transducing phage.

It is unlikely that changes in elements physically unlinked to DNA are responsible for the density changes since no phenotypic mixing was observed between the genetic determinants of density and the density itself. Thus, no transducing phages of normal density or active phages of a typical density were found.

To estimate the increment in density corresponding to a change in DNA content by a fraction α from that present in ordinary λ, we assume first that the volume and mass of phage material is increased or decreased by the volume and mass of the DNA added or subtracted, and, second, that the partial specific volume of phage DNA in cesium chloride solution is independent of the particular phage from which it might be isolated. It follows that the change $\Delta\rho$ in density of the phage is given by

$$\Delta\rho = \rho_0\, \alpha\, \frac{F_m - F_v}{1 + \alpha\, F_v} \tag{1}$$

where F_m and F_v are respectively the mass and volume fraction of DNA in ordinary λ with buoyant density ρ_0. The measured value for ρ_0 is $1\cdot51$ g cm^{-3}; for λ DNA the density is $1\cdot71$ g cm^{-3}, and we take the value $1\cdot30$ g cm^{-3} for protein. Then F_m and F_v are found to be $0\cdot57$ and $0\cdot50$ respectively and equation (1) becomes

$$\Delta\rho = \frac{0\cdot21\,\alpha}{2 + \alpha} \cong 0\cdot10\,\alpha \ . \tag{2}$$

Accordingly, the least dense λdg has 14% less DNA than λ and the most dense contains 8% more. If we take the total DNA content of the virus to be 7×10^7 molecular weight units (MWU) (Stent & Fuerst, 1955), then the greatest alteration entails a loss of about 10^7 MWU. The smallest density differences observed between pairs of λdg correspond to a DNA increment of 5×10^7 MWU, about 800 nucleotide pairs. If the amounts of DNA which are gained or lost in the formation of λdg possess a common denominator, this elementary unit cannot be larger than 5×10^5 MWU.

Unequal genetic exchanges of the type described here seem to take place only during the original event leading to the formation of transducing phages. Thus genetic exchanges within the *Gal* region between transducing phage and bacterial chromosome, between normal and transducing phage outside the *Gal* region, and among ordinary λ during multiplication in induced cultures of doubly lysogenic cells, all had no demonstrable effect on the density of the participating phages. It is possible that the occurrence of unequal genetic exchange is a consequence of incomplete pairing between partially homologous chromosomes.

One must remember that the model described above has not been demonstrated to be correct. The possibilities that the density differences between different lambda phages are due to changes in their protein coat, or to changes in both the DNA and the protein components of the particles, are being tested.

REFERENCES

Adams, M. H. (1950). *Methods of study of bacterial viruses*, in *Methods in Medical Research*, Vol. 2, pp. 1–73. Chicago.

Arber, W., Kellenberger, G. & Weigle, J. (1957). *Schweiz. Z. Allg. Path.* **20**, 659.

Arber, W. (1958). *Arch. Sci. Phys. Nat.* **11**, 259.

Baldwin, R. L. (1959). *Proc. Nat. Acad. Sci., Wash.* **45**, 939.

Campbell, A. (1958). *Virology*, **4**, 366.

Kaiser, A. D. (1955). *Virology*, **1**, 424.

Lederberg, E. M. & Lederberg, J. (1953). *Genetics*, **38**, 51.

Meselson, M., Stahl, F. & Vinograd, J. (1957). *Proc. Nat. Acad. Sci., Wash.* **43**, 581.

Morse, M. L., Lederberg, J. & Lederberg, E. M. (1956). *Genetics*, **41**, 142.

Stent, G. S. & Fuerst, C. R. (1955). *J. Gen. Physiol.* **38**, 441.

Wollman, E. L. (1953). *Ann. Inst. Pasteur*, **84**, 281.

PRINTED IN GREAT BRITAIN AT THE UNIVERSITY PRESS ABERDEEN

Reprinted from *Proc. Natl. Acad. Sci. (U.S.)*, **49**(5), 748–755 (1963)

COHESION OF DNA MOLECULES ISOLATED
FROM PHAGE LAMBDA

By A. D. Hershey, Elizabeth Burgi, and Laura Ingraham

GENETICS RESEARCH UNIT, CARNEGIE INSTITUTION OF WASHINGTON, COLD SPRING HARBOR,
LONG ISLAND, NEW YORK

Communicated March 25, 1963

Aggregation of DNA is often suspected but seldom studied. In phage lambda we found a DNA that can form characteristic and stable complexes. A first account of them is given here.

Materials and Methods.—DNA was extracted from a clear-plaque mutant (genotype cb^+) of phage lambda[1] by rotation[2] or shaking[3] with phenol. Sodium dodecylsulfate, ethylenediaminetetraacetate, citrate, or trichloroacetate was sometimes included in the extraction mixture without effect on the properties of the DNA. Phenol was removed by dialysis, with or without preliminary extraction with ether, against 0.1 or 0.6 M NaCl.

Sedimentation coefficients were measured[4] at 10 μg DNA/ml in 0.1 and 0.6 M NaCl in aluminum cells at 35,600 rpm with consistent results, and are reported as $S_{20,w}$.

Zone sedimentation[5] of labeled DNA's[6] was observed in 0.1 M NaCl immobilized by a density gradient of sucrose. A sample, usually containing less than 0.5 μg of DNA in 0.15 ml of 0.1 M NaCl, was placed on 4.8 ml of sucrose solution, and the tube was spun for 5 or 6 hr at 28,000 rpm in an SW39L rotor of a Spinco Model L centrifuge at 10°C.

Solutions containing 5–40 μg DNA/ml in 0.1 or 0.6 M NaCl were stirred on occasion for 30 min at 5°C with a thin steel blade turning in a horizontal plane.[7] Since we used two stirrers of different capacities, stirring speeds given in this paper are comparable only within a context.

Salt solutions were buffered at pH 6.7 with 0.05 M phosphate.

Results.—Disaggregation and breakage: Solutions containing 0.5 mg/ml of lambda DNA in 0.1 M NaCl acquire an almost gel-like character on standing for some hours in a refrigerator. Diluted to 10 μg/ml, the DNA exhibits in the optical centrifuge an exceedingly diffuse boundary sedimenting at 40–60 s (Fig. 1*A*). If the diluted solution is aged for several days, the sedimentation rate may fall somewhat (not below 40 s), but the boundary remains diffuse and often appears double.

Stirring the diluted solution at 1,300 to 1,700 rpm yields a single component sedimenting at 32 s (Fig. 1*B*). The product so obtained is stable for a week or more in the cold in 0.1 M NaCl. We call this process disaggregation by stirring.

If samples of the diluted solution are stirred at increasing speeds between 1,800 rpm and 2,100 rpm, one sees a stepwise transition from 32 s to 25.2 s components, each by itself exhibiting a sharply sedimenting boundary (Figs. 1*C* and 1*D*). We call this phenomenon breakage. Broken DNA can form aggregates, but the characteristic 32 s species cannot be regained.

Aggregation: Disaggregation, in contrast to breakage, is reversible, as shown by the following experiment. Lambda DNA at 40 μg/ml in 0.6 M NaCl was

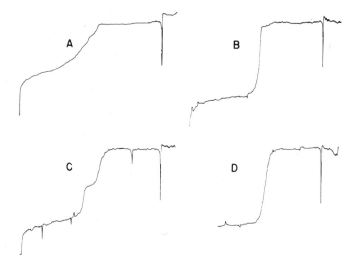

Fig. 1.—Sedimentation pattern of initially aggregated DNA after stirring at several speeds. *A*, unstirred; *B*, 1,600 rpm; *C*, 1,900 rpm; *D*, 2,000 rpm. The meniscus shows at the right.

disaggregated by stirring at 1,700 rpm, and samples at either 40 μg/ml or 10 μg/ml in the same solvent were warmed to 45°C. After measured time intervals, the tubes were chilled and their contents diluted to 10 μg/ml, if necessary, with cold 0.6 M NaCl. Sedimentation coefficients were measured over the course of some hours. Unheated samples showed the same sedimentation rate at the beginning and end of the series of measurements. The heated samples were analyzed in random order, so that the results reflect mainly the duration of heating, not the duration of subsequent storage.

The results, presented in Figure 2, show that the sedimentation rate of the DNA increases rapidly on heating at 40 μg/ml, and less rapidly on heating at 10 μg/ml. The reversibility of disaggregation, and the dependence of rate of aggregation on concentration of DNA, justify our choice of language.

Similar experiments showed that heating in 0.1 M NaCl under the same conditions does not cause appreciable aggregation. Aggregation occurs in that solvent at higher DNA concentrations, however. Thus, aggregation is accelerated by high DNA concentrations, high temperatures, and high salt concentrations.

Linear molecules: According to the description given above, aggregated lambda DNA can be reduced under shear to a uniform 32 s product, which is evidently the structure subject to breakage at higher rates of shear. The maximum stirring speed withstood by 32 s lambda DNA is 1,800 rpm at 10 μg/ml. When

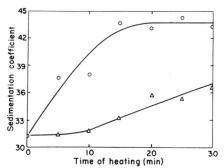

Fig. 2.—Aggregation at 45°C. Circles, 40 μg DNA/ml in 0.6 M NaCl; triangles, 10 μg/ml. The scale on the ordinate refers to observed sedimentation coefficients.

T2 DNA is stirred under the same conditions, it is reduced to fragments sedimenting at 31 s.   Thus, lambda DNA exhibits a fragility under shear that is appropriate to linear molecules[7] sedimenting at about 32 s.   We therefore conclude that 32 s lambda DNA consists of linear molecules.   These and other DNA structures are best identified by zone centrifugation, as illustrated below.

Linear molecules can also be prepared (irrespective of the initial state of aggregation of the DNA) by heating a solution in 0.1 or 0.6 M NaCl to 75°C for 10 min and cooling the tube in ice water (Fig. 3A, solid line).   This procedure is effective at concentrations up to 10 μg/ml at least.

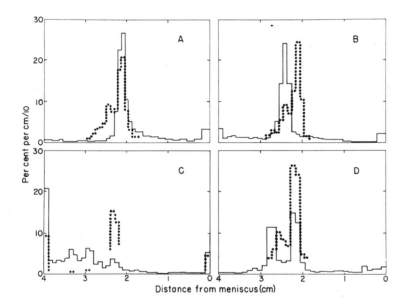

FIG. 3.—Zone sedimentation of several molecular forms.   Solid lines: P[32]-labeled linear molecules (A), folded molecules (B), aggregates (C), dimers and linear monomers (D).   Broken lines, a preparation of H[3]-labeled marker DNA containing linear and some folded molecules.

Linear molecules are obtained directly by extracting the DNA (by rotating, not shaking, the tubes)[6] at 2 μg/ml into 0.1 M NaCl (Fig. 3A, dotted line).   Control experiments showed that the mechanical operations involved in the extraction do not destroy previously formed complexes in solutions diluted to 2 μg/ml.

We conclude that the 32 s form of lambda DNA is analogous to more conventional phage DNA's and is a typical double-helical molecule.

Folded molecules: Another form of lambda DNA we usually prepare by heating a dilute solution (5 μg/ml or less) in 0.6 M NaCl to 75°C for 1 min, and allowing the container to cool slowly (0.4° per min at 65°) in the heating bath with the heater disconnected.   The resulting product sediments as a narrow band moving 1.13 times faster than linear molecules in zone centrifugation (Fig. 3B).   The expected sedimentation coefficient is 32 × 1.13 = 36.2 s.   Material prepared as described and then concentrated by dialysis against dry sucrose followed by 0.6 M NaCl shows in the optical centrifuge a sharp boundary at 37 s.

The formation of 37 s material is equally efficient at several DNA concentrations between 5 μg/ml and 0.1 μg/ml (at higher concentrations it is obscured by simultaneous aggregation). The 37 s product, therefore, is composed of monomers that we shall call folded molecules.

When a dilute solution containing either linear or folded molecules in 0.6 M NaCl is heated to 75°C, one gets only linear molecules by rapid cooling and only folded molecules by slow cooling. Partial conversion of linear to folded molecules occurs on heating to 45°C for 30 min followed by rapid cooling, and nearly complete conversion at 60°C. Thus, at 75°C linear molecules are the stable form of lambda DNA. At low temperatures, folded molecules are more stable but the conversion is slow. The slow cooling from 75°C serves to find a temperature near 60°C at which the conversion to folded molecules is rapid and the product is stable.

Folded molecules are formed on heating and slow cooling in 0.1 M NaCl as well as in 0.6 M NaCl, but the conversion is not complete at the lower salt concentration. Some molecular folding also occurs when linear molecules are stored at low concentration and low temperature for a few weeks in 0.1 M NaCl or a few days in 0.6 M NaCl. This is the origin of the faster-sedimenting component of the tritium-labeled marker DNA whose sedimentation pattern appears in Figure 3.

Folded molecules can be converted back into linear molecules by stirring as well as by heating, though the margin between the stirring speed required to accomplish this and the speed sufficient to break linear molecules is rather narrow.

It should be added that heating DNA at 10 μg/ml and 45°C in 0.6 M NaCl produces many folded molecules whose formation competes with the simultaneous aggregation. For this reason the dependence of rate of aggregation on DNA concentration is not truly represented in Figure 2.

Folded molecules themselves do not aggregate. Solutions concentrated for analytical centrifugation continue to yield sharp boundaries after aging in 0.6 M NaCl. Neither do folded molecules form complexes with linear molecules. This was shown by mixing P^{32}-labeled folded molecules with unlabeled linear molecules (20 μg/ml) and aging the mixture in 0.6 M NaCl for 4 days at 5°C. A similar mixture containing labeled linear molecules served as control. Zone centrifugation of each mixture with added H^3-labeled marker DNA showed that the labeled linear molecules but not the folded molecules had formed complexes with the unlabeled DNA.

The similarity between the conditions, other than DNA concentration, controlling formation and destruction of folded molecules, and formation and destruction of aggregates, suggests that similar cohesive forces are involved in both phenomena. The folding implies that each molecule carries at least two mutually interacting cohesive sites, which join to form a closed structure. The uniformity of structure of folded molecules, indicated by the narrow zone in which they sediment, suggests that there are not more than two cohesive sites, and that these are identically situated on each molecule.

Dimers and trimers: Aggregated DNA often shows multiple boundaries in the optical centrifuge and always shows multiple components in zone centrifugation. An example, prepared by heating linear molecules for 30 min at 45°C and 40 μg/ml in 0.6 M NaCl, is shown in Figure 3C. Since the characteristic folding seen in monomers is incompatible with aggregation, as already described, it is likely that some of the differently sedimenting products of aggregation are polymers differing in mass rather than configuration.

One form of aggregate can be obtained in moderately pure state by allowing aggregation to occur during a day or so in the cold at 100 μg/ml in 0.1 M NaCl (Fig. 3D). Such material contains a fraction of the molecules in linear form, and presumably contains in addition mainly the smaller and more stable aggregates. One of these, as shown in the figure, always predominates, and we assume that it is a dimer. It sediments 1.25 times faster than linear molecules.

In a study of zone centrifugation to be reported separately, we found a relation

$$\frac{D_2}{D_1} = \left(\frac{L_2}{L_1}\right)^{0.25} \qquad (1)$$

between molecular lengths (L) and distances sedimented (D) of two DNA's, which is valid for linear molecules. According to this relation, dimers are about twice as long as linear molecules of lambda DNA. The only alternative compatible with the sedimentation rate is a second form of folded monomer, which is ruled out by the requirement for high DNA concentrations during formation. Therefore, dimers are tandem or otherwise open structures. (For definitions of "open" and "closed," see hereafter.)

In more completely aggregated material (Fig. 3C) one sees few or no linear molecules, a very few folded molecules (fewer the more concentrated the solution in which aggregation occurred), a considerable fraction of dimers sedimenting 1.25 times faster than linear molecules, and another characteristic component sedimenting 1.43 times faster than linear molecules. According to its sedimentation rate, the last component could be a tandem trimer or a folded or side-by-side dimer. We believe that it is an open trimer for the following reasons.

A folded dimeric structure is ruled out because material sedimenting at rate 1.43 does not form when a dilute solution containing dimers (similar to that shown in Fig. 3D) is aged for two weeks in the cold in 0.1 or 0.6 M NaCl, or is heated in 0.6 or 1.0 M NaCl at 45°C. At high DNA concentrations, trimers do form under these conditions. At low DNA concentrations, dimers and trimers are stable and one sees only the conversion of linear to folded monomers.

A side-by-side dimeric structure can be ruled out on the basis of susceptibility to hydrodynamic shear. Figure 4 shows the result when samples of a mixture of trimers, dimers, and folded and linear monomers are stirred at increasing speeds. Trimers disappear first, being converted to dimers or linear molecules or both. Next to go are dimers. Folded monomers are much more resistant, but can be reduced to linear monomers at stirring speeds just insufficient to break the molecules. Thus, trimers, as expected if they are open structures, are more fragile than open dimers, whereas closed dimers should be more stable. We note, however, that a small amount of the material sedimenting at the rate of trimers is relatively resistant to stirring and could signify a minority of closed dimers.

We note also that destruction of dimers and trimers does not liberate any folded monomers, a result consistent with the evidence from sedimentation rates for an open polymeric structure, and with our finding that folded monomers do not form complexes. The fact that aggregation and folding are mutually exclusive processes implies that both utilize the same limited number of cohesive sites, which must be small in size to account for the open polymeric structure. As already suggested by the unique configuration of folded monomers, there may be only two sites per molecule.

Specificity of aggregation: If tracer amounts of P[32]-labeled lambda DNA are mixed with unlabeled lambda DNA at 25 μg/ml in 0.6 *M* NaCl, and the mixture is brought to 75°C for 1 min and allowed to cool slowly, subsequent zone sedimentation with added H[3]-labeled marker shows that most of the P[32]-labeled DNA has been converted to aggregates and a small remainder to folded molecules. When the same procedure is followed with H[3]-labeled or unlabeled T5 DNA substituted for the unlabeled lambda DNA, the T5 DNA sediments (at its normal rate) 1.20 times faster than the P[32]-labeled lambda DNA, which now consists entirely of folded molecules. Thus, lambda DNA shows no tendency to form complexes with T5 DNA, T5 DNA itself does not form stable aggregates, and T5 DNA does not inhibit molecular folding in lambda DNA. The cohesive sites in lambda DNA are therefore mutually specific, as our model requires.

Role of divalent cations: Divalent cations probably do not play any specific role in the phenomena described in this paper. In NaCl solutions, molecular folding and aggregation are not inhibited by added citrate or ethylenediaminetetraacetate. Neither are these processes appreciably accelerated, in the presence of NaCl, by added calcium or magnesium ions. In a solution of 0.01 *M* MgCl$_2$, 0.01 *M* CaCl$_2$, and 0.01 *M* tris (hydroxymethyl) aminomethane, pH 7.2,[8] linear monomers at 10 μg/ml are about as stable as they are in NaCl solutions.

Interpretation of sedimentation rates: Equation (1) shows that if two identical DNA molecules were joined end to end their sedimentation rate would increase by the factor 1.27, evidently owing to the loss of independent mobility. Perhaps the result would be about the same whether they were joined end to end or to form a V, a T, or an X. Thus, we are led to the definition of an *open dimeric structure* as one formed by the joining

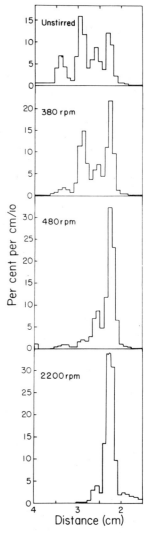

FIG. 4.—Successive destruction by stirring of trimers, dimers, and folded molecules, as seen by zone centrifugation. Linear molecules began to break at 2,400 rpm in this series. The starting material was prepared by heating a sample of DNA, already containing some spontaneously formed folded molecules and dimers, for 30 min at 45°C and 40 μg/ml in 1.0 *M* NaCl, and diluting to 5 μg/ml in buffered water.

of two linear molecules at a single point, recognizable by a 1.27-fold increase in sedimentation rate. The principle of independent mobility of parts suggests that, as the structure departed from the tandem arrangement, its sedimentation rate could only increase, not decrease, and in the order V, T, X.

Our results also show that molecules of lambda DNA undergo some sort of folding, apparently as the result of bonding between two cohesive sites lying at some distance from each other on each molecule. If that interpretation is correct, it would appear that when the molecule (regarded as two halves joined end to end) forms an additional point of attachment between its parts, the sedimentation rate increases by an additional factor of 1.13, evidently owing to a further loss of independent mobility of parts. Thus, we are led to the definition of a *closed dimeric structure* as one formed by joining two linear molecules at two points. Such a structure ought to sediment 1.27 × 1.13 or 1.43 times faster than the linear monomer. We have not found closed dimers, but the question remains how the factor 1.13 would depend on the point of closure of a threadlike molecule. The principle of independent mobility of parts suggests that the sedimentation rate would approach or pass through a maximum as the fraction of the molecular length contained in the loop increased. In some measure it may be possible to answer such questions empirically by determining the locations of cohesive sites on the molecules.

Discussion.—Lambda DNA can exist in at least four characteristic forms that we call linear monomers, folded monomers, open dimers, and open trimers, which sediment respectively at the rates 1.0, 1.13, 1.25, and 1.43, expressed in arbitrary units. These structures are interconvertible with certain restrictions according to the scheme

$$\text{open polymers} \leftrightarrows \text{linear monomers} \leftrightarrows \text{folded monomers}$$

As the scheme indicates, linear monomers are subject to two distinct processes: aggregation, seen at DNA concentrations exceeding 10 μg/ml, and folding, seen at any concentration but forced to compete with aggregation at high DNA concentrations. Both processes are accelerated as the temperature is raised to about 60°C, beyond which only linear monomers are stable, and as the salt concentration is raised from 0.1 to 1.0 M. Both processes are rapidly reversed at 75°C or by hydrodynamic shear. All four structures are stable at low temperatures, low DNA concentrations, and low salt concentrations, except for a slow conversion of linear to folded molecules.

Since folded molecules exist in only one stable configuration, and since molecular folding and aggregation are mutually exclusive processes, we postulate that each molecule carries two cohesive sites in prescribed locations, and that these are responsible for both processes. To account for the considerable effect of molecular folding on sedimentation rate, the sites must lie rather far apart along the molecular length. To account for the moderate effect of dimerization on sedimentation rate, the cohesive sites must be small compared to the total molecular length.

According to the proposed model, one might anticipate two dimeric forms, open (that is, joined by one pair of cohesive sites) and closed (joined by two). We find only open polymers, though a minority with closed structures is not excluded. Failure to detect closed polymers may be explained, at least in part, by the fact that the rate of folding must decrease as the length of the linear structure increases.[9]

Whether all details of our model are correct or not, it is clear that lambda DNA forms a limited number of characteristic complexes, not the continuously variable series that might be expected if the molecules could cohere at random. The

limited number of mutually specific cohesive sites implied thereby suggests a specialized biological function, one that remains to be identified.

Summary.—The DNA of phage lambda undergoes reversible transitions from
linear to characteristically folded molecules, and from linear monomers to open
polymers. Some conditions favoring one state or another have been defined. It
may be surmised that each molecule carries two specifically interacting cohesive
sites.

This work was aided by grant CA-02158 from the National Cancer Institute, National Institutes
of Health, U.S. Public Health Service. Its direction was determined in part in conversation with
Dr. M. Demerec about heterochromatin, synapsis, deletions, and speculations to be pursued.
Professor Bruno Zimm contributed useful suggestions about the manuscript.

[1] Kellenberger, G., M. L. Zichichi, and J. Weigle, these PROCEEDINGS, **47**, 869 (1961).
[2] Frankel, F. R., these PROCEEDINGS, **49**, 366 (1963).
[3] Mandell, J. D., and A. D. Hershey, *Anal. Biochem.*, **1**, 66 (1960).
[4] Burgi, E., and A. D. Hershey, *J. Mol. Biol.*, **3**, 458 (1961).
[5] Hershey, A. D., E. Goldberg, E. Burgi, and L. Ingraham, *J. Mol. Biol.*, **6**, 230 (1963).
[6] Burgi, E., these PROCEEDINGS, **49**, 151 (1963).
[7] Hershey, A. D., E. Burgi, and L. Ingraham, *Biophys. J.*, **2**, 423 (1962).
[8] Kaiser, A. D., *J. Mol. Biol.*, **4**, 275 (1962).
[9] Jacobson, H., and Stockmayer, W. H., *J. Chem. Phys.*, **18**, 1600 (1950).

Reprinted from *Cold Spring Harbor Symp. Quant. Biol.*, **33**, 729–734 (1968)

Structure and Function of DNA Cohesive Ends

A. D. Kaiser and Ray Wu

*Department of Biochemistry, Stanford University, Stanford, California and Section of
Biochemistry and Molecular Biology, Cornell University, Ithaca, New York*

Since the discovery of cohesive ends on the DNA of phage lambda by Hershey, Burgi, and Ingraham in 1963, a number of other types of DNA molecules have been shown to possess cohesive ends. They are summarized in Table 1. There are as yet no

TABLE 1. TYPES OF DNA MOLECULES WITH COHESIVE ENDS

Origin	Specificity family	Test	Ref.
Bacteriophages λ	λ		1
ϕ80	λ	P, B	2, 4
21	λ	P, B	3, 4
424	λ	P, B	3, 4
434	λ	P, B	3, 4
82	λ	B	4
186	186		3, 4
P2	186	P, B	5, 6
N1			7
Yeast Mitochondrial DNA			8
χ			9

References: 1 is Hershey et al., 1963; 2 is Yamagishi et al., 1965; 3 is Baldwin et al., 1966; 4 is Table 2 of this paper; 5 is Mandel, 1967; 6 is Mandel and Berg, 1968; 7 is Wetmur et al., 1966; 8 is Shapiro et al., 1968; 9 is Mitani and Lang, pers. commun. P stands for the physical test for homology—the formation of mixed dimers; B stands for the biological test—the ability to help in DNA infection.

exceptions to the rule that temperate phages whose DNA molecules have a nonpermuted base sequence possess cohesive ends. Cohesive ends permit a linear DNA molecule packaged inside the head of a phage particle to become circular inside the host cell. Thus, the presence of cohesive ends on the DNA molecules of temperate phages fits into the scheme for recombination between a circular phage DNA and the host chromosome leading to insertion of the phage DNA into the chromosome as proposed by Campbell (1962). However, circular DNA molecules and cohesive ends must serve other purposes because at least two virulent phages, χ and N1, and yeast mitochondrial DNA also have them.

From a more general point of view, cohesive ends are a device for joining two DNA molecules with the proper specificity. This joining can be rapid and efficient since it has been observed that 90% of the λ DNA molecules injected into a bacterium become covalently joined within 5 minutes (Bode

and Kaiser, 1965). Beyond their established role in the formation of circular DNA molecules, cohesive ends might be involved in the break and join steps of genetic recombination. With this in mind, we have investigated the detailed structure of the cohesive ends of λ DNA.

After discussing the way cohesive ends can be detected and their specificity analyzed, we will turn to the evidence that the cohesive ends of λ DNA are short, protruding, single, complementary strands of DNA with a well-defined base sequence.

SPECIFICITY OF COHESIVE ENDS

Cohesive ends can be detected in vitro by observing a reversible conversion of linear monomers to circular monomers or to linear polymers (Fig. 1) as demonstrated by electron microscopy or by sedimentation velocity measurements. To test whether the cohesive ends on two different kinds of DNA molecules are the same, one can also look for the formation of mixed dimers. There is a second test for the presence and specificity of cohesive ends applicable to viral DNA molecules: to determine their infectivity for helper-infected bacteria. Although the way free DNA penetrates these bacteria is not yet understood, it is nevertheless clear that infection depends upon homology between cohesive ends on the infecting DNA and

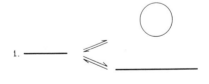

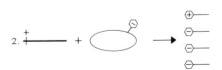

FIGURE 1. Tests for detection of cohesive ends.

TABLE 2. EFFICIENCY OF DNA INFECTION

Helper-phage	\(\lambda\)	434	424	21	82	\(\phi80\)	186
				DNA from bacteriophage			
\(\lambda\)	1.0	0.4	6.4	0.5	0.8	0.9	$<3 \times 10^{-7}$
434	2.4		2.6		3.8		$<3 \times 10^{-6}$
424	0.03	0.08		0.1	4.2		$<3 \times 10^{-6}$
21	0.72	0.05	0.03		0.3		3×10^{-6}
82	5.7	4.4	0.3	1.4			3×10^{-6}
\(\phi80\)		0.01	0.05	0.03	0.02	0.5	1.1×10^{-5}
186	4.8×10^{-6}	1.8×10^{-6}	5×10^{-6}	4×10^{-5}	5×10^{-7}	1.5×10^{-5}	0.002
none	5×10^{-8}						5×10^{-8}

Helper-infected bacteria were prepared as described by Kaiser and Inman (1965). An efficiency of 1 is 10^7 plaques/μg of added DNA. In each case where the DNA and helper phage are different, immunity was used as the selective marker. In cases where the DNA and helper phage have the same immunity, *sus* mutations were used as selective markers.

those on the helper phage. The data of Table 2 show that any phage in the set containing λ, 434, 424, 21, 82, and $\phi80$ can serve as helper for the infection by DNA isolated from any other member of the set. Phage 186, however, belongs to a different set since 186 phage will help its own DNA but no member of the first set. Phage P2 belongs to the 186 set by the helper test (Mandel, 1967).

The physical test for homology of cohesive ends, namely the formation of mixed dimers, defines exactly the same two sets. That is, two DNAs which can form mixed dimers can also help each other for DNA infection. Conversely, two DNAs which cannot form mixed dimers, such as λ and 186, also fail to help each other in DNA infection. The data on cohesive end specificity is summarized in Table 1. The set of nine phage DNAs tested reveal only two types (families) of specificity. We will return later to the question whether two cohesive ends belonging to the same family are identical.

MOLECULAR STRUCTURE OF COHESIVE ENDS

Hershey et al. (1963) discovered the thermal reversibility of the cohesion of lambda DNA. The equilibrium constant relating cohered and separated molecules depends on temperature just as that of a DNA helix-coil transition does, except that the transition for cohesion occurs at a lower temperature than the denaturation of any high mol wt DNA (Kaiser and Inman, 1965). This suggests that the tendency of free bases to pair is the cohesive force. Four possible structures for ends which could cohere by base pairing are shown in Fig. 2. Discrimination among the four structures is possible with the aid of two enzymes: *E. coli* DNA polymerase and *E. coli* exonuclease III. Exonuclease III removes mononucleotide units from the 3' terminus of double-stranded DNA. DNA polymerase adds mononucleotide units to the 3' terminus of a primer strand but requires another strand opposite to act as a template. Thus, the two enzymes have similar specificities but carry out opposing reactions.

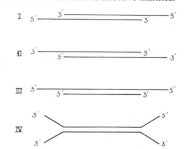

FIGURE 2. Four possible structures for the cohesive ends. 5' and 3' refer to the terminal hydroxyl group in the polydeoxynucleotide chain.

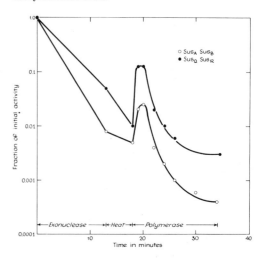

FIGURE 3. Inactivation of the two ends of λ DNA by *E. coli* exonuclease III and their reactivation by *E. coli* DNA polymerase. λ DNA 33 μM was incubated at 37°C with 20 units/ml of exonuclease for 13 min, then heated to 75° for 5 min to destroy exonuclease activity. DNA polymerase, 200 units/ml, and the four deoxynucleoside triphosphates, each at 1 μM, were then added and incubation continued at 15°. At the indicated times aliquots were removed, diluted 50-fold, stirred to break the molecules in half, and assayed for $sus_A^+ sus_B^+$ and $sus_Q^+ sus_R^+$ activity. Additional details may be found in Wu and Kaiser (1968).

The need for cohesive ends in the DNA infection of helper-infected bacteria provides a sensitive test for the ability to cohere (Kaiser and Inman, 1965). If λ DNA is exposed to exonuclease III, it loses its ability to transfer markers to the progeny of helper-infected bacteria. The data are presented in Fig. 3. Exonuclease III-catalyzed destruction of the ability to cohere was also observed by Wang (this volume), who measured the equilibrium between linear and circular molecules by sedimentation velocity. When exonuclease III-treated DNA serves as primer template for polymerase, some of the lost biological activity is restored. The observed restoration of activity by polymerase rules out a structure with protruding 3′-terminated strands, because polymerase could not regenerate the original state without an opposed 5′-terminated template strand. In this experiment the two cohesive ends were assayed separately by shearing the product DNA into half-molecules at each assay time point and measuring separately the activity of two markers near the left end (*sus A sus B*) and two markers near the right end (*sus Q sus R*). The two ends respond similarly. To account for these results both ends of λ DNA must either be double-stranded (structure IV) or single-stranded with protruding 5′-terminated strands (structure I).

The two remaining possibilities can be distinguished by applying the same two enzymes in reverse order. As shown in Fig. 4, polymerase-

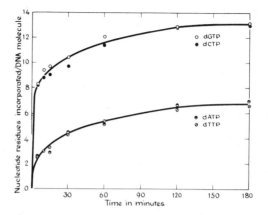

FIGURE 5. Nucleotide incorporation into the cohesive ends. Each reaction mixture contained 0.15 mM λ DNA, 100 units DNA polymerase/ml and 0.02 mM of each of the four deoxynucleoside triphosphates. In one mixture both dGTP and dCTP were radioactively labeled, one with ³H and the other with ³²P. In another mixture dATP was labeled and in the third dTTP was labeled.

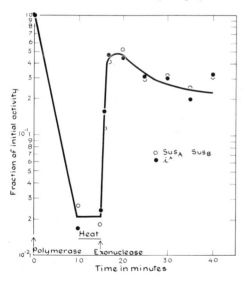

FIGURE 4. Inactivation of the two ends of λ DNA by *E. coli* DNA polymerase and their reactivation by *E. coli* exonuclease III. Experimental details are given in Wu and Kaiser (1968).

catalyzed synthesis inactivates both cohesive ends, and subsequent treatment with exonuclease III restores activity. Therefore, native λ DNA must have two protruding single strands which are 5′-terminated (structure I). This structure is supported by other evidence. For example, a DNA exonuclease induced by infection with phage lambda (Korn and Weissbach, 1963; Radding, 1964; Little, Lehman, and Kaiser, 1967) acts specifically at 5′-termini and acts faster on native than on denatured DNA (Little, 1967). Lambda exonuclease attacks native λ DNA slowly and with a lag, but λ exonuclease attacks λ DNA to which nucleotides have been added by prior treatment with DNA-polymerase rapidly and without lag (Little, 1967). This is the expected behavior toward DNA molecules with protruding 5′-terminated single strands.

If native λ DNA has protruding 5′-terminated single strands then the polymerase-catalyzed incorporation of nucleotides should occur until the ends of the growing 3′-terminated strands come into register with the ends of the corresponding 5′-terminated template strands. As shown in Fig. 5, incorporation reaches a limit when 13 residues each of G and C and 7 residues each of A and T are incorporated. Since the incorporation of G and C are equal and the incorporation of A and T are equal, the two protruding single strands must have complementary base compositions. This is consistent with the idea that their sequences are complementary. A total of 40 ± 4 residues are incorporated into the two ends. Assuming the

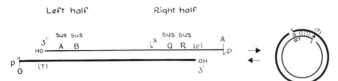

FIGURE 6. The structure of linear and hydrogen-bonded circular molecules is shown, including the known 5′-terminal nucleotides and their relation to the vegetative recombination map (Wu and Kaiser, 1967). The strand indicated by the thicker line is, by convention, designated the *l*-strand and the other the *r*-strand.

two ends to be of equal length, each protruding single strand would contain 20 nucleotide units.

Two kinds of experiments show that the two single strands do have the same length. If the two single strands have complementary sequences and are of equal length, then, as shown in Fig. 6, the circular molecules will have the 5′ terminus of each strand immediately adjacent to its own 3′ terminus with no nucleotide 'teeth' missing. Because the right 5′ terminal nucleotide is A, base-pairing implies the presence of a T in the first position beyond the 3′OH terminus in the left cohesive end; and because the left 5′ terminal nucleotide is G, there must be a C in the first position beyond the 3′OH terminus in the right cohesive end (Wu and Kaiser, 1967). These assignments then predict that native λ DNA should prime two specific limit reactions: one with dGTP alone at the right end and one with dATP alone at the left end.

Both limit reactions are observed, and no limit reaction is observed with dCTP or dTTP (Wu and Kaiser, 1968). The second experiment supporting the structure shown in Fig. 6 used the DNA-joining enzyme of Olivera and Lehman (1967). Both strands of λ DNA were observed to join with purified DNA-joining enzyme in the absence of added polymerase or nucleoside triphosphates (Wu and Kaiser, 1968).

Continuation of the limit reaction technique with pairs of nucleotides may permit the elucidation of the sequence of bases in the cohesive ends. The

current state of the sequence analysis is summarized in Fig. 7. Starting at the 3′ terminal position of the right cohesive end there is a run of 4 G-C pairs with homo G in one strand and homo C in the other, followed by 6 or 7 more G-C pairs in approximately alternating sequence. This G-C run is terminated by an A-T pair and is followed by 5 more A-T pairs and 2 or 3 G-C pairs in an as yet undetermined sequence. The cohesive end sequence terminates with an A-T pair in the orientation shown. The sequence is striking in that it is divided into a pure G-C half and an A-T rich half. It seems likely that this sequence singularity is recognized by the enzyme(s) which create the cohesive ends specifically at this point in the 50,000 base pair sequence which makes up one molecule of λ DNA.

RECOGNIZING THE COHESIVE END SEQUENCE

Different DNA molecules which belong to the same family of cohesive end specificity have similar if not identical base sequences in their protruding single strands. The heterodimers of λ and 21 DNA dissociate at approximately the same pH as homo-λ dimers (Baldwin et al., 1966). And the efficiency with which one member of a family helps DNA infection by another member of the same family is always 3 or 4 orders of magnitude greater than helping between members of different families as shown in Table 2.

The identity or near identity of sequence within one family creates a paradox. On the one hand, phage λ and phage 21, to take a specific example, though related are sufficiently diverse to have evolved different nucleotide sequences in two other protein-DNA recognition systems: immunity repression and prophage integration. The sequences recognized by the immunity repressors are different because bacteria lysogenic for lambda are sensitive to 21 and vice versa. The sequences recognized by the prophage integration proteins are different because prophage λ and prophage 21 occupy different, specific sites on the *E. coli* chromosome (Liedke-Kulke and Kaiser, 1967). On the other hand, the sequence in the cohesive ends of λ and 21 DNA seems to be the same.

How can this paradox be resolved? One possibility is that there exists only one sequence of 20 base pairs which can join and disjoin at 37° in the

Base Sequence in λ Cohesive Ends

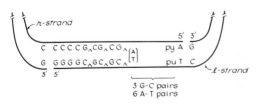

FIGURE 7. Nucleotide sequence in the cohesive ends of λ DNA. The two ends are shown in the cohered state. A *G* residue is located at one of the three carat (∧) marks in the *l*-strand and a *C* residue at the corresponding site in the *r*-strand.

ionic environment of the host cell. Although it seems unlikely that two sequences differing only in the orientation of a single base pair would have very different thermal stabilities, this has not been established. This possibility can be tested by comparing the sequences and lengths of the λ and 186 DNA cohesive ends. A second possibility is that within one family the sequences are not identical, but only very similar. More detailed studies of the relative stabilities of homo- and heterodimers would test this explanation. A third possibility is that the cohesive end sequence is involved in some other function which requires recognition of the same sequence by another protein. Evolution of base sequence would be slowed by the need for parallel evolution of two proteins. A final possibility is that prophage integration and immunity repression each have several specific, distinct, but interdependent recognition components, whereas cohesive end synthesis has only one, permitting greater variability in the former than the latter. Ptashne's (1967) finding of a single, low molecular weight λ repressor protein argues against this possibility.

One way in which two cohesive ends having exactly the same length and complementary base sequences might be created on each λ DNA molecule is by an endonuclease type enzyme or enzyme complex which would introduce two single strand scissions on opposite strands, 20 base pairs apart in a region with the correct base sequence. This is, of course, not the only possible mechanism for cohesive end synthesis (Wu and Kaiser, 1968) but we raise it here because it permits one to imagine that cohesive ends function in other processes.

A Cohesive End Mechanism for Prophage Integration

For instance, a specific set of cohesive ends could insert a circular phage chromosome as a linear element into the bacterial chromosome as suggested in Fig. 8. In this scheme we assume the existence of an enzyme which recognizes a particular sequence of bases (A through V in Fig. 8) found both in the phage chromosome and in the bacterial chromosome. This enzyme, like the one postulated to account for the synthesis of the vegetative cohesive ends, would make two single strand breaks on opposite strands 10 to 20 base pairs apart. The two pairs of mutually complementary cohesive ends, thus generated, could exchange pairing partners. Finally, formation of the missing phosphodiester bonds would complete the insertion process. Attempts to test this mechanism are in progress.

A considerable body of evidence that phage and bacterium have different sequences in their attachment regions has been reviewed by Signer (1968).

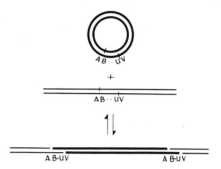

FIGURE 8. The double circle represents the DNA molecule of an infecting phage and the double line below it represents that portion of the bacterial DNA which contains the prophage attachment site. AB $\cdots$ UV represents a particular sequence of base pairs. The double line in the lower part represents the bacterial DNA after the phage DNA has been inserted but before the four phosphodiester bonds have been made.

These findings might seem to contradict a cohesive end mechanism which requires that the phage and bacterium have identical sequences within the segment bounded by the two, hypothetical, specific single strand breaks. However, the evidence can also be explained by an identical sequence in phage and bacterium which is flanked by sequences differentiating one from the other (Signer, 1968).

Summary

The ends of molecules of DNA from bacteriophage λ and, possibly, all temperate phages with nonpermuted base sequences possess protruding, 5'-terminated, single strands. The sequences of bases in the two protruding single strands of one DNA molecule are complementary and permit the two ends to join by base pairing. There are at least two families of cohesive end sequences and cohesive ends with different sequences may also be involved in other processes, such as the specific insertion of phage DNA into the bacterial chromosome.

Acknowledgments

This work was supported by research grants GB-7287 and GB-7029 from the National Science Foundation and AI-04509 from the National Institutes of Health.

REFERENCES

BALDWIN, R. L., P. BARRAND, A. FRITSCH, D. A. GOLDTHWAIT, and F. JACOB. 1966. Cohesive sites on the deoxyribonucleic acids from several temperate coliphages. J. Mol. Biol. *17*: 343.

BODE, V. C., and A. D. KAISER. 1965. Changes in the structure and activity of λ DNA in a superinfected immune bacterium. J. Mol. Biol. *14*: 399.

CAMPBELL, A. M. 1962. Episomes. Adv. Genet. *11:* 101.

HERSHEY, A. D., E. BURGI, and L. INGRAHAM. 1963. Cohesion of DNA molecules isolated from phage lambda. Proc. Nat. Acad. Sci. *49:* 748.

KAISER, A. D., and R. B. INMAN. 1965. Cohesion and the biological activity of bacteriophage lambda DNA. J. Mol. Biol. *13:* 78.

KORN, D., and A. WEISSBACH. 1963. The effect of lysogenic induction on the deoxyribonucleases of *Escherichia coli* K12λ. I. Appearance of a new exonuclease activity. J. Biol. Chem. *238:* 3390.

LIEDKE-KULKE, M., and A. D. KAISER. 1967. Genetic control of prophage insertion specificity in bacteriophages λ and 21. Virology *32:* 465.

LITTLE, J. W. 1967. An exonuclease induced by bacteriophage λ. II. Nature of the enzymatic reaction. J. Biol. Chem. *242:* 679.

LITTLE, J. W., I. R. LEHMAN, and A. D. KAISER. 1967. An exonuclease induced by bacteriophage λ. I. Preparation of the crystalline enzyme. J. Biol. Chem. *242:* 672.

MANDEL, M. 1967. Infectivity of Phage P2 DNA in the presence of helper phage. Molec. Gen. Genet. *99:* 88.

MANDEL, M., and A. BERG. 1968. Cohesive sites and helper phage function of P2, Lambda, and 186 DNAs. Proc. Nat. Acad. Sci. *60:* 265.

OLIVERA, B. M., and I. R. LEHMAN. 1967. Linkage of polynucleotides through phosphodiester bonds by an enzyme from *Escherichia coli*. Proc. Nat. Acad. Sci. *57:* 1426.

PTASHNE, M. 1967. Isolation of the λ phage repressor. Proc. Nat. Acad. Sci. *57:* 306.

RADDING, C. M. 1964. Nuclease activity in defective lysogens of phage λ, II. A hyperactive mutant. Proc. Nat. Acad. Sci. *52:* 965.

SHAPIRO, L., L. I. GROSSMAN, J. MARMUR, and A. K. KLEINSCHMIDT. 1968. Physical studies on the structure of yeast mitochondrial DNA. J. Mol. Biol. *33:* 907.

SIGNER, E. R. 1968. Lysogeny: the integration problem. Ann. Rev. Microbiol. *22:* 451.

WETMUR, J., N. DAVIDSON, and J. SCALETTI. 1966. Properties of DNA of bacteriophage N1, a DNA with reversible circularity. Biochem. Biophys. Res. Comm. *25:* 684.

WU, R., and A. D. KAISER. 1967. Mapping the 5′-terminal nucleotides of the DNA in bacteriophage λ and related phages. Proc. Nat. Acad. Sci. *57:* 170.

—, —. 1968. Structure and base sequence in the cohesive ends of phage lambda DNA. J. Mol. Biol. *35:* 523.

YAMAGISHI, H., K. NAKAMURA, and H. OZEKI. 1965. Cohesion occurring between DNA molecules of temperate phages φ80 and lambda or φ81. Biochem. Biophys. Res. Comm. *20:* 727.

26

Reprinted from *Proc. Natl. Acad. Sci. (U.S.)*, **57**(2), 306–313 (1967)

ISOLATION OF THE λ PHAGE REPRESSOR*

By Mark Ptashne

DEPARTMENT OF BIOLOGY, HARVARD UNIVERSITY

Communicated by J. D. Watson, December 27, 1966

A bacterium can carry within it, integrated in its chromosome, the genome of a potentially lethal phage (prophage), because the genes of the prophage are prevented from functioning by a repressor. This lysogenic bacterium, as it is called, will lyse and produce phage if the repressor is inactivated. This same repressor, which is made by a gene on the prophage, is also responsible for the immunity of lysogenic cells to superinfection by phages similar to the prophage. Such superinfecting phages inject their DNA but the newly introduced phage genes neither function nor replicate.

These facts were elucidated largely by studies of the phage λ which grows on *E. coli*. In 1957, Kaiser and Jacob showed that the prophage gene C_1, required for the maintenance of lysogeny, is also responsible for the immunity against superinfecting phages.[1] On the basis of this and other observations, Jacob and Monod[2] proposed that the C_1 gene produces a repressor molecule (often referred to as the immunity substance) which blocks lytic development of both prophage and superinfecting phages by selectively repressing the expression of one or more phage genes. This is exactly analogous to their model for the action of the *i* gene of the lactose operon. Although both the *lac* and λ repressors have been the objects of extensive studies *in vivo*, their mechanism of action is unknown. Only recently has a repressor, the *lac* repressor, been detected *in vitro*.[3] This paper describes the specific labeling and partial purification of the C_1 product of phage λ.

Preliminary Results and Considerations.—In an ordinary λ-lysogen, phage repressor synthesis probably constitutes on the order of only one part in 10^4 of the cell's total protein synthesis. In order to label the λ phage repressor specifically, I sought conditions in which the rate of synthesis of repressor is a significant fraction (5–10%) of the total protein synthesis of the cell. To achieve the necessary differential increase in repressor synthesis, I attempted (1) to inhibit the synthesis of host proteins while maintaining the capacity to synthesize phage proteins; (2) to inhibit the synthesis of most or all of the phage proteins other than that of the repressor; and (3) to maximize the number of functioning λ repressor genes in the cell. The first objective was achieved by irradiating cells with massive doses of ultraviolet (UV) light, a treatment which damages the host DNA and dramatically decreases the level of cellular protein synthesis. Figure 1 shows that lytic infection of phage in sufficiently damaged cells stimulates about tenfold the incorporation of labeled leucine. This incorporation presumably represents the synthesis of numerous phage proteins, only one of which is the C_1 product. The second objective is achieved in principle by infecting λ-lysogens with other λ phages of the same immunity. Although most of the genes of these superinfecting λ phage chromosomes are repressed, one newly injected gene which *does* function is the C_1 gene itself.[4] The third objective might be achieved simply by increasing the multiplicity of λ phages infecting the lysogenic cell. However, at multiplicities above 10–15, immunity to superinfection breaks down, and phage proteins other than the repressor are synthesized. I have found that in lytic infection the production of many of the proteins of phage λ, perhaps all of them except for the C_1 product, is blocked if the phage carries a mutation in the early gene N, a result consistent with results of others.[5] Phages carrying mutations in gene N fail to stimulate the large increase in labeled leucine incorporation when added to UV-irradiated cells. This result is specifically confirmed by visualization of the labeled λ phage proteins made in irradiated cells by electrophoresis and autoradiography in polyacrylamide gels.[6] Although 8–10 prominent phage bands appear in extracts of λ-sensitive irradiated cells infected with wild type λ, these bands all disappear when the infecting λ phage carries a mutation in gene N. We know from genetic experiments that mutations in gene N do not block synthesis of the repressor. Therefore, high multiplicities of these mutant phages should stimulate the production of C_1 product without stimulating the production of most of the other λ proteins. This greatly simplifies the task of finding the C_1 product itself.

These considerations lead to the following experiment: a strain of *E. coli* lysogenic for λ is subjected to heavy UV irradiation. (The prophage used is λind^-, a mutant of λ which, unlike wild type, is not inducible by UV light.) One portion of the cells is infected with 30–35 λ phages carrying mutations in gene N, the other with λ phages carrying in addition a mutation in the C_1 gene which prevents synthesis of the C_1 product. One culture then receives H^3-leucine, the other C^{14}-leucine. After a period of labeling, the cells are mixed and sonicated, and the extracts are fractionated to look for a single protein marked with one label but not with the other. Figure 2 shows the location and names of the C_1 mutants used in these experiments. At each stage, I ran parallel experiments with the labels reversed to avoid being misled by the artifacts which can arise in double-label experiments.

The block provided by single mutations in gene N is noticeably incomplete on some bacterial strains, and so, in the experiments described below, *all* the phages carry double suppressor-sensitive mutations (ambers) in the N gene. Only the name of the C_1 gene, be it wild type, amber mutant, or temperature-sensitive mutant, will be explicitly used.

TABLE 1

DETECTION OF THE C_1 PRODUCT BY RATIO COUNTING

Cell fraction	H^3 cpm/- 0.1 ml	C^{14} cpm/- 0.1 ml	Ratio H^3/C^{14}
Sonicate	14,653	7,290	2.01
	15,079	7,654	1.97
Supernatant	6,852	3,201	2.14
	6,853	3,248	2.11
Pellet	7,007	3,747	1.87
	7,641	3,589	1.85

Counts and ratios are given for duplicate samples from each fraction. See text and *Materials and Methods* for experimental details.

Detection of the C_1 product: When a double-label experiment is performed comparing wild-type λ with the amber C_1 mutant $C_1 sus34$, a differential fractionation of the labels can be observed following sonication and high-speed centrifugation. Table 1 shows the results of a typical experiment. In this case the cells infected with wild-type phage were labeled with H^3-leucine, the others with C^{14}-leucine. There is approximately a 15 per cent increase in the ratio of H^3 to C^{14} in the supernatant fraction compared to that found in the pellet. This means that there is a component which is labeled with H^3 but not with C^{14} comprising about 15 per cent of the H^3 label in the supernatant. A ratio difference of 10–15 per cent is consistently observed when wild-type λ is compared with this amber mutant, whether the labels are in the configuration described in Table 1 or reversed.

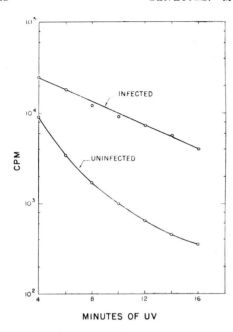

MINUTES OF UV

FIG. 1.—Stimulation of labeled leucine incorporation into irradiated cells by lytic phage infection. The cells were grown and irradiated as described in *Materials and Methods.* One-ml portions were labeled with $1/2$ μc C^{14}-leu for 1 hr with or without added phage at a multiplicity of 30 phage/cell. The cells are lysogenic for λind^-. In this particular experiment the infecting phage is λimm^{434}, a phage similar to λ but of different immunity which grows normally on λ-lysogens. The incorporation was terminated by the addition of TCA. The precipitated cells were collected and washed on a Millipore filter and counted in a gas-flow counter. These curves do not extrapolate to the correct unirradiated values. The uninfected cell synthesis is decreased about 5,000-fold at 14 min of irradiation.

DEAE chromatography: The ratio counting experiment suggests that the supernatant fraction contains a protein made by the wild-type repressor gene but not by the amber C_1. This protein is isolated by fractionation on a DEAE column. The results of a typical experiment are presented in Figure 3. In this case the H^3 label is again in cells infected with λ wild type, and the C^{14} label in cells infected with the C_1 amber mutant C_1sus10. The gradient elution profile shows a distinct peak present in the H^3 label with no corresponding peak in the other label. This experiment has also been performed using the amber C_1 mutants C_1sus34, C_1sus14, and C_1sus80. In each case the results are as seen in Figure 3, that is, none of these mutants produces detectable amounts of the major protein peak. These mutants are also lacking the much smaller peak which is usually seen immediately following the major one. Although other minor peaks preceding and trailing the major peak at tube 20 often appear, there are no other major peaks (excluding the flowthrough) in either label throughout a gradient run from 0.07 M KCl to 0.5 M KCl. Furthermore, there is no increase in the ratio of H^3/C^{14} in the flowthrough nor in the acid and base washings of the column, nor is there any significant differential loss of label. Finally, the percentage of supernatant counts appearing in the major peak is large enough to account for the 10–15 per cent ratio difference described in Table 1. The remaining 85–90 per cent of the label in the supernatant represents

NAMES AND APPROXIMATE MAP
POSITIONS OF λ C₁ MUTANTS

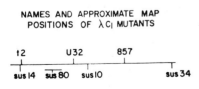

FIG. 2.—Mutants shown above the line are temperature sensitives (C_1ts), and those below the line are ambers (C_1sus). Mutants C_1sus34 and C_1sus14 are located at the extremities of the C_1 gene. Mutant C_1sus80 is only known to lie between C_1sus14 and C_1ts857. This map is a composite of two maps constructed by Drs. M. Lieb and F. Jacob. (These C_1sus mutants should not be confused with amber mutants of the same *sus* numbers located in other λ genes.)

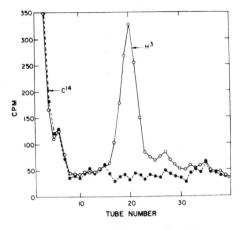

Fig. 3.—DEAE elution profile of an extract of a mixture of H³-leu-labeled λ wild-type infected cells and C¹⁴-leu-labeled λC₁-sus10 infected cells. The cells are lysogenic for λind⁻. The gradient begins around tube 4. The counts are normalized by multiplying the measured C¹⁴ values by the ratio of H³/C¹⁴ in the sample applied to the column.

Fig. 4.—DEAE elution profile of an extract of H³-labeled λC₁ls857 infected cells and C¹⁴-labeled λ wild-type infected cells. Reconstituted protein hydrolysate is the label used in this experiment. Other details as in Fig. 3.

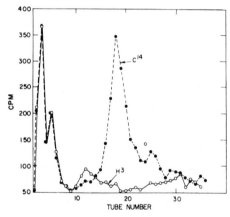

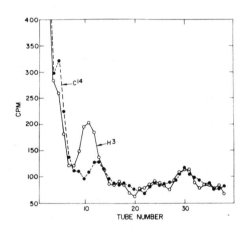

Fig. 5.—DEAE elution profile of a mixture of H³-leu-labeled λC₁lsU32 infected cells and C¹⁴-leu-labeled λ wild-type infected cells. Other details as in Fig. 3.

Fig. 6.—DEAE elution profile of a mixture of H³-leu-labeled λC₁lsU32 infected cells and C¹⁴-leu-labeled λC₁sus34 infected cells. Other details as in Fig. 3.

the synthesis of some bacterial proteins and possibly the limited synthesis of a few phage proteins other than the repressor.

Temperature-sensitive C₁ mutants: The absence of the isolated protein from C₁ amber mutant infected cells strongly suggests, but does not prove, that the structural gene for this protein is the C₁. To confirm this directly, three temperature-sensitive C₁ mutants (C₁ls857, C₁lsU32, and C₁lst2) have been examined to determine whether a modified form of the protein is produced by these mutants. In each case the repressors are labeled at temperatures at which they are functional *in vivo.*

Figure 4 shows the DEAE elution profile of a mixture of H³-labeled C₁ls857

TABLE 2

DIFFERENTIAL FRACTIONATION OF $\lambda C_1 ts2$ AND WILD-TYPE C_1 PRODUCTS

Fraction	Ratio H³/C¹⁴	Fraction	Ratio H³/C¹⁴
(1) Sonicate	2.39	(2) Sonicate	3.03
	2.39		3.00
Supernatant	2.27	Supernatant	3.13
	2.30		3.19
Pellet	2.52	Pellet	2.85
	2.49		2.78

In (1), H³-leu-labeled $\lambda C_1 ts2$ infected cells were mixed with C¹⁴-leu-labeled $\lambda C_1 sus34$ infected cells. In (2), H³-leu-labeled wild-type infected cells were mixed with an aliquot of the same C¹⁴-leu-labeled $\lambda C_1 sus34$ infected cells used in (1). Ratios are given from duplicate samples from each fraction. About half the label appears in the pellet following centrifugation. The experiment was performed as described in *Materials and Methods*, except that the infected cells were labeled at 32° C.

repressor and C¹⁴-labeled wild-type repressor. The mutant $C_1 ts857$ repressor chromatographs identically to wild type. The result of a similar experiment with mutant $C_1 tsU32$ gives a strikingly different result (Fig. 5). No peak corresponding to the $C_1 tsU32$ repressor is seen under the major C¹⁴ peak. A much smaller H³ peak of variable height which is probably the $C_1 tsU32$ gene product is observed around tube 12. This small peak is seen more easily in an experiment in which the $C_1 tsU32$ and $C_1 sus34$ gene products are chromatographed together (Fig. 6). The most likely reason that only a small amount of the $C_1 tsU32$ gene product is seen is that this material is partially insoluble. Another temperature-sensitive mutant, $C_1 ts2$, produces an even less soluble product. No significant differentially labeled H³ peak is seen when the experiment described in Figure 6 is performed with this mutant. The $C_1 ts2$ repressor is labeled in these experiments, but it precipitates and appears in the pellet upon sonication and subsequent centrifugation (Table 2). Although H³-labeled wild-type repressor produces a H³/C¹⁴ ratio increase in the supernatant when tested against C¹⁴-labeled $C_1 sus34$, H³-labeled $C_1 ts2$ repressor produces a ratio increase in the pellet when tested against this mutant. The fact that the magnitude of the ratio changes is about equal in the two cases suggests that $C_1 sus34$ produces only a small C_1 fragment, if any.

The important finding is that two mutations which produce modified repressors *in vivo* also alter the major protein synthesized in the present experiments.

Electrophoresis of the C_1 product: The migration of the wild-type C_1 product in a 7.5 per cent polyacrylamide gel is shown in Figure 7. The material applied to the gel was collected and concentrated from the peak fractions of a DEAE run. A single major H³ band is observed, with no corresponding C¹⁴ peak. This shows that the differentially labeled peak recovered from DEAE consists of a single-labeled protein of high isotopic purity. The C_1 product migrates toward the anode at pH 8.7 as would be expected for an acidic protein.

A mixture of H³-labeled $C_1 ts857$ repressor and C¹⁴-labeled wild-type repressor was isolated on

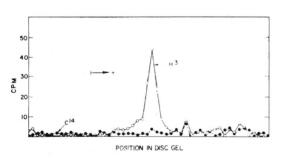

FIG. 7.—Polyacrylamide gel electrophoresis of the protein isolated from DEAE columns. See *Materials and Methods* for experimental details.

DEAE and subjected to electrophoresis as in Figure 7. A single superimposable major band appears in both labels, showing that the C_1ts857 product bears the same charge as the wild-type C_1 product.

Sedimentation of the C_1 product: Sedimentation of the wild-type C_1 product from the major DEAE peak was followed on sucrose gradients. Figure 8 shows the approximate s value of the C_1 product as $2.7–2.8S$. This corresponds to a molecular weight of approximately 30,000 if the label is in a spherical protein.

Discussion.—The behavior of the temperature-sensitive mutants is particularly interesting because of the observation[7,8] that, *in vivo*, temperature-sensitive mutations mapping on the left side of C_1 cause a more drastic alteration of the repressor than do temperature-sensitive mutations mapping on the right side of C_1. It has been suggested that right-hand side mutants such as C_1ts857 reversibly denature when heated; repression is restored when the temperature is lowered. In contrast, left-hand side mutants, such as C_1tst2 and C_1tsU32, apparently are irreversibly denatured by a pulse of heat. Attempts to attribute this difference to the existence of two C_1 products have so far failed: no complementation has been detected between amber mutants from opposite sides of the C_1 gene, nor between any amber mutants and temperature-sensitive mutants located anywhere in the gene.[9] In

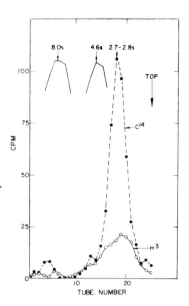

FIG. 8.—Sedimentation of the protein isolated from DEAE columns. The peak fractions from a DEAE run were pooled and concentrated, and a sample was layered on a 5–30% sucrose gradient in T-M buffer plus 0.15 M KCl. The gradient was spun 60,-000 rpm for 10.5 hr at 8°C. Parallel tubes were run carrying a mixture of the markers aldolase and hemoglobin. Fractions were collected and assayed for radioactivity or for absorbance at 230 and 410 mμ.

experiments reported here, no difference has yet been detected between the wild type and C_1ts857 repressors *in vitro*, but the C_1tst2 and C_1tsU32 repressors are found to differ markedly from wild type. Under standard conditions, the unmutated repressor appears in the supernatant following high-speed centrifugation, the C_1tst2 repressor invariably precipitates, and the C_1tsU32 repressor shows an intermediate solubility. The fraction of C_1tsU32 repressor which remains in the supernatant following centrifugation chromatographs separately from the wild type on DEAE columns. Furthermore, the *in vitro* distinction between the C_1tst2 and C_1tsU32 repressors is consistent with the observation that *in vivo* the latter is more heat-stable than the former.

The repressor isolated from DEAE has an apparent molecular weight of approximately 30,000. The molecular weight of another repressor, the *lac i*-gene product, has been estimated at 150,000–200,000.[3] There is indirect evidence that the functional *i*-gene product is an oligomer.[10] It is possible that the λ repressor can also exist as an oligomer, and that I have isolated it in some subunit form. In this regard it may be significant that the sedimentation of the *lac* repressor was effected at high concentrations (about 10^{-6} M), whereas in the experiments described here, the concentrations are much lower (about 10^{-8} to 10^{-9} M). There is a small peak of high s value in the sedimentation profile shown in Figure 8, but I have not attempted to characterize this fraction or the small peak which consistently emerges immediately behind the major peak on DEAE columns. Either of these components might represent alternate forms of the repressor.

Summary.—A product of the λC$_1$ (repressor) gene has been labeled with amino acids and isolated free from any other labeled components of the cell. The identification is based on the fact that this molecule is not made by amber mutants and is made in a modified form by temperature-sensitive mutants of the C$_1$ gene. The product electrophoreses in a single band, moving as an acidic protein. It sediments at 2.8S, which corresponds to a molecular weight of approximately 30,000.

Materials and Methods.—*Bacteria:* The bacterial strain used in all these experiments is a UV-sensitive mutant isolated by Dr. M. Meselson from the *E. coli* strain W3102.[11] This strain is prototrophic and nonpermissive (su$^-$) for amber mutants. The UV^s locus mutated in this strain is unknown. This strain is used because, as with some other UV^s strains,[12] its capacity to support λ phage growth is higher than that of wild-type strains following exposure of the host to the same physical dose of UV irradiation.

Bacteriophages: All the λ phages carrying double N mutations were prepared by recombination with the phage λNosus$_7$sus$_{53}$ isolated by Dr. D. Hogness. The derivative of phage λimm^{434} used here was described previously.[13] Phage λind$^-$ was isolated and described by Jacob and Campbell.[14] Two amber mutants,[15] C$_1$sus34 and C$_1$sus80, were supplied by Dr. F. Jacob. One of the temperature-sensitive mutants, C$_1$ts857, was isolated and described by Sussman and Jacob.[16] The other temperature-sensitive[7] and amber mutants were isolated and donated by Dr. M. Lieb. All the phages used in these experiments carry the *h* marker from the phage λh of Kaiser[17] to ensure efficient absorption. Phage stocks were grown in strain C600 by the agar-overlay method, pelleted, resuspended in phage buffer, and dialyzed against the same buffer.

Media and buffers: Cells were grown in A medium[18] with 4 gm/liter maltose as carbon source. Preconditioned medium was prepared by filtration of A medium which had supported the growth of cells to 2–3 × 10^8 cells/ml. T-M buffer is 0.01 M Tris pH 7.4, 0.005 M MgSO$_4$. Phage buffer is T-M buffer made 0.1 M NaCl.

Radioactivity: H^3- and C^{14}-labeled leucine and reconstituted protein hydrolysates were purchased from Schwartz BioResearch, Inc., and from New England Nuclear Corp. These labeled compounds were used without further dilution with cold amino acids. Numerous different batches of isotopes were used during the course of these experiments. The specific activity of the C^{14}-leu is about 200 mc/mmole, and that of the H^3-leu is about 2.0 c/mmole. All radioactive isotope counting, unless specified otherwise, was performed according to the method described by Fox and Kennedy,[19] except that the ethanol wash was replaced with another TCA wash.

Electrophoresis: Disc electrophoresis was performed at pH 8.7 as described by Davis.[20] After electrophoresis the gels were frozen and cut into slices 1 mm thick using a slicing apparatus designed and built by Dr. C. Levinthal. The slices were distributed into counting vials and dissolved in $1/2$ ml of 15% hydrogen peroxide by heating at 60–70°C for a few hours. One-half ml of hydroxide of hyamine (Packard Inst. Co.) was added to the vials, and the mixture was incubated at 57°C for 3 min. After cooling, 10 ml of Bray's[21] solution was added, and the samples were counted in a Packard scintillation counter.

Chromatography: DEAE-cellulose chromatography was performed using a 15-ml column, 1 cm in diameter, with Whatman DEAE-cellulose #DE52. After applying the sample and washing the column with about 10 ml of T-M buffer containing 0.07 M KCl, a 150-ml linear KCl gradient was run to 0.2 M KCl, and fractions containing about 4 ml were collected. Aliquots from each fraction were assayed for radioactivity.

Irradiation: Cells were irradiated at a distance of 30 cm from 2 G.E. 15-w germicidal lamps. The incident dose at this distance is about 75 ergs mm^{-2}sec^{-1}. Eighty ml of cells at concentration 3 × 10^8 cells/ml were irradiated at 0°C in a Petri dish of diameter 11 cm, with swirling.

A typical experiment: Cells grown to a concentration of 10^9 cells/ml in medium A are chilled and diluted to a concentration of 3 × 10^8 cells/ml with cold preconditioned medium. The cells are irradiated for 12 min, and the $MgSO_4$ concentration is then raised from $10^{-3} M$ to $2 \times 10^{-2} M$. Two 40-ml portions are distributed into flasks, and phage are added at a multiplicity of 30–35 phage/cell. After swirling 5 min at 37°C, 0.4 mc H^3-leu is added to one flask, and 0.04 mc C^{14}-leu to the other, and the cells are aerated for 1 hr at 37°C. The cells are then chilled, centrifuged, mixed together, washed twice in T-M buffer, and finally resuspended in 2 ml of T-M buffer containing 0.4 M KCl and a few micrograms of DNase. The cells are disrupted while on ice by several 30-sec pulses delivered by an MSE sonicator. Small aliquots are taken to determine the original ratio of H^3/C^{14}, and the extract is immediately spun at 350,000 g for 36 min in an International B-60 centrifuge. The supernatant is withdrawn and the pellet is resuspended in 2 ml of T-M buffer. After removing samples for ratio counting, the supernatant is dialyzed overnight against T-M buffer plus 0.07 M KCl.

It is of critical importance that the Mg^{++} concentration be raised before addition of the phage. Phage λ absorbs well to cells grown in medium A with maltose as the carbon source, even at low Mg^{++} concentrations. However, if the experiment described in Figure 1 is performed in the presence of only $10^{-3} M$ Mg^{++}, no stimulation of labeled leucine incorporation is observed. In fact, even at high UV doses, the addition of 35 phage/cell causes a further 10–20-fold decrease in the incorporation of labeled leucine. Others have noted a strong inhibition in protein and RNA synthesis following infection of unirradiated cells with high multiplicities of phage λ.[22] The discovery that high concentrations of Mg^{++} can reverse this inhibition, at least in the irradiated strain used here, made possible the success of these experiments.

Throughout the course of this work I have had the advice and encouragement of Walter Gilbert and James Watson, for which I am very grateful. I would also like to thank Matthew Meselson for providing laboratory space and many facilities necessary for these experiments; Nancy Hopkins and Louise Rogers for excellent assistance; and Margaret Lieb, Racquel Sussman, and François Jacob for phage mutants.

* These experiments were performed while the author was a Junior Fellow of the Society of Fellows, Harvard University, and were supported by grants from the National Science Foundation and the National Institutes of Health.

[1] Kaiser, A. D., and F. Jacob, *Virology*, **4**, 509 (1957).

[2] Jacob, F., and J. Monod, *J. Mol. Biol.*, **3**, 318 (1961).

[3] Gilbert, W., and B. Müller-Hill, these PROCEEDINGS, **56**, 1891 (1966).

[4] Lieb, M., *Virology*, **29**, 367 (1966); and Horiuchi, T., and H. Inokuchi, *J. Mol. Biol.*, **15**, 674 (1966).

[5] Thomas, R., *J. Mol. Biol.*, in press; and Protass, J., and D. Korn, these PROCEEDINGS, **55**, 1089 (1966).

[6] Autoradiography was performed as described in Fairbank, G., Jr., C. Levinthal, and R. H. Reeder, *Biochem. Biophys. Res. Commn.*, **20**, 393 (1965).

[7] Lieb, M., *J. Mol. Biol.*, **16**, 149 (1966).

[8] Naona, S., and F. Gros, *J. Mol. Biol.*, in press.

[9] Lieb, M., personal communication.

[10] Sadler, J. R., and A. Novick, *J. Mol. Biol.*, **12**, 305 (1965).

[11] Hill, C. W., and H. Echols, *J. Mol. Biol.*, **19**, 38 (1966).

[12] Devoret, R., and T. Coquerelle, personal communication.

[13] Ptashne, M., *J. Mol. Biol.*, **11**, 90 (1965).

[14] Jacob, F., and A. Campbell, *Compt. Rend.*, **248**, 3219 (1959).

[15] Jacob, F., R. Sussman, and J. Monod, *Compt. Rend.*, **254**, 4214 (1962).

[16] Sussman, R., and F. Jacob, *Compt. Rend.*, **254**, 1517 (1962).

[17] Kaiser, A. D., *J. Mol. Biol.*, **4**, 275 (1962).

[18] Meselson, M., and F. W. Stahl, these PROCEEDINGS, **44**, 671 (1958).

[19] Fox, C. F., and E. P. Kennedy, these PROCEEDINGS, **54**, 891 (1965).

[20] Davis, B. J., *Ann. N. Y. Acad. Sci.*, **121**, 404 (1964).

[21] Bray, G. A., *Anal. Biochem.*, **1**, 279 (1960).

[22] Howes, W. V., *Biochim. Biophys. Acta*, **103**, 711 (1965); Tertzi, M., and C. Levinthal, *J. Mol. Biol.*, in press.

27

DEVELOPMENTAL PATHWAYS FOR THE TEMPERATE PHAGE: LYSIS VS LYSOGENY

Harrison Echols

Department of Molecular Biology, University of California, Berkeley, California

1. INTRODUCTION

This article is divided into two major sections: first, a general summary of current beliefs concerning molecular events during the life cycle of the temperate phage; second, one aspect of this life cycle in which there has been a considerable amount of recent progress—the mechanism by which the temperate phage provides itself with a choice between the alternative life styles of lysis or lysogeny. The aim of the second section is to give some insight into experimental approaches and therefore provide an introduction to current literature. I will consider primarily the extensively studied phage λ. The emphasis of this review is complementary to that of an earlier selective review (1) in which I stressed other aspects of the lysogenic pathway. The reader who wishes a more complete view of λ should consult *The Bacteriophage* λ (2). Other recent reviews have dealt in depth with the temperate phages P22 (3) and P2 (4). A valuable historical discussion of many of the points in this article may be found in the reviews by Lwoff (5) and Jacob & Wollman (6).

2. DIVERGENT LIFE STYLES: THE LYTIC AND LYSOGENIC PATHWAYS

The Life Cycle of the Temperate Phage

Definitions and evolutionary considerations.—When a temperate phage infects a bacterium, typically there are two developmental pathways open to the virus: the *lytic pathway*, which culminates in production of a large number of new virus particles and lysis of the cell; and the *lysogenic pathway*, in which the cell survives with the lytic capacity of the virus turned off, and the viral DNA integrated into the host DNA and replicating as part of the host chromosome. Once established, the lysogenic state is quite stable: the viral DNA (or prophage) remains repressed and integrated. However, most temperate phages have evolved the option for divorce; a variety of agents which interfere with host DNA replication induce the prophage to undergo lytic development. Rarely, lytic growth also ensues in the absence of an external inducing agent (10^{-2}–10^{-5} per cell generation).

There are clearly two areas of viral decision-making: whether to carry out

157

the lytic or lysogenic pathway after infection; and whether to maintain lysogeny or undergo induction. The evolutionary basis for lysogeny and induction seems clear. Lysogeny provides a solution to the problem of exhaustion of the host supply. The virulent phage must seek more uncertain means to avoid exhaustion of its natural resources: perhaps through bacterial mutation to virus resistance, followed by viral mutation to a new host range; perhaps through a continual supply of fresh cells administered by a friendly sewer system or molecular biologist. Induction prevents the demise of a prophage when the host chromosome of which it is part loses the capacity to replicate. The evolutionary basis for the typical occurrence of both lytic and lysogenic responses in an infected bacterial population is less clear. Most temperate phage devotees take as an article of faith the premise that the lytic response is favored under conditions when the cell population is growing well and thus able to propagate a successful lytic infection. This point will be considered in more detail at the end of this article.

In order to present an overall picture of the life cycle of the temperate phage, I will first summarize the essential features of the lytic and lysogenic pathways which are common to a variety of temperate phages. I will then consider in more detail for phage λ the series of molecular events that defines each pathway and provides for the capacity of the virus to switch from one to the other.

Essential features of effective lytic growth.—The basic developmental problem of the lytic pathway is to maximize virus yield. All large DNA phages, temperate or virulent, accomplish this by sequential expression of viral genes such that the proteins for DNA replication and genetic recombination are synthesized first, and the proteins involved in head and tail structures and cell lysis are not made until a later time. One can thus define an early and a late stage of lytic development.

This sequential developmental pattern provides for a temporal assembly line: the concentration of the cell energy and resources first on maximal replication of viral DNA and then on the coating of the DNA and viral release. In addition, this temporal assembly line prevents premature destruction of the host cell before a maximal viral yield is obtained and probably aids in DNA encapsulation, which may typically proceed from a very long multimeric piece of DNA (7). For the temperate phage, sequential development also aids in the potentiality for the lysogenic response.

In every case examined so far, the primary mechanism for sequential regulation of lytic growth appears to involve positive regulation (activation) of the viral genes for late proteins by a phage coded protein or proteins from the early stages of development. The absence of this positive regulatory protein is characterized by a failure of RNA synthesis from the late gene region of the viral DNA; thus the mechanism for activation of late genes is likely to involve an alteration of the host machinery for DNA transcription (8, 9). An additional feature of regulation may be provided by the transcription time required for an RNA polymerase molecule to traverse a long region of DNA (10, 11).

The necessity for a phage coded protein for late protein synthesis allows for a

302

clear experimental separation of the early and late stages of lytic development. For phage λ and possibly for phage T4, there is a similar experimental separation of early events into an immediate-early and a delayed-early stage, in which the delayed-early stage is the period of maximal viral DNA replication and genetic recombination (10, 12). The activation of the delayed-early stage of lytic development depends on a phage coded protein or proteins from the immediate-early stage.

In summary, lytic growth of temperate phage (and large virulent phage) is characterized by sequential positive regulation of the viral genome, probably at the level of RNA synthesis. Lytic development begins with the transcription of a limited portion of the injected viral DNA by the host RNA polymerase. Development proceeds through one or two stages of new transcription, dependent upon phage coded regulatory proteins. This sequential expression of viral genes provides for a temporal division of the lytic cycle into a period for viral DNA synthesis and recombination and a period for encapsulation of the viral DNA and cell lysis.

Essential features of an effective lysogenic response and its reversal.—The basic developmental problem of the lysogenic pathway is to provide an effective solution for three fundamentally different problems: establishment, maintenance, and induction.

The establishment of lysogeny in an infected cell generally requires two functionally separate events: repression of the genes for lytic functions and integration of the viral DNA into the host DNA. Integration occurs through a site-specific genetic recombination event at a specific region of the phage and host DNA which provides for a covalent insertion of the viral DNA into the host DNA (Figure 1). The establishment stage of the lysogenic pathway requires coordination between repression and integration; an uncertain future faces a viral DNA that is completely repressed but fails to integrate or vice-versa. An exceptional case is phage P1, which does not integrate, but duplicates in synchrony with the host as a typical plasmid (13); this may involve a judicious degree of partial repression (14).

The maintenance of lysogeny is inherently simpler than establishment. The repression of viral genes must be maintained, but no site-specific recombination events are involved; on the contrary, stability of the prophage state requires that such occurrences be avoided. In addition to maintenance of its own repression, a prophage often protects itself against other phages either through direct repression, abortion of lytic growth, or alteration of receptor sites on the cell wall.

The maintenance of repression generally involves a complete turnoff of all genes for lytic functions. At first consideration, this appears to be a formidable task. However, since a phage coded protein is required to activate the later stages of lytic development, the maintenance of repression requires a turnoff of only those genes active during the earliest stage. The action of a single repressor protein at one or two sites prevents the host RNA polymerase from initiating this earliest stage of viral development. The establishment of repression involves a

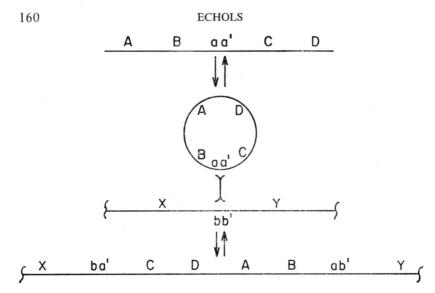

FIG. 1. Integration and excision by phage DNA. *A*, *B*, *C*, and *D* denote phage genes; *X* and *Y* denote bacterial genes. The linear DNA molecule injected from the phage (top of figure) forms a circle by joining ends. Site-specific recombination between phage and host attachment regions (*aa'* and *bb'*, respectively) provides for insertion of the viral DNA into the host DNA. Recombinant prophage attachment regions (*ba'* and *ab'*) result at the ends of the prophage DNA. Excision occurs through a recombination between prophage attachment regions to regenerate the original viral DNA molecule.

more complex regulatory problem because phage coded proteins are required for integration. Thus to establish lysogeny the phage must not impose repression before these proteins are synthesized, but must shut off viral development before an irreversible commitment to lytic growth. Not surprisingly, more regulatory genes are generally required to establish than to maintain lysogeny.

As for establishment, induction from the lysogenic state requires two functionally separate events: release of repression and excision of the viral DNA from the host DNA. The release of repression follows treatment of the cell with agents which inhibit DNA replication (e.g. ultraviolet light, mitomycin C, thymine deprivation). An "inducing substance" is generated that antagonizes the repressor responsible for maintenance of the lysogenic state. The lytic development that ensues is identical to that found after infection, except for the additional requirement for excision of the viral DNA. Excision occurs through a site-specific recombination event that regenerates the free viral DNA (Figure 1). Not all temperate phages have the capacity for induction (at least by typical agents).

The simplicity of the maintenance system serves to make induction relatively uncomplicated—typically only one protein must be inactivated by the inducing treatment. Thus sequential positive regulation of viral development serves not only the needs of efficient lytic growth, but also the complex set of requirements for the establishment and maintenance of lysogeny and for induction from the lysogenic state.

In summary, lysogeny by temperate phage generally exhibits two characteristic features: negative regulation of the viral genome and site-specific recombination. The lysogenic response to infection requires the establishment of repression for lytic genes and an integrative recombination event. Once established, stable lysogeny requires repression of genes concerned with lytic growth and with site-specific recombination. Induction from the lysogenic state to lytic development requires the release of repression and an excisive recombination event.

The Lysogenic and Lytic Pathways for Phage λ

Introductory comments.—The study of λ regulation has involved three levels of analysis: genetic experiments, which define the regulatory genes and suggest their functional targets; measurements of RNA and protein synthesis in vivo, which also indicate functional targets and in addition suggest possible molecular mechanisms; and in vitro experiments with separated components, which establish the biochemical basis for the regulatory phenomena. The first two levels of analysis provide a likely working hypothesis; the in vitro experiments carry this hypothesis to the stage of biochemical reality. The actual experimental procedures used have been summarized before (12). Some examples will be given in the more detailed discussion of Section 3.

Only for the maintenance of repression has the study of λ regulation progressed to the stage of in vitro reconstruction with purified components and only at this stage can firm biochemical conclusions be drawn. The remaining areas of λ regulation remain at the level of the working hypothesis. Most recent progress has been concerned with the development of such a working hypothesis for the establishment of repression—this topic will be considered in much more depth in Section 3. Other reviews in the past two years have considered in some detail lytic regulation (12), the maintenance of repression (1, 15), and integrative and excisive recombination (1, 16). In my opinion, the level of understanding in these areas has not changed substantially since these articles were written.

The lytic pathway.—Lytic development by phage λ is regulated through sequential activation of transcription by two λ proteins—the products of the N and Q genes. The general concept of N and Q regulation can be followed by reference to Figure 2.

λ DNA enters the cell and forms a circular molecule (17, 18) through joining of single-stranded complementary ends ("cohesive sites") (19, 20). Development begins by transcription from gene N and to some extent from the region including the genes O and P which specify λ replication proteins. This immediate-early stage of RNA synthesis is carried out by the host RNA polymerase (21–23). The delayed-early stage of RNA synthesis is dependent on N protein, which activates transcription from the recombination genes, the replication genes, and gene Q (21–27). The late stage of RNA synthesis is dependent on Q protein, which activates transcription from the genes for head, tail, and lysis proteins (21, 28, 29). Q-activated transcription probably begins from a site between the Q and lysis genes and proceeds sequentially around the circular DNA molecule through the tail region (30, 144).

The maintenance of repression.—The maintenance of lysogeny is the simplest stage of lysogenic development—it involves only the maintenance of viral repression. For phage λ, this stage is also the best understood. Repression is maintained through the action of a single phage protein—the product of the cI gene—which acts at sites to the left and right of the cI gene (Figure 2). The cI protein inhibits the immediate-early stage of RNA synthesis and provides for its own continued synthesis.[1] The repression of immediate-early transcription is sufficient to shut off the lytic capacity of the virus because the N protein is required for the delayed-early and late stages of lytic development. The cI protein not only maintains prophage repression, but also provides immediate repression of the DNA of a λ phage that infects a cell lysogenic for λ. Thus a lysogenic cell is "immune" to further infection by additional λ phage.

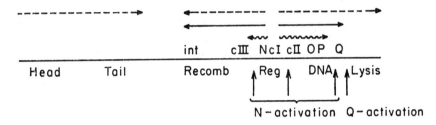

FIG. 2 Transcription events during lytic development by phage λ. Approximate DNA regions transcribed during the different stages of lytic growth are shown: (∿∿➤) represents the immediate-early stage of RNA synthesis, performed solely by the host transcription machinery; (⮂➤) represents the delayed-early stage of RNA synthesis, in which N protein activates transcription of the cIII to *int*, and cII to Q regions; (----➤) represents the late stage of RNA synthesis, in which Q protein activates transcription of the lysis, head, and tail regions. Since λ DNA exists in a circular form during much of its intracellular life, it is likely that the actual unit of transcription is the circular DNA with the lysis region joined to the head region, rather than the linear molecule extracted from phage and indicated here. The probable sites at which N and Q-activation occurs are indicated by the vertical arrows (↑). No effort has been made to show individual RNA chains nor indicate quantity of RNA produced. The DNA strand transcribed from right to left is denoted l; the strand transcribed from left to right is denoted r. The genetic organization of the λ DNA molecule is indicated by the generic designation below the λ DNA: *Head* for genes concerned with the structure of the phage head; *Tail* for genes concerned with tail structure; *Recomb* for genes involved with general and site-specific recombination; *Reg* for genes exerting a regulatory function in lytic development or lysogeny; *DNA* for genes specifying replication proteins; and *Lysis* for genes concerned with cell lysis. Specific genes of the "regulation region"—$cIII$, N, cI, cII—are indicated above the "λ DNA", as are the integrative recombination gene *int*, the DNA replication genes OP, and the late regulatory gene Q.

[1] The lysogenic cell retains another function—the ability to inhibit growth of certain strains of phage T4 (rII mutants) (136). Mutants of λ have been isolated—termed rex^-—which have lost this "rII exclusion" function, but which retain the repression capacity of a typical lysogen (137). Thus, the protein required for rII exclusion may be distinct from the cI protein; however, this point is not yet well established.

The capacity of the cI protein to repress immediate-early RNA synthesis is the best established aspect of λ regulation. The evidence for this activity is derived from all three levels of analysis noted earlier. Genetic studies and measurements of in vivo macromolecular synthesis have provided the working hypothesis for the role of the cI protein and its target genes (24, 31–36); in vitro experiments on DNA binding (35, 37–39) and RNA synthesis (38–42) have shown that the cI protein binds to specific operator sites on λ DNA and inhibits RNA synthesis from the immediate-early initiation sites.

This level of understanding seems profound only in contrast to the dearth of biochemical understanding in other areas of the λ life cycle. Many problems remain unsolved concerning the maintenance of repression. The actual molecular mechanism for the repression of RNA synthesis has not been completely defined. In vitro the cI protein can inhibit binding or initiation by RNA polymerase if repressor is bound first (42, 143); the cI protein can also partially inhibit RNA synthesis by a prebound RNA polymerase (143). Thus in vivo the cI protein may be capable of acting at two points in the RNA polymerase reaction.

The biochemical mechanism for the maintenance of cI protein synthesis has not been determined. Evidence has been presented for a "self-activation mechanism" in which the presence of the cI protein is necessary to provide for transcription of the cI gene. This problem will be considered further as part of a discussion of the transition from the establishment of repression to its maintenance.

The establishment of repression.—The primary regulatory elements for the establishment of repression are the protein products of the cII and cIII genes. These proteins function in concert to activate synthesis of the cI protein and repress synthesis of certain lytic proteins. The functional activity and time of synthesis of the cII and cIII proteins provide for the properly timed establishment of repression—after the requirements for integration have been satisfied, but before the late stage of lytic growth.

For phage λ, integration requires the protein product of the *int* gene; integration is probably also enhanced by viral DNA replication, because the presence of many DNA copies is likely to favor the insertion direction of the reversible integration reaction. Thus repression should not be imposed before the delayed-early stage of viral development, which includes maximal transcription of the *int* and *OP* genes. Since the cII and cIII genes are part of the delayed-early transcription units (Figure 2), synthesis of the cII and cIII proteins should be coordinated with that of the *int* and replication proteins; the cII and cIII genes are sensors of effective delayed-early transcription. Once the requirements for integration have been satisfied, successful lysogeny requires a rapid establishment of repression before an irreversible commitment to lytic growth has occurred. The cII and cIII proteins presumably provide for this by an activation of rapid synthesis of cI protein and an ancillary repression of synthesis of lytic proteins.

A possible mechanism for the activity of the cII and cIII proteins is the following (Figure 3). The cII and cIII proteins form a regulatory oligomer which acts at a site between the cI and cII genes to activate leftward (*l*-strand) transcription and repress rightward (*r*-strand) transcription. The experimental justification

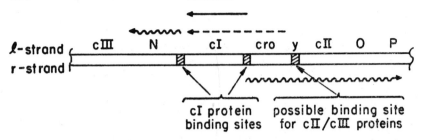

FIG. 3. Possible mechanism for the action of the cII and cIII proteins. The cII/cIII regulatory oligomer acts at a site between the cI and cII genes ("y-region") to provide for activation of l-strand transcription toward the cI gene (◄----) and inhibition of r-strand transcription toward the cIIOP genes (〜〜➤). Once the supply of cI protein becomes sufficient, the lytic capacity of the virus is completely shut off by the ability of the cI protein to inhibit immediate-early RNA synthesis, represented by (〜〜〜) (Figure 2). The maintenance of cI protein synthesis probably involves a different transcription start from that provided by cII/cIII activation (◄—).

for this working hypothesis and some possible biochemical mechanisms are considered in the section beginning on page 165.

Another feature of this rapid establishment of repression is probably provided by an additional regulatory capacity of the cI protein. This protein inhibits the "activity" of the N and replication proteins—the N and OP proteins will not act effectively on a λ DNA molecule to which the cI protein is bound. Thus the presence of excess N and replication proteins in an infected cell need not impede lysogeny once cI-mediated repression is established. Possible mechanisms for this activity of the cI protein are considered in the section beginning on page 168.

An additional regulatory element that can influence the establishment of repression is the product of the cro gene (Figure 3). The role of the cro gene is the least understood of λ regulatory phenomena. The cro product[2] can inhibit the synthesis of cI protein and thus antagonize the establishment of lysogeny. The physiological function for this cro activity after infection might be to prevent cII/cIII-mediated repression from channeling all infected cells to lysogeny. Such an anti-repression activity is of more obvious use at the other end of the lysogenic cycle when a stable prophage is induced to lytic growth, and reestablishment of repression is particularly undesirable.

The study and interpretation of the role of cro is complicated by the fact that the cro product probably also turns off synthesis of recombination proteins during the late stage of lytic growth. If this turnoff of delayed-early protein synthesis is critical for a successful lytic response, the primary target for cro activity might be one or more delayed-early genes, and the effect on cI might be a by-product that is unimportant in a normal infection.

[2] Nonsense mutations have been found for the cI, cII, cIII, N, and Q genes; however, no nonsense mutations in the cro gene have yet been reported. For this reason, the products of the cII, cIII, N, and Q genes are called "proteins" even though the gene products have not been isolated; the creation of the cro gene is referred to as a "product."

The best available working hypothesis at present for the role of the *cro* gene seems to be the following (Figure 3). The *cro* gene is transcribed from the *r*-strand during both the immediate-early and delayed-early stages of viral development. The *cro* product acts at or near the same sites as the *c*I protein to exert the same basic repression activity: inhibition of immediate-early RNA synthesis. This represses synthesis of *c*II and *c*III proteins because *N* protein cannot be synthesized and even if present cannot effectively turn on the transcription of the *c*II and *c*III genes. The action of *cro* to inhibit gene *N* transcription initiated at the immediate-early promoter site also provides for the turnoff of synthesis of recombination proteins noted above. In addition, *cro* product may prevent the normal maintenance stage of *c*I protein synthesis by interfering with transcription of the *c*I gene initiated at or near the right-hand operator site. Evidence for this working hypothesis is presented in the section beginning on page 182.

A paradox is apparent with such an hypothesis even at this rudimentary level. Repression of *N* protein synthesis can peril the lytic response as well as the lysogenic. In a normal infection, this repression effect should not handicap lytic development because the action of *cro* as a repressor is not maximal until the late stage of lytic growth begins (as judged by repression of synthesis of recombination proteins). However, the way in which the phage provides for this solution is not obvious, since transcription of the *cro* gene should be at least as early and efficient as that of the *c*II gene.

The release of repression.—The cellular signal for the release of repression is an inhibition of host DNA synthesis (6, 43). This initiates a sequence of events that occupies about 30 minutes (44–47) and culminates in the inactivation of the *c*I protein (48). Thus viral repression is released and the prophage is induced to commence lytic growth. The actual cellular events that lead to repressor-inactivation are unknown. The protein product of the *rec*A gene is probably an essential element of the "induction pathway", since *rec*A⁻ mutants are defective in release of repression (49, 50). Lambdologists usually avoid the inconvenience of normal induction procedures by the use of phage with a mutation in the *c*I gene which renders the *c*I protein thermolabile; repression can be released for such strains simply by raising the temperature (51, 142).

The possible importance of the *cro* product in consummating the induction process has been noted earlier. No data are yet available as to whether *cro* function is more effective under conditions of prophage induction than infection.

Integration and excision.—The establishment of stable lysogeny for phage λ requires both repression of lytic functions and insertion of the viral DNA into the host DNA (Figure 4, also 1, 16, 52, 53). Integration occurs through a site-specific recombination event (54, 55), which requires the protein product of the λ *int* gene (56–58). The induction of prophage λ from the lysogenic state requires both release of repression and excision of the viral DNA from the host DNA. Excision occurs also through a site-specific recombination event (59), which requires the protein products of the λ *int* and *xis* genes (60–63).

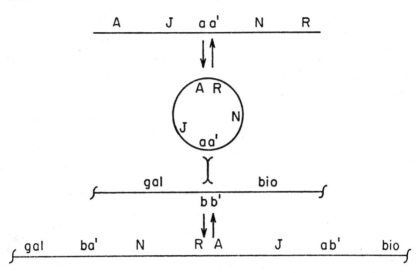

FIG. 4. Integration and excision by phage λ DNA. Figure 1 has been redrawn to show specific λ genes and the specific bacterial attachment region for λ between the *gal* and *bio* operons. *A, J, N,* and *R* denote particular λ genes in the head, tail, regulation, and lysis regions, respectively. The linear λ DNA molecule at the top of the figure forms a circle through base-pairing between complementary single-stranded ends. A reciprocal recombination between phage and host attachment regions (*aa'* and *bb'*, respectively) inserts the λ DNA into its prophage home in the host DNA. Excision occurs through a recombination between prophage attachment regions (*ba'* and *ab'*) to yield free viral DNA and the original host attachment region.

From its functional and structural requirements (Figure 4), the integration-excision reaction can thus be written formally as follows:

$$aa' + bb' \overset{int}{\underset{\substack{int \\ xis}}{\rightleftarrows}} ba' + ab'$$

Integration is an efficient event in the establishment of lysogeny; excision is typically highly efficient when a prophage is induced to lytic growth. The way in which this reaction is controlled is clearly important to effective lysogenic or lytic development. The basis for this control is not known. The number of viral copies (*aa'*) is probably important in driving the integration reaction (64); relative quantity of *int* and *xis* proteins and/or state of aggregation also may be important (63, 65–67). Virtually nothing is known in biochemical terms concerning the enzymes and DNA substrates for the integration-excision reaction, nor of the sequence of molecular events that occur during this reaction. An understanding of the molecular basis for the integration-excision reaction probably requires the isolation of both enzymes and DNA intermediates; progress in this area has been slow.

Summary of the λ life cycle.—Phage λ has the capacity for a choice of alternative life-styles: lytic or lysogenic. These two pathways of development and their regulation are indicated schematically in Figure 5. After injection into a nonlysogenic bacterial cell, λ DNA first enters an immediate-early stage, in which both pathways are open (upper left of Figure 5). Synthesis of the *N* protein provides for the delayed-early stage, characterized by maximal synthesis of proteins concerned with replication, recombination, and regulation. At this point, the pathways diverge, and a choice is made for lysis or lysogeny.

Lytic development into the late stage is activated by *Q* protein, which provides for rapid synthesis of head, tail, and lysis proteins. Lysogenic development is activated by *c*II and *c*III proteins, which provide for rapid synthesis of *c*I protein and reduced synthesis of head, tail, and lysis proteins. Lysogenic development may be opposed by the *cro* gene product through its capacity to repress synthesis of the *c*II and *c*III proteins.

The establishment of *c*I-mediated repression and the integration of the viral DNA consummate the establishment of stable lysogeny. Once established, the lysogenic state is maintained through repression of lytic genes by the *c*I protein, which must also provide for its own continued synthesis. The *c*I protein maintains repression by blocking the immediate-early stage of development.

The developmental process may be started anew by release of *c*I-mediated repression, normally accomplished by inactivation of the *c*I protein through some event that follows inhibition of host DNA replication. Again a choice can presumably ensue. The delayed-early stage can diverge to either pathway: excision and transition to the late stage along the lytic pathway; or re-establishment of repression and integration along the lysogenic pathway. If the release of *c*I-mediated repression is accomplished by an irreversible inactivation of *c*I protein, the lytic pathway will prevail.

The schematic diagram of Figure 5 encompasses the major features of the λ

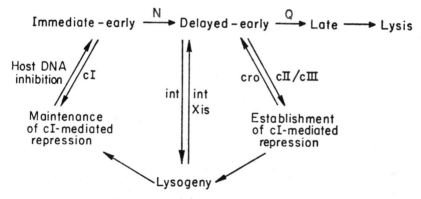

FIG. 5. Pathways of λ development and their regulation (see text). Both the establishment of repression and integrative recombination are required for the establishment of stable lysogeny; both the release of repression and excisive recombination are required for successful lytic development from the lysogenic state.

life cycle, but should not be taken too seriously in detail. A completely accurate presentation probably requires both divine guidance and three dimensions.

3. THE ESTABLISHMENT OF REPRESSION BY PHAGE λ
The cii and ciii Proteins Provide for the Establishment of Repression

Genetic analysis.—As for other regulatory phenomena, the study of the establishment of repression has depended heavily on the isolation and characterization of mutants. Repression-defective mutants of λ are recognized by their capacity to form plaques with few or no surviving bacteria ("clear plaques") (31, 68), rather than the typical plaques characteristic of wild-type λ, which contain a rather dense population of surviving lysogenic bacteria ("turbid plaques").[3]

Repression-defective mutants isolated in this way fall into three major classes —cI^-, cII^-, and $cIII^-$—which differ in mutational site, frequency of lysogenization, and capacity to maintain lysogeny. cI^- mutants generally fail to lysogenize at all, and cannot form stable single lysogens even if the cI function is supplied transiently by another co-infecting λ phage (31). cII^- and $cIII^-$ mutants lysogenize rarely (frequency approximately 10^{-4} and 10^{-2} respectively); once formed the lysogen is stable (31). Each mutational class defines a gene for a cytoplasmic product, as judged by complementation analysis (31).

The genetic analysis thus defines three regulatory products required for effective development along the lysogenic pathway—the proteins specified by the cI, cII, and $cIII$ genes. Since lysogens of cII^- and $cIII^-$ mutants are stable, the cII and $cIII$ proteins presumably are needed for the efficient establishment of repression, but not its maintenance. Since cI^- mutants cannot achieve a stable lysogenic state,[4] the cI protein presumably is the essential regulatory element for the maintenance of repression.

An additional genetic region important for the establishment of repression is indicated by a minor class of mutations—termed cy^-—which are located in the "y-region" between the cI and cII genes (69, Figure 3) and confer a phenotype similar to cII^- mutations in terms of lysogenization frequency (31). In contrast to cII^- mutants, cy^- mutants do not exhibit efficient complementation for lysogeny after mixed infection with cI^-; this *cis* effect of cy^- mutation on cI gene function suggests the possibility that the cy^- mutations might affect a site concerned with cI protein synthesis (69).

From the genetic analysis, a knowledge of the function of the cII and $cIII$ genes and of the "y-region" is clearly central to an understanding of the estab-

---

[3] Integration-defective mutants are essentially the converse of repression-defective mutants. Repression can be established, so that the infected cell often survives; however, the repressed viral DNA cannot integrate or replicate and is diluted out as the cell population grows. Integration-defective mutants thus form turbid plaques, but most of the surviving cells in the population do not contain prophages (138).

[4] As discussed later, stable lysogeny is possible for a cI^- mutants if additional mutations block the target functions for the cI protein—gene N and the genes for viral DNA replication.

lishment of repression. Two general mechanisms have been considered for the action of the *c*II and *c*III proteins: a repression of lytic function, in addition to that provided by the *c*I protein (70); an activation of synthesis of *c*I protein (70–72, 145). The utility of both mechanisms seems to have been evident to the phage as well as to the lambdologist—both appear to exist.

Repression of lytic genes.—The possibility that the *c*II and *c*III proteins might provide for an ancillary repression of lytic genes has been investigated mainly through comparative kinetic studies of the synthesis of λ-specified proteins (70). The experiments have involved assays for λ proteins in cell extracts. The generality of this approach is limited by the paucity of available assays. The only proteins studied so far in this regard are λ-exonuclease, which is a delayed-early recombination protein, and endolysin and a tail fiber protein, which are both late proteins. The results of the protein assays show that the late proteins begin synthesis at an earlier time after infection by *c*II⁻ or *c*III⁻ mutants than after infection by *c*I⁻. No such kinetic difference is discernable for λ-exonuclease.

These results indicate that the *c*II and *c*III proteins participate in a repression of late lytic proteins. This repression effect of *c*II and *c*III is independent of the *c*I protein, because *c*I⁻ *c*II⁻ and *c*I⁻ *c*III⁻ double mutants also show advanced synthesis of endolysin with respect to *c*I. These results do not imply that *c*II and *c*III proteins have no effect on *c*I protein synthesis; in fact, the identity of *c*I⁻ *c*II⁻ and *c*I⁺ *c*II⁻ phage in terms of endolysin repression indicates that *c*I⁺ *c*II⁻ mutants may be phenotypically *c*I⁻ as well (70). More rigorous evidence for this conclusion has been obtained from the direct measurements of *c*I protein discussed in the next section.

Activation of the cI gene.—The concept that the synthesis of *c*I protein might be subject to regulation is critical for much of the remaining discussion. This concept has been derived mainly from the finding that the capacity for *c*I-mediated repression can be lost by a particular class of mutant prophage strains (71, 73). These phage possess two properties: the activity of the mutationally altered *c*I protein can be controlled in a reversible way by raising and lowering the temperature [*c*I857 mutation (51, 74)]; lytic development is blocked through mutational inactivation of the target functions of the *c*I protein—gene *N* and the *O* or *P* genes for viral DNA replication.

For such strains, elevated temperature releases repression without any effect on cell viability. If the temperature is lowered soon after elevation, repression is rapidly restored, presumably because the *c*I protein can "renature" to an active form. However, prolonged growth at an elevated temperature produces an inability to regain repression rapidly when the temperature is lowered. The latter property suggests that the *c*I protein is no longer being produced and therefore that synthesis of *c*I protein is not constitutive, but is regulated (71, 73).

The nuances of this experiment will be pursued in more detail under the topic of *cro* regulation. In the context of *c*II/*c*III regulation, the important point is that this capacity for *c*I gene regulation might be useful in the establishment of repression.

The major evidence for the role of the *c*II and *c*III proteins in *c*I protein synthesis has come from measurements of the kinetics of *c*I protein production in infected cells. These measurements have utilized either a DNA-binding assay (75, 76) or an antibody assay (76) for *c*I protein in cell extracts. The basic kinetic result is shown in Figure 6a. Substantial production of *c*I protein does not begin until several minutes after infection; there follows a period of rapid synthesis, followed by a shut-off (75, 76). *c*II⁻ or *c*III⁻ mutants are extremely defective in production of *c*I protein (75, 76). Thus the *c*II and *c*III proteins probably activate *c*I protein synthesis.

For comparison, the repression effect of the *c*II and *c*III proteins for endolysin

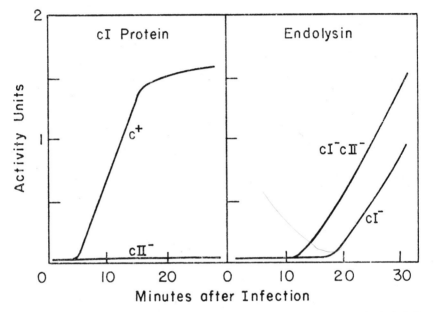

Fig. 6. The action of the *c*II and *c*III proteins. Figure 6a shows the kinetics of *c*I protein synthesis after infection of nonlysogenic cells by wild-type (c^+) phage or a *c*II⁻ mutant. The assay for *c*I protein is either specific binding to λ DNA (75, 76) or specific reaction with antibody to purified *c*I protein (76). The effect of *c*III⁻ mutation is similar to that found with *c*II⁻, but less severe. Since mutational inactivation of the *c*II or *c*III proteins results in under-production of *c*I protein, the normal role of *c*II and *c*III proteins is likely to involve an activation of *c*I protein synthesis.

Fig. 6b shows the kinetics of endolysin synthesis after infection of nonlysogenic cells by *c*I⁻ or *c*I⁻ *c*II⁻ phage (70). A similar advanced synthesis of endolysin is found for *c*I⁻ *c*III⁻ phage. The *c*I⁻ mutation is used to dissociate the repression effect of *c*II and *c*III from that of *c*I protein. Since mutational inactivation of the *c*II or *c*III proteins results in over-production of endolysin, the normal role of the *c*II and *c*III proteins is likely to involve a repression of synthesis of endolysin and possibly other lytic proteins.

synthesis is shown in Figure 6b. The concept that these two apparent activities result from the same basic mechanism rests mainly on two additional findings: both activities depend on the number of phages per infected cell ("multiplicity of infection") and both are eliminated by the same presumptive "site-mutation" in the y-region of λ DNA (77).

Possible site of action of the cII and cIII proteins.—cy^- phage are an additional class of mutants defective in the establishment of repression. cy^- mutants differ from cII^- by their inability to establish lysogeny after co-infection with cI^-. The nature of the cy^- mutation can be analyzed further by a *cis-trans* study at the level of cI protein synthesis. Co-infection by both cI^+cy^- and cI^-cy^+ phage (*cis* configuration) does not lead to synthesis of cI protein; co-infection by both cI^+cy^+ and cI^-cy^- (*trans* configuration) leads to synthesis of cI protein (75, 76). The *cis*-dominant, *trans*-recessive characteristic of the cy^- mutation is most simply interpreted as mutational inactivation of a site essential for turn-on of cI protein synthesis.

Since the cII and $cIII$ proteins provide for activation of cI protein synthesis, the location of cy^- mutations may define the site at which the cII and $cIII$ proteins act. If this hypothesis is correct, the repression activity of the cII and $cIII$ proteins should also be eliminated by cy^- mutation; this expectation has been verified for endolysin synthesis (77). Thus the complete role of the cII and $cIII$ proteins in the establishment of repression may involve bifunctional regulation at a single site: positive regulation of the cI gene and negative regulation of lytic genes.

The cellular level of both regulatory activities is likely to be RNA synthesis. The presence of cII^- or $cIII^-$ mutations increases RNA synthesis from the head and tail genes (70) and decreases l-strand RNA synthesis from the x-region (Figure 3) (78). The late gene RNA is assayed by the difference in DNA-RNA hybridization between normal λ DNA and λ DNA lacking the head and tail genes through a deletion-addition mutation (λgal transducing phage) (21). l-strand RNA from the x-region is assayed by the formation of an RNA-RNA hybrid with r-strand RNA from the x-region (78).

Possible molecular mechanism for cII/cIII regulation.—The transition from the general definition of a regulatory activity to an understanding of the molecular mechanism requires a biochemical analysis which is not yet available for any regulatory phenomenon discussed in this section. However, biochemical postulates are useful (if not taken too seriously) as a guide for new experiments and as a way to remember the old ones. A possible mechanism for $cII/cIII$ action is that summarized earlier in Figure 3: the cII and $cIII$ proteins form a regulatory oligomer that acts at a site in the y-region to activate leftward transcription through the cI gene and inhibit rightward transcription of lytic genes from y through the head and tail genes.

How might such a bifunctional regulatory mechanism work in biochemical terms? Possible mechanisms for the activation function are obvious but not unique. Most simply, the $cII/cIII$ oligomer might provide for new initiations by RNA polymerase at a previously inaccessible promoter site in the y-region. Possible mechanisms for the repression function are more subtle. Transcription from the y-site leftward might inhibit initiation of delayed-early r-strand transcription; alternatively binding of cII and $cIII$ at the y-site might provide for "downstream" repression of r-strand transcription. In turn, inhibited delayed-early transcription might lead to repression of late genes in several ways. In the most obvious mechanism, synthesis of Q protein would be insufficient and thus late gene activation inhibited. Alternatively (or in addition), each prior stage of RNA synthesis might be required to activate fully the promoter sites for the next stage. Thus inhibited delayed-early transcription would also inhibit late transcription.

Clearly a number of possible molecular mechanisms exist, and in vitro biochemical analysis will be required to decide how $cII/cIII$ regulation works. The possible mechanisms discussed above are merely a partial list. I have discussed before the problems inherent in such biochemical postulates from in vivo data (12) and will repeat only one cautionary note here. A site essential for the activity of a regulatory protein (e.g. y-region) need not define a site at which the regulatory protein carries out a biochemical function. For example, host RNA polymerase might begin transcription from a promoter site defined by the cy^- mutation, and be modified at another site by the cII and $cIII$ proteins to provide for RNA synthesis from the cI gene.

Whatever the mechanism, an unraveling of the biochemical basis for the bifunctional regulatory activity of the cII and $cIII$ proteins is not likely to be a trivial occupation.

ESTABLISHMENT VS MAINTENANCE—THE TRANSITION TO cI-MEDIATED REPRESSION

Inhibition of the "activity" of the N and replication proteins.—The cII and $cIII$ proteins provide for the properly timed establishment of repression through rapid synthesis of cI protein during the delayed-early stage of development and through a delay in the transition to the late stage of lytic growth. However, the total potential for lytic development is not quelled by these events because the previously synthesized N and replication proteins are presumably still present in the infected cell. This problem has been solved by the phage through inhibition mechanisms that prevent these proteins from working efficiently on repressed λ DNA. Thus, as for many bacterial regulatory circuits, control occurs through an inhibition of both synthesis and activity of the target enzymes.[5]

The inhibition of activity is most readily studied after co-infection of a lysogenic cell by two different phage—one sensitive and one insensitive to the cI protein. The repression-insensitive phage is generally $\lambda imm434$; $\lambda imm434$ is identical to λ except for the region of the cI gene and its immediate surroundings, which are

_____

[5] Indirect evidence has been presented for still another level of control—metabolic instability of the N protein (79, 80). Such a mechanism of course eliminates N activity after repression blocks further synthesis.

derived from a phage (434) with a different *c*I protein and different binding sites for the *c*I protein (32, 33, 37) (Figure 7). In such a coinfection experiment, λ*imm*434 will replicate and produce a normal burst of phage, whereas λ will not replicate extensively and hence will produce a small burst of phage (81–83, 146). Thus the *O* and *P* replication proteins supplied by λ*imm*434 do not act effectively on the repressed λDNA. In similar studies of *N* function, *N* protein supplied by λ*imm*434 will not activate repressed λ DNA for the synthesis of recombination proteins (80, 84) or probably for the normal synthesis of replication proteins (24, 34, 36); other types of experiments indicate a similar inhibition of *N*-mediated activation of *Q* protein synthesis in the presence of *c*I protein (85, 86).

The mechanism for this inhibition of *N* and *OP* activity is unknown at present. The effect is particularly interesting because the sites at which the *c*I protein binds are probably distinct from the site at which DNA synthesis is activated (87–89), and the sites at which *N*-mediated activation of RNA synthesis occurs (Figure 7). Therefore the inhibition probably does not involve direct interference by the *c*I protein. A number of possible mechanisms exist; these include "transcription-activation", "primer", and "physical accessibility."

In a transcription-activation mechanism, RNA synthesis up to or through an initiation site is required to activate the site. Thus initiation of DNA replication might depend on rightward transcription, which is blocked by the *c*I protein (89); initiation of delayed-early RNA synthesis might depend on immediate-early transcription, which is blocked by the *c*I protein (30, 85, 90).

In a primer mechanism, RNA or DNA synthesis might terminate at a site unless the proper phage protein is supplied to allow continuation of the poly-nucleotide chain. Thus activation of DNA replication might depend on RNA or

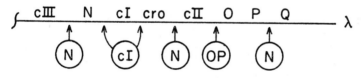

FIG. 7. Inhibition of *N* and *OP* activity by *c*I protein. If phage λ infects a cell carrying a λ prophage, the superinfecting λ DNA is immediately repressed by the *c*I protein produced by the prophage. If phage λ*imm*434 infects a cell carrying a λ prophage, λ*imm*434 can carry out lytic growth because λ*imm*434 lacks the binding sites for the λ *c*I protein [the *c*I gene and its surroundings (notched line in figure) are derived from phage 434]. If both λ and λ*imm*434 infect, the *N* and *OP* proteins provided by λ*imm*434 will not work effectively on the repressed λ DNA, even though the sites at which *N* or *OP* activity is demonstrable are not close to the binding sites for *c*I protein. The relevant sites are indicated by vertical arrows.

DNA synthesis from the *c*I protein binding site to the point at which *N* or *OP* proteins act to elongate the chain. For *N*-activated RNA synthesis, the primer would be the immediate-early chain (91), which is blocked by the *c*I protein; for *OP*-activated DNA synthesis, the primer might be a DNA chain ("Okazaki piece") or even an immediate-early RNA chain, as suggested for phage M13 (92).

In a physical accessibility mechanism, the availability of the activation site is altered by the presence of *c*I protein. For example, *N* and *OP* activity might depend on membrane association by λ DNA, which might be directly blocked by *c*I protein (93). One aspect of DNA structure which might provide sensitive regulation of binding capacity is superhelical density (94).

At present the inhibition function of *c*I protein is clearly much better utilized by the phage than understood by the lambdologist.

Self-regulation by the cI protein.—When the supply of active *c*I protein becomes sufficient for the number of λ DNA molecules, the synthesis of the *c*II and *c*III proteins will cease, and another mechanism must provide for continued synthesis of *c*I protein. There are two obvious possibilities: the *c*I protein might be synthesized at a much lower constitutive rate adequate for the maintenance of repression; or there might be a "maintenance mode" of *c*I gene activation in a lysogen. The *c*I protein itself is the only likely candidate to provide for such a maintenance mode; thus the interesting possibility arises that the *c*I protein might activate its own continued synthesis. Since the *c*I gene is probably transcribed from the *l*-strand (22, 90, 95), the "right-side" operator site o_{I_r} is the likely site for such an activation function (Figure 8).

An experimental distinction between the two possibilities for maintenance could be made if the rate of synthesis of *c*I protein (or *c*I gene RNA) in a lysogen could be compared to the "constitutive" rate in the absence of any regulatory elements. The difficulty arises in defining the constitutive rate of synthesis of *c*I protein amid a plethora of regulatory influences. The closest approximation to a "nonregulated" *c*I gene so far analyzed is probably the situation found after thermal induction of an N^-x^-cI857 prophage. Such a prophage is defective in *N*-mediated positive regulation because of an *N* mutation and is extremely defective in the residual *r*-strand transcription because of a "promoter-type" mutation in the *x*-region (22, 91, Figures 3 & 8). After thermal induction an N^-x^-cI857 prophage should be extremely defective in the synthesis of all known λ proteins required for viral development along either the lytic or lysogenic pathways; in particular, the *c*II, *c*III, and *cro* gene products should not be made. Furthermore, the *c*I protein will be inactive.

Efforts to measure *c*I gene RNA in such prophage strains have led to contradictory results. Two groups have reported strong reduction in *c*I gene RNA synthesis after thermal induction of an N^-x^-cI857 prophage (90, 95); a third group has found no reduction (96). The hybridization assays for *c*I gene RNA have differed slightly; the experimental discrepancy may arise from the difficulty of assaying RNA from a single gene, a procedure which requires multiple hybridization steps.

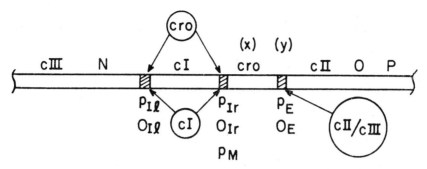

FIG. 8. Summary of genes and sites (proven or postulated) concerned with the establishment and maintenance of repression. To maintain repression, the cI protein acts at o_{ll} and o_{lr} to repress leftward and rightward transcription initiated at the immediate-early promoter sites p_{ll} and p_{lr}, respectively; the cI protein may also activate leftward transcription of the cI gene initiated at a maintenance promoter p_M. To establish repression, the cII and $cIII$ proteins are presumed to act at an operator site o_E to provide for cI gene transcription initiated at an establishment promoter p_E. The two putative promoters for cI gene transcription are separated by the cro gene, which is presumed to be transcribed as part of the r-strand RNA initiated at p_{lr} (Figure 3). The cro product is likely to act at or near o_{ll} and o_{lr} to repress RNA synthesis initiated at p_{ll} and p_{lr} and possibly at p_M. The region between the cI and cII genes has been traditionally called xy—x to denote the area between the right end of the cI gene and the right end of the imm434 region, y to denote the area between the right end of the imm434 region and the left end of the cII gene (135).

Measurements of cI protein production before and after thermal induction are possible by the use of antibody determinations of inactive cI protein at the elevated temperature. Such experiments show a 12-fold lower amount of cI protein in cells grown at the high temperature (76). From these experiments, the constitutive rate of synthesis of cI protein is probably much less than that found in a lysogen. More indirect genetic experiments are consistent with this result (71, 140). Thus there is likely to be a "maintenance" mode of cI gene activation.

Another series of quantitative measurements has compared the rate of synthesis of cI protein after infection of a lysogen to the rate found after infection of a nonlysogen by normal or cII^- phage. Production of cI protein under the conditions of prior repression is intermediate between the $cII/cIII$-activated and the cII^- ("no activation") rates found after infection of a nonlysogen (shown in Figure 6a) (75, 76). This finding is readily interpretable in terms of a maintenance mode of cI gene activation which is independent of the cII and $cIII$ proteins. This interpretation is supported by the fact that the synthesis of cI protein in a lysogen is unaffected by cII^-, $cIII^-$, or cy^- mutations (75, 76, 104), but is markedly decreased by mutations that reduce the ability of cI protein to bind to its "right-side" binding site o_{lr} (76, 104) (Figure 8). The latter finding in particular is consistent with the idea that the cI protein itself activates the maintenance mode of cI gene expression.

From the preceding discussion, the complete role of the cI protein in the maintenance of repression may involve a bifunctional regulatory activity: positive regulation of the cI gene and negative regulation of lytic genes. The repression activity of the cI protein at the level of RNA synthesis has been established in vitro. The nature of the maintenance mode of cI gene activation and the role of the cI protein is less clear. If the cI protein activates cI gene transcription, a number of molecular mechanisms are possible. The cI protein might activate directly a promoter site for RNA synthesis (p_M) at or near o_{Ir}. Alternatively, the promoter site for cI gene transcription may be unavailable to RNA polymerase unless RNA synthesis initiated at the adjacent immediate-early promoter site (p_{Ir}) is blocked by the cI protein. The latter idea is presumably incorrect if the x-mutation described above affects RNA synthesis through a promoter type mutation, since x^- mutants appear to exhibit "self-regulation" by the cI protein (76, 90, 95). However, it is possible that x-mutations are typically mutations that produce early termination of RNA chains rather than blocked initiation. As for the other regulatory phenomena considered, molecular interpretations are difficult without biochemical data.

The CRO Gene Product Opposes the Establishment of Repression

Genetic analysis.—Repression-defective mutants of λ can be isolated by their clear-plaque phenotype; the genetic and biochemical analysis of such mutants has defined the central role of the cII and $cIII$ genes in the establishment of repression and provided a plausible working hypothesis for the biochemical basis of $cII/cIII$ action. Genetic analysis of another type of repression-defective mutant has led to the identification of a gene that can function to oppose the establishment of repression—the cro gene; the physiological role and biochemical basis for cro activity are at present less well-defined than is the case for $cII/cIII$.

The basic experimental system for the "anti-repression" concept is the highly defective prophage described earlier for which lytic development is blocked by mutation in the N gene and in one or both of the O and P genes for DNA replication. For such an N^-DNA^- prophage which also carries the $cI857$ mutation, an extended period of cell growth at high temperature produces a "repression-defective" phenotype: repression is not regained rapidly when the temperature is lowered. Although potentially reversible, this repression-defective phase may persist for many cell generations[6] (71, 73, 97, 98). A repression-defective phase is also produced by an N^-DNA^- prophage that has a normal cI gene, but carries a mutation in the right-side operator site for the cI protein (99, 100). This $v1v3$ mutation allows the genes to the right of the mutation to escape cI-mediated repression (35, 99); the maintenance mode of cI protein synthesis is also inhibited (76).

At least three regulatory influences might (and presumably do) contribute to

---

[6] Since the typical biological assay for cI-mediated repression in a lysogen is the ability to repress a superinfecting cI^- phage ("superinfection immunity"), the repression-defective phase has also been termed the immunity-defective or Imm^- phase (97).

the inferred defectiveness of these prophage strains in the synthesis of cI protein: a loss of positive self-regulation by the cI protein; an indirect loss of positive regulation by the cII and $cIII$ proteins as a result of the N^- mutation; an acquisition of negative regulation through derepression of the gene for a "repression-antagonist." The genetic reasoning for such a repression-antagonist stems from the observation that a normal λ phage is unable to establish lysogeny efficiently after infection of a cell harboring an N^-DNA^- prophage in the repression-defective phase. This channeling of the superinfecting phage to the lytic pathway suggests the prior presence in the cell of a regulatory element that can actively oppose the establishment of repression (73).

A further analysis of the repression-defective phase has provided additional evidence for an anti-repression activity and some indication of the location of the regulatory gene. The approach can be followed by reference to Table 1. N^-x^- $cI857$ prophage[7] in the repression-defective phase reestablish cellular repression much more rapidly when the temperature is lowered than do N^-O^- (or N^-P^-) prophage (73, 101) (Table 1, lines 2 & 3). An F′ partial diploid cell carrying both prophage types remains in the repression-defective phase when the

TABLE 1. Genetic Evidence for a Repression-Antagonist

Prophage	Capacity for repression after growth at	
	40°C	40°C, then 30°C
$N^-O^-cI^+$	Yes	Yes
N^-O^-cI857	No	No
N^-x^-cI857	No	Yes
$N^-O^-v1v3cI^+$	No (37°C)	—
$N^-O^-cro^-cI857$	No	Yes
N^-O^-cI857/N^-x^-cI857	No	No
$N^-O^-v1v3cI^+/N^-O^-cI^+$	No (37°C)	—
$N^-x^-cI857/N^-O^-imm434cIts$	{No—immλ {No—imm434	{Yes—immλ {No—imm434
$N^-O^-cI857/N^-x^-imm434cIts$	{No—immλ {No—imm434	{No—immλ {Yes—imm434

For each experiment involving $cI857$ prophage, cells were grown at 40°C and then tested for capacity to repress an infecting cI^- phage, as judged by plaque formation at 30°C. The results of column 1 refer to immediate infection by cI^- after growth at 40°C; the results of column 2 refer to approximately four generations of growth at 30°C before infection by cI^-. The experiments involving $v1v3$ prophage are independent of temperature because the prophage is cI^+; the assay for cellular repression was ability to repress an infecting cI^+ phage in liquid culture. For details of the $cI857$ experiments, see reference 101. For details of the $v1v3$ experiments, see reference 99.

[7] x^- prophages are phenotypically O^- and P^- because the x^- mutation exerts a pleiotropic effect on genes to the right of the mutation (Figure 8) (139).

temperature is lowered (Table 1, line 6) (101). These findings indicate that a regulatory gene that can antagonize *c*I protein synthesis is inactivated by the x^- mutation and that this gene produces a cytoplasmic product. Since the x^--mutation abolishes *r*-strand transcription from the *x-P* region (Figure 8), the implied location of the regulatory gene is in this region of λ DNA. $N^-v1v3O^-P^-$ prophage confer a stable repression-defective phase, either as a single prophage or in a partial diploid cell carrying an additional $N^-O^-c^+$ prophage (99) (Table 1, lines 1, 4, 7). These data imply that a regulatory gene that can antagonize *c*I protein synthesis is rendered insensitive to the *c*I protein by the *v1v3* mutation. Again the inferred location of the regulatory gene is in the *x-P* region of λ DNA.

In an analogous situation to the first diploid experiment, the anti-repression activity of an N^-O^- *imm*434 prophage can prevent the rapid reestablishment of repression by an *imm*434 prophage, but not by an *imm*λ prophage (Table 1, line 8) (101). Thus the location of the regulatory gene is likely to be inside the *imm*-region. Taken together, all of these results are most consistent with the existence of a gene in the *x*-region that can antagonize the establishment of repression. This gene has been termed *cro* (or *Ai*) (100–102). An additional diploid experiment leads to the inference that the site of action of the *cro* product is within the *imm*-region: the anti-repression activity of an N^-O^- *imm*λ prophage cannot prevent the rapid reestablishment of repression by an N^-x^- *imm*434 prophage (Table 1, line 9) (101).

Starting from an N^-O^- *c*I857 prophage, additional mutations have been found that confer the capacity to reestablish repression rapidly (Table 1, line 5) (100, 101). These mutations—termed cro^- or Ai^-—are located in the *x*-region, but lack the DNA-defective phenotype of the previously defined pleiotropic x^- mutations; thus these mutations are likely to define the *cro* gene. However, no nonsense mutations have yet been reported with the phenotype of the cro^- mutations, and so the location of the anti-repression gene and the nature of the gene product remain somewhat uncertain at present.

Based upon the earlier discussion about the role of the *c*I, *c*II, and *c*III genes, there are at least four general ways in which the *cro* gene might antagonize the establishment of repression: (*a*) a direct inhibition of *c*I protein activity; (*b*) a direct inhibition of *c*I protein synthesis; (*c*) an inhibition of the activity of the *c*II and *c*III proteins; (*d*) an inhibition of the synthesis of either or both of the *c*II and *c*III proteins. The last stated mechanism almost certainly exists; the second is likely to occur in addition.

Repression of cI protein synthesis through turnoff of the cII and cIII genes.— The repression activity of *cro* on *c*I protein synthesis is readily observable in two types of experiments: infection of a nonlysogen or infection of one of the repression-defective lysogenic strains described in the preceding section: N^-O^- *c*I857 or N^-O^-v1v3. In the case of infection of a nonlysogen, the action of *cro* is not apparent until relatively late after infection; cro^- phage fail to shut off synthesis of *c*I protein (76) (Figure 9a). In the case of prior presence of the *cro* product in a repression-defective lysogen, effective synthesis of *c*I protein never begins (103,

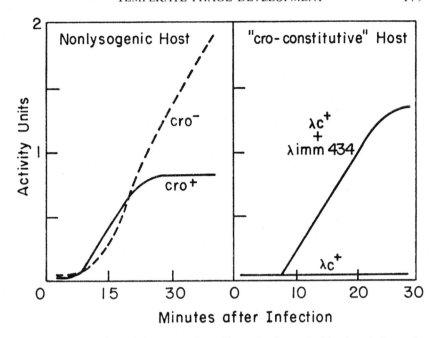

FIG. 9. The action of the *cro* product. Figure 9a shows the kinetics of *c*I protein synthesis after infection of nonlysogenic cells by *c*I857 phage (solid line) or *cro⁻c*I857 phage (dashed line) (76). Since the *cro⁻* mutation causes a failure of the normal turnoff of *c*I protein synthesis (see also Figure 6a), the *cro* product is likely to be responsible for this turnoff. The time scale is longer because the experiment was performed at 30°C, rather than the normal 37°C, to provide for an active *c*I857 protein. Figure 9b shows the kinetics of *c*I protein synthesis after infection of "*cro*-constitutive" cells by *c⁺* phage (lower curve) or by both *c⁺* and λ*imm*434*c⁺* phage (upper curve) (103). The presumptive "*cro*-constitutive" host is the N⁻DNA⁻*v1v*3 lysogen described in the text. Since coinfection by λ*imm*434 allows the λ*c⁺* phage to synthesize *c*I protein, the λ*c⁺* phage is probably unable to produce adequate amounts of *c*II and *c*III proteins in the presence of *cro* product.

104) (Figure 9a). The regulation involves *c*I protein synthesis and not activity because the effect is observable by either antibody or DNA-binding activity (103, 104). A strong inference can be made that the *cro* product acts through turnoff of *c*II/*c*III activation. If the *c*II and *c*III products are supplied by co-infection with a *cro*-insensitive λ*imm*434*c⁺*phage, *c*I protein synthesis proceeds at an approximately normal rate (Figure 9b). A λ*imm*434 *c*III⁻ or a λ*imm*434 *c*II⁻ phage cannot activate *c*I protein synthesis in this situation (103). Thus the *cro* product presumably acts through turnoff of *c*II and *c*III protein synthesis. If the *cro* product inhibited directly the synthesis or activity of the *c*I protein or the activity of the *c*II and *c*III proteins, the complementation for *c*I protein production by the λ*imm*434 *c⁺* phage would not have occurred. At a more biological level,

complementation for lysogeny by $\lambda imm434c^+$ has also been observed in similar experiments (147).

The ability of *cro* to inhibit *cIII* protein synthesis is not surprising because synthesis of the recombination protein λ-exonuclease is turned off after some minutes of synthesis (105), and the gene responsible for turnoff has been located in the *x*-region (106); *cro⁻* mutants have been shown to lack capacity for turnoff (101). In fact, mutations with the *cro⁻* phenotype (designated *fed⁻*) have been isolated by a selection for enhanced synthesis of proteins of the recombination region (141). Thus, the ability of the *cro* product to repress synthesis of *cI* protein probably results at least in part from a general capacity to repress the *cIII* to *int* region of λ DNA (Figure 8). Since *x⁻* and *cro⁻* prophages overproduce RNA from the recombination region of λ DNA (90, 96), the repression effect probably occurs at the level of RNA synthesis.

The capacity of the *cro* product to inhibit the synthesis of the *cIII* and *cII* proteins suggests that the *cro* product may carry out a repression function very similar to that provided by the *cI* protein: an inhibition of the immediate-early state of transcription. However, the *cro* and *cI* products presumably differ in their capacity to regulate *cI* gene RNA synthesis. The *cI* protein probably activates transcription of the *cI* gene to maintain its own synthesis. The *cro* product does not activate expression of the *cI* gene.

Possible direct effect of cro on the cI gene.—The anti-repression activity of the *cro* product has led to the suggestion that the *cro* product not only fails to activate but can repress directly the synthesis of *cI* protein (96, 101). If it exists, the direct repression activity does not inhibit *cII/cIII* activation of the *cI* gene because λ*imm*434 can complement a *cro*-repressed λ phage for the synthesis of *cI* protein. Such a direct repression might affect self-activated synthesis of *cI* protein or constitutive synthesis or both.

There is experimental evidence that seems to argue both for and against a direct effect of *cro* product on the *cI* gene. Efforts to study this possibility are complicated by the effect of the *cro* product on *cII/cIII* already noted.[8] In principle, the best experimental system is probably thermal induction of a *cI857* prophage that carries additional *cII⁻*, *cIII⁻* or *cy⁻* mutations. Effects of *cro* on the constitutive expression of the *cI* gene can be determined at 42°C in the probable absence of self-activation by the thermolabile *cI* protein; effects of *cro* on the self-activated expression of the *cI* gene can be determined if the denatured *cI* protein is renatured by lowering the temperature after the *cro* product has been synthesized. For these experiments, *cI* gene expression can be estimated by measurements of either *cI* gene RNA or *cI* protein (by antibody at 40°C).

---

[8] Even *N⁻* phage probably produce enough residual *cII* and *cIII* proteins to provide some *cI* gene activation (103, 104). This residual *cII/cIII* activity is presumably enhanced by *cro⁻* mutation because some *cro⁻* mutants (*fed⁻*) have been isolated by capacity to produce more recombination region proteins under *N⁻* conditions (141).

Although an isogenic $cro^{\pm}$ comparison has not yet been done in the absence of all possible $cII/cIII$ effects, the results of cI protein assays so far provide a strong indication that both the constitutive and self-activated syntheses of cI protein are less for a cro^+ situation than for cro^-. N^-O^- $cI857$ prophage which are cro^+ produce tenfold less cI protein at 40°C than the cro^- derivative, and the cro^- level is also found for an N^- $cI857$ prophage that has deleted the cro, cII, and cy region (104). After heating N^-O^- cro^+ $cI857$ prophage to allow cro product synthesis and cooling to regain active cI protein, the synthesis of additional cI protein by an infecting cy^- phage is fivefold less in a cro^+ lysogen than in a cro^- (the normal cI protein of the infecting cy^- phage can be distinguished from the altered thermolabile cI protein already present by a DNA-binding assay) (104). The former experiment implies that cro product represses constitutive cI gene expression; the latter experiment suggests that cro product represses self-activated synthesis as well.

The results of RNA measurements after thermal induction are conflicting, but the experiments done so far may have been complicated by $cII/cIII$ effects and technical limitations. One report notes a sharp drop in production of cI gene RNA after thermal induction of either cro^+ or cro^- prophage and finds no effect of cro^- or x^- mutation on the residual "constitutive" rate of synthesis (95). In terms of the protein measurement just discussed, these results are understandable if the drop in cI gene RNA results from a loss of cI-activated transcription of the cI gene, and the effect of cro on the low residual level of cI gene RNA synthesis was not observed because of technical problems or because cro product was not yet active at the early time studied. However, another report finds that the drop in cI gene RNA after thermal induction is eliminated by cro^- or x^- mutation (96). These conflicting reports appear to be readily reconciled only if residual $cII/cIII$ activation raised both the cro^+ and cro^- levels of cI gene RNA in the latter set of experiments. Further protein and RNA measurements should clarify the situation.

Possible sites of action of the cro *product.*—From the similarity between cro and cI-mediated repression, the cro product might be expected to act also to the left and right of the cI gene within the *imm*-region (Figure 8). A possible left-side site can be inferred from studies of exonuclease turnoff. Exonuclease repression is not abolished by a deletion into the cI gene from some distance to the right (when cro product is supplied by another phage) (107); exonuclease repression is impaired by the $v2$ mutation (99), which diminishes cI protein binding at the left-side operator site (35). Thus the cro product may act at or near this operator site. If this concept is correct, infection of a cro-repressed cell by $\lambda v2$ and $\lambda imm434$ $cIII^-$ should lead to some escape of the λ cI gene from cro-mediated repression, since $\lambda v2$ should supply $cIII$ protein and $\lambda imm434$ $cIII^-$ should supply cII protein. This anticipated effect has been found (103, 104).

The presumptive right-hand site for cro activity is less well defined. The $v1v3$ mutation impairs cI protein binding at the right-side operator site; however, in the analogous experiment to that just described, infection of a *cro*-repressed cell by λ*v*1*v*3 and λ*imm*434 *c*II⁻ does not lead to substantial escape of the λ *c*I gene from *cro*-mediated repression (103, 104). This may mean that *cro* product acts at a different right-side from the *c*I protein. However, this result is difficult to evaluate because there is some residual repression of the mutant operator site by *c*I protein (35, 42), and partial repression by *cro* product or by *c*I protein might yield different results for the indirect parameters studied so far in vivo.

Possible molecular mechanism for cro *regulation.*—From the available evidence, the best working hypothesis for *cro* regulation appears to be the following. The *cro* product binds at sites to the left and right of the *c*I gene to repress the immediate-early stage of transcription. The binding sites for *cro* are close to and might be identical with the binding sites for the *c*I protein. In addition to a repression of *N* gene transcription, the *cro* product mimics the activity of the *c*I protein as an inhibitor of *N* protein activity and thus provides for a highly efficient turnoff of the early gene region of λ DNA. *cro*-mediated repression does not normally become effective until the onset of the late stage of lytic development. Thus a cell clearly destined for lysis will not continue to expend energy and materials on the proteins required for repression or on the proteins needed for DNA replication and recombination.

Inherent in the preceding description is the assumption that the *cro* product does not inhibit the late state of transcription or probably the late stage of DNA replication. It is conceivable that the *cro* protein might also mimic the activity of the *c*I protein as an inhibitor of new initiation events for DNA replication from a circular molecule and thus actually potentiate a late phase of replication involving a rolling circle and concatemer formation. One major problem for this picture of *cro* function is how the action of *cro* is deferred until well into the delayed-early stage of development. Clearly there is much to be learned about *cro* regulation.

WHAT CONTROLS THE FREQUENCY OF THE LYSOGENIC CHOICE?

General considerations.—The frequency with which phage λ develops along the lysogenic pathway is subject to environmental influences. Before I discuss the little that is known of these influences, I will consider briefly some of the regulatory elements that might be expected to affect the choice of lytic or lysogenic development. When phage λ infects a cell, there is an immediate-early stage of development, followed by an *N*-activated delayed early stage. The delayed-early stage is the decision point—lytic development through the *Q*-activated late stage or lysogenic development through repression and integration. The establishment of repression provides for the lysogenic response, but does not ensure it. The integrative recombination event is required for stable lysogeny.

The establishment of repression in an infected cell can clearly be influenced in two general ways: the rate or timing of *c*I protein synthesis; and the rate or timing of protein synthesis from the late genes, particularly those concerned with cell lysis. Thus lysogeny will be favored by intracellular events that facilitate the synthesis of cI protein or antagonize the synthesis of late lytic proteins. There are several obvious potential targets for regulatory influences of this nature: enhanced synthesis or activity of the cII and $cIII$ proteins; diminished synthesis or activity of cro product; diminished synthesis or activity of Q protein; directly enhanced synthesis or activity of the cI protein. Even without a consideration of subtle regulatory features, there are clearly a large variety of possibilities for environmental regulation of the lysogenic response. So far the nature of the environmental "effectors" and their targets are unknown.

Phage influences.—One environmental effect that can influence the establishment of repression is the number of phage particles which infect a cell. Lysogeny is generally more frequent after infection by many phage than after infection by few (108). At an intracellular level, this "multiplicity effect" is associated with a discernable influence on the activities attributed to the cII and $cIII$ proteins: activation of cI protein synthesis and repression of lytic protein synthesis. The synthesis of cI protein is enhanced (77, 104) and that of endolysin is repressed at a high multiplicity of infection (77). A possible biochemical basis for this multiplicity effect is the more rapid attainment of a concentration threshold for $cII/cIII$ oligomer formation when more copies of phage DNA are present initially as templates for RNA synthesis.

Host influences.—The influence of host physiology on λ development is often discussed (mainly as a blame for experimental vagaries) but has seldom been studied systematically. The lysogenic response is often favored if the cells are grown past the exponential phase of growth or incubated in medium lacking a carbon source ("old and starved cells") (109). These physiological effects on lysogeny probably result from a more effective establishment of repression, rather than an enhanced frequency of integration, but there is no information available as to which of the potential regulatory features are involved. The physiological usefulness of this "old and starved" regulation may be to prevent lytic growth under conditions when the cells are ill-prepared for the demands of large scale production of new viral particles.

One host influence on lysogeny clearly affects integrative recombination rather than repression—poor lysogeny at high temperature ($>40°C$). After infection by λc^+ at high temperature cell survival is normal, but stable lysogeny rare (110, 111). Under these conditions normal integrative recombination is virtually eliminated (112, 113); surprisingly, normal excisive recombination is not markedly affected (113).

The study of host mutants that can influence the frequency of lysogenization constitutes one potentially promising approach to an understanding of environmental regulation of the lysogenic response. Host mutants have been isolated that enhance lysogenization by λc^+, λcII^-, and $\lambda cIII^-$ phage (114, 115). In one case, the effect of the mutation on lysogenization by λcIII$^-$ is particularly strong (114). There is no information yet as to which regulatory element is affected by the mutations.

Host mutants with impaired capacity for lysogenization by λc^+ are also available. Somewhat reduced lysogenization frequencies have been reported for host mutants defective in the cyclic AMP-mediated activation of catabolic operons (adenyl cyclase$^-$ or CAP$^-$) (116). The cyclic AMP activation system is not essential for the establishment of repression because a cII$^-$ phage mutation reduces lysogenization by a factor of 10^3, whereas a cyclase$^-$ or CAP$^-$ host mutation reduces lysogenization only three- to four-fold. The interpretation of such relatively small decreases in the frequency of lysogenization is difficult in the absence of some knowledge of intracellular effects on viral development. Although these findings imply some interaction between the cyclic AMP system and λ development, the effect may be quite indirect.

Such an interaction is also suggested by the inference that phage λ may cause a reduction in the intracellular level of cyclic AMP during the early stages of λ development (41). Phage λ development is associated with a repression of the host lac and gal operons (41, 117–119); this repression requires the synthesis of recombination region proteins (119), and is reversed by the addition of exogenous cyclic AMP (41). Thus one or more λ proteins may produce a lowered cyclic AMP level (or less active cyclic AMP activation system). Since synthesis of recombination region proteins is repressed by cro product, the cro gene may regulate indirectly the intracellular level of cyclic AMP. The significance of such regulation, if it exists, is unclear at present.

There are other host mutants in which λ may be repression-defective, as judged by a clear-plaque phenotype for λc^+ (e.g. rep^- mutants unable to grow phage ϕx-174) (120). The study of physiological influences on the lysogenic response is obviously on its infancy.

Comparison of Phages λ, P22, and P2

Two temperate viruses that have been extensively studied in parallel to λ are P22 and P2. Since these phages have been the special subjects of recent review articles (3, 4), I will not consider them in detail here. The striking similarities between λ, P22, and P2 imply that the set of solutions evolved by phage λ is not unique, but may have elements in common with other bacterial viruses, with animal viruses, and perhaps with chromosomal segments which must alternate between active and quiescent states in a temporal pattern.

Phage P22.—Phage P22 is superficially very different from λ. P22 differs in natural host (*Salmonella*), in virus morphology, and in structure of the DNA molecule. P22 DNA is smaller than λ DNA, contains a long region of identical sequence at each double-stranded end, and a population of P22 DNA molecules contains a permuted rather than a unique sequence with respect to end points (121). However, in terms of gene organization and intracellular behavior, λ and

P22 are strikingly similar. Like λ, P22 DNA forms a circular molecule inside an infected cell, exercises a choice between lytic or lysogenic development, integrates by site-specific recombination, and can reverse the lysogenic choice through an induction mechanism triggered by inhibitors of host DNA replication (3).

A more detailed comparison reveals a striking similarity in the way these events are carried out (Figure 10). Phage P22 has a cluster of three genes required for the establishment and maintenance of repression: the c3 and c1 genes are required for establishment, the c2 gene for maintenance (122). There is also a gene termed cly, which is a possible analog to cro; cly⁻ mutations establish lysogeny with a very high frequency, and thus there may be a cly product that opposes the establishment of repression (123). The regulatory region of P22 is flanked by genes for DNA replication to the right and genes for recombination to the left; as for λ, the establishment genes c3 and c1 begin these presumptive operons. The genes for late proteins are also clustered and follow the λ pattern: lysis, head, tail (3). No genes clearly analogous to N and Q have yet been identified, although there is a P22 gene in the "proper place" for Q, which might have a similar function.

From the genetic similarity, it is possible to consider the establishment of lysogeny by phage P22 in terms of the same sequence of events postulated for λ. However, there is not yet sufficient biochemical evidence to decide for or against such a possibility. For a λ analogy, the c3 and c1 proteins of P22 might act at a site to the left of the c1 gene to activate the c2 gene and repress the DNA and late lytic genes. The cly protein might oppose the establishment of repression through an inhibition of synthesis of c1 or c3 or both. There is one apparent difference, which need not indicate a fundamental distinction in mechanism—the c1 (and possibly c3) proteins provide for a repression of P22 DNA synthesis (124). This may only indicate that the repression activity of the P22 c1 and c3 proteins is earlier or more effective than the repression activity of the λ cII and cIII proteins [which do not show a clear repression of DNA synthesis, presumably because enough of the requisite proteins are made before cII and cIII act (70)]. Alternatively, there may be a basic difference in mechanism.

Once repression is established, the c2 protein presumably represses RNA synthesis from the DNA region and from the recombination region. The major

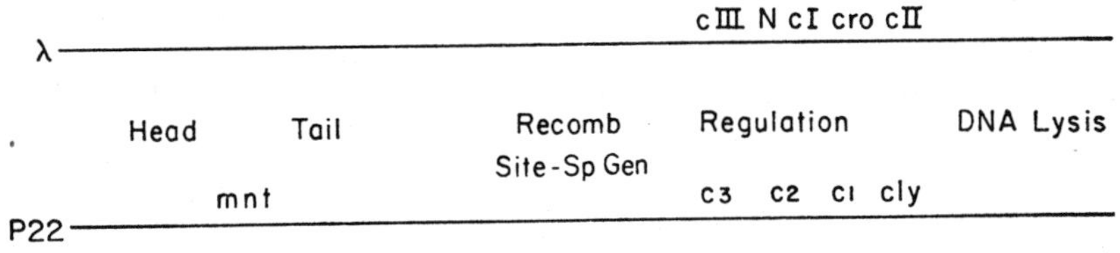

FIG. 10. Comparative gene order for the DNA of phage λ and P22 (3). The DNA of phage P22 is permuted with respect to end points—an example of P22 is chosen in which the head genes are at the left end. The λ and P22 genes of the regulatory region are described in the text. The other genes are shown only generically.

difference between λ and P22 is the existence of an additional gene—termed *mnt*—required for the maintenance of repression (125, 126). This may indicate that the *c2* gene of P22 does not activate its own synthesis, as seems likely for λ, but an additional protein is required for the maintenance of *c2* protein synthesis. Such a function would explain the delayed induction associated with thermal denaturation of a temperature-sensitive *mnt* product (127); the release of repression would not be direct, but would occur only as the supply of *c2* protein became inadequate because of loss of the capacity for further synthesis.

The physiological influences on the choice of lysis or lysogeny appear to be similar for P22 and λ, as far as they have been studied. P22 lysogenization is enhanced by infection with many phage per cell (122, 128). P22 lysogenization is somewhat reduced in host mutants defective in the cyclic AMP activation system (123). There are a large variety of other host mutants on which P22 exhibits a clear-plaque phenotype (123, 129), and these may be repression-defective as well.

Phage P2.—Phage P2 is also superficially very different from λ in virus morphology and DNA molecule, although P2 has unique, single-stranded ends that provide cohesive sites for end-joining (4). In terms of gene organization and intracellular behavior, P2 is similar in many respects to λ. Like λ, P2 DNA forms a circular molecule inside an infected cell, can develop along either a lytic or lysogenic pathway, and integrates by site-specific recombination. P2 differs from λ in two major features: capacity to integrate efficiently at more than one chromosomal site; and inability to undergo induction to lytic growth after inhibition of host DNA replication (4).

The repression system for phage P2 may be simpler than that for λ. Only one known gene—termed *C*—is clearly required to establish and maintain repression in *E. coli* (130); another gene—termed *Z*—is required in the case of Shigella (131) and may have some influence in *E. coli* (132). As for λ, the *C* gene is flanked on the right by genes required for DNA replication and late gene activation and on the left by the *int* gene required for site-specific recombination (133). Evidence has been presented that the *int* gene is not repressed directly by the C product (134). The *int* gene might be inactivated through separation from its promoter site as a consequence of the integrative recombination (134). Alternatively, the *int* gene might be controlled by a separate regulatory element (e.g. a positive regulator active only during the establishment of lysogeny).

At present λ appears to differ from P2 much more than from P22. However, this may only mean that fewer P2 regulatory genes have been identified and studied.

4. CONCLUDING REMARKS

The temperate phage provides a useful experimental system for the study of many phenomena of current interest to molecular geneticists: positive and negative regulation of RNA synthesis, DNA replication and its regulation, general and site-specific recombination. The temperate phage provides an equally useful experimental system for the study of the integration of these phenomena into a temporal sequence which can be termed a developmental pathway. For the temperate phage, there are two developmental pathways—lytic and lysogenic. For each pathway, the phage chromosome is active for a defined sequence of events and then becomes quiescent—in one case in a phage particle, in the other embedded in the host chromosome.

The extent to which temperate phage development is similar to other developmental systems can be only a subject for conjecture at present. In my opinion, a radical difference is unlikely in view of the near identity between microorganisms and "higher" organisms in basic metabolic pathways, in the genetic code, and in the mechanism for protein synthesis. Regardless of the universality of the conclusions, however, the temperate phage provides a delightful array of regulatory mechanisms that show no signs of diminished interest for someone interested in the molecular basis for gene regulation.

Acknowledgment

The concepts of temperate phage development presented in this article have evolved from the ideas and experiments of many people. This communal nature of scientific progress is embraced by some and fought by others, but in my view remains the enduring feature of a scientific discipline. I thank my fellow lambdologists for their contributions to this article and phage λ for revealing itself to Esther Lederberg in 1951.

My research has been supported by U. S. Public Health Service Research Grant GM 17078 from the National Institute of General Medical Sciences.

LITERATURE CITED

1. Echols, H. 1971. *Ann. Rev. Biochem.* 40:827–54
2. Hershey, A. D., Ed. 1971. *The Bacteriophage Lambda.* Cold Spring Harbor, New York: Cold Spring Harbor Laboratory
3. Levine, M. 1971. *Current Topics in Microbiology and Immunology.* In press
4. Bertani, L. E., Bertani, G. 1971. *Advan. Genet.* 16:200–37
5. Lwoff, A. 1953. *Bacteriol. Rev.* 17: 269–337
6. Jacob, F., Wollman, E. L. 1961. *Sexuality and the Genetics of Bacteria.* New York: Academic
7. Thomas, C. A., Jr. 1967. *J. Cell. Physiol.* 70, Sup. 1:13–34
8. Calendar, R. 1970. *Ann. Rev. Microbiol.* 24:241–96
9. Losick, R. 1972. *Ann. Rev. Biochem.* In press
10. Brody, E., Sederoff, A., Bolle, A., Epstein, R. H. 1970. *Cold Spring Harbor Symp. Quant. Biol.* 35: 203–11
11. Schleif, R., Greenblatt, J., Davis, R. W. 1971. *J. Mol. Biol.* 59:127–50
12. Echols, H. 1971. See Ref. 2, pp. 247–70
13. Ikeda, H., Tomizawa, J. 1968. *Cold Spring Harbor Symp. Quant. Biol.* 33:791–98
14. Scott, J. R. 1970. *Virology* 41:66–71
15. Ptashne, M. 1971. See Ref. 2, pp. 221–38
16. Gottesman, M. E., Weisberg, R. A. 1971. See Ref. 2, pp. 113–38
17. Young, E. T., Sinsheimer, R. L. 1964. *J. Mol. Biol.* 10:562–64
18. Bode, V. C., Kaiser, A. D. 1965. *J. Mol. Biol.* 14:399–417
19. Hershey, A. D., Burgi, E., Ingraham, L. 1963. *Proc. Nat. Acad. Sci. USA* 49:748–55
20. Strack, H. B., Kaiser, A. D. 1965. *J. Mol. Biol.* 12:36–49
21. Skalka, A., Butler, B., Echols, H. 1967. *Proc. Nat. Acad. Sci. USA* 58:576–83
22. Taylor, K., Hradecna, Z., Szybalski, W. 1967. *Proc. Nat. Acad. Sci. USA* 57:1618–25
23. Kourilsky, P., Marcaud, L., Sheldrick, P., Luzzati, D., Gros, F. 1968. *Proc. Nat. Acad. Sci. USA* 61: 1013–20
24. Thomas, R. 1966. *J. Mol. Biol.* 22: 79–95
25. Radding, C. M., Echols, H. 1968. *Proc. Nat. Acad. Sci. USA* 60: 707–12
26. Kumar, S. et al 1969. *Nature* 221: 823–25
27. Heinemann, S. F., Spiegelman, W. G. 1970. *Cold Spring Harbor Symp. Quant. Biol.* 35:315–18
28. Dove, W. F. 1966. *J. Mol. Biol.* 19: 187–201
29. Oda, K., Sakakibara, Y., Tomizawa, J. 1969. *Virology* 39:901–18
30. Herskowitz, I., Signer, E. R. 1970. *J. Mol. Biol.* 47:545–56
31. Kaiser, A. D. 1957. *Virology* 3:42–61
32. Kaiser, A. D., Jacob, F. 1957. *Virology* 4:509–21
33. Isaacs, L. N., Echols, H., Sly, W. S. 1965. *J. Mol. Biol.* 13:963–67
34. Pereira da Silva, L. H., Jacob, F. 1967. *Virology* 33:618–24
35. Ptashne, M., Hopkins, N. 1968. *Proc. Nat. Acad. Sci. USA* 60: 1282–87
36. Thomas, R. 1970. *J. Mol. Biol.* 49: 393–404
37. Ptashne, M. 1967. *Nature* 214:232–34
38. Chadwick, P., Pirrotta, V., Steinberg, R., Hopkins, N., Ptashne, M. 1970. *Cold Spring Harbor Symp. Quant. Biol.* 35:283–94
39. Wu, A. M., Ghosh, S., Echols, H., Spiegelman, W. G. 1972. *J. Mol. Biol.* In press
40. Echols, H., Pilarski, L., Cheng, P. Y. 1968. *Proc. Nat. Acad. Sci. USA* 59:1016–23
41. Wu, A. M., Ghosh, S., Willard, M., Davison, J., Echols, H. 1971. See Ref. 2, pp. 589–98
42. Steinberg, R., Ptashne, M. 1971. *Nature New Biol.* 230:76–80
43. Echols, H., Joyner, A. 1968. *Molecular Basis of Virology*, ed. H. Fraenkel-Conrat, New York: Reinhold, pp. 556–59
44. Joyner, A., Isaacs, L. N., Echols, H., Sly, W. S. 1966. *J. Mol. Biol.* 19: 174–86
45. Naono, S., Gros, F. 1967. *J. Mol. Biol.* 25:517–36
46. Green, M. 1966. *J. Mol. Biol.* 16: 134–48
47. Tomizawa, J.-I., Ogawa, T. 1967. *J. Mol. Biol.* 23:247–63
48. Chadwick, P., Reichardt, L., Danielson, G. Personal communications

49. Brooks, K., Clark, A. J. 1967. *J. Virol.* 1:283–93
50. Hertman, I., Luria, S. E. 1967. *J. Mol. Biol.* 23:117–33
51. Sussman, R., Jacob, F. 1962. *C. R. Acad. Sci. Paris* 254:1517–19
52. Campbell, A. 1962. *Advan. Genet.* 11:101–45
53. Campbell, A. 1969. *Episomes*, New York: Harper & Row, 193 pp.
54. Echols, H., Gingery, R., Moore, L. 1968. *J. Mol. Biol.* 34:251–60
55. Weil, J., Signer, E. R. 1968. *J. Mol. Biol.* 34:273–79
56. Zissler, J. 1967. *Virology* 31:189
57. Gingery, R., Echols, H. 1967. *Proc. Nat. Acad. Sci. USA* 58:1507–14
58. Gottesman, M. E., Yarmolinsky, M. B. 1968. *J. Mol. Biol.* 31:487–506
59. Echols, H. 1970. *J. Mol. Biol.* 47:575–83
60. Gingery, R., Echols, H. 1968. *Cold Spring Harbor Symp. Quant. Biol.* 33:721–27
61. Gottesman, M. E., Yarmolinsky, M. B. 1968. *Cold Spring Harbor Symp. Quant. Biol.* 33:735–43
62. Guarneros, G., Echols, H. 1970. *J. Mol. Biol.* 47:565–74
63. Kaiser, A. D., Masuda, T. 1970. *J. Mol. Biol.* 47:557–64
64. Brooks, K. 1965. *Virology* 26:489–99
65. Signer, E. R. 1970. *Virology* 40:624–33
66. Echols, H., Court, D. 1971. See Ref. 2, pp. 701–10
67. Weisberg, R. A., Gottesman, M. E. 1971. See Ref. 2, pp. 489–500
68. Jacob, F. Wollman, E. L. 1954. *Ann. Inst. Pasteur* 87:653–73
69. Brachet, P., Thomas, R. 1969. *Mutat. Res.* 7:257–60
70. McMacken, R., Mantei, N., Butler, B., Joyner, A., Echols, H. 1970. *J. Mol. Biol.* 49:639–56
71. Eisen, H. A., Pereira da Silva, L., Jacob, F. 1968. *C. R. Acad. Sci. Paris* 266:1176–78
72. Signer, E. R. 1969. *Nature* 223:158–60
73. Calef, E., Neubauer, Z. 1968. *Cold Spring Harbor Symp. Quant. Biol.* 33:755–64
74. Naono, S., Gros, F. 1967. *J. Mol. Biol.* 25:517–36
75. Echols, H., Green, L. 1971. *Proc. Nat. Acad. Sci. USA* 68:2190–94
76. Reichardt, L., Kaiser, A. D. 1971. *Proc. Nat. Acad. Sci. USA* 68:2185–89
77. Court, D., Green, L., Echols, H. In preparation
78. Eisen, H. A., Yanif, M., Spiegelman, W. G., Heinemann, S. F., Reichardt, L. In preparation
79. Konrad, M. W. 1968. *Proc. Nat. Acad. Sci. USA* 59:171–78
80. Schwartz, M. 1970. *Virology* 40:23–33
81. Thomas, R., Bertani, L. E. 1964. *Virology* 24:241–53
82. Ptashne, M. 1965. *J. Mol. Biol.* 11:90–96
83. Dove, W. F., Hargrove, E., Ohashi, M., Haugli, F., Guha, A. 1969. *Jap. J. Genet.* 44:11–22
84. Luzzati, D. 1970. *J. Mol. Biol.* 49:515–20
85. Butler, B., Echols, H. 1970. *Virology* 40:212–22
86. Couturier, M., Dambly, C. 1970. *C. R. Acad. Sci. Paris.* 270:428
87. Schnös, M., Inman, R. B. 1970. *J. Mol. Biol.* 51:61–73
88. Stevens, W. F., Adhya, S., Szybalski, W. 1971. See Ref. 2, pp. 515–33
89. Dove, W. F., Inokuchi, H., Stevens, W. 1971. See Ref. 2, pp. 747–71
90. Kourilsky, P., Bourguignon, M.-F., Bouquet, M., Gros, F. 1970. *Cold Spring Harbor Symp. Quant. Biol.* 35:305–14
91. Roberts, J. W. 1969. *Nature* 224:1168–74
92. Brutlag, D., Schekman, R., Kornberg, A. 1971. *Proc. Nat. Acad. Sci. USA* 68:2826–29
93. Hallick, L., Boyce, R. P., Echols, H. 1969. *Nature* 223:1239–42
94. Botchan, P., Wang, J. In preparation
95. Heinemann, S. F., Spiegelman, W. G. 1970. *Proc. Nat. Acad. Sci. USA* 67:1122–29
96. Kumar, S., Calef, E., Szybalski, W. 1970. *Cold Spring Harbor Symp. Quant. Biol.* 35:331–39
97. Neubauer, Z., Calef, E. 1970. *J. Mol. Biol.* 51:1–13
98. Spiegelman, W. G. 1971. *Virology* 43:16–33
99. Sly, W. S., Rabideau, K., Kolber, A. 1971. See Ref. 2, pp. 575–88
100. Calef, E., Avitabile, A., del Giudice, L., Marchelli, C., Menna, T., Neubauer, Z., Soller, A. 1971. See Ref. 2, pp. 609–20
101. Eisen, H., Pereira da Silva, L., Jacob, F. 1970. *Proc. Nat. Acad. Sci. USA* 66:855–62
102. Oppenheim, A. B., Neubauer, Z., Calef, E. 1970. *Nature* 226:31–32
103. Echols, H., Green, L. In preparation

104. Reichardt, L. 1972. *Control of Repressor Synthesis and Early Gene Expression by Bacteriophage λ.* Thesis, Stanford University, Stanford, Calif.
105. Radding, C. M. 1964. *Proc. Nat. Acad. Sci. USA* 52:965-73
106. Pero, J. 1970. *Virology* 40:65-71
107. Pero, J. 1971. See Ref. 2, pp. 599-608
108. Fry, B. A. 1959. *J. Gen. Microbiol.* 21:676-84
109. Séchaud, J. 1960. *Arch. Sci. Phys. Nat., Geneve* 13:428-74
110. Lieb, M. 1953. *Cold Spring Harbor Symp. Quant. Biol.* 18:71-73
111. Campbell, A., Killen, K. 1967. *Virology* 33:749-52
112. Signer, E. R., Weil, J. 1968. *Cold Spring Harbor Symp. Quant. Biol.* 33:715-19
113. Guarneros, G., Echols, H. In preparation
114. Belfort, M., Wulff, D. L. 1971. See Ref. 2, pp. 739-42
115. Strack, H. B. Personal communication
116. Grodzicker, T., Arditti, R. R., Eisen, H. A. 1972. *Proc. Nat. Acad. Sci. USA* 69:366-70
117. Starlinger, P. 1963. *J. Mol. Biol.* 6: 128-36
118. Willard, M., Echols, H. 1968. *J. Mol. Biol.* 32:37-46
119. Cohen, S. N., Chang, A. C. Y. 1970. *J. Mol. Biol.* 49:557-71
120. Denhardt, D. T., Dressler, D. H., Hathaway, A. 1967. *Proc. Nat. Acad. Sci. USA* 57:813-20
121. Rhoades, M., MacHattie, L. A., Thomas, C. A. 1968. *J. Mol. Biol.* 37:21-40
122. Levine, M. 1957. *Virology* 3:22-41
123. Hong, J.-S., Smith, G. R., Ames, B. N. 1971. *Proc. Nat. Acad. Sci. USA* 68:2258-62
124. Smith, H. O., Levine, M. 1964. *Proc. Nat. Acad. Sci. USA* 52:356-63
125. Gough, M. 1970. *J. Virology* 6:320-25

126. Bezdek, M., Amati, P. 1968. *Virology* 36:701-03
127. Gough, M., Scott, J. V., Malik, V. S., de la Rosa, O. 1972. *Virology* 47: 276-84
128. Boyd, J. S. K. 1951. *J. Pathol. Bacteriol.* 63:445-57
129. Smith, G. R. Personal communication
130. Bertani, L. E. 1968. *Virology* 36:87-103
131. Bertani, L. E. 1960. *Virology* 12: 553-69
132. Sunshine, M. G., Calendar, R. C. Personal communication
133. Lindahl, G. 1969. *Virology* 39:839-60
134. Bertani, L. E. 1970. *Proc. Nat. Acad. Sci. USA* 65:331-36
135. Eisen, H. A., Fuerst, C. R., Siminovitch, L., Thomas, R., Lambert, L., Pereira da Silva, L., Jacob, F. 1966. *Virology* 30:224-41
136. Benzer, S. 1957. *The Chemical Basis of Heredity* Baltimore: John Hopkins Press, pp. 70-93
137. Howard, B. D. 1967. *Science* 158: 1588-89
138. Kellenberger, G., Zichichi, M. L., Weigle, J. 1961. *J. Mol. Biol.* 3: 399-408
139. Brachet, P., Green, B. R. 1970. *Virology* 40:792-99
140. Oppenheim, A. B., Slonim, Z. 1971. *Mol. Gen. Genet.* 112:255-62
141. Court, D., Campbell, A. 1972. *J. Virology* In press
142. Lieb, M. 1966. *J. Mol. Biol.* 16:149-63
143. Wu, A. M., Ghosh, S., Echols, H. 1972. *J. Mol. Biol.* In press
144. Toussaint, A. 1969. *Mol. Gen. Genet.* 106:89-92
145. Strack, H. B., Cox, J. H. 1970. *Virology* 41:562-63
146. Green, M. H., Gotchel, B., Hendershott, J., Kennel, S. 1967. *Proc. Nat. Acad. Sci. USA* 58:2343-50
147. Oppenheim, A. B., Oppenheim, A., Honigman, A. In preparation

VI
The Spartans

Editor's Comments on Papers 28 Through 31

In contrast to the phages discussed in previous sections of this book, the ones to be considered here are small and structurally simple. Because of their smallness, these phages might be expected to reveal something close to the ultimate minimum requirements for a replicating biological entity, uncomplicated by the luxuries possessed by the more sophisticated phages. Table 1 compares the sizes of the large and small phages and gives an estimate of the number of genes they contain. As noted from the table, the small phages have genetic material which differs radically from that of the large phages and from most of the rest of the biological world. The DNA of phages S13 and φX174 does not have the Watson – Crick double-helical form but is, instead, a circle of single-stranded DNA, while for phages f2, MS-2, and Qβ the genetic material is not DNA at all, but single-stranded linear RNA. Recently, the symmetry of the phage world has been completed with the discovery of a *Pseudomonas* phage, φ6, which contains double-stranded RNA *(190)*.

Table 1

Phage	Kind of nucleic acid	Particle weight (daltons)	Nucleic acid weight (daltons)	No. genes
T4	DS-DNA	200×10^6	130×10^6	100–150
λ	DS-DNA	62×10^6	31×10^6	35–50
S13, φX174	SS-DNA	6.2×10^6	1.7×10^6	8–10
f2, MS-2, Qβ	SS-RNA	4×10^6	1.1×10^6	3

Single-Stranded DNA Phage

The minuteness of S13, first isolated by Burnet in 1927, soon became apparent from centrifugation and filtration studies, but not until 1957 did suspicion arise that its DNA is unusual. The first indications of this fact came from radiobiological studies, originally utilizing X-rays and later the unstable isotope of phosphorus, ^{32}P.

The decay of ^{32}P atoms incorporated into the DNA of a population of phage results in the inactivation of the phage particles at a rate that depends both on the number of

^{32}P atoms incorporated and on the efficiency or probability with which these atoms inactivate the phage when they disintegrate. When ^{32}P inactivation measurements were made for a number of different phage strains, it was found that the rate of inactivation is proportional to the DNA content of the phage *(92, 172)*. This would be expected if the probability of inactivation per decay is the same for all phage strains. From such experiments it could be deduced that this probability is about 0.1; that is, only about 1 of 10 disintegrations results in inactivation. This low efficiency of killing could be understood on the basis of the duplex structure of DNA if it were assumed that nine-tenths of the disintegrations break only one of the two polynucleotide strands. Such breaks might, in general, be nonlethal because the duplex would be held together by the complementary unbroken strand. Only those disintegrations would be lethal which are of such direction and energy that both strands are broken at sites very close to one another.

As implied by the above, the amount of DNA contained in a phage particle can be determined, with an appropriate calibration, by measuring its rate of ^{32}P inactivation. This is a useful method for viruses that cannot be purified sufficiently or grown in great enough quantity for direct chemical determinations. In applying this method to phage S13, which had not been purified at the time, Tessman, Tessman, and Stent *(183)* deduced a value for the DNA content that was unreasonably large. If S13 had the volume it was thought to have, based on indirect measurements, their value for the DNA content predicted that the phage particle would have a density close to that of metallic lead. It was clear that either the measurements of the size of the phage were grossly inaccurate or else this phage was an exception to the rule that the efficiency of killing per ^{32}P decay is invariant.

As described in Paper 28, Tessman reinvestigated this question by comparing the radiosensitivity of S13 to that of the closely related phage, ϕX174, whose DNA content had recently been determined chemically by Sinsheimer *(163)*. Tessman found both phages to have essentially the same high radiosensitivity from which he concluded that, unlike all other phages then known, almost every ^{32}P disintegration results in inactivation. On the basis of this observation, Tessman suggested that these phages might possess single-stranded DNA. At about the same time Sinsheimer published direct chemical evidence for this idea, showing that the usual rule of molar equality of adenine and thymine and of guanine and cytosine, characteristic of double-stranded DNA, is violated by the DNA of ϕX174 *(164)*.

The existence of single-stranded DNA immediately raised questions about its mode of replication. Did it replicate in a manner radically different from that of double-stranded DNA or did it become double-stranded upon entering a cell by acting as a template for the synthesis of a complementary strand? Sinsheimer found the latter to be the case *(165)*. Immediately after entrance into the cell, the single-stranded ϕX174 DNA is converted into a double-stranded ring called the RF (replicative form) composed of the original infecting (V) strand and a complementary (C) strand. The question remaining was how circular single-stranded progeny DNA, identical to the parental DNA, is generated and exclusively packaged into phage particles.

During seven more years of research, the details of the replication process gradually emerged, the essential features of which are summarized in Fig. 1 of Paper 29. After the first RF molecule forms, it becomes attached to one of a limited number of sites on the host cell membrane. When so attached, the RF replicates to form progeny RF molecules which, if additional membrane sites exist, occupy these available sites and continue to replicate, resulting in the accumulation of unattached RF molecules. At about 12 minutes after infection, when some 15 unattached RF molecules have accumulated per cell, RF replication ceases, the V strands are displaced from the unattached RF duplex rings in a process coupled to packaging, and new V strands are synthesized on the freed C-strand rings of the RF. Release of the V strand from the RF is made possible by an enzyme that nicks the V strand exclusively. Thus, after the formation of the first RF molecule, the process can be characterized as symmetric replication of the membrane-bound RF (in which V and C strands are synthesized in equal numbers) followed by asymmetric replication (in which V strands are synthesized exclusively). The RF molecules bound to the membrane are never released and do not participate in the synthesis of the V strands destined to be packaged. It is assumed that the linear progeny V strands released from the unattached RFs somehow recircularize during packaging, since this is the form of the DNA in mature phage particles.

In the same volume of the Cold Spring Harbor Symposium in which Paper 29 appeared, Gilbert and Dressler (72) further elaborated on the replicative process of φX174 and described how the scheme of Fig. 1 could be incorporated into their rolling-circle model of DNA replication. According to this model, progeny V strands are generated by cyclic replication around the RF ring, C-strand growth being prevented by complexing of the phage coat proteins to the V strand. The nicking enzyme, acting at a specific site on the V strand, then releases unit length V strands from the replicating complex. Gilbert and Dressler also suggested a mechanism for the recircularization of the V strand during packaging.

The isolation and characterization of conditional lethal nutants of φX174 by Sinsheimer's group (9, 10) and of S13 by Tessman's group (103, 180, 182) have revealed the roles played by many of the phage genes in the replication process. At the present time, nine genes have been identified for these phages (see the Appendix). Five of these genes specify proteins that are incorporated into the mature phage particle, three are involved in RF- and SS-DNA synthesis, and one is responsible for lysis of the host cell. If synthesis of a coat protein is blocked by mutation, SS-DNA synthesis is also blocked, suggesting that SS-DNA synthesis is coupled to packaging (145). It has been shown that all the genes of φX174 are transcribed from the strand of the RF complementary to the V strand (81). Thus, until RF is formed by the action of the host DNA polymerase, no phage genes are expressed.

Another group of phage also with single-stranded DNA but quite different from S13 and φX174 in morphology and biological properties are the phage fd, f1, and M13, recently reviewed by Marvin and Hohn (134). In contrast to the icosahedral S13 and φX174, these phage are filamentous and attack only male bacteria, that is, bacteria that possess the fertility factor F. The reason for this specificity is that these phage do not adsorb to the surface of the cell but only to the ends of the sex pili or fimbriae characteristic of male bacteria. When seen in the electron microscope the long phage filaments appear as extensions of the pili, having very closely the same diameter. It is presumed, although not proved, that the phage DNA enters the cell by passing through the hollow pilus. A unique feature of these phage is that replication is not accompanied by cell lysis. Instead, the phage filaments are extruded through the cell wall and are probably assembled during the process of extrusion since no pool of intracellular phage can be demonstrated. These interesting phage underscore the warning stated by Luria long ago that ". . . no conclusion based on the study of one virus can *a priori* be generalized as valid for other viruses" *(130)*.

Single-Stranded RNA Phage

The identification of DNA as the genetic material of phage, bacteria, and higher organisms during the 1950s provided a unifying thread to all of biology, but at the same time made conspicuous the few obvious exceptions. Certain plant and animal viruses had long been known to contain RNA rather than DNA. Cells infected with such viruses must somehow copy RNA sequences, but the mechanism by which this is accomplished was a mystery. In the past several years it has been established that some RNA viruses contain an enzyme capable of converting RNA sequences to DNA sequences, whereas others synthesize an enzyme that replicates the viral RNA directly without the intervention of DNA *(179)*, but prior to about 1960, the experimental difficulties associated with plant and animal viruses precluded any detailed analysis of their mode of replication and there was no known RNA phage to serve as a model system.

This situation changed in 1961 when, as described in Paper 30, Loeb and Zinder isolated an RNA-containing coliphage called f2. Subsequently, a number of other RNA phages (MS-2, R17, Qβ) were isolated and studied in various laboratories. All these phages are similar in size and morphology to each other and to the DNA icosahedral phage ϕX174. They also share with the filamentous phage, such as f1, the property of being specific for male bacteria. The mode of attachment is different, however, the RNA phage attaching to the side of the sex pili rather than to its end as do the filimentous phage. Cells infected with an RNA phage can, under certain conditions, produce enormous numbers of progeny—up to 20,000 per cell—making these phage, as Zinder points out *(215)*, probably the most populous of biological entities, yet they were unknown little more than ten years ago.

Early studies on RNA phage replication sought to determine whether transfer of information to DNA was a possible step in the replication process. The results were uniformly negative. Almost complete suppression of DNA synthesis in f2-infected cells failed to reduce the yield of the virus *(38)*. Furthermore, no DNA hybridizable to phage RNA could be detected in phage-infected cells *(47)*. The RNA phage thus appeared to be capable of replicating its RNA directly without a DNA intermediate.

Since there was no evidence that the RNA of uninfected cells replicated directly, it was reasoned that the phage RNA probably codes for the enzymatic machinery necessary to perform this task. This enzyme, or RNA replicase as it was termed, must be highly specific, since it would have to recognize phage RNA among all the other kinds of RNA present in the cell.

The search for the phage RNA replicase was undertaken by the groups of August, Spiegelman, and Weissmann, and all three reported success at about the same time. Summary reviews of their findings were published in the 1968 Cold Spring Harbor Symposium volume *(6,199)*, one of which (Paper 31) is reproduced here. As expected, the enzyme is highly specific. It was found to catalyze the incorporation of ribonucleotide triphosphates into RNA polymers only if the reaction mixture contained RNA isolated from the same phage strain used to produce the replicase. No variety of host RNA was active in stimulating incorporation. Characterization of the enzyme showed that it is composed of four dissimilar subunits only one of which is specified by the phage RNA, the other three being derived from the host.

In the early in vitro studies, RNA synthesis was defined simply as the incorporation of labeled ribonucleotides into acid-insoluble material. An essential question to be answered was whether the product RNA was identical to the template RNA or represented some biologically irrelevant polymerization occurring in the test tube but not in the intact cell. It was thus necessary to show that the product RNA is biologically competent.

A particularly advantageous property of the RNA phage is that free RNA isolated from phage particles can infect properly treated cells (protoplasts) and produce progeny phage particles. Spiegelman and colleagues used this protoplast assay to show that infectious phage RNA was synthesized in vitro at the same rate as total RNA and, moreover, that the infectious RNA produced was more than could possibly be accounted for by the RNA added to initiate the reaction. However, as stated in Paper 31, "None of the experiments thus far described *proved* that the RNA synthesized in this system is, in fact, the self-duplicating entity. Minute quantities of contaminating RNA present in the enzyme preparation or even the replicase itself could be the instructive agent. What was required was a rigorous demonstration that the RNA, and not the replicase preparation, was the progenitor." Such a proof was provided by incubating RNA marked by a mutation with replicase prepared from cells infected with nonmutant phage. The product RNA was found to give rise to mutant phage when used to infect protoplasts, thus showing that the initiating RNA itself is the instructive agent.

The details of the replicative process were investigated by several groups *(3, 16, 68)* who sought to identify the intermediate structures formed both in vivo and in vitro. Both approaches pointed to the existence of intermediate structures composed of parental (+) strands and complementary (−) strands. The picture was complicated, however, by the fact that, upon isolation, these intermediate structures exhibited several forms. One was a duplex structure called HS and another was a multistranded structure called FS containing single-stranded segments, as revealed by its partial sensitivity to ribonuclease. A suggested scheme that includes these intermediates in the replicative cycle of Qβ is given in Fig. 10 of Paper 31. The multistranded structures are pictured as arising by the initiation of synthesis of additional strands on the template RNA before the first strand has been completed. Aside from these special features, the mechanism of single-stranded RNA replication closely parallels that of single-stranded DNA replication, i.e., the phage (+) strand directs the synthesis of a complementary (−) strand, which in turn serves as the template for progeny (+) strand formation.

The ability to synthesize Qβ RNA in vitro permitted the demonstration that (−) strands isolated from infected cells can serve as the template for replicase-mediated RNA synthesis as well as (+) strands. In fact, (−) strands are an even more effective template than (+) strands *(6, 199)*. This may be related to the fact that certain host factors are required for (−) strand synthesis from a (+) strand template but are not required for the converse, and may explain the predominant synthesis of (+) strands in vivo during the latter part of the latent period.

In vitro replication of RNA also provided an opportunity to mimic evolutionary changes in nucleotide sequences which might have occurred in the precellular past. Spiegelman reasoned (see Paper 31, Part VIII) that in a population of Qβ-phage RNA molecules, any rare variant RNA molecule which has lost part of its nucleotide sequence but which has retained the recognition site for the replicase should replicate in less time than normal molecules and should have a selective advantage over the normal molecules. To select for these more rapidly replicating molecules, the replicating RNA was diluted into a fresh sample of replicase at an early time, when only a few molecules had been completed. The process was repeated serially, over and over again, using at each transfer a progressively shorter incubation period. After four such transfers, synthesis of infectious RNA (but not total RNA) ceased, indicating that the species now replicating had lost information necessary to produce phage. By the eighth transfer, the rate of RNA synthesis showed a sharp increase. At the 75th transfer, the molecular weight of the product RNA was found to be only 17 percent that of the original Qβ RNA. This RNA was undoubtedly heterogeneous but by diluting the RNA so as to obtain samples containing only one RNA molecule and then replicating these single molecules, "clones" of homogeneous RNA could be obtained.

Very recently Mills, Kramer, and Spiegelman *(137)* have determined the complete nucleotide sequence of an even smaller Qβ variant which contains only 218 nucleotides. The most interesting feature of this sequence is the occurrence of stretches of nucleotides which are complementary to neighboring stretches of nucleotides, thus allowing the formation of intrastrand double-stranded regions. Although several structures could be formed using the known sequence, the simplest predict that as many as 80 percent of the nucleotides are involved in intrastrand base pairing. Extensive secondary structure has also been deduced for the wild-type R17-phage RNA from partial sequences *(1)*. Mills, Kramer, and Spiegelman argue that intrastrand complementarity would be expected to arise in the course of evolution of replicating RNA molecules. For example, if a sequence advantageous to the replication of the RNA molecules occurs in the (+) strand, this advantage might be enhanced by creating the same sequence in the (−) strand. Since the (+) and (−) strands are themselves complementary, this would automatically result in intrastrand complementary sequences in both strands. These authors further point out that the evolution of secondary structure in RNA, in contrast to uniformly double-stranded DNA, which probably arose later in evolution, gives such molecules biological potentialities in addition to that contained in the nucleotide sequence alone. An example of how the RNA phage might exploit this secondary structure is discussed below.

The phage RNA is some 3500 nucleotides long and, although it is still a formidable task, recent advances in nucleotide sequence analysis give reason to expect that the entire sequence will be known for one of the phages in the near future. The RNA phage have only three genes. One of these specifies the replicase, a second determines the structure of the major coat protein (CP), and a third, the A gene, codes for a minor coat protein responsible for adsorption of the phage to the host cell. All three proteins can be formed in vitro using the (+) strand as template *(118)*. These phage do not undergo genetic recombination, precluding gene ordering by genetic means, but, using RNA fragments as messenger, the gene order has been determined to be 5'-A-CP-Rep-3' *(118)*.

Partial nucleotide sequence determinations of RNA phages, particularly R17, have revealed the interesting fact that regions at each end of the molecule and regions between the genes are not translated *(169)*. It is thought that such intergenic regions may function as regulators of translation by forming double-stranded duplexes with neighboring complementary sequences. If this hairpin duplex region includes a ribosome binding site, ribosome attachment at this site would be prevented until the secondary structure is temporarily removed by the passage of a ribosome arriving from the 5' direction. Such a mechanism has been proposed to explain the finding that translation of the replicase gene requires prior translation of the coat protein gene. The current state of knowledge of the replication, translation, and structure of the RNA of these phages has been summarized in several recent articles *(93, 118, 168, 188, 200)*.

In reviewing the status of RNA phage several years after their discovery, Zinder *(215)* made the prediction that: "In time, it is certain that the whole story of these phages will be encompassed and, with this, there will be one independent piece of genetic material about which we can say—we know and we understand." That time has not yet arrived, but it can at least be said that since the day phage were first identified in the electron microscope, the progress has been remarkable.

28

Reprinted from *Virology,* **7**(3), 263 – 275 (1959)

Some Unusual Properties of the Nucleic Acid in Bacteriophages S13 and φX174

Irwin Tessman[1, 2]

Laboratory of Nuclear Studies, Cornell University, Ithaca, New York, and Biology Department, Massachusetts Institute of Technology, Cambridge, Massachusetts

Accepted November 18, 1958

Phages φX174 (and S13), labeled with P^{32}, are inactivated by decay of their incorporated radioactive atoms with approximately 100% efficiency, even at $-196°$ C (assuming the DNA content of φX174 to be as determined by Sinsheimer). The high sensitivity to P^{32} decay suggests an unusual structure and function of the DNA in these phages. One possibility is a single-stranded DNA.

The efficiency of inactivation of φX174 is measured at temperatures up to 37°. The rate of death of the phage at the elevated temperatures is too great to be entirely due to decay of the incorporated P^{32} atoms. This suggests a technical difficulty in studying the effects of temperature on the lethality of phage-incorporated P^{32} atoms.

The resemblance of S13 and φX174 to some plant and animal viruses is discussed.

INTRODUCTION

It is generally recognized that deoxyribonucleic acid (DNA) is an important, and probably the most important, carrier of genetic information. The Watson-Crick double-stranded helical structure for DNA has been well established for DNA derived from many sources under a wide variety of biological conditions (Watson and Crick, 1953; Langridge *et al.*, 1957), and this structure has been serving as the model for ideas concerning the functioning mechanism of DNA. The experiments to be

[1] Aided by a postdoctoral fellowship grant from the American Cancer Society. The investigation was supported, in part, by a grant from the American Cancer Society to Dr. Philip Morrison and, in part, by research grant E-2128 (C1) from the National Institute of Allergy and Infectious Diseases, United States Public Health Service.

[2] A preliminary report of this work is given in *Program and Abstracts. The Biophysical Society*, 1958 Meeting, p. 42.

263

reported here describe the two related bacteriophages, S13 and ϕX174, and suggest that in comparison with other phages, these two viruses contain DNA that is unusual in its structure, or in its function, or possibly in both these respects.

The conclusions will be based on the high rate of death of these phages when labeled with P^{32}. It was originally shown by Hershey, Kamen, Kennedy and Gest (1951) that P^{32}-labeled T2 and T4 are inactivated by decay of their incorporated radioactive phosphorus atoms. The survival is given by $s = 10^{-kD}$, where s is the fraction of phage particles surviving after a dose D, and D is the fraction of P^{32} atoms decayed by the time of assay. The rate constant, k, depends on the number of radioactive atoms in the phage and the effectiveness of these atoms in killing the phage when they decay. k equals $1.48 \times 10^{-6} A_0 \alpha N$, the numerical factor arising from the particular units chosen. A_0 is the specific activity of the phage phosphorus in millicuries per milligram, N is the total number of phosphorus atoms in a phage, and α, the efficiency of killing, is the probability that a P^{32} decay will inactivate the phage in which the decay occurs. The finding of a strictly exponential survival curve indicates a *single-hit* inactivation process, which means that each inactivation is caused by a single P^{32} decay, although not every decay may inactivate. The average number of decays for each inactivation is $1/\alpha$, and only when α equals 1.0 does every decay inactivate the phage in which it occurs.

By measuring k, A_0, and N, it was found for phages T1, T2, T3, T4, T5, T7, and P22 (Hershey *et al.*, 1951; Stent and Fuerst, 1955; Garen and Zinder, 1955), that the rate of inactivation was proportional to the amount of phosphorus in the phage, i.e., that α was very nearly constant and equal to 0.10 at 4° for the variety of phages tested. A small deviation from $\alpha = 0.10$ in the case of T7 can be explained by the presence of non-DNA phosphorus (Lunan and Sinsheimer, 1956), and one may assume the same explanation for the small deviation in the related phage T3. Therefore, the interesting conclusion has been that the sensitivity of phages to inactivation by P^{32} decay is proportional to the DNA content of the phages.

An experiment was performed by Tessman, Tessman, and Stent (1957) to determine the DNA content of phage S13 by measuring its rate of inactivation by P^{32} decay (and X-rays) relative to T2. On the assumption that the radiosensitivity is proportional to the DNA content, the result obtained by them was almost ten times larger than were unpublished chemical and light-scattering determinations by Dr. R. L. Sinsheimer of the DNA content of a related phage, ϕX174. As an alternative to the possibility that these two phages simply have different amounts of DNA, it was suspected that α might be very much higher than 0.1 for these phages. If instead of only one in ten P^{32} decays being effective ($\alpha = 0.10$), every P^{32} decay inactivated the S13 particle in which it occurred ($\alpha = 1.0$), the particle weight of S13-DNA would be 1.8×10^6 ($= 3.0 \times 10^{-18}$ g). This is the lower limit for the DNA content of S13 on the basis of its sensitivity to P^{32} decay.

The recent value obtained by Sinsheimer (personal communication) for the particle weight of DNA in ϕX174 is 1.6×10^6 (recalculated as the free acid), in striking agreement with the lower limit predicted for the S13-DNA on the basis of its P^{32} sensitivity. But, to argue conclusively that these phages have an unusual radiosensitivity, it is necessary to have both the DNA content and P^{32} sensitivity for the same phage. For this reason the P^{32} experiment has been repeated for ϕX174, and temperature studies have been included to examine the P^{32} sensitivity in greater detail.

MATERIALS AND METHODS

The method used previously (Stent and Fuerst, 1955) for making radioactive phage has been modified only by the replacement of the H medium with "tris-glucose" (Hershey, 1955). In experiment 1, *Escherichia coli*, strain C, was grown for about three generations in a radioactive medium containing P^{32} at a specific activity of 210 mc per milligram phosphorus, inferred from the inactivation rate of T2 labeled in this medium (Stent and Fuerst, 1955). The culture was then divided into three equal aliquots for growth of three phage strains: T2hr_1, ϕX174 (obtained from Dr. R. L. Sinsheimer) and S13ch_b (a host-range mutant derived in two steps from the previously used wild strain). Each aliquot of cells was inoculated with approximately 1×10^6 particles per milliliter of the appropriate phage. The phage concentrations increased about 4000-fold in the lysates in about 3 hours. Each lysate was immediately diluted in the glycerol-casamino acid medium (designated 3XD) and samples were stored at 4° in the refrigerator and at −196° under liquid nitrogen. Nonradioactive phage stocks stored at these temperatures showed no inactivation.

In experiment 2 only ϕX174 was labeled and no comparison was made with T2. The specific activity of the tris-glucose medium was increased and inferred to be 360 mc per milligram phosphorus by comparing the inactivation rate at −196° with the corresponding rate in experiment 1. The titer increased about 400-fold in the lysate. In addition to frozen samples, the phage were stored at various temperatures up to 37°, as will be described in more detail later.

The possibility of lethal effects from β-particles emitted by P^{32} atoms in the medium during the storage period seems unlikely. Hershey *et al.* (1951) found that 13 μc/ml was a safe initial P^{32} concentration for storing the much larger phage T2 at 5° for 30 days. In both experiments described here the concentration of P^{32} in the tubes used to store the labeled phage was considerably lower than these values (although the concentration of P^{32} in the lysate medium was approximately 3 mc/ml, approximately equivalent to a dose of 2×10^3 rads per hour from the β-particles). The following values apply to experiment 2; the values in experiment 1 are only about 20% lower. In the frozen tubes the concentration was 0.25 μc/ml (equivalent to a dose of approximately 5 rads per day) and all assays from these tubes were made by diluting and plating samples immediately after thawing. The tubes stored at temperatures above freezing contained tenfold serial dilutions, and assays were made by plating directly from the appropriate tubes. Thus, as the phage became inactivated it was necessary to use the more concentrated dilution tubes. As a result, the P^{32} concentration varied from 0.0025 μc/ml in the dilution tubes used at early times to 0.025 μc/ml in the dilution tubes used at the end of the experiment, and in experiment 1 a few of the latest points were from tubes that had an initial concentration of 0.20 μc/ml. The significant evidence that the concentration of radioactivity in the stored dilution tubes had no effect on the survival of the phage is that in the curves of Figs. 1 and 2, to be described below, the initial points join smoothly with the final points even though the initial and final assays in the above-freezing experiments were made from stored tubes with at least a tenfold difference in P^{32} concentrations.

For all curves, except those in Fig. 3, straight lines are fitted to the experimental points by the method of least squares (Bacon, 1953). In Fig. 3 the straight lines are fitted visually.

RESULTS

The survival curves of S13, φX174, and T2 from experiment 1 are displayed in Fig. 1. The slopes obtained from these curves are given in Table 1. The results can be summarized as follows:

1. The sensitivity of φX174 to the decay of incorporated P^{32} is essen-

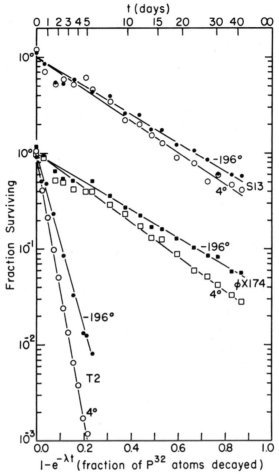

FIG. 1. *Experiment 1.* Survival at −196° and at 4° of T2, S13, and φX174 containing equal specific activities of P³² as a function of the fraction of P³² atoms decayed by the time, t, of assay. λ is the fraction of P³² atoms decaying per day and equals 0.048 per day. The S13 curves have been displaced to avoid confusion with the φ X174 curves.

tially equal to that of S13. That this equality holds more exactly at −196° than at 4° is considered significant and will be discussed below.

2. The sensitivity of the mutant strain S13ch_b, is very nearly the same as that of the wild type used previously (Tessman *et al.*, 1957). The difference can be attributed to variability in the 4° experiments.

TABLE 1

SLOPES OF SURVIVAL CURVES AND EFFICIENCY OF KILLING CALCULATED
FROM FIG. 1

| Phage | Slope of survival curve (k) | | slope (−196°) / slope (4°) | Efficiency of killing[a] (α) | |
	−196°	4°		−196°	4°
ϕX174	−1.47	−1.78	0.82	0.94	1.14
S13ch_b	−1.48	−1.65	0.90	0.94[b]	1.04[b]
T2hr_1	−9.06	−13.8	0.65		

The following data are from Stent and Fuerst (1955)		
T1	0.56	0.12
T2	0.66	0.10
T3	0.52	0.07
T5	0.55	0.12
T7	0.56	0.08
λ	0.54	?

[a] A_0 = 210 mc/mg.

[b] Assuming the DNA content of S13 is the same as that of ϕX174.

3. The sensitivities of ϕX174 and S13 at −196° are not the same as at 4°, but it is important to note that they are more nearly the same than in the cases of the many other phages shown.

4. All survival curves are single hit.

The serological interrelationship and common host-range characteristics of ϕX174 and S13 (Zahler, 1958) and their identical ultraviolet (UV) sensitivities (Tessman, unpublished) have suggested that these phages are very closely related. This conclusion is now reinforced by their equal sensitivity to P^{32} decay. It is therefore assumed in making calculations based on the amount of DNA in S13 that this value is equal to the amount of DNA in ϕX174.

Using the survival curve of T2 as a measure of A_0 and using the value of 1.6×10^6 (Sinsheimer, personal communication) for the DNA content of ϕX174, the efficiency of killing is calculated and is given in the last columns of Table 1. It can be concluded that, within an error of approximately $\pm 20\%$, every P^{32} decay in ϕX174 or in S13 inactivates the particle. Clearly this cannot be true at both 4° and −196°, for the slopes of the survival curves are not the same at these two temperatures. (The ratios of the slopes in any one experiment are far more accurately determined than their absolute values.) But the differences between

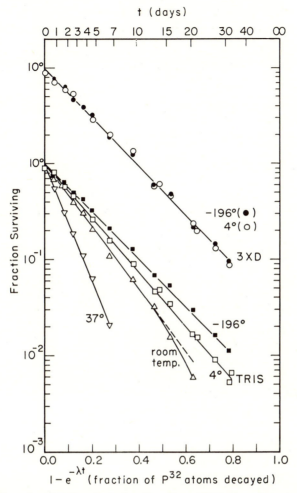

FIG. 2. *Experiment 2.* Survival of radioactive ϕX174 at various temperatures as a function of the fraction of P^{32} atoms decayed by the time, t, of assay. The curves are all relative to the survival of nonradioactive controls stored at the same temperature. The controls are plotted in Fig. 3. The 3XD curves have been displaced to avoid confusion with the tris curves.

α at $-196°$ and α at $4°$ are too small to permit a decision as to the temperature at which α is more nearly equal to 1.0.

It was expected that if every P^{32} decay were lethal there would probably not be any subtle effects depending on the temperature. This expec-

TABLE 2

SLOPES OF SURVIVAL CURVES AND EFFICIENCY OF KILLING CALCULATED FROM FIG. 2

Storage medium		$-196°$	$4°$	Room temp. (corrected for control)	37° (corrected for control)
3XD	Absolute slope (k)	-2.56	-2.56		
	Slope relative to $-196°$ slope	1.00	1.00		
	Efficiency of killing (α)	0.94	0.94		
Tris	Absolute slope (k)	-2.46	-2.82	-3.27	-5.98
	Slope relative to $-196°$ slope	1.00	1.15	1.33	2.43
	Efficiency of killing (α)	0.90	1.03	1.20	2.2

To calculate α the specific activity A_0 is obtained by comparing the survival curves of ϕX174 in 3XD at $-196°$ in experiments 1 and 2. $A_0 = 360$ mc/mg.

tation was only partially fulfilled, as can be seen from Fig. 1 and Table 1. The following experiment clarifies this by showing that the small temperature effect probably arises from extra inactivations unrelated to the P^{32} decays. It also reinforces the main conclusions about the high radiosensitivity of S13 and ϕX174.

In experiment 2, dilutions of the ϕX174 lysate were made in 3XD and also in 0.1 M trishydroxymethylaminomethane (tris) at pH 9.8. In both cases samples were stored at $-196°$ under liquid nitrogen and at $4°$ in the refrigerator. In addition, dilutions of phage in tris were stored at room temperature (about 25°) and at 37°. Tris was chosen for the elevated temperatures because it was believed that the phage is relatively stable against spontaneous inactivation in this medium. Control tubes of nonradioactive phage were made in each case. The survival curves for the radioactive phage are shown in Fig. 2 with the slopes tabulated in Table 2; the survival curves for the nonradioactive phage are shown in Fig. 3.

In this experiment the slopes for the radioactive phage are identical at $-196°$ and at $4°$ in 3XD, but not in tris medium. Significantly, the major difference is in the $4°$ slope. At these temperatures the controls are completely stable, but at room temperature and at 37° the controls are slowly inactivated, the rates being greater at the higher temperature. At these two elevated temperatures the inactivation rate of the radio-

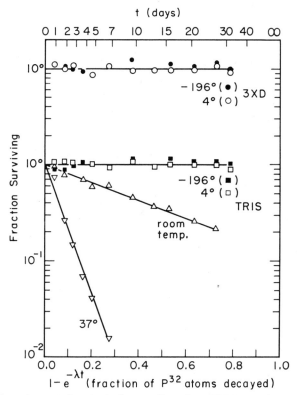

FIG. 3. *Experiment 2.* Survival of nonradioactive φX174 stored at various temperatures as controls for the data shown in Fig. 2. The abscissa is the same as in Figs. 1 and 2, even though the phages are nonradioactive.

active phage, corrected for the spontaneous inactivation rate of the control phage, is significantly greater than can be explained by the number of incorporated P^{32} atoms, i.e., α is definitely greater than 1, and this is referred to as an extrasensitivity. Again, the explanation is unlikely to be radiation effects of the P^{32} in the stored dilution tubes because of the agreements among assays from tubes containing different concentrations of P^{32}. One possible explanation is that the radioactive phage, while in the highly radioactive lysate medium, may have suffered sublethal effects, which, even after susequent dilution to a low radiation level, rendered the radioactive phage more temperature sensitive than the controls. This explanation might be tested by examining the stability of nonradioactive phage grown with a high concentration of P^{32} but with a large excess of P^{31} to give a comparatively low specific activity (Hershey *et al.*, 1951).

Whatever the explanation, extrasensitivity is definitely established at 37° and to a lesser extent at about 25°. It is then suggestive that this might also be the case at 4°, though to an even lesser extent. It is then understandable that the 4° results should vary from time to time, depending in part on the radiation level in the growth solution, and on the particular merits of the storage media. Thus it seems that α equals 1.0 at all temperatures but that because of the extrasensitivity of the phage, the true inactivation rate from P^{32} decays is measured only at very low temperatures.

Clearly in these experiments not all the inactivations were due to P^{32} decays within the DNA of the particles being inactivated, because α was greater than 1 at the elevated temperatures. If α were much less than 1, the extra inactivations could be overlooked. This suggests that other findings (Castagnoli and Graziosi, 1954; Castagnoli *et al.*, 1955; Stent and Fuerst, 1955) of the effects of temperature on P^{32} inactivations be re-examined, in particular, the finding that for phage T5 α rises from 0.066 at −196° to as high a value as 0.31 at 65° (Stent and Fuerst, 1955). The interpretation of the temperature effect bears upon the argument of Stent and Fuerst that inactivation requires a scission of the double-stranded DNA molecule, and that breaks in only one strand do not seriously prevent the DNA from functioning.

CONCLUSIONS

The fine points concerning the extrasensitivity should not obscure the major result that for ϕX174 and S13 the decay of essentially every incorporated P^{32} atom is lethal (with the qualification that this has been shown rigorously only for ϕX174). This is a sensitivity significantly greater than has been found for any other bacteriophage or bacterium. It suggests novel features for the DNA in these phages.

It was argued by Stent and Fuerst (1955) that the combination of a low and constant value of α at a given temperature for a varied collection of bacteriophages probably reflected the common basic structure of the DNA macromolecule. Their subsequent study of bacterial death by P^{32} decay strengthened their argument (Fuerst and Stent, 1956). It is certainly difficult in the light of their evidence to conceive of an explanation for the low value of α that involves a DNA heterogeneous in either structure or function such that part of the DNA is insensitive to P^{32} decay while another part is sensitive. It would be unusual that DNA from such a large variety of sources should have roughly the same proportion of sensitive and insensitive regions. The particular model proposed by Stent and Fuerst (1955) to explain the relative insensitivity of the particles to P^{32} decays was that the DNA could function even with the individual strands severed in various places. Its ability to function could be destroyed only by a decay that managed to sever both strands, breaking the molecule.

These arguments imply a different DNA structure as the explanation for the unusual sensitivity of φX174 and S13. A simple hypothesis is that of a single-stranded molecule which could be completely cut by every P^{32} decay. There is, in fact, evidence from light-scattering, spectrophotometry, and chemical reactivity (Sinsheimer, personal communication) that the DNA of φX174 is unusual in its structure.

Another important possibility must be considered. Whereas a common radiosensitivity implies a common DNA structure, an exceptional radiosensitivity does not imply an exceptional structure; for the DNA in these phages may have the standard Watson-Crick structure (Watson and Crick, 1953), but yet have some unusual function which cannot be performed whenever one strand of the double helix is severed. The latter possibility is in keeping with the observation of Lerman and Tolmach (in press) that the apparent severing of only a single strand of transforming principle DNA with deoxyribonuclease is sufficient to destroy the biological activity; for it is generally believed that transforming principle DNA has the structure of a Watson-Crick double helix (Zamenhof, 1956) (although it is difficult to rule out the possibility that part of the transforming activity resides in a small fraction of the DNA having an unusual structure). It may be unnecessary to distinguish carefully the structural and functional implications of the unusual radiosensitivity of these two phages because, since structure and function are intimately connected, it is very possible that if φX174 and S13 are unique in the one respect they are unique in the other respect also.

It is interesting to speculate on the resemblance of these small bacteriophages to the small "spherical" plant and animal viruses such as bushy stunt and poliomyelitis virus (Williams, 1954; Crick and Watson, 1957), which contain ribonucleic acid (RNA). Electron micrographs of S13 (Tessman, unpublished) and of φX174 (Hall, Maclean, and Tessman, unpublished) show the similarity to the plant and animal viruses in regard to size and shape, and the chemical determinations by Sinsheimer of the DNA content of ϕX174 show the similarity in regard to nucleic acid content. The resemblance is intriguing because it raises the possibility that the difference in the kind of nucleic acid in these viruses (DNA and RNA) may be minimized by the unusual structure of the DNA in the very small phages. In fact, the DNA may well have a structure resembling the structure of viral RNA.

ACKNOWLEDGMENTS

It is a pleasure to thank Mr. R. Hede, Drs. C. Levinthal, S. E. Luria, F. Rothman, N. Symonds, and E. S. Tessman for their criticisms, and Drs. L. S. Lerman and L. J. Tolmach, and Dr. R. L. Sinsheimer for communicating their results before publication.

REFERENCES

BACON, R. H. (1953). The "best" straight line among the points. *Am. J. Phys.* **21,** 428–446.

CASTAGNOLI, C., and GRAZIOSI, F. (1954). Effect of temperature on the inactivation of phage labelled with phosphorus-32. *Nature* **174,** 599–600.

CASTAGNOLI, C., DONINI, P., and GRAZIOSI, F. (1955). Indagine biofisica del complesso virus-cellula ospite durante i primi stadi dell'infezione. *Giorn. Microbiol.* **1,** 52–64.

CRICK, F. H. C., and WATSON, J. D. (1957). Virus Structure: General principles. In *The Nature of Viruses* (G. E. W. Wolstenholme and E. C. P. Millar, eds.), pp. 5–13. Churchill, London.

FUERST, C. R., and STENT, G. S. (1956). Inactivation of bacteria by decay of incorporated radioactive phosphorus. *J. Gen. Physiol.* **40,** 73–90.

GAREN, A., and ZINDER, N. D. (1955). Radiological evidence for partial genetic homology between bacteriophage and host bacteria. *Virology* **1,** 347–376.

HERSHEY, A. D. (1955). An upper limit to the protein content of the germinal substance of bacteriophage T2. *Virology* **1,** 108–127.

HERSHEY, A. D., KAMEN, M. D., KENNEDY, J. W., and GEST, H. (1951). The mortality of bacteriophage containing assimilated radiophosphorus. *J. Gen. Physiol.* **34,** 305–319.

LANGRIDGE, R., SEEDS, W. E., WILSON, H. R., HOOPER, C. W., WILKINS, M. H. F., and HAMILTON, L. D. (1957). Molecular structure of deoxyribonucleic acid (DNA). *J. Biophys. Biochem. Cytol.* **3,** 767–778.

LERMAN, L. S., and TOLMACH, L. J. Genetic transformation. II. The significance of damage to the DNA molecule. *Biochim. et Biophys. Acta* in press.

LUNAN, K. D., and SINSHEIMER, R. L. (1956). A study of the nucleic acid of bacteriophage T7. *Virology* **2,** 455–462.

STENT, G. S., and FUERST, C. R. (1955). Inactivation of bacteriophages by decay of incorporated radioactive phosphorus. *J. Gen. Physiol.* **38,** 441–458.

TESSMAN, I., TESSMAN, E. S., and STENT, G. S. (1957). The relative radiosensitivity of bacteriophages S13 and T2. *Virology* **4,** 209–215.

WATSON, J. D., and CRICK, F. H. C. (1953). A structure for deoxyribose nucleic acid. *Nature* **171,** 737–738.

WILLIAMS, R. C. (1954). Electron microscopy of viruses. *Advances in Virus Research* **2,** 183–239.

ZAHLER, S. A. (1958). Some biological properties of bacteriophages S13 and φX-174. *J. Bacteriol.* **75,** 310–315.

ZAMENHOF, S. (1956). Biology and biophysical properties of transforming principles. *Prog. in Biophys. and Biophys. Chem.* **6,** 85–119.

Reprinted from *Cold Spring Harbor Symp. Quant. Biol.*, **33**, 443–447 (1968)

Stages in the Replication of Bacteriophage φXI74 DNA in vivo

ROBERT L. SINSHEIMER, ROLF KNIPPERS, AND TOHRU KOMANO

Biology Division, California Institute of Technology, Pasadena, California

Research with bacteriophage φX174 over the past few years has led to the recognition of three distinct stages in the replication of the viral DNA. An outline of the replicative process is formulated in Fig. 1 (Sinsheimer et al., 1962; Lindqvist and Sinsheimer, 1968; Knippers et al., 1968; Komano et al., 1968).

Upon entry into the cell, the single-stranded DNA of the virus, which is a ring DNA, is quickly converted to a double-stranded DNA ring, the replicative form or RF. Two centrifugal forms of the double-stranded ring are found in the cell: the supercoiled form with both strands covalently closed (RFI), and the open form with one strand open (RFII).

The conversion from single- to double-stranded ring is performed by pre-existent host enzymes. No viral functions are needed for this step. It can occur in the presence of high concentrations of chloramphenicol, or in amino acid auxotrophs starved for the essential amino acid. This step has been closely duplicated in vitro with the aid of the host DNA polymerase and the host polynucleotide ligase (Goulian et al. 1967).

After formation, the RF becomes associated with a pre-existent essential site within the bacterial cell. The number of such sites is limited and depends upon the physiology of the cell. In previously starved cells it is usually one (Yarus and Sinsheimer, 1967). Unless an RF becomes associated with such a site it cannot replicate or give rise to progeny.

After association with such a site the RF replicates semiconservatively (Denhardt and Sinsheimer, 1965). This replication is such that the input viral DNA strand always remains associated with the site, exchanging partners at each replication. This site sediments with the membrane fraction of the cell lysate (Knippers and Sinsheimer, 1968).

This replication of the RF requires both host and viral function (Tessman, 1966; Lindqvist and Sinsheimer, 1967). It is blocked in certain host mutants, and by prior ultraviolet irradiation of the

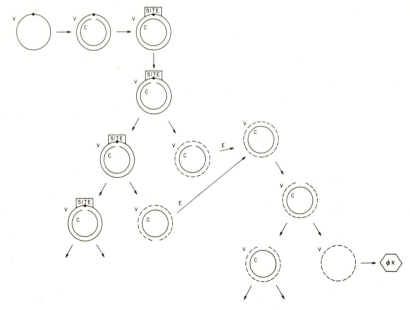

FIGURE 1. Schematic outline of the replication cycle of bacteriophage φX174.

443

host. The product of viral cistron VI is needed for RF replication (Sinsheimer, 1968).

Pulse-labeling experiments indicate that the RF which is replicating on the site is in the open form, with one strand open. In virtually all of such open RF molecules the viral strand is closed while the complementary strand is open.

Similarly the nascent daughter RF molecules released from the site and sedimenting free of the membrane fraction are initially in the open form. Again, in approximately half of these the viral strand is open; in the other half the complementary strand. The significance of this distribution is unclear. Conceivably it represents an alternation of open strands in the replicative cycle.

RF replication is initiated about 2–3 min after infection (at 37°C) and continues until about 12 min after infection. In this time some 15 daughter RF molecules accumulate. Although they first appear in the open form they are converted within a minute to the closed or supercoiled form, so that by 12 minutes the large majority of RF molecules are in the closed form.

The daughter RF molecules remain free of the membrane fraction unless the cell has more than one site; in this case the secondary site may be colonized by a daughter RF which can in turn start to replicate.

The second phase of RF replication ends at about 12 min after infection. At this time host DNA synthesis which had continued at its preinfection rate ceases. Net RF synthesis also ceases although a low level of RF replication at the site may persist for a time. Synthesis of progeny single-strand DNA begins at a rate equivalent to 5–10 times the previous rate of RF synthesis.

To initiate this third phase of DNA replication (progeny DNA synthesis) the RF molecules which have accumulated are converted within a few minutes to an open form, in all of which the viral strand is open and the complementary strand closed. All of these RF molecules are then used in an asymmetric semiconservative synthesis in which the viral strand is displaced from the RF into a virus particle while a new viral strand is laid down on the complementary ring. This process proceeds at a rate of approximately one new viral strand per RF per minute. Obviously the displaced viral strand must be closed at some time, as it is known to be closed in the mature virus particle.

Free single-strand rings of DNA are never found. Unless functional coat proteins of cistrons III and IV are both present the single-stranded progeny DNA is not made. The coat protein evidently plays a regulatory role as well as a structural role, and mutants are known in the coat protein cistron which

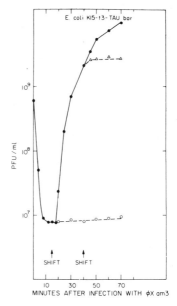

FIGURE 2. A culture of *E. coli* strain 15t3-TAU bar was grown at 30°C to 4.5×10^8/ml, infected with ϕX am3 (phage multiplicity, $m = 2$) and divided into three portions. Infective centers were measured (by artificial lysis) at various times after infection in each portion.
●– –● This portion was maintained at 30°C.
○– –○ The temperature of this portion was raised to 42°C at 12 min after infection (prior to the close of the eclipse period).
△– –△ The temperature of this portion was raised to 42°C at 40 min after infection (during formation of progeny phage).

produce progeny DNA and particles at 2–3 times the rate of wild type.

It has been a long-standing question whether the viral DNA actually penetrates the cell as a single-stranded ring or whether it might be in effect pulled into the cell as it became an RF. With the aid of a host mutant known to be temperature-sensitive for DNA synthesis it has been possible to demonstrate that the free viral single-stranded rings are indeed released from the virus in the absence of DNA synthesis.

If a temperature-sensitive mutant of the strain 15 TAU-bar, originally isolated by Rasmussen and Weywadt and given to us by Dr. Phillip Hanawalt, is infected with ϕX at the high temperature, no phage are produced (Fig. 2). If, during the period of phage production in this host, the temperature is raised to the restrictive level, ϕX synthesis abruptly stops. Upon infection of such a strain with ϕX at the high temperature no RF is formed (Fig. 3). (An analogous result was shown earlier in a Bonhoeffer mutant by Dr. David Denhardt.)

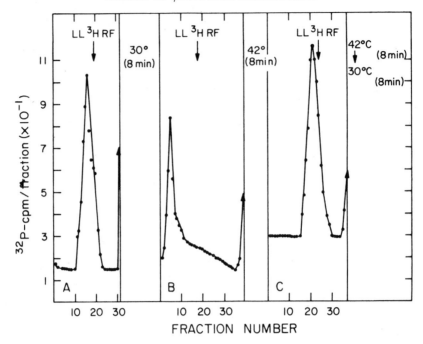

FIGURE 3. A culture of *E. coli* strain 15t3-TAU bar was grown at 30°C to 4 × 10⁸/ml and divided into three portions. Each portion was separately infected with ¹³C¹⁵N³²P-φX (*m* = 5).

(A) This portion was maintained at 30°C. At 8 min after infection, cells were lysed with lysozyme-EDTA. The lysate was sheared by passage through a syringe and then centrifuged to equilibrium in a neutral CsCl density gradient.

(B) The temperature of this portion was raised to 42°C before infection. At 8 min after infection, a lysate was prepared as in part A and centrifuged to equilibrium in a neutral CsCl density gradient.

(C) The temperature of this portion was raised to 42°C before infection. At 8 min after infection the temperature was reduced to 30°C. At 16 min after infection, a lysate was prepared as in part A and centrifuged to equilibrium in a neutral CsCl density gradient.

In each gradient the arrow marks the location of ³H-labeled RF (without density label).

However, under these conditions free single-stranded DNA is present in the cell. If the temperature is dropped these are then converted to RF. It is interesting to speculate upon what cellular component is involved here, since such cells are known to perform DNA repair at the restrictive temperature.

The result of another experiment to demonstrate the failure of daughter RF molecules to replicate during the period of RF replication is presented in Fig. 4 (see also Stone, 1967). Cells were infected with ³²P labeled phage in heavy (¹³C¹⁵N) medium in the presence of ³H-thymine. After four minutes a sample was taken for DNA extraction and the remainder transferred to light (¹²C¹⁴N) medium in an excess of cold thymine. After seven more minutes an extract of DNA was made from these cells.

After four minutes fully heavy progeny DNA containing ³H and hybrid density parental DNA containing both ³H and ³²P are present. After transfer to light medium and seven minutes' further incubation, about half of the parental DNA

has shifted (the input multiplicity was 3) to the fully light position. Infectivity data indicate a synthesis of fully light progeny DNA. But the thymine-labeled progeny DNAs, made in heavy medium, remain at the full heavy density indicating their failure to replicate.

Experiments carried out in our laboratories and others some years ago, intentionally done at a low multiplicity of infection, indicated a failure of transfer of parental DNA to progeny (Sinsheimer, 1961; Kozinski, 1961). This result is now explained in our model as a consequence of the persistence of the input viral strand at the membrane site. If, however, a considerable multiplicity of input phage were used, one might expect that at least some of the RF molecules formed would join the progeny RF pool and be utilized, thereby contributing their viral strands to progeny. Indeed this can happen. In multiplicities of 10 or above, transfer of parental strands through RF to progeny can be demonstrated. That such high multiplicity is needed is a consequence of a poorly understood but

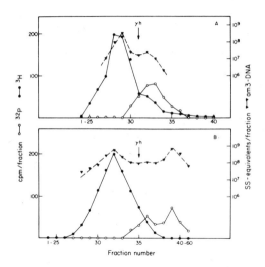

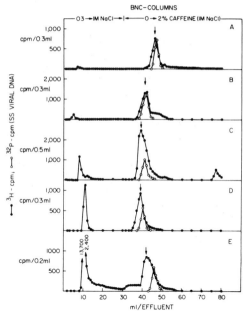

FIGURE 4. Cells of *E. coli* strain HF4704 were grown to 3×10^8/ml in $^{13}C^{15}N$ medium and infected in this medium with ^{32}P-ϕX ($m = 3$) in the presence of ^3H-thymine. At four min after infection, DNA was extracted from one portion of the culture. Cells from the other portion were transferred to $^{12}C^{14}N$ medium with an excess of unlabeled thymine and incubated for another seven min. After this, DNA was extracted from this portion.

(A) Distribution in a neutral CsCl density gradient of ^{32}P, ^3H and infectivity of the DNA extracted after 4 min.

(B) Distribution in a neutral CsCl density gradient of ^{32}P, ^3H and infectivity of the DNA extracted after the additional 7 min incubation.

In both gradients the arrow indicates the location of marker single-stranded viral DNA.

FIGURE 6. The fractions of Fig. 5 were pooled into five groups, A–E. The DNA in each group was then fractionated on a column of benzoylated, naphthoylated DEAE-cellulose (2 cm × 1.5 cm diam.) and the distribution of ^3H-label determined. A marker of ^{32}P-labeled ϕX viral DNA was added to each DNA sample before fractionation.

The fractions indicated by arrows in each gradient were further analyzed as shown in Figure 7.

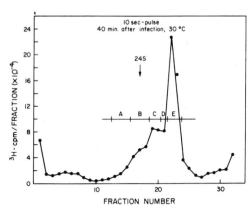

FIGURE 5. Cells of *E. coli* strain CH502 (uvr⁻, DNase I⁻, thy⁻) were grown at 30°C to 1.2×10^8/ml and infected with ϕX $am3$ ($m = 6$). After 40 min of infection, ^3H-thymidine was added. Ten seconds later the culture was abruptly chilled. DNA was extracted, sedimented through a neutral sucrose gradient (0.3 M NaCl, Tris-EDTA) and the distribution of ^3H-label determined.

The arrow indicates the portion of a marker of infective single-stranded ϕX DNA.

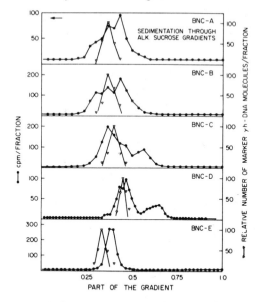

FIGURE 7. Peak fractions from the column eluate of Fig. 6 (indicated by arrow) were sedimented through a sucrose gradient at pH 12.5 (measured) and the distribution of ^3H (pulse-label) and of ^{32}P (added ϕX DNA marker) determined.

446

359

well-observed phenomenon whereby parental RF molecules formed from input DNA are rather firmly attached to the membrane fraction even if they are not at a site at which they can replicate. This is in contrast to progeny RF, which rarely attach to membrane. Such parental RF only occasionally and slowly become detached to join the progeny RF pool which can contribute DNA to progeny phage. Thus at an input multiplicity of 33 (light ^{32}P-phage infecting cells in heavy medium), after 10 minutes 90% of the ^{32}P of the parental phage was in RF. After 3 hr this was reduced to 60% and ^{32}P corresponding to 11 viral DNAs was found in virus particles of hybrid density (light DNA, heavy coat).

More recently we have attempted to observe possible intermediates in the synthesis of the progeny single-strands by reducing the effective length of pulse both by decreasing the absolute time and by lowering the temperature. Thus Fig. 5 indicates the sedimentation pattern of the radioactivity incorporated during a 10-sec pulse of ^{3}H-thymine at 30°C at 40 min after infection. The arrow indicates the position of single-stranded DNA.

The fractions from the gradient were divided into five groups, A-E as shown. These were then analyzed on a column of benzoylated-napthoylated DEAE cellulose which has previously been shown to fractionate nucleic acids according to their content of exposed purine and pyrimidine rings— or roughly according to their content of single-stranded DNA (Tener et al., 1966; Sedat et al., 1967).

As can be seen in Fig. 6, the fastest sedimenting fractions cling most tenaciously to the column (almost as tightly as does a viral φX DNA marker, suggesting the presence of a free single-strand portion comparable in length to a viral DNA) while the slower sedimenting portions split into two fractions—an increasing proportion of pure RF and a decreasing proportion of a fraction that behaves as though it contained shorter single-stranded regions.

When the peak fractions from each of the columns were analyzed by sedimentation in alkali (Fig. 7), it became clear that almost all of the original A–E fractions contained labeled components which after denaturation sedimented faster than either viral rings or linear single-stranded DNA of viral DNA length. This might be expected from a mode of replication which added on to the 3′ end and displaced the prior viral strand as it grew.

However, several fractions also clearly contain well-defined components smaller than the open viral single-strands. While one can formulate ad hoc hypotheses to account for these results they contain just about as many assumptions as facts. It is clear that more detailed experiments are needed.

ACKNOWLEDGMENTS

This research has been supported, in part, by grants RG-6965 and GM-13554 from the United States Public Health Service.

REFERENCES

DENHARDT, D., and R. L. SINSHEIMER. 1965. The process of infection with bacteriophage φX174. IV. Replication of the viral DNA in a synchronized infection. J. Mol. Biol. 12: 647.

GOULIAN, M., A. KORNBERG, and R. L. SINSHEIMER. 1967. Enzymatic synthesis of DNA. XXIV. Synthesis of infectious phage φX174 DNA. Proc. Nat. Acad. Sci. 58: 2321.

KNIPPERS, R., T. KOMANO, and R. L. SINSHEIMER. 1968. The process of infection with bacteriophage φX174. XXI. Replication and fate of the replicative form. Proc. Nat. Acad. Sci 59: 577.

KNIPPERS, R., and R. L. SINSHEIMER. 1968. The process of infection with bacteriophage φX174. XX. Attachment of the parental DNA of bacteriophage φX174 to a fast-sedimenting cell component. J. Mol. Biol. 34: 17.

KOMANO, T., R. KNIPPERS, and R. L. SINSHEIMER. 1968. The process of infection with bacteriophage φX174. XXII. Synthesis of progeny single-stranded DNA. Proc. Nat. Acad. Sci. 59: 911.

KOZINSKI, A. W. 1961. Uniform sensitivity to P^{32} decay among progeny of P^{32}-free phage φX174 grown on P^{32} labeled bacteria. Virology 13: 377.

LINDQVIST, B. H., and R. L. SINSHEIMER. 1967. The process of infection with bacteriophage φX174. XV. Bacteriophage DNA synthesis in abortive infections with a set of conditional lethal mutants. J. Mol. Biol. 30: 69.

——, ——. 1968. The process of infection with bacteriophage φX174. XVI. Synthesis of the replicative form and its relationship to viral single-stranded DNA synthesis. J. Mol. Biol. 32: 285.

SEDAT, J., R. B. KELLY, and R. L. SINSHEIMER. 1967. Fractionation of nucleic acid on benzoylated-naphthoylated DEAE cellulose. J. Mol. Biol. 20: 537.

SINSHEIMER, R. L., 1961. Replication of bacteriophage φX174. Proc. R. A. Welch Conf. on Chem. Research, V, Houston, Texas (The Robert A. Welch Foundation). 227.

SINSHEIMER, R. L. 1968. Bacteriophage φX174 and related viruses. Prog. Nucl. Acid. Res. and Mol. Biol. 8: 115.

SINSHEIMER, R. L., B. STARMAN, C. NAGLER, and S. GUTHRIE. 1962. The process of infection with bacteriophage φX174. I. Evidence for a replicative form. J. Mol. Biol. 4: 142.

STONE, A. B. 1967. Some factors which influence the replication of the replicative form of bacteriophage φX174. Biochem. Biophys. Res. Commun. 26: 247.

TENER, G. M, I. GILLIAM, M. VON TIGERSTROMM, S. MILLWARD, and E. WINNER. 1966. Purification of tRNAs on benzoylated DEAE-cellulose. Fed. Proc. 25: 519.

TESSMAN, E. 1966. Mutants of bacteriophage S13 blocked in infectious DNA synthesis. J. Mol. Biol. 17: 218.

YARUS, M. and R. L. SINSHEIMER. 1967. The process of infection with bacteriophage φX174. XIII. Evidence for an essential bacterial "site". J. Virol. 1: 135.

30

Reprinted from *Proc. Natl. Acad. Sci. (U.S.)*, **47**(3), 282–289 (1961)

A BACTERIOPHAGE CONTAINING RNA

By Tim Loeb and Norton D. Zinder

THE ROCKEFELLER INSTITUTE

Communicated by E. L. Tatum, January 17, 1961

Although viruses may in general contain either deoxyribonucleic acid (DNA) or ribonucleic acid (RNA), there have been no previous reports of a bacteriophage containing RNA. The single-stranded DNA of phage ϕX-174 is the most similar in structure to RNA.[1] This report describes a bacteriophage which contains RNA but no DNA as its nucleic acid.

Loeb reported the isolation of a number of phages which would grow only on *E. coli* K-12 donor strains (Hfr and F+, males).[2] The inability of the phage to grow on female bacteria is due to its failure to attach to them.[3] By serological criteria, these phage strains fell into three groups, of which there was a single representative of the first group (f1) and several of the second group (f2). The third group cross-reacts with f2 and might be deemed a subgroup of f2.[3] For a variety of technical reasons, f2 was chosen for detailed study.

Materials and Methods.—Although f2 grows on all tested *E. coli* K-12 males, its plaques were found to be clearest and to give the highest efficiency of plating on an Hfr strain (which transfers methionine as its first marker) provided by Dr. A. Garen. This strain was used for all of the experiments to be described.

The medium used for growth of bacteria and phage contained in grams per liter: Bacto-tryptone 10, yeast extract 1, glucose 1, and NaCl 8. To this basal medium was added $M/500$ CaCl$_2$. In the absence of added calcium, the phage infections are abortive, although the calcium is not needed for attachment.[3] Plaques were obtained by the agar-layer method[4] by the appropriate addition of agar to the medium described above.

One step of the phage purification utilized the technique of gradient centrifugation as described by Meselson *et al.*[5]

DNA was measured by the diphenylamine reaction[6] and the modified diphenylamine reaction as described by Burton.[7] The assay was standardized against both deoxyadenosine and a sample of calf thymus nucleic acid supplied by Dr. Muriel Roger.

RNA was measured with the orcinol reaction as standardized against ribose.[8] It was assumed that only purine bound ribose would be detected and that the purine-to-pyrimidine ratio in the samples was unity.

Phage nucleic acid was prepared as described by Gierer and Schramm.[9] Hydrolysis of phage nucleic acid and chromatography of the products were done according to the methods of Smith and Markham.[10] The sugar present was identified as described by Partridge.[11]

The synthesis of nucleic acid following infection was measured by the Schmidt-Tannhauser procedure as modified by Hershey.[12]

Experimental.—*The single cycle growth of f2:* Bacteria at a density of 2×10^8 cells per ml were infected with f2. A portion of the culture was assayed for free phage as a function of time, while another portion was lysed by the addition of cyanide to stop growth and chloroform and lysozyme to break open the bacteria. Figure 1 shows intra- and extracellular growth curves of f2 when the bacteria were infected with less than one phage particle per bacterium and diluted to prevent loss of phage by resorption. Figure 1 also shows the intracellular growth of a multiply infected culture maintained at the original density. The most pertinent point to note is the large yield per bacterium amounting to 2,000 to 4,000 in the dilute culture and to better than 9,000 in the dense culture. This latter figure has been as high as 20,000 plaque-forming units (P.F.U.) in some similar experiments, which indicates that the bacteria are probably lysis-inhibited in dense culture.

An electron micrograph of f2: Figure 2 shows an electron micrograph of f2. The phage appears to have about the same size as ϕX-174. It is not unlikely that the two phages have similar amounts of nucleic acid per particle, about $3 \times 10^{-12}\,\mu$g (see further evidence below).

The synthesis of nucleic acid following infection: Using large volumes of culture, it was possible to measure the synthesis of nucleic acid following infection. At appropriate time intervals, 10 ml samples were precipitated with trichloracetic acid

(TCA) and subjected to the Schmidt-Tannhauser (S-T) procedure as modified by Hershey to fractionate the nucleic acids.   Figure 3 shows the results of one such experiment.   We may note that there is a net synthesis of DNA and RNA amounting in both instances to about 2.5 times that originally present.   In this experiment, the number of P.F.U. synthesized per bacterium was 12,000.   Using our previous

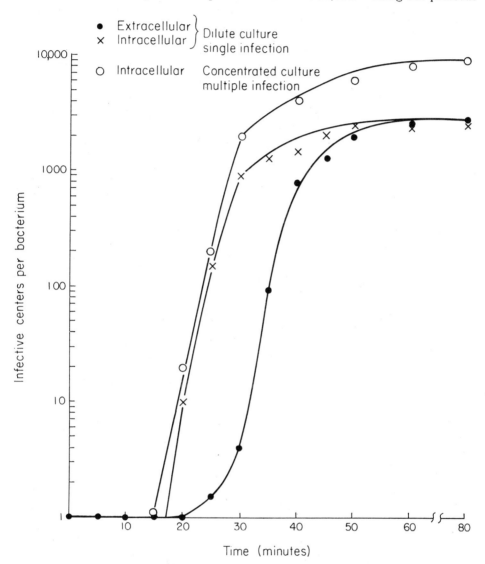

Fig. 1.—The intra- and extracellular growth of f2.

figure of 3×10^{-12} μg of nucleic acid per phage and the figure 6.0×10^{-8} μg of DNA present per bacterium, it is apparent that only if almost all of the DNA were in f2 particles could it account for the number of P.F.U.

An experiment demonstrating the nucleic acid associated with f2:   To ascertain which type of nucleic acid was associated with f2, the following experiment was

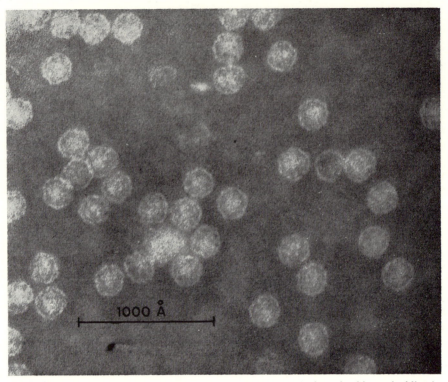

Fig. 2.—An electron micrograph of 2.   The phage were negatively stained by embedding in neutral phosphotungstate.   This micrograph was kindly taken by Dr. W. Stoeckenius.

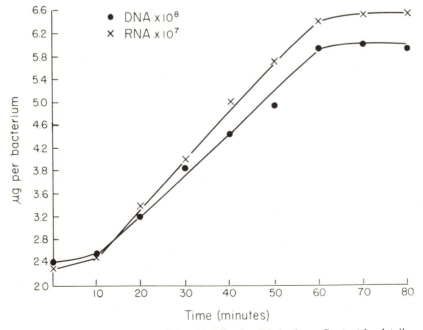

Fig. 3.—The synthesis of the nucleic acids following f2 infection.   See text for details.

done.   A culture of 1.4×10^8 bacteria per ml was multiply infected and allowed to lyse (chloroform and lysozyme were added to complete lysis).   An uninfected culture of equivalent volume with 3.4×10^8 bacteria per ml was lysed with chloroform and lysozyme.   Aliquots of each culture were immediately precipitated with TCA and fractionated by the S-T procedure.   Deoxyribonuclease (DNAase) and ribonuclease (RNAase) were added to the rest of both cultures at a concentration of 10 μg per ml each, and the cultures were incubated for one hour.   This treatment has no effect on the number of P.F.U. that are present other than occasionally raising it from its initial value.   In this instance, there were 1.2×10^{12} P.F.U. per ml.   At the end of this period, another aliquot was subjected to S-T fractionation.   Further aliquots were centrifuged in the Spinco centrifuge at 26,000 g, a force in-

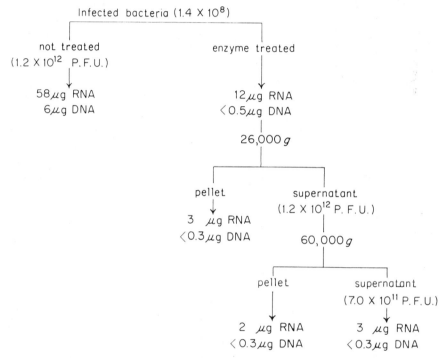

Fig. 4.—The distribution of the nucleic acids synthesized by a phage-infected culture.   See text for details.

sufficient to sediment any phage.   The pellets were dissolved in dilute ammonia and analyzed for pentose and deoxypentose.   The supernatants were centrifuged at 60,000 g for one hour.   This sedimented not quite half of the phage.   The final supernatants were precipitated with TCA.   These precipitates and pellets were analyzed for nucleic acid as above.   The data are presented in the flow diagrams (Figs. 4 and 5) and are calculated as nucleic acid equivalents per ml of the original lysates.

We may note that in both instances the DNA essentially disappears following the enzyme treatment.   The limit was set by the amount of culture used for the assay and the sensitivity of the test, which in this instance was of the order of 5 μg total DNA.   In the infected culture, about 20 per cent of the RNA survives treatment while only about 3 per cent survives in the uninfected culture, although there were initially almost equivalent amounts of RNA in both cultures.   The RNA-like material which is not lost following enzyme treatment of the uninfected culture appears in the 26,000 g pellet as does an equivalent amount in the infected culture. This may represent pentose in the bacterial debris.   RNA appears in both the 60,000 g pellet and the TCA precipitate of the 60,000 g supernatant, while there is none in the pellet and precipitate of the uninfected culture.   Phage titers are presented for the supernatants, since sedimenting the phage causes a large loss of viability, amounting to about 90 per cent as determined in many experiments. However, as will be described, there is reason to believe that the particles are still intact although they can no longer form a plaque.   The figure of 4×10^{-12} μg RNA per P.F.U. calculated from the above data is consistent with the previous estimate.

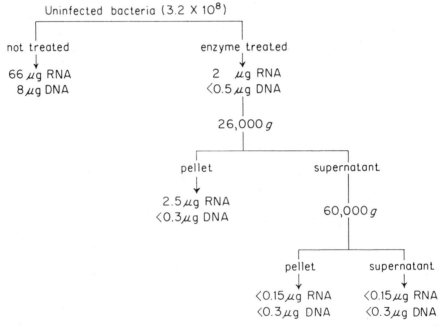

FIG. 5.—The distribution of the nucleic acids from an uninfected culture that was lysed with chloroform and lysozyme.   See text for details.

The isolation of f2 in large quantities: The following procedures were empirically developed for the isolation of large amounts of phage.   A 15-liter culture (approximately 2×10^8 bacteria per ml) was infected and allowed to go to lysis.   Chloroform and lysozyme were added to complete the lysis.   The yield was 7×10^{11} P.F.U. per ml.   The lysate was made 2 molar with ammonium sulfate and the precipitate collected by centrifugation.   This precipitate was suspended in 200 ml of water, brought to pH 8 with NaOH, and then stirred in a blendor.   DNAase and RNAase were added (each at a final concentration of 2 μg per ml).   The solution was centrifuged at 13,000 g and then at 26,000 g, giving a supernatant containing a total of 8×10^{15} phage.   The supernatant was filtered through a filter candle and the phage again precipitated with ammonium sulfate.   The major contaminants to this point of the purification were constituents of the broth medium that had also been precipitated by the ammonium sulfate. There was only a minor loss of phage viability. The precipitate was resuspended in pH 8 phosphate buffer and the solution centrifuged at 13,000 g. The supernatant was centrifuged at 60,000 g for two hours. Although the phage sedimented almost quantitatively (by the analysis described below), there was a marked loss of viable titer, about 90 per cent. The pellet was resuspended in saline, made to a density of 1.4 with CsCl, and centrifuged in the swinging bucket rotor of the Spinco Model L centrifuge at 125,000 g for 20 hours.[4] A heavy band about 2 mm in width was found in the center of the centrifuge tube and this was carefully removed. This fraction contained all of the infective material and showed no further loss of viable titer. The material was diluted in water and then recentrifuged in the Spinco at 60,000 g. A translucent pellet which was easily resuspended was obtained and was again centrifuged at 12,000 g to remove any further debris. These latter procedures resulted in no further loss of viability. The final solution (of 11 ml) had a viable titer of 5×10^{13} per ml of a total of 5.5×10^{14} P.F.U. This represents a recovery of about five per cent of viable particles.

The type of nucleic acid present in f2: The preparation described above had an optical density at 260 mμ of 55 per ml, which corresponds to about 2.8 mg of nucleic acid per ml. The 260 mμ to 280 mμ ratio was 1.78, and the 230 mμ to 260 mμ ratio was 1.05. There was found to be 2.5 mg of orcinol reacting material per ml (presumably RNA) and less than 2 μg of DNA per ml as determined by the modified diphenylamine reaction.

It is apparent that the quantity of orcinol reacting material corresponds well with the quantity of nucleic acid as determined by optical density. On the other hand, the amount of DNA, assuming the limit of the sensitivity of the analysis to be the true value, divided by the actual infectivity would give a value of about 4×10^{-14} μg of nucleic acid per particle of f2. This value would be equivalent to about 2×10^{-4} of a coliphage T2 particle (2×10^{-10} μg per particle[13]) or about 20,000 M.W.U. However, the above assumption of about 3×10^{-12} μg of nucleic acid per f2 particle would indicate that the final solution had an infectivity of less than 0.1. Thus the quantity of DNA per phage would have to be reduced a further factor of 10. The evidence certainly indicates that f2 contains RNA and no significant amount of DNA.

A value of 3×10^{-12} μg of nucleic acid per P.F.U. gives for the original lysate a value of 30 mg of nucleic acid (by viable titer) before purification. The optical density at 260 mμ indicates that virtually all of the nucleic acid was recovered after purification and that the loss of viability in the Spinco still leaves the particles intact.

The base and sugar composition of f2 nucleic acid: A sufficient quantity of f2 to provide approximately 1,400 μg equivalents of nucleic acid determined by optical density was deproteinized by shaking with aqueous phenol.[9] The resulting nucleic acid suspension was suspended in 10 μl of 1 N HCl and heated for one hour at 100°C in a sealed tube.[10] The hydrolysate and an appropriate control solution were chromatographed on Whatman #41 filter paper using a mixture of butanol and HCl. Four spots were found having the Rf and absorption spectra of guanine, adenine, cytidylic acid, and uridylic acid.   The molar ratios were 1.17 ± 0.03, 1.00, 1.21 ± 0.06, and ± 0.07, respectively.

The sugar present was determined by chromatographing the above HCl hydrolysate and a control mixture of sugars on Whatman #1 paper using a mixture of butanol, ethanol, and ammonia.[11]   The sugar had an Rf characteristic of ribose.

Summary.—The evidence presented indicates that the bacteriophage f2 contains RNA and not DNA as its nucleic acid.   The evidence is based primarily on the analysis of purified material but also on the distribution of the two types of nucleic acid synthesized after infection.   Although f2 is an extremely small phage, there is a compensating large yield per bacterium (about 10,000 P.F.U.).   Therefore, the synthesis of phage materials can be followed and the phage itself readily purified. In its general features such as adsorption and intracellular growth, f2 resembles the DNA bacteriophages.   Further studies on the biology and chemistry of f2 are in progress.

We gratefully acknowledge our indebtedness to Dr. M. Jesaitis for advice and aid in the preparation and analysis of large quantities of f2.   We also thank Miss Doris Degen, Mr. S. Cooper, and Mr. M. Estrin for technical aid.

[1] Sinsheimer, R. L., *J. Mol. Biol.*, **1**, 43 (1959).
[2] Loeb, T., *Science*, **131**, 932 (1960).
[3] Unpublished data.
[4] Adams, M. H., *Methods in Med. Research*, **2**, 1 (1950).
[5] Meselson, M., F. W. Stahl, and J. Vinograd, these PROCEEDINGS, **43**, 581 (1957).
[6] Dische, Z., *Mikrochemie*, **8**, 4 (1930).
[7] Burton, K., *Biochem. J.*, **62**, 315 (1956).
[8] Mejbaum, W. Z., *Z. physiol. Chem.*, **258**, 117 (1939).
[9] Gierer, A., and G. Schramm, *Nature*, **177**, 702 (1956).
[10] Smith, J. D., and R. Markham, *Biochem. J.*, **46**, 509 (1950).
[11] Partridge, S. M., *Biochem. J.*, **42**, 238 (1948).
[12] Hershey, A. D., *J. Gen. Physiol.*, **37**, 1 (1953).
[13] Hershey, A. D., J. Dixon, and M. Chase, *J. Gen. Physiol.*, **36**, 39 (1952).

Erratum

In line 2 of this page, ± 0.07 should read 1.13 ± 0.07.

31

Reprinted from *Cold Spring Harbor Symp. Quant. Biol.*, **33**, 101–124 (1968)

The Mechanism of RNA Replication

S. Spiegelman, N. R. Pace, D. R. Mills, R. Levisohn, T. S. Eikhom,[*] M. M. Taylor,
R. L. Peterson, and D. H. L. Bishop

Department of Microbiology, University of Illinois, Urbana, Illinois

I. INTRODUCTION

The discovery of the RNA-containing coliphage f2 by Loeb and Zinder (1961) was rapidly followed by the isolation of related phages such as MS2 (Strauss and Sinsheimer, 1963), R17 (Paranchych and Graham, 1962) and others. The armamentarium of techniques accumulated during several decades of T-phage technology thus became available to those concerned with RNA viruses. Because some of the difficulties and disadvantages inherent in the use of plant and animal systems could not be easily obviated, a number of laboratories turned their attentions to the bacterial systems.

Initial concern centered on the possibility that RNA viruses employ the 'DNA to RNA to protein' pathway of information flow. In an attempt to settle this issue, Doi and Spiegelman (1962) employed the specific DNA-RNA hybridization test (Hall and Spiegelman, 1961; Yankofsky and Spiegelman, 1962) and were unable to detect hybrids between the RNA of phage MS2 and DNA derived from infected cells. The negative outcome of this hybridization test implied that the DNA-to-RNA pathway is not employed in the life cycle of the RNA phages. Further support for this view came from the experiments of Cooper and Zinder (1962) who showed that infection of thymine-requiring *E. coli* under conditions in which DNA synthesis was suppressed to the extent of 97 % resulted in undiminished yields of virus.

It must be recognized that in a logically rigorous sense, negative evidence cannot be used to eliminate a proposed mechanism. Nevertheless, the absence of any evidence of DNA involvement was generally accepted to imply that RNA bacteriophage had evolved a mechanism of generating RNA copies from RNA templates. The next obvious step was to identify and then isolate the new type of RNA-replicating enzyme system predicted by this line of reasoning.

A. The Search for an RNA Replicase

The search for an RNA-dependent RNA polymerase (replicase) unique to cells infected with an

RNA virus is complicated by the presence in the host cell of a variety of enzymes capable of mediating RNA synthesis. It was clear from the onset that a claim for a new type of polymerase would ultimately have to be supported by a demonstration that the enzyme possesses some unique characteristic which differentiates it from known polymerases.

In addition to enzymological difficulties, we recognized a potential source of another complication inherent in the fact that an RNA virus must always operate in a heterogenetic environment replete with strange RNA molecules. It seemed possible that the viral replicase had evolved some means of recognizing its genome and ignoring all host-specific RNA molecules. As a practical consequence, this meant that we could only use intact Qβ RNA as the challenging template at all stages of enzyme purification.

Despite these potential obstacles, our first success was achieved in 1963 with *E. coli* infected by the phage MS2 (Haruna et al., 1963). A procedure involving protamine fractionation combined with column chromatography yielded what seemed to be the relevant enzyme. Most important of all, the preparation exhibited a virtually complete dependence on added RNA, permitting a test of the expectation of specific template requirement. The response of MS2-replicase to various kinds of nucleic acids revealed a striking preference for its own RNA. No significant activity was observed with tRNA or ribosomal RNA of the host cell. Our intuitive guess was apparently confirmed. By producing a polymerase which ignores cellular RNA components, a guarantee is provided that replication is focused on the single strand of parental viral RNA, the ultimate origin of viral progeny.

B. Confirmation of Specific Template Requirement with Qβ Replicase

The announcement of the specific template requirement of the MS2 replicase was greeted with what may best be described as "well controlled enthusiasm." The acceptance of template specificity clearly required further evidence. It seemed important, therefore, to examine another virus unrelated to the MS2 group. The Qβ phage of

* Present address: Dept. Biochemistry, University of Bergen, Bergen, Norway.

Watanabe (1964) was chosen because of its serological and other chemical differences (Overby et al., 1966a,b).

The isolation and purification of the $Q\beta$ replicase (Haruna and Spiegelman, 1965a) followed, with slight modifications, the procedures devised earlier for the MS2 replicase. The general properties of the $Q\beta$-replicase were similar to those observed with MS2 replicase, including requirement for all four triphosphates and Mg^{++}. Further, the response of the $Q\beta$ replicase to various RNA primers was in accord with that reported for the MS2 replicase, the preference being clearly for its own template. The heterologous viral RNA's, MS2, and STNV (Satellite Tobacco Necrosis Virus) were completely inactive as were the ribosomal and transfer RNA species of the host cell. Furthermore, it was found (Haruna and Spiegelman, 1965b) that fragments of $Q\beta$ viral RNA were unable to stimulate the replicase to significant activity.

The inability of the replicase to employ fragments of its own genome as templates argues against a recognition mechanism involving only one beginning sequence. The enzyme can apparently sense when it is confronted with an intact RNA molecule, implying that another element of structure is involved. A plausible formal explanation can be proposed in terms of 'functional circularity.' Thus, a decision by the enzyme on the intactness of a linear heteropolymer could be readily made if it examined both ends for the proper sequences. An examination of this sort would be physically aided by forming a circle using terminal sequences of overlapping complementarity. The enzyme could then recognize the resulting double-stranded region. This 'amphora' model was offered only as an example of how an enzyme could simultaneously distinguish both sequence and intactness.

II. NATURE OF PRODUCT SYNTHESIZED

By 1965, the purification of $Q\beta$ replicase had been brought to a stage at which it had been largely freed of contaminating nucleases and host polymerase activities. It could mediate virtually unlimited synthesis of RNA over long periods of time. It seemed desirable to begin a study of the product of extensive synthesis. It was quickly established that the synthetic material was indistinguishable from viral RNA in both size and base composition. Furthermore, the $Q\beta$ replicase was shown to recognize equally well both natural viral RNA and synthetic RNA as templates (Haruna and Spiegelman, 1965c).

The next question concerns the extent of similarity between product and template. *Have identical duplicates been in fact produced?* The most

decisive test would be to determine whether the product contains all the information required to program the synthesis of complete virus particles in vivo. The first simple observation that infectious units accumulate during the enzyme reaction at the same rate as total RNA synthesis was not considered sufficiently conclusive, since the possibility could not be eliminated that the agreement observed was fortuitous. One could argue, however implausibly, that the enzyme was 'activating' the *input* RNA to higher levels of infectivity while synthesizing new, noninfectious RNA.

The issues raised by such arguments were answered by a serial transfer experiment (Spiegelman et al., 1965). The product of a standard replicase reaction was used to initiate synthesis in a second reaction which, in turn, after a suitable synthetic period, was used to initiate a third reaction, and so on, until the template RNA of the first tube was diluted to an insignificant level. Aside from controls, 15 transfers were involved, each resulting in a 1 to 6 dilution. By the eighth tube there was less than one infectious unit ascribable to the initiating RNA and the 15th tube contained less than one strand of the initial input. *Nevertheless, every tube showed an increment in infectious units corresponding to the radioactive RNA found.* It was clear, therefore, that the $Q\beta$ replicase generates biologically competent viral RNA.

III. A RIGOROUS PROOF THAT THE ADDED RNA IS THE SELF-DUPLICATING ENTITY

None of the experiments thus far described *proved* that the RNA synthesized in this system is, in fact, the self-duplicating entity. Minute quantities of contaminating RNA present in the enzyme preparation or even the replicase itself could be the instructive agent. What was required was a rigorous demonstration that the RNA, and not the replicase preparation, was the progenitor. A definitive decision would be provided by an experimental answer to the following question, "If the replicase is challenged with an altered RNA molecule, is the product produced always identical to the initiating template?" A positive outcome would establish that the RNA template is directing its own synthesis and would simultaneously completely eliminate any remaining possibility of 'activation' of pre-existing RNA.

The discriminating selectivity of the replicase for its own genome as template makes it impossible to employ heterologous RNA in the test experiments and recourse was had, therefore, to mutants. For ease in isolation and simplicity in distinguishing between mutant and wild type, temperature sensitive (ts) mutants were chosen. Their diagnostic phenotype is poor growth at 41°C as compared with 34°C. The wild type grows equally well at both temperatures.

We should, therefore, be able to determine whether the product produced by a normal replicase primed with ts-Qβ-RNA is mutant or wild type. As in previous investigations, this was done by a serial transfer experiment to avoid the ambiguity of examining reactions containing significant quantities of the original initiating RNA.

Accordingly (Pace and Spiegelman, 1966a) the first reaction tube of a series containing normal Qβ replicase was initiated with ts-Qβ-RNA, and the product transferred through the series as described previously. Figure 1 summarizes the outcome of the experiment in a cumulative plot of the RNA synthesized and the plaque formers at the two test temperatures of 34°C and 41°C. It is clear that the RNA synthesized has the ts-phenotype; plaque formation at 34°C increases in parallel with the new RNA synthesized; no such increase is seen when the tests are carried out at 41°C. It should be noted from the upper panel of Fig. 1 that no significant synthesis of either RNA or infectious units was observed in a control series of tubes to which initiating templates were not added.

The experiments just described demonstrate that normal replicase can produce distinguishably different though genetically related RNA molecules. The genetic type produced is completely determined by the RNA used to start the reaction and is always identical to it. The following two conclusions would appear to be inescapable from these findings: (1) the RNA *is the instructive agent* in the replicating process and, therefore, satisfies the operational definition of a self-duplicating entity; (2) it is not some cryptic contaminant of the enzyme, but rather the input RNA that multiplies.

IV. PROPERTIES OF THE Qβ REPLICASE

Our initial attempts (Haruna and Spiegelman, 1965a) at purifying Qβ replicase were concerned with removal of ribonuclease and the DNA-dependent-RNA-polymerase (transcriptase). Subsequently, our interest centered on removal of the residue of virus particles. Chemically these were trivial contaminants; however, they prevented simple direct assays for infectivity of newly synthesized RNA. A purification scheme (Pace and Spiegelman, 1966b) was designed to bring the virus particle content of enzyme preparations to insignificant levels by taking advantage of expected disparities in size and density between the replicase and the RNA-containing virus particles. The

enzyme protein was banded in CsCl gradients, and residual contamination was lowered to acceptable levels by subsequent sedimentation through sucrose gradients. Replicase purified in this manner retained the ability to generate infectious viral RNA from Qβ template RNA.

A. IDENTIFICATION OF TWO PROTEIN COMPONENTS OF Qβ REPLICASE

At each purification stage replicase activity appeared to act as a single entity. However, we observed some persistent peculiarities in the properties of the enzyme, suggesting the possible existence of two components. We may list some of these hints as follows: (1) during fractionation on DEAE it was noted that exhaustive washing with 0.12 M NaCl removed a component necessary for activity; (2) the enzyme was quite sensitive to dilution and required high ionic strength for stability; (3) it was routinely observed that pooling the active fractions from a sucrose gradient

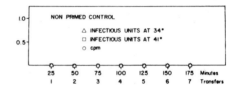

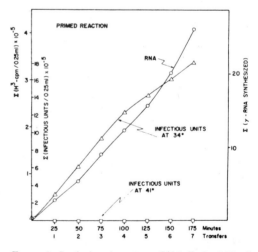

FIGURE 1. *Synthesis of mutant RNA.* Each 0.25 ml reaction contained 60 μg of Qβ replicase purified by CsCl and sucrose centrifugation. The first reaction was initiated by addition of 0.2 μg of ts-RNA. Each reaction was incubated at 35°C for 25 min, whereupon 0.02 ml was withdrawn for counting and 0.025 ml used to prime the next reaction. All samples were stored frozen at −70°C until infectivity assays were performed at both 41°C and 34°C. Control series were incubated with no initiating RNA added (Pace and Speigelman, 1966b).

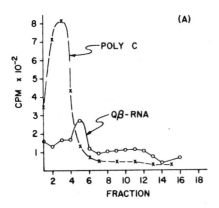

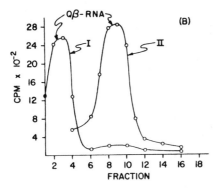

FIGURE 2. *Extensive centrifugation of Qβ replicase in sucrose gradients.* Replicase fractions after sucrose purification were pooled, concentrated, dialyzed, and rerun on 5–20% sucrose gradients in SW-39 swinging buckets; centrifugation time was 36 hr at 39,000 rpm. Nineteen fractions were collected.

(A) Assays were performed using 20 μl enzyme and 1 μg Qβ-RNA with ^{32}P-UTP or 10 μg poly C with ^{32}P-GTP to monitor synthesis.

(B) Curve II: Tubes 1–3 were pooled. 20 μl aliquots of this combined pool were added to 20 μl samples of each fraction. The combinations were assayed with 1 μg Qβ RNA as template.

(B) Curve I: Tubes 8–10 were pooled and 20 μl aliquots of this pool added to 20 μl samples of each fraction. The combinations were then assayed with 1 μg Qβ RNA as template (Eikhom and Spiegelman, 1967).

resulted in preparations with activities *exceeding* the sum of the individual fractions; (4) the enzyme band was much narrower in both CsCl and sucrose gradients than would be expected from its estimated molecular weight; (5) finally, it was noted that at all stages of purification, even after complete removal of transcriptase, the replicase contained a poly-C-dependent poly-G-polymerase activity.

All of these peculiarities could be readily explained if the DEAE and sucrose steps of the

purification scheme did, in fact, partially separate two components of the Qβ replicase. If so, this would imply that neither component could carry out the reaction in isolation and that one of them possesses the poly-G-polymerase activity.

It was decided to explore these possibilities. Our previous experience indicated that sucrose gradients should be effective in achieving a separation. This, in fact, turned out to be the case (Eikhom and Spiegelman, 1967). Uninfected cells were found to contain only a poly-G-polymerase activity *associated with the transcriptase,* whereas infected cells contained *in addition* a more slowly sedimenting poly-G-polymerase activity, sedimenting slightly more rapidly than the replicase activity.

If the poly-G-polymerase of the infected cells was, in fact, one of the necessary components of Qβ replicase, then a more complete separation should lead to the virtual disappearance of the response to Qβ RNA without loss of the poly-G-polymerase activity. Figure 2A shows that this occurs. Extended and repeated sedimentation in sucrose gradients led to the loss of Qβ-dependent activity, although the poly-G-polymerase activity was retained. This finding is consistent with the notion that the Qβ replicase has been separated into two components, one of which can carry out the poly-G polymerization. If the hypothesis is true, the other component was likely to be found in the upper part of the gradient. Further, the presence of this lighter component should be revealed by adding aliquots *from the poly-G-polymerase region* to the other fractions of the gradient and challenging the mixture with Qβ RNA. Profile II of Fig. 2B shows that active Qβ replicase can, in fact, be reconstituted in this manner.

To confirm that the restoration of activity in the light region was indeed due to reconstitution of the enzyme from two necessary components, the reverse experiment (i.e., adding pooled light fraction to the other gradient fractions) should also yield replicase activity and thus locate the 'heavy' component of the replicase. Profile I of Fig. 2B shows that this was obtained, and that the novel poly-G-polymerase was coincident in the sucrose gradient with the heavy fraction of the *replicase* activity. From their relative positions in the gradient, it may be estimated that the 'heavy' component is approximately 130,000 in molecular weight and the 'light' about 80,000.

B. EVIDENCE THAT THE 'LIGHT' COMPONENT IS A HOST PROTEIN

The size of each of these components raises an interesting issue related to the information available in the viral genome. The RNA of Qβ has been shown to be about 1×10^6 daltons and can, therefore, code for about 1×10^5 dalton equivalents of protein. Of this, 0.47×10^5 is accounted for by the viral coat and A proteins. This leaves only 0.53×10^5 equivalents to be shared between the two replicase units which add up to almost four times (2.1×10^5 daltons) that quantity.

Logically, this situation could be explained if both components of the active replicase were constructed of several identical subunits, all defined by the host genome. A second and perhaps more likely way to ease the apparent pressure for genetic information is to suggest that one of the replicase components is a host-specified protein.

The previous experiments demonstrated that the poly-G-polymerase, apparently associated with the 'heavy' replicase component, is in fact *induced* in the infected cells. Accordingly, equivalent 'replicase preparations' from uninfected cells were examined (Eikhom et al., 1968) for 'light' content on sucrose gradients by assaying for $Q\beta$ replicase activity in the presence of added, authentic 'heavy' fraction purified from infected cells. Light component activity was readily identified in the expected position of the sucrose gradient and possessed an activity comparable to similar preparations derived from infected cells. Further, the 'light' component was found not only in uninfected *E. coli* Q13, but also in cells of *E. coli* B, which cannot be infected by $Q\beta$ or any of the other known RNA phages. Most important, the 'light' components from uninfected cells can catalyze the synthesis of infectious viral RNA in the presence of the 'heavy', viral-specific component from infected cells and $Q\beta$ RNA template.

In addition to repeated sucrose gradient runs, agarose gel filtration has been utilized (Wong, Leichtling, Stockley, and Spiegelman, in prep.) to separate the two components of replicase activity on a preparative scale. Resolution by Biorad A 0.5 m agarose filtration allows the preparation of milligram quantities of the two components, and has the added advantage of separating RNase II and III from the replicase fractions.

The precise role played by each component in the replicase reaction remains to be delineated. The most obvious possibility that one mediates the first and the other the second stage in a two-step reaction is not supported by available data. Both must be present simultaneously for the initiation of the reaction stimulated by $Q\beta$ RNA.

V. ANALYSIS OF EARLY INTERMEDIATES

When it became certain that infectious RNA was being generated we felt that the system had been brought to a stage where one could, without ambiguity, undertake to analyze the mechanism of RNA replication. Since the reaction starts with viral RNA (plus strands) and ends up with more of the same, every necessary intervening stage must be represented in the reaction.

Investigations prior to 1965 had led to the postulation of double-stranded intermediates involving the plus templates and their complements (minus strands). This possibility was implied by experiments with mutants (Lodish and Zinder, 1966) and suggested by the presence in virus-infected cells of structures partially resistant to ribonuclease (Montagnier and Sanders, 1963; Ammann et al., 1964; Weissmann et al., 1964; Kelly and Sinsheimer, 1964; Nonoyama and Ikeda, 1964; Fenwick et al., 1964).

The terms 'replicative form' (RF) and 'replicative intermediate' (RI) have been commonly used to designate the two distinguishable RNase resistant complexes. We preferred a more neutral terminology and suggested the use of 'Hofschneider Structures' (HS) to designate the material isolated by the procedure of Hofschneider and his colleagues (Ammann et al., 1964; Francke and Hofschneider, 1966a,b) and "Franklin structures" (FS) to denote the physically different complexes isolated by Franklin (Franklin, 1966). We felt that this terminology would not prejudge the nature and function of the complexes while identifying them with the investigators who operationally defined their existence. In view of the doubts still existing (see below, Weissmann and colleagues, this volume) on whether these complexes, as isolated, are the actual intermediates, it seems desirable to retain this terminology.

Our initial search (Haruna and Spiegelman, 1966) during the early stages of synthesis for evidence of duplex intermediates failed to provide unambiguous support for their intermediary function during *in vitro* RNA replication. We cautioned against accepting as established that the resistant structures were intermediates but noted that negative evidence could not logically eliminate a proposed mechanism.

A. SEDIMENTATION ANALYSIS OF EARLY INTERMEDIATES

Certain technical advances permitted a more certain and detailed examination of the early events occurring in reaction mixtures. Pycnographic and sucrose density purification of the replicase (Pace and Spiegelman, 1966a) made it possible to do direct infectivity assays of the reaction mixture, thus permitting a detailed analysis of any interval of synthesis with regard to the genetic integrity of replicase and product. In addition, we had developed a simple treatment

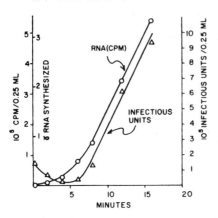

FIGURE 3. In vitro *synthesis of labeled and infectious RNA*. A 2.0 ml standard reaction mixture contained in *μ*moles: Tris-HCl, pH 7.4, 160; MgCl$_2$, 24; ATP, CTP, GTP, UTP, 1.6 each, UTP-α-^{32}P 4.05 × 10^7 count/min per 0.2 *μ*M UTP; 1.6 *μ*g Q*β* ^{3}H-RNA; 400 *μ*g enzyme protein and incubated at 35°C. Zero time was obtained by the addition of template to the prewarmed mixture. The left-hand ordinate gives the ^{32}P count/min in TCA-precipitable product in each 0.25 ml. The right-hand ordinate gives the number of infectious units per 0.25 ml observed in the spheroplast assay (Mills et al., 1966).

of the reaction mixture which avoids phenol purification and permits direct examination in sucrose gradients of templates and products with complete recovery of both.

With the aid of these technical advances, we demonstrated (Mills et al., 1966) the existence of a latent period in the reaction prior to the appearance of new infectious RNA. A search during this latent period should maximize the chances of finding replicative complexes if they are mandatory intermediates. Further, at least some of these complexes should be noninfectious and release infectious plus strands on heat denaturation as reported by Hofschneider and his colleagues (Ammann et al., 1964; Francke and Hofschneider, 1966a,b). Conversion of templates into Hofschneider structures (HS) should result in their disappearance as plaque forming units (pfu). Thus, if the time required to complete the first plus strands is significant in the time scale of the experiment, one should observe a latent period accompanied by an *apparent eclipse* of the initiating infectious RNA.

Figure 3 describes data which show that these expectations were realized. It is evident that a considerable loss (∼75%) of pfu is observed by the fourth minute of synthesis providing clear evidence of the eclipse and latent periods in terms of the pfu assay. Both end at about six minutes, which appears

to correspond to the time required to complete the first new mature strands.

If the disappearance of pfu is associated with the formation of HS RNA, a peak containing ^{3}H template and ^{32}P-product should appear in the 15 S region of a sucrose gradient. Since, with time, the loss of pfu is extensive, the shift of ^{3}H to the 15 S region should be considerable by 2 min but negligible at 15 sec. Further, this peak should yield infectious material only after heat denaturation. Accordingly, aliquots of the samples taken in the experiment of Fig. 3 at 15 sec, 2, 4, and 6 min were subjected to analysis in sucrose gradients. The number of pfu found at 15 sec agreed with the initial input of RNA and the amount of ^{32}P incorporated was negligible. The two-minute sample (Fig. 4) shows a dramatically different picture. A rather large proportion (62%)

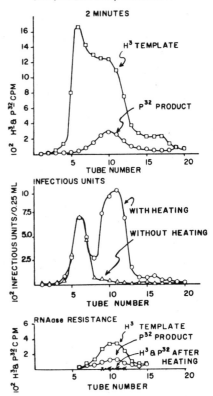

FIGURE 4. *Sedimentation analysis of the two-minute sample*. 0.25 ml of reaction mixture (Fig. 3) was resolved by a sucrose gradient centrifugation. 0.25 ml fractions were collected and diluted fivefold, and aliquots (heated or unheated) were taken for determining the incorporation of ^{32}P and ^{3}H, for assay of infectious units, and for ribonuclease digestion (Mills et al., 1966).

of the ³H-template has been shifted from the 28 S to the 15 S region, a movement which is accompanied by the appearance of ³²P-product which is virtually confined to the 15 S region. Assays for infectious RNA *without* prior heating yield pfu *only* in the region of mature plus strands (28 S). However, heat denaturation uncovers a large peak of activatable pfu at 15 S. Finally, it is evident that both the tritiated template and the ³²P-product in the 15 S region show resistance (about 50%) to ribonuclease. This resistance disappears completely when the samples are heated for 2 min at 100°C in 0.003 M EDTA.

The 4- and 6-min samples exhibited essentially the same sedimentation profile, except that some 28 S ³²P product had clearly been produced. As at 2 min, both ³H-template and ³²P-product, as well as heat-activatable plaque-forming units, were resistant in the 15 S region of the gradients.

The striking fact to emerge from the data described is that a set of experimental conditions had been devised which permits the in vitro demonstration of a structure found in infected cells. It has all the characteristics of HS RNA including the key one of yielding infectious plus strands on denaturation.

It is clear from these results that HS has the following relations to the evolution of the synthetic reaction. It contains both initiating ³H-template and newly synthesized ³²P-product. Further, the appearance of HS as a component in the latent period is accompanied by loss of pfu on direct assay. Finally, it is synthesized before any new 28 S product is made and prior to emergence from the latent period. While these are all *necessary* characteristics of a replicative intermediate, *they are not logically sufficient to establish that HS is playing this role.*

B. ELECTROPHORETIC ANALYSIS OF THE REPLICATION REACTION

It became apparent as our investigations progressed that centrifugal analysis in sucrose gradients did not possess the resolving power necessary to identify all the components of the replicase reaction and unravel the temporal relations obtaining among them. It was particularly difficult to separate the Qβ-HS RNA from the multistranded structures (FS) identified by Franklin (1966) in cells infected with R17.

We turned our attention, therefore, to the use of electrophoretic separation of RNA on polyacrylamide gels (Leoning, 1967) and found (Bishop et al., 1967a) that separation of different species of single-stranded RNA could be achieved proportional to their molecular weight differences. Using purified HS and FS RNA from infected cells

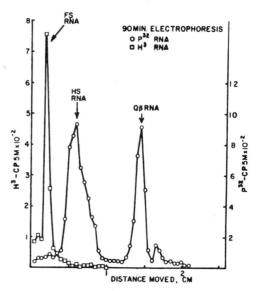

FIGURE 5. *Separation of Qβ-FS, Qβ-HS, and Qβ-RNA.* ³H-labeled Franklin structure RNA (Qβ-FS) and ³²P-labeled Hofschneider structure RNA (Qβ-HS) were mixed with ³²P-labeled Qβ viral RNA and subjected to electrophoresis on a polyacrylamide gel (0.7 cm diameter) for 90 min under standard conditions (Bishop et al., 1967a). The gels were sliced, dried, and counted (Pace et al., 1967a).

it was possible to demonstrate (Fig. 5) that each species could be separated from the other and from single-stranded Qβ RNA as well as from small molecular weight (SMW) materials. It was immediately evident that electrophoresis through polyacrylamide gels possessed evident advantages over the sucrose gradient procedure (Pace et al., 1967a). Further, quantitative estimation of the amounts of each component (FS, HS, Qβ, and SMW RNA) present in the reaction at any given time could be obtained by summation of the relevant gel regions. A kinetic description of the replication process in terms of the sequential appearance of each component thus became possible.

1. *A kinetic analysis of the appearance of intermediates.* Since true replicative intermediates must involve input template molecules, we first undertook a detailed analysis of the fate of the template during the early stages of RNA replication (Bishop et al., 1967b). A reaction was initiated with ³²P-Qβ RNA and aliquots were removed at intervals for resolution by polyacrylamide gel electrophoresis. Three representative profiles are shown in Fig. 6.

The course of events is apparent. With time, the template moves first into the HS region and subsequently into the FS region of the gels. Both of these

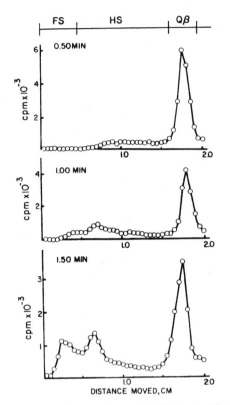

FIGURE 6. *Involvement of template RNA.* A 0.75-ml reaction mixture containing unlabeled triphosphates was incubated at 38°C with 0.9 μg ^{32}P-labeled Qβ-RNA. Aliquots were removed at the indicated time intervals, mixed with sucrose (10% final concentration) and SDS (0.2% final concentration), and subjected to polyacrylamide electrophoresis (0.9 cm diameter gels) under standard conditions. The gels were frozen, sliced, dried, and counted to determine the distribution of radioactivity. The regions of gel in which Qβ-FS, Qβ-HS, and Qβ are situated are indicated at the top of each panel (Bishop et al., 1967b).

events occur prior to the appearance of new infectious units in the reaction (see Fig. 3).

Summation of radioactive template RNA found in each region of the gels provided a complete kinetic picture of the progress of the reaction. It was clear that 70% of the initiating template molecules were involved in complex formation. They were first converted into HS, the RNA duplex structure, followed by their appearance in FS, the multi-stranded complex.

A similar reaction was carried out with ^{3}H-Qβ-RNA as template, and α-^{32}P-UTP marking products. At intervals, samples were withdrawn and submitted to gel electrophoresis. As before, summation of ^{32}P-radioactivity in products within

each region of the gel permits a quantitative representation of the course of events. Figure 7 graphically summarizes the relative accumulation of ^{32}P-product in the various components as the replicase reaction proceeds (Pace et al., 1967b).

As with the involvement of template molecules, ^{32}P-product RNA appears first in the HS region of the gels, followed by its accumulation in FS. New, mature Qβ RNA appears after the two duplex structures accumulate. It should be noted that the RNA in the SMW region begins to increase simultaneously with the production of mature RNA molecules. The fragments do not, therefore, play a role in the events occurring in the latent period.

The data described provide information about the replicative process. It is clear that the sequence of events which emerges is the same whether attention is focused on template involvement or product synthesized. The temporal order of their appearance and the fact that the duplexes contain both initiating template and newly synthesized product are consistent with a mechanism of synthesis which involves the following sequence of steps:

$$Q\beta\text{-Template} \rightarrow HS \rightarrow FS \rightarrow Q\beta \text{ Product.}$$

2. *The immediate physical precursor of Qβ RNA.* The scheme just outlined implies that the FS structures are those which directly generate new Qβ RNA strands. Suitably designed "pulse-chase" experiments could test this prediction.

To be maximally informative, the 'pulse chase'

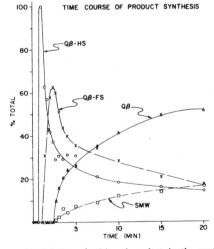

TIME COURSE OF PRODUCT SYNTHESIS

FIGURE 7. *Relative quantities of product in the reaction.* A standard reaction mixture containing α-^{32}P-UTP was initiated at 35°C with Qβ-RNA. Aliquots were removed at intervals, and RNA species were resolved by gel electrophoresis. The data for each species are considered as a percentage of the total ^{32}P-product recovered from the respective gels (Pace et al., 1967b).

experiment had to be carried out during a period in which mature plus strands were being generated. A latent period of 3–5 min precedes the appearance of the first newly synthesized plus strands. The pulse with α-^{32}P-UTP was, therefore, started at 10 min and the chase with nonradioactive UTP initiated one minute later. After the chase, samples were removed at intervals and submitted to electrophoretic separation in polyacrylamide gels. Summations were made of ^{32}P-radioactivity in each of the gel regions, and the results are shown in Fig. 8 (Pace et al., 1968).

It is evident that nearly 70% of the RNA synthesized between 10 and 11 min is located in the FS region after 15 sec of chase. Virtually all of the ^{32}P-RNA is subsequently displaced from the FS region to appear in the region characteristic of mature Qβ RNA. It therefore seems clear that the most rapidly labeled species are the FS complexes and that they are the immediate physical precursors of mature Qβ RNA. The small amount of ^{32}P which enters the HS complexes during the pulse does not leave during the chase, suggesting that a portion of the HS complexes which accumulate in the reaction do not participate in the complete replication process.

The kinetics of their appearance and the results of the pulse chase experiment are all *consistent* with a mechanism invoking the HS and FS structures as mandatory intermediates in the replication of RNA. The data do not, however, *establish* the mechanism and there are reasons to doubt its validity in its simplest form. In the first place, none of the duplex intermediates, HS or FS, have been successfully employed as initiators of synthesis (Weissmann et al., 1967; Mills, Pace, and Spiegelman, 1966). Secondly, as pointedly emphasized by Weissmann and his colleagues (Borst and Weissmann, 1965; Feix et al., 1967) it was, and still is necessary to entertain the possibility that the HS and FS structures *as isolated* do not correspond to the replicative intermediates functional *in situ*. The manipulations required for their isolation may well convert them into the more or less complete duplexes they appear to be in the purified state. If, as seems likely, they are direct derivatives of the true replicative intermediates, the kinetics of their appearance and labeling properties still provide us with information relevant to the replicative process. However, a detailed molecular understanding of the synthesis will require a knowledge of the nascent operative structures themselves.

Finally, there are the complications, to be discussed below, introduced by the early generation of free negative strands and the fact that negative strands are actually superior to plus strands as templates; features which were first reported by

Weissmann and his collaborators (Weissmann et al., 1967; Feix et al., 1968). As will be seen, we were led to very similar observations by a somewhat different route and have confirmed them by electrophoretic analyses of the reaction (Pace, Taylor, Eikhom, and Spiegelman, 1968, in prep.). We now consider in further detail these findings and their implications for the mechanism of replication.

3. *The accumulation of free negative strands and its implications.* Two particularly puzzling features emerged from our detailed electrophoretic and biological characterization of the early products. First, although 28 S product RNA begins to appear in the Qβ region of gels by 2.5 min, new infectious RNA (i.e., plus strands) does not arise until 5 min have elapsed. Secondly, the first FS complexes

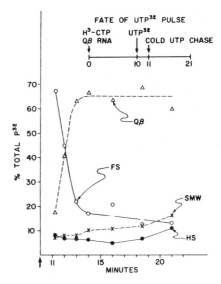

FIGURE 8. *Kinetic fate of UT^{32}P pulse.* A 1.25 ml standard reaction mixture contained 15 μmoles MgCl$_2$, 100 μmoles Tris·HCl (pH 7.4), 1.0 μmole each ATP, GTP, ^{3}H-CTP (1 mc/μmole), 0.2 μmole UTP, 600 μg replicase protein, and 3.75 μmoles Mg·EDTA. The reaction was initiated at 38°C by addition of 15 μg Qβ·RNA, and allowed to incubate for 10 min, at which time a 0.25 ml aliquot was removed and added to a second tube containing 0.02 μmole UTP-α-^{32}P (0.7 mc/μmole). To the remaining 1.0 ml of reaction mixture were added additional cold UTP (4.4 μmoles) and MgCl$_2$ (3.5 μmoles). The pulse-labeling period was continued for 1 min and then chased by adding the reaction mixture containing excess cold UTP to the 0.25 ml aliquot containing UTP-α-^{32}P. The final content of the unlabeled UTP equaled at 75-fold chemical dilution of the radioactive UTP. After the addition, the reaction at 38°C was continued. At intervals after the chase, 0.15-ml aliquots were adjusted to contain 0.2% SDS, and submitted to electrophoresis through polyacrylamide gels. Counts were summed over respective regions of the gel profiles, expressed as a fraction of total amounts recovered, and plotted as a function of reaction time (Pace et al., 1968).

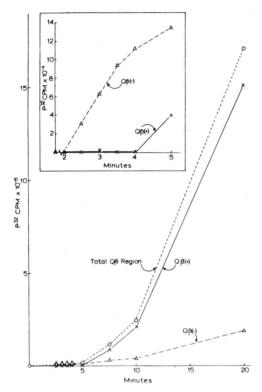

FIGURE 9. *Accumulation of single-strand Qβ-RNA in the replicase reaction.* A standard reaction mixture containing α-^{32}P-UTP was incubated at 35°C with Qβ-RNA. Aliquots were removed at intervals, and RNA species resolved by gel electrophoresis. Gel slices were eluted, 0.3 ml aliquots from the Qβ-RNA regions of gels adjusted to 0.4 M NaCl, 4.5 μg of self-annealed Qβ-RNA added to give a final volume of 1 ml, and the mixture annealed at 70°C for 1 hr. After the incubation, 1 ml of a solution of 0.003 M EDTA, RNase A 20 μg/ml and RNase T₁ 10 μg/ml were added, and the samples digested at 35°C for 1 hr. The RNase core always found with single-stranded viral RNA (usually 1%) was subtracted from the resistance observed prior to calculation of the per cent Qβ(−) content. The relative amounts of Qβ(−) or Qβ(+) are expressed as the proportion of the total counts recovered in the Qβ region of the gel.

observed appeared to contain no *completed* product 28 S RNA. Since, as will be seen below, the principal direction of synthesis of both Qβ(+) and Qβ(−) strands proceeds from the 5′ toward the 3′ terminus, it would be predicted that the Qβ(−) in the duplex must be completed *before* the new Qβ(+) strand can be initiated.

These observations suggested the possibility that the first FS to appear is actively generating free Qβ(−) strands, and that the earliest 28 S RNA to appear is, in fact, free single-strand Qβ(−) RNA.

To examine this possibility, a reaction was sampled at early times, and the resolved products

were isolated from gels by elution. The respective reaction components were then examined for single-strand Qβ(−) content by their conversion to ribonuclease resistance after annealing to purified Qβ(+) RNA.

The accumulation of single-strand Qβ(−) in the 28 S RNA region of the gels is considered in Fig. 9. As is evident, Qβ(+) strands are the major products of the replicase reaction, although Qβ(−) strands continue to accumulate slowly. However, during the first few minutes of the reaction (inset to Fig. 9), the single strand product is exclusively Qβ(−). Qβ(+) accumulation does not begin until two minutes after the onset of Qβ(−) appearance, at about the same time that new infectious RNA production begins.

Since Qβ(−) strands accumulate *before* the appearance of Qβ(+) product, they must be considered as potential templates for the subsequent production of Qβ(+) RNA, especially since Qβ(−) strands serve as excellent templates for the replicase (Feix et al., 1968).

VI. A FIVE-INTERMEDIATE MODEL OF RNA REPLICATION

The generation of single-strand Qβ(−) RNA and its possible subsequent use as template in generation of Qβ(+) strands, requires the identification of five intermediates intervening between the initial plus template and the final infectious product. Four of these would be duplexes containing either intact plus or minus strands and the fifth would be free negatives. These classes and the possible course of events are illustrated in Fig. 10.

Thus, Qβ(+) RNA serves as template for construction of a Qβ(−) strand, forming the duplex

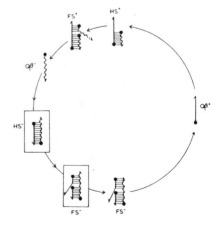

FIGURE 10. *A possible replicating scheme of Qβ-RNA.* For details see text.

HS(+); here the intact (+) serves as the template. Even *before* the completion of the first negative complement, HS(+) could act as template in generation of another Qβ(−) strand, forming FS(+) the multistranded structures involved in generation of complete, free Qβ(−) strands. Free Qβ(−) strands would then direct the synthesis of Qβ(+), forming the duplex HS(−), which upon further Qβ(+) strand initiation would be converted into FS(−) the structure actively generating infectious viral RNA.

It must be emphasized that the mechanism schematized in Fig. 10 is offered only as a working model to aid in the design of future experiments. In addition to negative strands preceding new positive strands, it makes the following specific predictions open to experimental confirmation or denial; (1) the first FS structures formed will be of the FS(+) kind, identifiable by RNase sensitive 5′ tails of the nascent *negatives*; (2) the second FS structures to appear will be of the FS(−) type, identifiable by RNase sensitive 5′ tails of nascent *positives*.

It is evident that similar predictions with respect to RNase sensitivity of the 5′ ends of the template in HS(+) and HS(−) complexes can be made. Some of these features have been observed but more experiments will be required before this scheme can be accepted.

VII. DIRECTION OF RNA SYNTHESIS

It is evident that a complete understanding of the chemistry of replication requires that we know the direction in which minus and plus strands are synthesized. To provide a basis for understanding the nature of the experiments needed for a decision, we compare in Fig. 11 (first stage) the consequence of a 5′ to 3′ with that of 3′ to 5′ formation of a minus strand on a hypothetical plus template.

Two distinguishing features stand out; (1) if product synthesis is in a 5′ to 3′ direction, the product RNA would start with the sequence pppGpGpApCpGp . . . etc., complementary and antiparallel to the 3′ *sequence* of the template RNA. However, if the direction is 3′ to 5′, then the product sequence would start with CpCpUpGpAp . . . etc., complementary and antiparallel to the 5′ *sequence* of the template RNA; (2) in the instance of a 5′ to 3′ mode, only one triphosphate (i.e., pppG) would be found in the product RNA whether the strand be complete or incomplete. However, with a 3′ to 5′ mode, *all four triphosphates* would be recovered on incomplete product strands.

A. DIRECTION OF MINUS STRAND SYNTHESIS

Two sets of experiments were designed for the analysis of the direction of minus strand synthesis.

The first of these tests prediction (1) of the above section. Limited phosphodiesterase digestion of the 3′ end of Qβ-RNA revealed a uniquely high U content, being 42% for a 2% digest and 37% for an 8% digest, compared to 29% for the whole molecule. On the other hand, the base ratio of an early HS-RNA labeled with all four triphosphates gave values of 18% for U and 37% for A. The implication of this result is that the initial product is complementary to the 3′ end of Qβ-RNA, and hence that the negative is polymerized in the 5′ to 3′ direction. However, until the base ratio of a similar limited digest of the 5′ end of Qβ-RNA is obtained, this evidence cannot be taken as conclusive.

A second approach tests prediction (2). Here, we examine the incorporation of triphosphates labeled in only the β,γ positions into the incomplete and complete HS RNA. No appreciable β,γ-^{32}P-labeled ATP, CTP, or UTP were deposited in the HS RNA from late or early reactions. However, significant amounts of β,γ-^{32}P-GTP were recovered in the HS RNA from both late *and early* reactions. This result and the *lower* ^{3}H to ^{32}P ratio for GTP early reactions indicates a 5′ to 3′ direction of synthesis for minus strands (Bishop et al., 1967c).

B. DIRECTION OF PLUS STRAND SYNTHESIS

Analysis of the direction of plus strand synthesis can be attempted from three standpoints; (1) examination of the incorporation of β,γ-^{32}P-labeled triphosphates in *plus* primed reactions at times when predominantly *plus* strands are being synthesized; (2) a similar examination of reactions primed by single-strand *minus* strands; (3) end analysis as described in Section X.

1. *Late plus primed reactions.* An attempt was made (Bishop et al., 1967c) to examine late reactions templated with plus strands and labeled individually with each of the four β,γ-^{32}P-labeled riboside triphosphates. The results with β,γ-^{32}P-GTP and β,γ-^{32}P-ATP are given as the gel profiles in Fig. 12. It will be seen that *all* components contain β,γ-^{32}P-GTP, demonstrating that pppG is the 5′ terminus of both plus and minus strands. It was found for the single-strand Qβ RNA that β,γ-^{32}P-GTP was recovered in molar quantities (relative to the ^{3}H label) equivalent to 1 per strand of 1.3 × 10^6 mol wt. Neither β,γ-^{32}P-ATP, CTP, or UTP was deposited in the single strand Qβ RNA. Secondly, all four labeled triphosphates were recovered in the FS region of the gel in amounts approximately equivalent to that expected when one of two product strands is incomplete and growing in a 3′ to 5′ direction. Although the quantities incorporated were in reasonable agreement with that expected for a 3′ to 5′ synthesis, it does not *prove* that the

FIRST STAGE

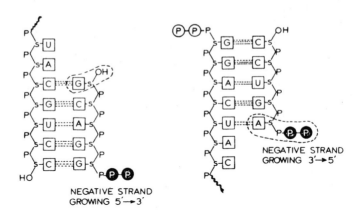

SECOND STAGE

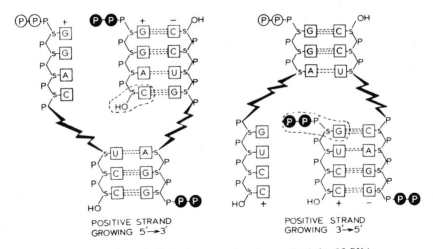

FIGURE 11. *Alternative methods for the synthesis of polymers complementary or identical to Qβ-RNA.*

First stage: Formation of duplex HS-RNA; synthesis of the negative strand. The synthesis of a complementary negative strand in a 5′ to 3′ direction (left-hand side) or 3′ to 5′ direction (right-hand side) using Qβ-RNA as template is shown. The 5′ phosphates of the new polymer derived from the β,γ-^{32}P-labeled triphosphate are depicted as P in blackened circles in order to differentiate them from the unlabeled phosphates of the template or the α-phosphates of the triphosphates. It is assumed that reactions are carried out with β,γ-^{32}P ribosidetriphosphates. The last nucleotide to be added to the new polymer is enclosed in a broken line.

Second stage: Formation of a multistranded FS-RNA: synthesis of the positive strands. Subsequent 5′ to 3′ or 3′ to 5′ synthesis of positive strands on the negative-strand template, which has been synthesized in a 5′ to 3′ direction (first stage, left-hand side) is shown.

The alternative methods of synthesis, and predictions therefrom are discussed in the text. Apart from the terminal bases the sequence of bases shown for the positive or negative strands is arbitrary (Bishop et al., 1967c).

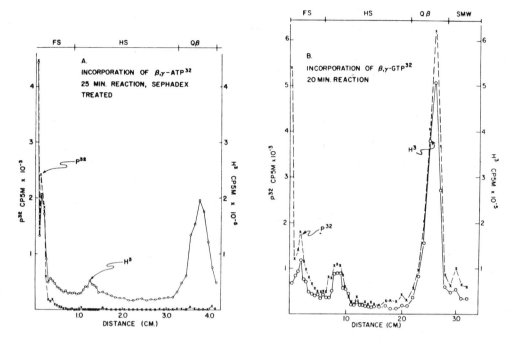

FIGURE 12. *Incorporation of β,γ-^{32}P-labeled GTP or ATP into the products of a $Q\beta$ replicase reaction.* Standard reaction mixtures containing ^{3}H-labeled UTP and β,γ-^{32}P-labeled ATP (A) or GTP (B) were incubated for 25 or 20 min, respectively, and either Sephadex treated (A) to remove triphosphates before electrophoresis, or subjected to electrophoresis (B). Subsequent to electrophoresis the gels were sliced, acid-washed, dissolved in H_2O_2 and counted (Bishop et al., 1967c).

label is incorporated into relevant structures or even into RNA. In order to examine this latter point, advantage was taken of the fact that alkali degradation of RNA gives rise to nucleoside monophosphates *except* for the terminal nucleoside triphosphate which is recovered as a tetraphosphate (pppXp).

DEAE column chromatography in 7 M urea of β,γ-^{32}P-labeled GTP, and alkali digests of $Q\beta$-FS, $Q\beta$-HS, and $Q\beta$-RNA, each labeled with β,γ-^{32}P-GTP are shown in Fig. 13. As expected, the guanosine tetraphosphate, because it has two extra phosphate charges, can be readily resolved from the guanosine triphosphate. The position of the labeled guanosine tetraphosphate in the second panel of Fig. 13 serves as the control for authentic pppGp, since it has been demonstrated independently by August and his co-workers that pppG is the 5′-terminus of $Q\beta$ RNA (Banerjee et al., 1967).

Similar alkali digests of β,γ-^{32}P-ATP or UTP-labeled $Q\beta$-FS-RNA yield the profiles shown in Fig. 14. However, the *recovery* of adenosine or uridine tetraphosphates is equivalent to only about 10% of that expected. Even though the existence of adenosine and uridine tetraphosphates

supports the 3′ to 5′ mode of plus strand RNA synthesis, it is not *sufficient* evidence *per se*. Two relevant points should be made; (1) there is no assurance that the adenosine, uridine, or cytidine triphosphate resident in the FS structures are the *actual growing points* for RNA synthesis, since they could represent false starts of a 5′ to 3′ polymerization; (2) the number of *active* FS species in the FS RNA which is recovered is unknown. In this regard, it should be emphasized that *relevant quantities of tetraphosphates can only be related to the amounts of active FS species* in the reaction and cannot be related to the *total* RNA present. Whether the amounts of tetraphosphates recovered are compatible with the active FS species cannot be determined by this type of analysis and, consequently, β,γ-triphosphate incorporation cannot be accepted as a sufficient criterion for the direction of RNA synthesis. Unfortunately, the low recovery of tetraphosphates for adenosine and uridine preclude a pulse-chase experiment of the type performed as described above (V, B-2).

C. MINUS STRAND PRIMED REACTIONS

As already noted, single-stranded minus strands isolated and purified from infected cells are capable

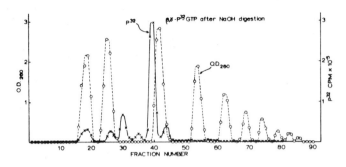

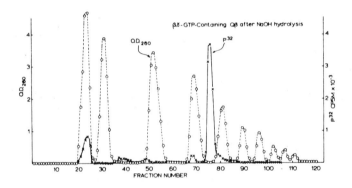

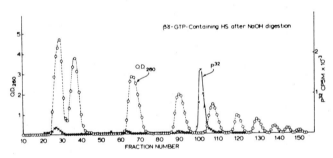

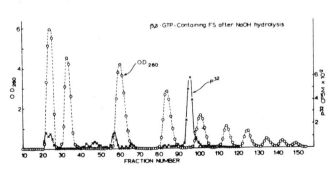

FIGURE 13. *DEAE chromatography of* β,γ-*^{32}P-GTP-labeled components.* A standard reaction mixture containing β,γ-^{32}P-GTP (6 c/mmole) was allowed to proceed at 38°C for 20 min, halted by addition of SDS to 1%, phenol extracted twice, and passed through a G-50 Sephadex column equilibrated in 0.01 M Tris, pH 7.9, 0.2 M NaCl, 0.003 M EDTA, 0.2% SDS (TSES). Excluded fractions were precipitated from 60% ethanol, dried under N_2, dissolved in 0.2 ml TSES and submitted to gel electrophoresis. Gel slices were eluted in TSES, and appropriate regions were pooled and precipitated from ethanol. Samples for hydrolysis including the β,γ-^{32}P-GTP were adjusted to 0.3 N NaOH, incubated at 35°C for 12 hr, neutralized with Dowex-50(H$^+$) and submitted to DEAE chromatography as described in Fig. 19.

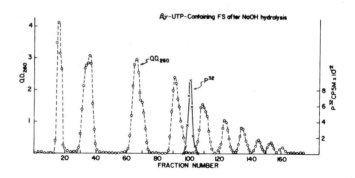

FIGURE 14. *DEAE chromatography of* β,γ-^{32}P-*UTP and ATP-labeled FS.* Replicase reaction products labeled with β,γ:^{32}P-UTP (9 c/mmole) or β,γ-^{32}P-ATP (9 c/mmole) were prepared and resolved by gel electrophoresis as described in Fig. 13. Material from the FS regions of gels were hydrolyzed with 0.3 N NaOH and chromatographed on DEAE as described in Fig. 13.

of priming the Qβ replicase and producing infectious plus strands earlier than similar *plus* primed reactions (Feix et al., 1967). The incorporation of β,γ-^{32}P-labeled GTP and UTP into early minus st·and-primed reactions are shown in Fig. 15. Only β,γ-^{32}P-GTP is incorporated in the FS, HS, or single-stranded product RNA. This result suggests a 5′ to 3′ direction of plus strand synthesis for reactions templated with purified minus strands. The same conclusion has been drawn from more detailed experiments with Qβ-variant minus strands (Mills et al., 1968).

In conclusion, it has been shown that both minus strand synthesis on plus templates and plus strand synthesis on minus templates proceed in the 5′ to 3′ direction. In light of these findings, it must be considered that the tetraphosphate evidence (Section VII-B) for 3′ to 5′ synthesis in reactions initiated with plus strands possibly represents the product of an abnormal synthesis. More definitive information comes from an end analysis of the replicative structures. The next section will show the technical feasibility of such an analysis.

D. END ANALYSIS

The third approach to defining the direction of plus strand synthesis is to determine which end sequences are synthesized during a pulse experiment. Consider an FS RNA which consists of a minus strand backbone with two plus strands, one complete and the other incomplete (Fig. 11, second stage). If the direction of plus strand synthesis is 3′ to 5′, then the *incomplete* strands will have a sequence complementary to the minus strand 5′ sequence, as illustrated in Fig. 11, beginning with CpCpUpGp . . . etc. However, for a 5′ to 3′ direction, the *incomplete* strand would have a sequence complementary to the 3′ sequence of the minus strand, i.e., pppGpGpApCp . . . etc. Provided that the 5′ and 3′ plus strand sequences may be identified, we can determine which is present within the *incomplete*, pulse-labeled plus strands.

The 5′ *plus* strand sequence can be selectively labeled with β,γ-^{32}P-GTP and the unique 5′ pancreatic ribonuclease oligonucleotide may be identified by DEAE chromatography as described for the variant RNA in Section X. Further, the 3′ plus strand ribonuclease T_1 oligonucleotide may

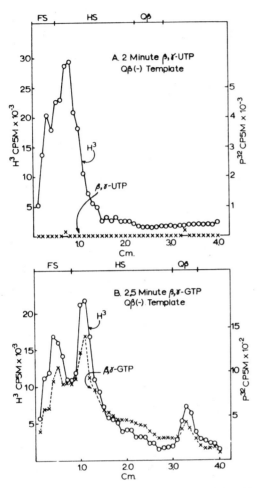

FIGURE 15. β,γ-^{32}P-GTP and UTP incorporation with $Q\beta(-)$ templates. Standard reaction mixtures containing ^3H-CTP (1 c/mmole), β,γ-^{32}P-UTP (8 c/mmole) or GTP (7 c/mmole) and templated by $Q\beta(-)$ RNA (0.5 μg/0.25 ml) prepared as described by Feix et al., (1968) were incubated at 38°C for the indicated time interval. Reaction mixtures were phenol extracted, passed through G-50 Sephadex in TSES, precipitated from 60% ethanol, and resolved by gel electrophoresis.

be identified by two dimensional paper electrophoresis (Dahlberg and Sanger, pers. commun.).

The profile of a β,γ-^{32}P-GTP pulse-labeled reaction is given in Fig. 16a. The electrophoretic pattern of heat-denatured total RNA indicated that the β,γ-^{32}P-GTP pulse-labeled material was exclusively small molecular weight and *not* in complete strands (Fig. 16b). Examination by DEAE chromatography of a pancreatic RNase

digest of the pulse-labeled reaction products revealed that 89% of the β,γ-^{32}P-GTP pulse was resident in plus strand terminal sequences and 11% was present as minus strand sequences as shown in Fig. 17 (see Section X).

These data support a 5′ to 3′ direction of plus strand synthesis during reactions initiated by $Q\beta$ plus strands. It must be emphasized that this β,γ-^{32}P-GTP pulse-labeled end analysis focuses on *metabolically active* FS. A similar end analysis, searching for a 3′ to 5′ direction of synthesis as outlined above, is being undertaken.

In conclusion, it has been shown that minus strand synthesis in plus primed reactions, plus strand synthesis in minus primed reactions, and plus strand synthesis in plus primed reactions proceed in a 5′ to 3′ direction.

VIII. EXTRACELLULAR DARWINIAN EXPERIMENTS WITH REPLICATING RNA MOLECULES

It was shown (Section III) that when $Q\beta$ replicase is presented with two genetically distinct $Q\beta$-RNA molecules, the RNA is identical to the initiating template. This specific response of the same enzyme preparation to the particular template added proved that the RNA is the instructive agent in the replicative process and hence satisfies the operational definition of a self-duplicating entity.

An opportunity is thus provided for studying the evolution of a self-replicating nucleic acid molecule outside a living cell. It should be noted that this situation mimics an early precellular evolutionary event, when environmental selection presumably operated directly on the genetic material. The comparative simplicity of the system and the accessibility of its known chemical components to manipulation permits the imposition of a variety of selection pressures during growth of the replicating molecules.

In the universe provided to them in the test tube, the RNA molecules are liberated from many of the restrictions derived from the requirements of a complete viral life cycle. The only restraint imposed is that they retain whatever sequences are involved in the recognition mechanism employed by the replicase. Thus, sequences which code for the coat proteins and replicase components *may* now be dispensable. Under these circumstances, it is of no little interest to design an experiment which attempts an answer to the following question: "What will happen to the RNA molecules if the only demand made on them is the biblical injunction, *multiply*, with the biological proviso that they do so as rapidly as possible?" The conditions

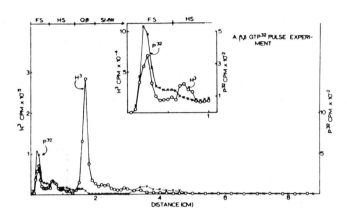

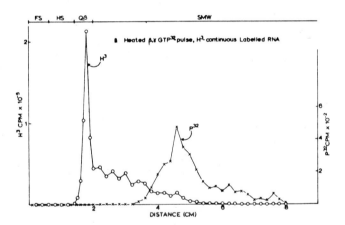

FIGURE 16. β,γ-^{32}P-GTP pulse-labeled reaction components. A standard reaction mixture containing ^{3}H-ATP (0.5 c/mmole) and Qβ(+) template was incubated at 38°C for 20 min, and then pulse-labeled with β,γ-^{32}P-GTP (7 c/mmole) for 1 min. The pulse was terminated by addition of SDS to 1%. After phenol extraction the mixture was passed through G-50 Sephadex in TSES. Aliquots after dialysis were resolved by gel electrophoresis without further treatment (Fig. 16A) and after heating to 100°C for 90 sec (Fig. 16B).

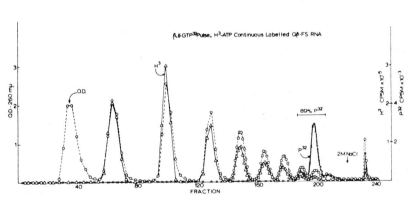

FIGURE 17. DEAE chromatography of β,γ-^{32}P-GTP pulse-labeled reaction. The β,γ-^{32}P-GTP pulse-labeled reaction products discussed in the legend to Fig. 16 were mixed with E. coli bulk RNA, digested with pancreatic ribonuclease and chromatographed on DEAE cellulose as described in Fig. 19.

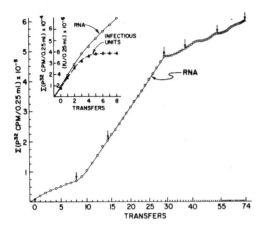

FIGURE 18. *Serial transfer experiment.* Each 0.25-ml standard reaction mixture contained 40 μg of Qβ replicase and (^{32}P) UTP. The first reaction (0 transfer) was initiated by the addition of 0.2 μg ts-1 (temperature-sensitive RNA) and incubated at 35°C for 20 min, whereupon 0.02 ml was drawn for counting and 0.02 ml was used to prime the second reaction (1 st transfer) and so on. After the first 13 reactions, the incubation periods were reduced to 15 min (transfers 14–29). Transfers 30–38 were incubated for 10 min. Transfers 39–52 were incubated for 7 min and transfers 53–74 were incubated for 5 min. The arrows above certain transfers (0, 8, 14, 29, 37, 53, and 73) indicate where 0.001–0.1 ml of product was removed and used to prime reactions for sedimentation analysis on sucrose. The inset examines both infectious and total RNA. The results show that biologically competent RNA ceases to appear after the 4th transfer (Mills et al., 1967).

required are readily attained by a serial transfer experiment in which the intervals of synthesis between transfers are adjusted to select the first molecules completed.

Mills et al. (1967) performed a series of 75 transfers (Fig. 18) during which each reaction mixture was diluted 12.5-fold into the next tube. The incubation intervals at 35°C were reduced periodically from 20 min to 5 min. As may be seen from the insert to Fig. 18, the synthesis of infectious RNA ceased after the fourth transfer. A dramatic increase in rate of RNA synthesis occurred after the eighth transfer. The products of several transfers were analyzed on a sucrose gradient. The product of the first reaction showed a 28 S peak, characteristic of Qβ-RNA as well as the peaks corresponding to the usual complexes observed during in vitro synthesis (Mills et al., 1966). Products of subsequent transfers showed a gradual shift of the RNA to smaller S values. By the 38th transfer, a single peak was found at about 15 S. By the 75th transfer, this peak moved to a value of 12 S.

The product of the 75th transfer, to be called variant-1 (V-1) was used for further analysis. Electrophoresis on polyacrylamide gel revealed

that the single stranded form of V-1 has a mol wt of 1.7×10^5 daltons. This corresponds to about 550 residues and represents a decrease of 83% in size compared to Qβ RNA.

It was of obvious interest to see whether phenotypically distinguishable variant molecules could be isolated. Levisohn and Spiegelman (1968) obtained a new, fast-growing mutant RNA by modifying the selection procedure. During a serial transfer experiment, the incubation interval at 38°C was held constant at 15 min. Selective pressure was applied by considerably increasing (up to 10^{-10}-fold) the dilution of reaction products at given stages during the transfer series. The variant RNA that evolved by the 17th transfer was selected for further studies and will be called variant-2 (V-2).

The properties of V-1 and V-2 were compared. One μμμg of V-2 RNA used as template was sufficient to allow synthesis of 0.13 μg RNA during a 15 min incubation while as much as 100 μμμg V-1 RNA led to no significant RNA synthesis (<0.005 μg) during this interval. A kinetic analysis revealed that during the exponential growth phase, V-2 had a doubling time of 0.403 min as compared with 0.456 min for V-1. Thus, V-2 can experience four more doublings (i.e., a 16-fold increase) than V-1 during a 15 min period of logarithmic growth. V-1 and V-2 had a similar electrophoretic mobility on polyacrylamide gels, suggesting that both have the same size.

IX. CLONING OF SELF-REPLICATING RNA MOLECULES

The Qβ replicase system should, in principle, be capable of generating clones descended from individual strands. The resulting clones would provide the sort of uniformity required for sequence and genetic studies.

Levisohn and Spiegelman (1968) obtained clones of RNA molecules in vitro using an approach that depends on a straightforward comparison of the observed frequency distribution with that expected from Poisson statistics in a series of repeated syntheses. Thus, if one strand is sufficient to initiate synthesis, then the proportion of tubes showing no synthesis should correspond to e^{-m}, m being the average number of strands per tube. Further, if the onset of synthesis in each tube is adequately synchronized, then one should be able to identify tubes that received initially one, two, or three strands and these tubes should appear with frequencies corresponding to me^{-m}, $(m^2/2!)e^{-m}$, and $(m^3/3!)e^{-m}$ respectively.

V-2, the fastest self-replicating RNA known, was employed in such a cloning experiment. A reaction mixture containing 0.29 $\mu\mu\mu g$ (corresponding to one RNA strand) V-2 RNA per 0.1 ml was distributed at 0°C in 0.1 ml portions into each of 82 tubes. All the tubes were placed simultaneously into a 38°C bath and after a 30 min interval, the reactions were stopped simultaneously. If the assumption underlying the Poisson distribution has been satisfied by the conditions of the experiment, 36.8% of the 82 tubes (i.e., 30) should show no synthesis. This is in excellent agreement with the 30 tubes found (Table 1).

Table 1 also shows good agreement between the actual results and the values expected from a Poisson distribution for tubes containing 1, 2, or 3 template strands. Thus, it is highly probable that a tube that exhibited an incorporation close to the value ascribable to a single template strand was, in fact, initiated by a single strand.

Our primary purpose in this section was to demonstrate the potentialities of the replicase system for examining the extracellular evolution

TABLE 1. POISSON ANALYSIS OF SYNTHESES
TEMPLATED WITH AN AVERAGE OF ONE STRAND

Strands per tube	P(r)	Number out of 82	
		Calc.	Found
0	0.368	30.2	30
1	0.368	30.2	29
2	0.184	15.1	19
3	0.061	5.4	3
4	0.015	1.3	1

A reaction mixture in which 13.2 μg variant-2 RNA were synthesized was diluted to a final concentration of 2.9 $\mu\mu g$/ml RNA into standard reaction mixtures containing 7.8×10^5 count/min ^{32}P-GTP per 0.25 ml. Aliquots (0.1 ml) were distributed into each of 82 tubes. Following 30 min incubation at 38°C the acid insoluble ^{32}P was determined. To identify tubes inoculated with 1, 2, 3 or more strands, the sum of the counts observed in all tubes was divided by the total number of template strands (82). The result, 309 count/min, is the average amount ascribable to a single template strand. The actual incorporation in each tube was divided by 309 and approximated to the nearest integer to give the number of strands that initiated synthesis in that tube. The observed distribution of templating strands per tube is compared to that expected for a Poisson distribution of the 82 templates (Levisohn and Spiegelman, 1968).

A

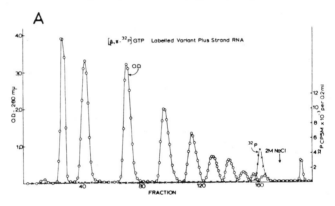

FIGURE 19. *DEAE Chromatography of pancreatic ribonuclease digests of (A) β,γ-³²P-GTP labeled variant plus RNA and (B) α-³²P-ATP, α³²P-UTP, α-³²P-GTP and α-³²P-CTP labeled variant plus RNA.* A pancreatic ribonuclease digest of unlabeled *E. coli* RNA with variant, plus strand RNA labeled with (A) β,γ-³²P-GTP and ³H-ATP or, (B) all four α-³²P triphosphates were resolved on DEAE column by a linear gradient of 0 to 0.35 M NaCl in 7 M urea-buffer. At the end of the gradient, the columns were stripped by passage of 2 M NaCl in 7 M urea, 0.003 M EDTA, 0.01 M Tris-HCl buffer pH 7.8. Samples were collected at 5 min intervals, the optical density measured, and radioactivity determined (Bishop et al., 1968).

It should be noted that the 5′ terminal oligonucleotide of the Qβ plus strand is coincident with the nonanucleotides, which is in contrast to the variant plus strand oligonucleotide. The 5′ terminal oligonucleotide of Qβ minus strand RNA can only be eluted in the 2 M NaCl wash under the conditions used (see Fig. 17).

B

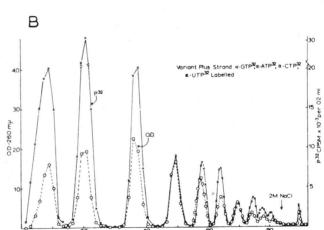

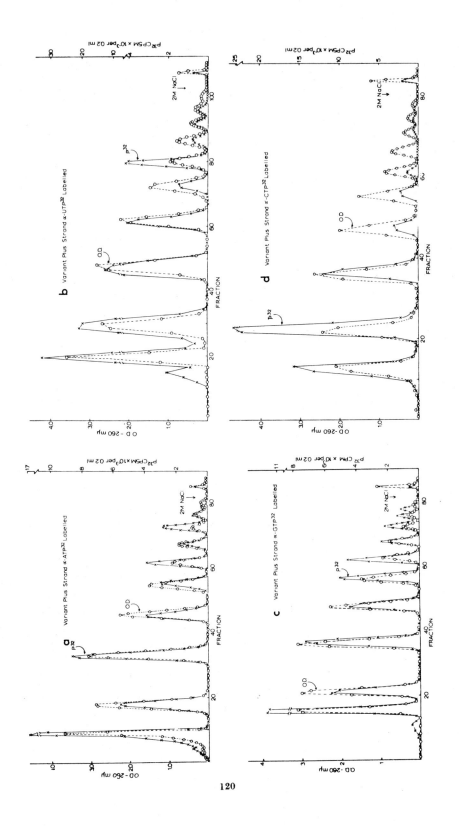

of a self-replicating nucleic acid molecule. Further, the experimental situation provides its own paleology; every sample is kept frozen and can be expanded at will to yield the components occurring at that particular evolutionary stage. While only 2 such samples are detailed here, they indicate that progress to a small size occurs in a series of steps.

The last products examined in the studies described are molecules which have eliminated 83% of their original length. Thus, neither the specific recognition nor the replication mechanisms require the complete original sequence. In this connection, it should be noted that although abbreviated, these variants are not equivalent to random fragments. The latter are unable to complete the replicative act (Haruna and Spiegelman, 1965b).

The availability of a molecule which has discarded large and unnecessary segments provides an object with obvious experimental advantages for the analysis of many aspects of the replicative process. Finally, these abbreviated RNA molecules have a very high affinity for the replicase but are no longer able to direct the synthesis of virus particles. This feature opens up a novel pathway toward a highly specific device for interfering with viral replication.

One should note that the situation described places at our disposal a completely novel method for the resolution of a variety of interesting problems. Potentially, other selective stresses can be imposed on the system to generate RNA entities which exaggerate other molecular features. The fact that they can be cloned increases their potentialities for both chemical and biological studies.

X. THE SEQUENCE AT THE 5′-TERMINUS OF A VARIANT OF Qβ-RNA

Since the ends of the replicating molecule are probably involved in the initiation and termination of replication, information on the sequences at the termini should help to illuminate mechanisms by which the replicase distinguishes one RNA molecule from another. The 'little' variants, isolated in a Darwinian selection experiment (Mills et al., 1967) contain only 550 nucleotides. They are of immediate interest because of their obviously greater amenability to sequence determination.

It has been shown (Mills et al., 1968; Bishop et al., 1967c) that the 5′ terminal nucleotide of both

Qβ RNA and the 'little' variant *plus* RNA is GTP. With this knowledge, we undertook the investigation of the 5′ terminal sequence of the 'little' variant plus strand with material containing β,γ-^{32}P-GTP to permit an easy identification of the 5′ end.

The DEAE-urea column profile (Tomlinson and Tener, 1963) of a pancreatic digest of unlabeled *E. coli* RNA and β,γ-^{32}P-GTP and ^{3}H-ATP labeled variant plus RNA is shown in Fig. 19a (Bishop et al., 1968). The majority of ^{32}P-label was recovered between the nona- and decanucleotides of *E. coli* digest.

Preliminary data on the approximate distribution of oligonucleotides labeled in vitro with each or all four α-^{32}P-ribonucleoside triphosphates were obtained for the variant plus RNA. As may be seen from Fig. 20, no decanucleotides are observed. Further, the septanucleotides were not labeled by α-^{32}P-CTP while the unique 5′ terminal oligonucleotide was not labeled by α-^{32}P-UTP.

To determine the number of nucleotides present in the 5′ terminal oligonucleotide, the effect of removing terminal phosphates with alkaline phosphatase was examined. A pancreatic digest of *plus* RNA labeled with all four ^{32}P-riboside triphosphates was resolved on DEAE as shown in Fig. 19b, digested with alkaline phosphatase, and again resolved by DEAE-urea chromatography. The subsequent fractionation of oligonucleotides showed that the dephosphorylated oligonucleotide was eluted on the leading side of the pentanucleotides, indicating a net −6 phosphate charge and hence a *septanucleotide* composition.

Knowledge of the base composition and nearest neighbor frequency of each component should provide the information required to deduce the sequence of the 5′-fragment, and the appropriate data are recorded in Table 2.

The last column (4) lists the outcome of a synthesis carried out with all four α-^{32}P-labeled riboside triphosphates. In agreement with the α-^{32}P-UTP of Fig. 20 we find no evidence for U residues in the 5′-fragment. The ^{32}P counts found in the various components indicate that the molar ratio of Cp:Ap:Gp:pppGp is 1:2:3:1. Note that pppGp, the 5′-terminus, carries two equivalents of ^{32}P as it should; one of its own and one from its neighbor. From column 3 we can deduce that every G has G as its neighbor and hence that all four are clustered at the 5′ end with no other residues intervening. From the synthesis with α-^{32}P-ATP

FIGURE 20. *DEAE chromatography of pancreatic ribonuclease digests of α-^{32}P-ATP, α-^{32}P-UTP, α-^{32}P-GTP or α-^{32}P-CTP labeled variant plus strand RNA.* RNA preparation, ribonuclease digestion, DEAE chromatography of the digests and monitoring of the eluants are identical to that described in Fig. 19. Plus strand variant RNA was labeled with either (a) α-^{32}P-ATP, (b) α-^{32}P-UTP, (c) α-^{32}P-GTP or (d) α-P^{32}-CTP (Bishop et al., 1968).

TABLE 2. BASE COMPOSITION OF [α-³²P] LABELED 5′ OLIGONUCLEOTIDE: VARIANT PLUS STRAND

Label: Nucleotide	[α-³²P] ATP Count Ratio		[α-³²P] CTP Count Ratio		[α-³²P] GTP Count Ratio		[α-³²P] 4 × TP Count Ratio	
Cp	89	0	3410	1	80	0	1812	1
Ap	2728	1	3750	1	309	0	3816	2
Gp	2930	1	33	0	5400	1	5826	3
Up	100	0	106	0	0	0	315	0
pppGp	103	0	0	0	4880	1	4200	2

Structure: pppGpGpGpGpApApCpCp . . .

Variant plus strand RNA labeled with (1) [α-³²P]ATP, (2) [α-³²P]CTP, (3) [α-³²P]GTP, or (4) all four [α-³²P] riboside triphosphates (4 × TP), were digested with pancreatic ribonuclease, and the oligo-nucleotides were resolved on DEAE and purified from urea and salt. After alkaline hydrolysis, the 2′, 3′ ribonucleoside monophosphates and guanosine tetraphosphate (pppGp) were resolved by pH 3.5 paper electrophoresis (Sanger et al., 1965), and the nucleotides were detected by autoradiography (Sanger et al., 1965) cut out and counted. The 'counts ratios' are ratios to the lowest significant value in each column rounded off to the nearest integer. Note that the counts for pppGp in columns 3 and 4 represent two equivalents of [³²P], one from its neighbor and one of its own, indicating a molar equivalence of one for the tetraphosphate. Further, the Gp in column 3 is represented by two equivalents of [³²P] although three molar equivalents are present, one having received an unlabeled P from its non-G neighbor (Bishop et al., 1968).

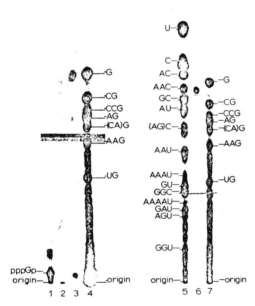

FIGURE 21. *Paper electrophoresis of digests of the 5′ terminal oligonucleotide from variant plus strand RNA.* β,γ-³²P-GTP and ³H-ATP-labeled variant plus strand RNA and the unique 5′ terminal oligonucleotide of the variant plus strand RNA labeled with α-³²P-GTP or α-³²P-CTP were obtained (Fig. 20). The assignment of oligonucleotides resolved from the digests by paper electrophoresis is based on the analyses of Sanger et al. (1965).
A comparison is made by electrophoresis between (1) an alkali digest of β,γ-³²P-GTP, ³H-ATP-labeled variant *plus* strand RNA; (2) the α-³²P-GTP-labeled variant *plus* strand 5′ terminal oligonucleotide; (3) a ribonuclease T₁ digest of the α-³²P-GTP variant *plus* strand 5′ fragment; (4) a ribonuclease T₁ of ³²PQβ viral RNA; (5) a pancreatic ribonuclease digest of ³²PQβ viral RNA; (6) a ribonuclease T₁ digest of the α-³²P-CTP labeled 5′ terminal fragment from the variant *plus* strand RNA; (7) a ribonuclease T₁ digest of ³²P-Qβ-RNA (Bishop et al., 1968).

(column 1) we find that A has A and G as neighbors with equal frequency. Finally, from column 2 we infer that C has A and C as neighbors, again with equal frequency. Finding C as a neighbor to the only C in the septanucleotide allows us to specify C as the base not recovered in the ribonuclease resistant 5′ fragment; a unique advantage conferred by the possibility of performing a nearest neighbor analysis. All of these facts leads us to write as the unique sequence of the 5′-end of the variant plus strand: pppGpGpGpGpApApCpCp . . .

Confirmation of the correctness of the assigned order of the residues can be obtained by examining T₁ digests of α-³²P-GTP and α-³²P-CTP labeled fragments. The results of such experiments are given in Fig. 21 where a comparison is made of the electrophoretic behavior on DEAE paper in 7% formic acid of (1) an alkali digest of [β,γ-³²P]GTP, [³H]ATP-labeled variant *plus* strand RNA to identify pppGp; (2) the [α-³²P]GTP-labeled variant *plus* strand 5′ terminal oligonucleotide; (3) a ribonuclease T₁ digest of the [α-³²P]GTP variant *plus* strand fragment; (4) a T₁ digest of ³²P-Qβ viral RNA to identify Gp; (5) a pancreatic digest of ³²P-labeled Qβ viral RNA to identify ApApCp; (6) a T₁ digest of the ³²P CTP-labeled 5′ pancreatic oligonucleotide exhibiting the recovery of the ApApCp expected from a T₁ cleavage of GpApApCp sequence; (7) another T₁ digest of the ³²P-labeled Qβ viral RNA to identify Gp. It is evident that the T₁ digests of the appropriately labeled 5′-fragment yielded in all instances the components predicted from the deduced sequence.

Haruna and Spiegelman (1965b) had earlier suggested that some sort of association of the 3′ and 5′ ends of the Qβ RNA, possibly through short regions of complementarity, was operating as an initiation and/or recognition site. Their reasons for proposing this model stemmed from the observation that fragmented viral RNA molecules would not serve as template for the replicase (Haruna and Spiegelman, 1965b). In light of reports from Weith et al. (1968) and Sanger and Dahlberg (pers. commun.) and also from our own unpublished observations that the 3' end of the $Q\beta$ and 'little' variant RNA molecules contain large numbers of pyrimidines (9 C's and 4 U's), it seems that the 5' terminal sequence with 4 G's and 2 A's could, in fact, associate in some manner with the 3' terminal sequence. However, we would like to point out that if the 3' terminal nucleotide sequence is the prime recognition requirement for the replicase, and if both plus and minus strands have 3' sequences with some degree of similarity, then in both cases the 5' and 3' sequences will be complementary.

IX. SUMMARY

The availability of a purified enzyme ($Q\beta$-replicase) which can mediate extensive synthesis of biologically competent viral nucleic acid has made possible a variety of informative experiments. The results described focus on our recent efforts to achieve further insight into the following aspects of the RNA replicating system:

(1) Further purification and properties of the viral induced and host determined protein components of $Q\beta$-replicase.

(2) A detailed analysis of the nature of the intermediates, the sequence of their appearance, and the direction of polymerization when *plus* and *minus* strands are used as initiating templates.

(3) The *selection* and *cloning* of phenotypically distinguishable mutant RNA molecules possessing predetermined competitive advantages over the original viral RNA molecule.

(4) The results of sequence studies at the 5' end of the 'little' variant.

ACKNOWLEDGMENTS

This investigation was supported by U.S. Public Health Service research grant CA-01094 from the National Cancer Institute and grant GB-4876 from the National Science Foundation.

N. R. P., R. L. P., and D. R. M. are postdoctoral trainees, USPH training grant 5-TO1-GM00-319. R. L. is a postdoctoral fellow—Damon Runyon fellowship. M. M. T. is a postdoctoral USPH fellow. D. H. L. B. holds a Wellcome Travel Scholarship.

REFERENCES

AMMANN, J., H. DELIUS, and P. H. HOFSCHNEIDER. 1964. Isolation and properties of an intact phage-specific replicative form of RNA phage M-12. J. Mol. Biol. *10:* 557.

BANERJEE, A. K., L. EOYANG, K. HORI, and J. T. AUGUST. 1967. Replication of RNA viruses. IV. Initiation of RNA synthesis by the $Q\beta$ RNA polymerase. Proc. Nat. Acad. Sci. *57:* 986.

BISHOP, D. H. L., J. R. CLAYBROOK, and S. SPIEGELMAN. 1967a. Electrophoretic separation of viral nucleic acids on polyacrylamide gels. J. Mol. Biol., *26:* 373.

BISHOP, D. H. L., J. R. CLAYBROOK, N. R. PACE, and S. SPIEGELMAN. 1967b. An analysis by gel electrophoresis of $Q\beta$-RNA complexes formed during the latent period of an in vitro synthesis. Proc. Nat. Acad. Sci. *57:* 1474.

BISHOP, D. H. L., D. R. MILLS, and S. SPIEGELMAN. 1968. The sequence of the 5'-terminus of a self-replicating variant of $Q\beta$-RNA. Biochem. *7:* 3744.

BISHOP, D. H. L., N. R. PACE, and S. SPIEGELMAN. 1967c. The mechanism of replication: A novel polarity reversal in the in vitro synthesis of $Q\beta$-RNA and its complement. Proc. Nat. Acad. Sci. *58:* 1790.

BORST, P., and C. WEISSMANN. 1965. Replication of viral RNA. VIII. Studies on the enzymatic mechanism of replication of MS2 RNA. Proc. Nat. Acad. Sci. *54:* 982.

COOPER, S., and N. D. ZINDER. 1962. The growth of an RNA bacteriophage: The role of DNA synthesis. Virology *18:* 405.

DOI, R. H., and S. SPIEGELMAN. 1962. Homology test between the nucleic acid of an RNA virus and the DNA in the host cell. Science *138:* 1270.

EIKHOM, T. S., and S. SPIEGELMAN. 1967. The dissociation of $Q\beta$-replicase and the relation of one of the components to a poly-C-dependent poly-G-polymerase. Proc. Nat. Acad. Sci. *57:* 1833.

EIKHOM, T. S., DARLENE STOCKLEY, and S. SPIEGELMAN. 1968. Direct participation of a host protein in the in vitro replication of viral RNA. Proc. Nat. Acad. Sci. *59:* 506.

FEIX, G., R. POLLET, and C. WEISSMANN. 1968. Replication of viral RNA. XVI. Enzymatic synthesis of infectious viral RNA with noninfectious $Q\beta$ minus strands as template. Proc. Nat. Acad. Sci. *59:* 145.

FEIX, G., H. SLOR, and C. WEISSMANN. 1967. Replication of viral RNA. XIII. The early product of phage RNA synthesis in vitro. Proc. Nat. Acad. Sci. *57:* 1401.

FENWICK, M. L., R. L. ERIKSON, and R. M. FRANKLIN. 1964. Replication of the RNA of bacteriophage R17. Science *146:* 527.

FRANCKE, B., and P. H. HOFSCHNEIDER. 1966a. Über infektiöse substrukturen aus *Escherichia coli* bakteriophagen. VII. Formation of biologically intact replicative form in ribonucleic acid bacteriophage (M12)-infected cells. J. Mol. Biol. *16:* 544.

——, ——. 1966b. Infectious nucleic acids of *E. coli* bacteriophages. IX. Sedimentation constants and strand integrity of infectious M12 phage replicative form RNA. Proc. Nat. Acad. Sci. *56:* 1883.

FRANKLIN, R. M. 1966. Purification and properties of the replicative intermediate of the RNA bacteriophage R17. Proc. Nat. Acad. Sci. *55:* 1504.

HALL, B. D., and S. SPIEGELMAN. 1961. Sequence complementarity of T_2-DNA and T_2-specific RNA. Proc. Nat. Acad. Sci. *47:* 137.

HARUNA, I., K. NOZU, Y. OHTAKA, and S. SPIEGELMAN. 1963. An RNA replicase induced by and selective for a viral RNA: isolation and properties. Proc. Nat. Acad. Sci. *50:* 905.

HARUNA, I., and S. SPIEGELMAN. 1965a. Specific template requirements of RNA replicase. Proc. Nat. Acad. Sci., *54:* 579.

——, ——. 1965b. Recognition of size and sequence by an RNA replicase. Proc. Nat. Acad. Sci. *54:* 1189.

——, ——. 1965c. The autocatalytic synthesis of a viral RNA in vitro. Science *150:* 884.

——, ——. 1966. A search for an intermediate involving a

complement during synchronous synthesis by a purified RNA replicase. Proc. Nat. Acad. Sci. 55: 1256.

KELLY, R. B., and R. I. SINSHEIMER. 1964. A new RNA component in MS2 infected cells. J. Mol. Biol. 8: 602.

LEVISOHN, R., and S. SPIEGELMAN. 1968. The cloning of a self-replicating RNA molecule. Proc. Nat. Acad. Sci. 60: 866.

LOEB, T., and N. D. ZINDER. 1961. A bacteriophage containing RNA. Proc. Nat. Acad. Sci. 47: 282.

LODISH, H., and N. ZINDER. 1966. Replication of the RNA bacteriophage f2. Science 152: 372.

LOENING, U. E. 1967. The fractionation of high molecular weight ribonucleic acid by polyacrylamide gel electrophoresis. Biochem. J. 102: 251.

MILLS, D. R., D. H. L. BISHOP, and S. SPIEGELMAN. 1968. The mechanism and direction of RNA synthesis templated by free minus strands of a "little" variant Qβ RNA. Proc. Nat. Acad. Sci. 60: 713.

MILLS, D., N. PACE, and S. SPIEGELMAN. 1966. The in vitro synthesis of a noninfectious complex containing biologically active viral RNA. Proc. Nat. Acad. Sci. 56: 1778.

MILLS, D. R., R. L. PETERSON, and S. SPIEGELMAN. 1967. An extracellular Darwinian experiment with a self-duplicating nucleic acid molecule. Proc. Nat. Acad. Sci. 58: 217.

MONTAGNIER, L., and F. K. SANDERS. 1963. Replicative form of encephalomyocarditis virus ribonucleic acid. Nature 199: 664.

NONOYAMA, M., and Y. IKEDA. 1964. Ribonuclease-resistant RNA found in cells of Escherichia coli infected with RNA phage. J. Mol. Biol., 9: 763.

OVERBY, L. R., G. H. BARLOW, R. H. DOI, MONIQUE JACOB, and S. SPIEGELMAN. 1966a. Comparison of two serologically distinct ribonucleic acid bacteriophages. I. Properties of the viral particles. J. Bacteriol. 91: 442.

——, ——, ——, ——, ——. 1966b. Comparison of two serologically distinct ribonucleic acid bacteriophages. II. Properties of the nucleic acids and coat proteins. J. Bacteriol., 92: 739.

PACE, N. R., D. H. L. BISHOP, and S. SPIEGELMAN. 1967a. Examination of the Qβ replicase reaction by sucrose gradient and gel electrophoresis. J. Virology 1: 771.

——, ——, ——. 1967b. The kinetics of product appearance and template involvement in the in vitro replication of viral RNA. Proc. Nat. Acad. Sci. 58: 711.

——, ——, ——. 1968. The immediate precursor of viral RNA in the Qβ replicase reaction. Proc. Nat. Acad. Sci. 59: 139.

PACE, N. R., and S. SPIEGELMAN. 1966a. The synthesis of infectious RNA with a replicase purified according to its size and density. Proc. Nat. Acad. Sci. 55: 1608.

——, ——. 1966b. The in vitro synthesis of an infectious mutant RNA with a normal RNA replicase. Science 153: 64.

PARANCHYCH, W., and A. F. GRAHAM. 1962. Isolation and properties of an RNA-containing bacteriophage. J. Cellular Comp. Physiol. 60: 199.

SANGER, F., G. G. BROWNLEE, and B. G. BARRELL. 1965. A two-dimensional fractionation procedure for radio-active nucleotides. J. Mol. Biol. 13: 373.

SPIEGELMAN, S., I. HARUNA, I. B. HOLLAND, G. BEAUDREAU, and D. R. MILLS. 1965. The synthesis of a self-propagating and infectious nucleic acid with a purified enzyme. Proc. Nat. Acad. Sci. 54: 919.

STRAUSS, J. H., and R. L. SINSHEIMER. 1963. Purification and properties of bacteriophage MS2 and of its RNA. J. Mol. Biol. 7: 43.

TOMLINSON, R. V., and G. M. TENER. 1963. The effect of urea, formamide and glycols on the secondary binding forces in the ion-exchange chromatography of poly-nucleotides on DEAE-cellulose. Biochemistry 2: 697.

WATANABE, I. 1964. Persistent infection with an RNA bacteriophage. Nihon Rinsho 22: 243.

WEISSMANN, C., P. BORST, R. H. BURDON, M. A. BILLETER, and S. OCHOA. 1964. Replication of viral RNA. IV. Properties of RNA synthesis and enzymatic synthesis of MS2 phage RNA. Proc. Nat. Acad. Sci. 51: 890.

WEISSMANN, C., G. FEIX, M. SLOR, and R. POLLET. 1967. Replication of viral RNA. XIV. Single stranded minus strands as template for the synthesis of viral plus strands in vitro. Proc. Nat. Acad. Sci. 57: 1870.

WEITH, H. L., G. T. ASTERIADIS, and P. T. GILHAM. 1968. Comparison of RNA terminal sequences of phages f2 and Qβ: Chemical and sedimentation equilibration studies. Science 160: 1459.

YANKOFSKY, S. A., and S. SPIEGELMAN. 1962a. The identification of the ribosomal RNA cistron by sequence complementarity. I. Specificity of complex formation. Proc. Nat. Acad. Sci. 48: 1069.

Appendix

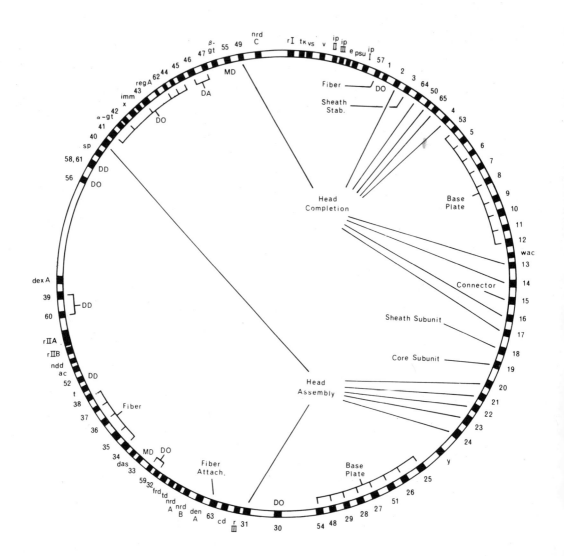

Figure 1. Genes of T4.

Genes of T4 (listed in map order)

Gene name	Gene product, gene function, or mutant defect	Key references
rI	Rapid lysis	12, 44
tk	Thymidine kinase	34
vs	Valyl-tRNA synthetase modifier	136
v	Ultraviolet sensitivity	80
ipII	Internal protein II	17, 95, 158
ipIII	Internal protein III	17, 158
e	Endolysin; lysis	175
psu$_1$	Serine tRNA; amber suppressor	76, 205
ipI	Internal protein I	17, 158
57	Tail fiber assembly (pleiotropic effects)	43, 51, 112
1	dHMP kinase; DNA negative	48, 57
2	Head completion	52, 57
3	Sheath stability	57, 111, 113, 114
64	Head completion	52
50	Head completion	52
65	Head completion	52
4	Head completion	52, 57
53	Baseplate	52, 113, 114
5	Baseplate	52, 57, 113, 114
6	Baseplate	52, 57, 113, 114
7	Baseplate	52, 57, 113, 114
8	Baseplate	52, 57, 113, 114
9	Baseplate	52, 57, 113, 114
10	Baseplate	52, 57, 113, 114
11	Baseplate modification	52, 57, 113, 114
12	Baseplate modification	52, 57, 113, 114
wac	Whisker antigen control (collar)	62
13	Head completion	52, 57
14	Head completion	52, 57
15	Connector for head attachment	57, 111, 113, 114
16	Head completion	52, 57
17	Head completion; quinacrine resistance	52, 57, 144
18	Sheath subunit	57, 111, 113, 114
19	Core subunit	57, 111, 113, 114
20	Head assembly (polyhead)	57, 60, 122
21	Head assembly (τ particle); protease (?)	57, 120, 158
22	Head assembly (polyhead); assembly core	57, 119, 158
23	Major head subunit	57, 155
24	Head assembly (τ particle); osmotic shock resistance	57, 120
y	Ultraviolet sensitivity	167
25	Baseplate	52, 57, 113, 114
26	Baseplate	52, 57, 113, 114
51	Baseplate	52, 113, 114
27	Baseplate	52, 57, 113, 114
28	Baseplate	52, 57, 113, 114
29	Baseplate	52, 57, 113, 114
48	Baseplate	52, 113, 114
54	Baseplate	52, 113, 114
30	DNA ligase; DNA negative	57, 58
31	Head assembly; random aggregation of P23	57, 121
rIII	Rapid lysis	12

Gene name	Gene product, gene function, or mutant defect	Key references
cd	dCMP deaminase	78, 79
63	Tail fiber attachment	207
denA	Endonuclease II	83, 149, 193
nrdB	Ribonucleotide reductase	77, 211, 212
nrdA	Ribonucleotide reductase	77, 211, 212
td	Thymidylate synthetase	161
frd	Dihydrofolate reductase (white halo)	79, 211
32	DNA recombination and replication; DNA negative	2, 57, 186
59	DNA negative	unpub.
33	Maturation defective (no late proteins)	18, 57, 65, 173
das	Suppressor of gene 46 and 47 mutations	82
34	Tail fiber subunit; A antigen	43, 51, 57, 112, 115
35	Tail fiber subunit	43, 51, 57, 112, 115
36	Tail fiber subunit; B antigen	43, 51, 57, 112, 115
37	Tail fiber subunit; C antigen (host range)	43, 51, 57, 112, 115
38	Tail fiber assembly	43, 51, 57, 112, 115
t	Lysis defect (phospholipase?)	104
52	DNA delay	210
ac	Acriflavine uptake	160
ndd	Nuclear disruption defect	unpub.
rIIB	Rapid lysis; restricted by λ-lysogens	11, 12, 85
rIIA	Rapid lysis; restricted by λ-lysogens	11, 12, 85
60	DNA delay	210
39	DNA delay	57, 210
dexA	Exonuclease A	194
56	dCTPase; DNA negative	192, 201
58(61)	DNA delay	210
sp	Suppressor of e mutations ("spackle")	56
40	Head assembly (polyhead)	57, 122
41	DNA negative	57
β-gt	β-glucosyl transferase	70, 150
42	Hydroxymethylase; DNA negative	57, 202
x	Ultraviolet sensitivity	80
imm	Superinfection immunity	189
43	DNA polymerase; DNA negative	42, 57
regA	Defective in shutoff of phage enzyme synthesis	203
62	DNA negative	unpub.
44	DNA negative	57
45	DNA negative	57
46	Fails to degrade host DNA; DNA arrest	57, 65, 201
47	Fails to degrade host DNA; DNA arrest	57, 201
α-gt	α-glucosyl transferase	70, 150
55	Maturation defective (no late proteins)	18, 173
49	Head completion (DNA packaging?)	52, 66, 128, 129
nrdC	Ribonucleotide reductase (thioredoxin)	181

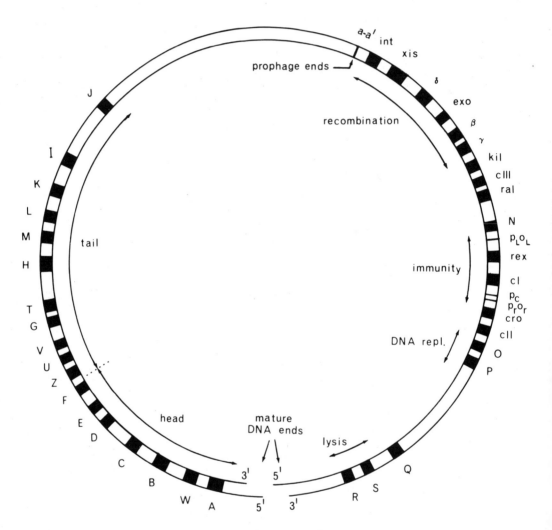

Figure 2. Genes of λ.

Genes of λ

Gene name	Gene product, gene function, or mutant defect
A	Formation of cohesive DNA ends
W	Controls synthesis of F protein
B	Head protein
C	Head protein
D	Head internal (?) protein
E	Major head protein
F	Head–tail joining
Z	Tail assembly
U	Tail assembly; abnormally long tails
V	Tail assembly; major tail protein
G	Tail assembly
T	Tail assembly
H	Tail assembly
M	Tail assembly
L	Tail assembly
K	Tail assembly
I	Tail assembly
J	Tail assembly; host range determinant
a-a′	Site of integrative recombination; prophage ends
int	Recombination enzyme for prophage integration and excision
xis	Recombination enzyme for prophage excision
δ	Interference with λ growth in P2 lysogens
exo	Exonuclease; general recombination
β	General recombination
Y	Mutants restricted in feb⁻ hosts
kil	Gene product kills host when N^+ lysogens are induced
cIII	Establishment of lysogeny; clear plaques
ral	Alleviates restriction of unmodified λ
N	Positive regulator of delayed-early transcription
P_LO_L	Leftward promoter-operator; binding site for cI protein
rex	Restriction of T4 rII mutants
cI	Immunity; clear plaques; structural gene for repressor
P_c	Promoter for transcription of cI
P_rO_r	Rightward promoter-operator; binding site for cI protein
cro	Negative regulator of immediate-early transcription
cII	Establishment of lysogeny; clear plaques
O	Initiation of DNA synthesis
P	Initiation of DNA synthesis
Q	Positive regulator of late gene transcription
S	Lysis
R	Lysis; endolysin

Note: Primary references and descriptions of gene functions are given in Hershey (88). The regulatory genes are discussed in detail in Paper 27.

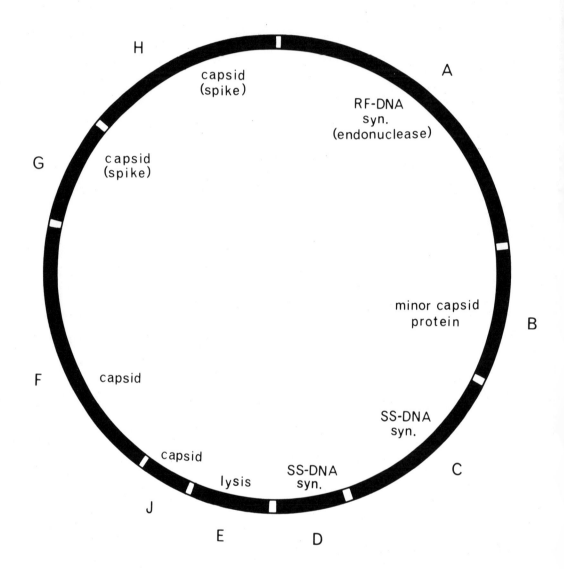

Figure 3. Genes of φX174 and S13.

Genes of φX174 and S13

Map order and gene identifications are from Benbow et al. *(9)*, Benbow et al. *(10)*, and Jeng et al. *(103)*, the designations being in accord with the revised nomenclature of *(10)*. Gene lengths have been drawn proportional to the molecular weights of their polypeptide products as determined by Benbow et al. *(10)*, with the exception of gene A, for which the larger value (62,000 daltons), reported by Godson *(73)* and Linney *(127)*, has been used.

References

1. Adams, J. M., P. G. N. Jeppersen, F. Sanger, and B. G. Barrell (1969). Nucleotide sequence from the coat protein cistron of R17 bacteriophage RNA. *Nature, 223,* 1009–1014.
2. Alberts, B. M., and L. Frey (1970). T4 bacteriophage gene 32: a structural protein in the replication and recombination of DNA. *Nature, 227,* 1313–1318.
3. Ammann, J., H. Delius, and P. H. Hofschneider (1964). Isolation and properties of an intact phage-specific replicative form of RNA phage M-12. *J. Mol. Biol., 10,* 557–561.
4. Anderson, T. M. (1953). The morphology and osmotic properties of bacteriophage systems. *Cold Spring Harbor Symp. Quant. Biol., 18,* 197–203.
5. Arber, W., G. Kellenberger, and J. Weigle (1957). La Défectuosité du phage λ transducteur. *Schweiz. Z. Allgem. Path. Bakteriol., 20,* 659–665.
6. August, J. T., A. K. Banerjee, L. Eoyang, M. T. F. De Fernandez, K. Hori, C. H. Kuo, U. Rensing, and L. Shapiro (1968). Synthesis of bacteriophage Qβ RNA. *Cold Spring Harbor Symp. Quant. Biol., 33,* 73–81.
7. Barnett, L., S. Brenner, F. H. C. Crick, R. G. Shulman, and R. J. Watts-Tobin (1967). Phase-shift and other mutants in the first part of the *r*IIB cistron of bacteriophage T4. *Phil. Trans. Roy. Soc. London, B252,* 487–560.
8. Bautz, E. K. F., and J. J. Dunn (1969). DNA-dependent RNA polymerase from phage T4 infected *E. coli:* an enzyme missing a factor required for transcription of T4 DNA. *Biochem. Biophys. Res. Commun., 34,* 230–237.
9. Benbow, R. M., C. A. Hutchison, III, J. D. Fabricant, and R. L. Sinsheimer (1971). Genetic map of bacteriophage φX174. *J. Virol., 7,* 549–558.
10. Benbow, R. M., R. F. Mayol, J. C. Picchi, and R. L. Sinsheimer (1972). Direction of translation and size of bacteriophage φX174 cistrons. *J. Virol., 10,* 99–114.
11. Benzer, S. (1955). Fine structure of a genetic region in bacteriophage. *Proc. Natl. Acad. Sci. U.S., 41,* 344–354.
12. Benzer, S. (1957). The elementary units of heredity. In W. D. McElroy and B. Glass (eds.), *The Chemical Basis of Heredity.* The Johns Hopkins Press, Baltimore, pp. 70–93.
13. Benzer, S. (1959). On the topology of the genetic fine structure. *Proc. Natl. Acad. Sci. U.S., 45,* 1607–1620.
14. Berns, K. I., and C. A. Thomas, Jr. (1961). A study of single polynucleotide chains derived from T2 and T4 bacteriophage. *J. Mol. Biol., 3,* 289–300.

15. Bernstein, H., and C. Bernstein (1973). Circular and branched circular concatenates as intermediates in phage T4 DNA replication. *J. Mol. Biol.*, 77, 355–361.
16. Bishop, D. H. L., J. R. Claybrook, N. R. Pace, and S. Spiegelman (1967). An analysis by gel electrophoresis of Qβ-RNA complexes formed during the latent period of an in vitro synthesis. *Proc. Natl. Acad. Sci. U.S.*, 57, 1475–1481.
17. Black, L. W., and C. Ahmad-Zadeh (1971). Internal proteins of bacteriophage T4D: their characterization and relation to head structure and assembly. *J. Mol. Biol.*, 57, 71–92.
18. Bolle, A., R. H. Epstein, W. Salser, and E. P. Geiduschek (1968). Transcription during bacteriophage T4 development: requirements for late messenger synthesis. *J. Mol. Biol.*, 33, 339–362.
19. Bradley, D. E. (1967). Ultrastructure of bacteriophages and bacteriocins. *Bacteriol. Rev.*, 31, 230–314.
20. Brenner, S., and J. R. Beckwith (1965). *Ochre* mutants, a new class of suppressible nonsense mutants. *J. Mol. Biol.*, 13, 629–637.
21. Brenner, S., F. Jacob, and M. Meselson (1961). An unstable intermediate carrying information from genes to ribosomes for protein synthesis. *Nature*, 190, 576–581.
22. Brenner, S., L. Barnett, F. H. C. Crick, and A. Orgel (1961). The theory of mutagenesis. *J. Mol. Biol.*, 3, 121–124.
23. Brenner, S., L. Barnett, E. R. Katz, and F. H. C. Crick (1967). UGA: a third nonsense triplet in the genetic code. *Nature*, 213, 449–450.
24. Brenner, S., G. Streisinger, R. W. Horne, S. P. Champe, L. Barnett, S. Benzer, and M. W. Rees (1959). Structural components of bacteriophage. *J. Mol. Biol.*, 1, 281–292.
25. Burnet, F. M. (1929). A method for the study of bacteriophage multiplication in broth. *Brit. J. Exptl. Pathol.*, 10, 109–115.
26. Cairns, J., G. S. Stent, and J. D. Watson (eds.) (1966). *Phage and the Origins of Molecular Biology*. Cold Spring Harbor Laboratory, Cold Spring Harbor, N.Y.
27. Calef, E., and G. Licciardello (1960). Recombination experiments on prophage host relationships. *Virology*, 12, 81–103.
28. Calendar, R. (1970). The regulation of phage development. *Ann. Rev. Microbiol.*, 24, 241–296.
29. Campbell, A. M. (1957). Transduction and segregation in *E. coli* K12. *Virology*, 4, 366–384.
30. Campbell, A. M. (1959). Ordering of genetic sites in bacteriophage λ by the use of galactose-transducing defective phages. *Virology*, 9, 293–305.
31. Campbell, A. M. (1961). Sensitive mutants of bacteriophage λ. *Virology*, 14, 23–32.
32. Campbell, A. M. (1963). Segregants from lysogenic heterogenotes carrying recombinant lambda prophages. *Virology*, 20, 344–356.
33. Campbell, A. M. (1969). *Episomes*. Harper & Row, Publishers, New York.
34. Chace, K. V., and D. H. Hall (1973). Isolation of mutants of bacteriophage T4 unable to induce thymidine kinase activity. *J. Virol.*, 12, 343–348.
35. Chamberlin, M., J. McGrath, and L. Waskell (1970). New RNA polymerase from *Escherichia coli* infected with bacteriophage T7. *Nature*, 228, 227–231.
36. Cohen, S. S. (1947). Synthesis of bacterial viruses in infected cells. *Cold Spring Harbor Symp. Quant. Biol.*, 12, 35–49.
37. Cohen, S. S. (1968). *Virus-induced Enzymes*. Columbia University Press, New York.
38. Cooper, S., and N. D. Zinder (1962). The growth of an RNA bacteriophage: The role of DNA synthesis. *Virology*, 18, 405–411.
39. Crick, F. H. C., J. S. Griffith, and L. E. Orgel (1957). Codes without commas. *Proc. Natl. Acad. Sci. U.S.*, 43, 416–421.
40. Davern, C. I., and M. Meselson (1960). The molecular conservation of ribonucleic acid during bacterial growth. *J. Mol. Biol.*, 2, 153–160.
41. Demerec, M., and U. Fano (1945). Bacteriophage-resistant mutants in *Escherichia coli*. *Genetics*, 30, 119–136.

42. DeWaard, A., A. V. Paul, and I. R. Lehman (1965). The structural gene for deoxyribonucleic acid polymerase in bacteriophages T4 and T5. *Proc. Natl. Acad. Sci. U.S., 54,* 1241–1248.
43. Dickson, R. C. (1973). Assembly of bacteriophage T4 tail fibers. IV. Subunit composition of tail fibers and fiber precursors. *J. Mol. Biol., 79,* 633–647.
44. Doermann, A. H. (1953). The vegetative state in the life cycle of bacteriophage: evidence for its occurrence and its genetic characterization. *Cold Spring Harbor Symp. Quant. Biol., 18,* 3–11.
45. Doermann, A. H. (1973). T4 and the rolling circle model of replication. *Ann. Rev. Genet., 7,* 325– 341.
46. Doermann, A. H., and L. Boehner (1963). An experimental analysis of bacteriophage T4 heterozygotes. I. Mottled plaques from crosses involving six *r*II loci. *Virology, 21,* 551–567.
47. Doi, R. H., and S. Spiegelman (1962). Homology test between the nucleic acid of an RNA virus and the DNA in the host cell. *Science, 138,* 1270–1272.
48. Duckworth, D. H., and M. J. Bessman (1967). The enzymology of virus-infected bacteria. X. A biochemical–genetic study of the deoxynucleotide kinase induced by wild-type and *amber* mutants of phage T4. *J. Biol. Chem., 242,* 2877–2885.
49. Eddleman, H. L., and S. P. Champe (1966). Components in T4-infected cells associated with phage assembly. *Virology, 30,* 471–481.
50. Edgar, R. S., and I. Lielausis (1964). Temperature-sensitive mutants of bacteriophage T4D: their isolation and genetic characterization. *Genetics, 49,* 649–662.
51. Edgar, R. S., and I. Lielausis (1965). Serological studies with mutants of phage T4D defective in genes determining tail fiber structure. *Genetics, 52,* 1187–1200.
52. Edgar, R. S., and I. Lielausis (1968). Some steps in the assembly of bacteriophage T4. *J. Mol. Biol., 32,* 263–276.
53. Edgar, R. S., G. H. Denhardt, and R. H. Epstein (1964). A comparative genetic study of conditional lethal mutations of bacteriophage T4D. *Genetics, 49,* 635– 648.
54. Eiserling, F. A., and R. C. Dickson (1972). Assembly of viruses. *Ann. Rev. Biochem., 41,* 476–502.
55. Ellis, E. L., and M. Delbrück (1939). The growth of bacteriophage. *J. Gen. Physiol., 22,* 365–384.
56. Emrich, J. (1968). Lysis of T4-infected bacteria in the absence of lysozyme. *Virology, 35,* 158–165.
57. Epstein, R. H., A. Bolle, C. M. Steinberg, E. Kellenberger, E. Boy de la Tour, R. Chevalley, R. S. Edgar, M. Susman, G. H. Denhardt, and A. Lielausis (1963). Physiological studies of conditional lethal mutants of bacteriophage T4. *Cold Spring Harbor Symp. Quant. Biol., 28,* 375–394.
58. Fareed, G. C., and C. C. Richardson (1967). Enzymatic breakage and joining of deoxyribonucleic acid. II. The structural gene for polynucleotide ligase in bacteriophage T4. *Proc. Natl. Acad. Sci. U.S., 58,* 665–672.
59. Farid, S. A. A., and L. M. Kozloff (1968). Number of polypeptide components in bacteriophage T2L contractile sheaths. *J. Virol., 2,* 308–312.
60. Favre, R., E. Boy de la Tour, N. Segrè, and E. Kellenberger (1965). Studies on the morphopoiesis of the head of phage T-even. I. Morphological, immunological, and genetic characterization of polyheads. *J. Ultrastruct. Res., 13,* 318–342.
61. Flaks, J. G., and S. S. Cohen (1959). Virus-induced acquisition of metabolic function. I.Enzymatic formation of 5-hydroxymethyldeoxycytidylate. *J. Biol. Chem., 234,* 1501–1506.
62. Follansbee, S., R. W. Vanderslice, L. G. Chavez, and C. D. Yegian (1974). A new set of adsorption mutants of phage T4D: identification of a new gene. *Virology, 58,* 180–199.
63. Frankel, F. R. (1963). An unusual DNA extracted from bacteria infected with phage T2. *Proc. Natl. Acad. Sci. U.S., 49,* 366–372.

64. Frankel, F. R. (1966). Studies on the nature of replicating DNA in T4-infected *Escherichia coli. J. Mol. Biol., 18,* 127– 143.

65. Frankel, F. R. (1966). Studies on the nature of the replicating DNA in *Escherichia coli* infected with certain amber mutants of phage T4. *J. Mol. Biol., 18,* 144– 155.

66. Frankel, F. R., M. L. Batcheler, and C. K. Clark (1971). The role of gene 49 in DNA replication and head morphogenesis in bacteriophage T4. *J. Mol. Biol., 62,* 439– 463.

67. Fraenkel-Conrat, H., and R. C. Williams (1955). Reconstitution of active tobacco mosaic virus from its inactive protein and nucleic acid components. *Proc. Natl. Acad. Sci. U.S., 41,* 690–698.

68. Franklin, R. M. (1966). Purification and properties of the replicative intermediate of the RNA bacteriophage R17. *Proc. Natl. Acad. Sci. U.S., 55,* 1504–1511.

69. Garen, A. (1968). Sense and nonsense in the genetic code. *Science, 160,* 149–159.

70. Georgopoulos, C. P. (1968). Location of glucosyl transferase genes on the genetic map of phage T4. *Virology, 34,* 364– 366.

71. Georgopoulos, C. P., R. W. Hendrix, S. R. Casjens, and A. D. Kaiser (1973). Host participation in bacteriophage lambda head assembly. *J. Mol. Biol., 76,* 45– 60.

72. Gilbert, W., and D. Dressler (1968). DNA replication: the rolling circle model. *Cold Spring Harbor Symp. Quant. Biol., 33,* 473–484.

73. Godson, N. G. (1971). Characterization and synthesis of ϕX174 proteins in ultraviolet-irradiated cells. *J. Mol. Biol., 57,* 541–553.

74. Goodman, H. M., J. N. Abelson, A. Landy, S. Brenner, and J. D. Smith (1968). Amber suppression: A nucleotide change in the anticodon of a tyrosine transfer RNA. *Nature, 217,* 1019–1024.

75. Gottesman, M. E., and R. A. Weisberg (1971). Prophage insertion and excision. In A. D. Hershey (ed.), *The Bacteriophage Lambda.* Cold Spring Harbor Laboratory, Cold Spring Harbor, N.Y., pp. 113–138.

76. Guthrie, C., and W. H. McClain (1973). Conditionally lethal mutants of bacteriophage T4 defective in production of a transfer RNA. *J. Mol. Biol., 81,* 137–155.

77. Hall, D. H. (1967). Mutants of bacteriophage T4 unable to induce dihydrofolate reductase activity. *Proc. Natl. Acad. Sci. U.S., 58,* 584–591.

78. Hall, D. H., and I. Tessman (1966). T4 mutants unable to induce deoxycytidylate deaminase activity. *Virology, 29,* 339–345.

79. Hall, D.H., I Tessman, and O. Karlström (1967). Linkage of T4 genes controlling a series of steps in pyrimidine biosynthesis. *Virology, 31,* 442–448.

80. Harm, W. (1963). Mutants of phage T4 with increased sensitivity to ultraviolet. *Virology, 19,* 66–71.

81. Hayashi, M., M. N. Hayashi, and S. Spiegelman (1963). Restriction of the in vivo genetic transcription to one of the complementary strands of DNA. *Proc. Natl. Acad. Sci. U.S., 50,* 664–672.

82. Hercules, K., and J. S. Wiberg (1971). Specific suppression of mutations in genes 46 and 47 by *das,* a new class of mutations in bacteriophage T4D. *J. Virol., 8,* 603–612.

83. Hercules, K., J. L. Munro, S. Mendelsohn, and J. S. Wiberg (1971). Mutants in a non-essential gene of bacteriophage T4 which are defective in the degredation of *E. coli* DNA. *J. Virol., 7,* 95–105.

84. d'Herelle, F. (1926). *The Bacteriophage and its Behavior.* English translation by G. H. Smith. Williams & Wilkins, Co., Baltimore.

85. Hershey, A. D. (1953). Inheritance in bacteriophage. *Advan. Genet., 5,* 89–106.

86. Hershey, A. D. (1955). An upper limit to the protein content of the germinal substance of bacteriophage T2. *Virology, 1,* 108–127.

87. Hershey, A. D. (1957). Some minor components of bacteriophage T2 particles. *Virology, 4,* 237–264.

88. Hershey, A. D. (ed.) (1971). *The Bacteriophage Lambda.* Cold Spring Harbor Laboratory, Cold Spring Harbor, N.Y.
89. Hershey, A. D., and E. Burgi (1965). Complementary structure of interacting sites at the ends of lambda DNA molecules. *Proc. Natl. Acad. Sci. U.S., 53,* 325–328.
90. Hershey, A. D., and M. Chase (1951). Genetic recombination and heterozygosis in bacteriophage. *Cold Spring Harbor Symp. Quant. Biol., 16,* 471–479.
91. Hershey, A. D., and M. Chase (1952). Independent functions of viral protein and nucleic acid in growth of bacteriophage. *J. Gen. Physiol., 36,* 39–56.
92. Hershey, A. D., M. D. Kamen, J. W. Kennedy, and H. Gest (1951). The mortality of bacteriophage containing assimilated radiophosphorus. *J. Gen. Physiol., 34,* 305–319.
93. Hindley, J. (1973). Molecular structure and function in RNA phages. *Brit. Med. Bull., 29,* 236–240.
94. Holland, J. J., and E. D. Kiehn (1968). Specific cleavage of viral proteins as steps in the synthesis and maturation of enteroviruses. *Proc. Natl. Acad. Sci. U.S., 60,* 1015—1022.
95. Howard, G. W., M. L. Wolin, and S. P. Champe (1972). Diversity of phage internal components among members of the T-even group. *Trans. N.Y. Acad. Sci., 34,* 36–51.
96. Ikeda, H., and J. Tomizawa (1965). Transducing fragments in generalized transduction by P1. I. Molecular origin of the fragments. *J. Mol. Biol., 14,* 85–109.
97. Jacob, F., and J. Monod (1961). Genetic regulatory mechanisms in the synthesis of proteins. *J. Mol. Biol., 3,* 318–356.
98. Jacob, F., and E. L. Wollman (1954). Étude génétique d'un bactériophage tempéré d'*Escherichia coli.* I. Le système génétique du bactériophage λ. *Ann. Inst. Pasteur, 87,* 653–673.
99. Jacob, F., and E. L. Wollman (1956). Sur les processus de conjugaison et de recombinaison génétique chez *Escherichia coli.* I. L'induction par conjugaison ou induction zygotique. *Ann. Inst. Pasteur, 91,* 486–510.
100. Jacob, F., and E. L. Wollman (1958). The relationship between the prophage and the bacterial chromosome in lysogenic bacteria. In G. Tunevall (ed.), *Recent Progress in Microbiology.* Charles C Thomas, Springfield, Ill., pp. 15–30.
101. Jacob, F., C. R. Fuerst, and E. L. Wollman (1957). Recherches sur les bactéries lysogènes défectives. II. Les Types physiologiques liés aux mutations du prophage. *Ann. Inst. Pasteur, 93,* 724–753.
102. Jacobson, M. F., and D. Baltimore (1968). Polypeptide cleavages in the formation of poliovirus proteins. *Proc. Natl. Acad. Sci. U.S., 61,* 77–84.
103. Jeng, Y., D. Gelfand, M. Hayashi, R. Shleser, and E. S. Tessman (1970). The eight genes of bacteriophage ϕX174 and S13 and comparison of the phage-specified proteins. *J. Mol. Biol., 49,* 521–526.
104. Josslin, R. (1970). The lysis mechanism of phage T4: mutants affecting lysis. *Virology, 40,* 719–726.
105. Kaiser, A. D. (1957). Mutations in a temperate bacteriophage affecting its ability to lysogenize *E. coli. Virology, 3,* 42–61.
106. Kaiser, A. D., and F. Jacob (1957). Recombination between related temperate bacteriophages and the genetic control of immunity and prophage localization. *Virology, 4,* 509–521.
107. Kaiser, A. D., and T. Masuda (1973). In vitro assembly of bacteriophage lambda heads. *Proc. Natl. Acad. Sci. U.S., 70,* 260–264.
108. Kellenberger, E., and W. Arber (1955). Die Structur des Schwanzes der Phagen T2 and T4 und der Mechanismus der irreversiblen Adsorption. *Z. Naturforsch., 10b,* 698–704.
109. Kellenberger, E., and E. Boy de la Tour (1964). On the fine structure of normal and "polymerized" tail sheath of phage T4. *J. Ultrastruct. Res., 11,* 545–563.
110. Kellenberger, E., and R. S. Edgar (1971). Structure and assembly of phage particles. In A.

D. Hershey (ed.), *The Bacteriophage Lambda*. Cold Spring Harbor Laboratory, Cold Spring Harbor, N.Y., pp. 271–295.

111. King, J. (1968). Assembly of the tail of bacteriophage T4. *J. Mol. Biol., 32*, 231–262.

112. King, J., and U. K. Laemmli (1971). Polypeptides of the tail fibers of bacteriophage T4. *J. Mol. Biol., 62*, 465–477.

113. King, J., and U. K. Laemmli (1973). Bacteriophage T4 tail assembly: structural proteins and their genetic identification. *J. Mol. Biol., 75*, 315–337.

114. King, J. and N. Mykolajewycz (1973). Bacteriophage T4 tail assembly: proteins of the sheath, core, and baseplate. *J. Mol. Biol., 75*, 339–358.

115. King, J., and W. B. Wood (1969). Assembly of bacteriophage T4 tail fibers: the sequence of gene product interaction. *J. Mol. Biol., 39*, 583–601.

116. Koch, G., and A. D. Hershey (1959). Synthesis of phage-precursor protein in bacteria infected with T2. *J. Mol. Biol., 1*, 260–276.

117. Kornberg, A., S. B. Zimmerman, S. R. Kornberg, and J. Josse (1959). Enzymatic synthesis of deoxyribonucleic acid. VI. Influence of bacteriophage T2 on the synthetic pathway in host cells. *Proc. Natl. Acad. Sci. U.S., 45*, 772–785.

118. Kozak, M., and D. Nathans (1972). Translation of the genome of a ribonucleic acid bacteriophage. *Bacteriol. Rev., 36*, 109–134.

119. Laemmli, U. K., and M. Favre (1974). Maturation of the head of bacteriophage T4. I. DNA packaging events. *J. Mol. Biol., 80*, 575–599.

120. Laemmli, U. K., and R. A. Johnson (1974). Maturation of the head of bacteriophage T4. II. Head-related, aberrant τ-particles. *J. Mol. Biol., 80*, 601–611.

121. Laemmli, U. K., F. Beguin, and G. Gujer-Kellenberger (1970). A factor preventing the major head protein of bacteriophage T4 from random aggregation. *J. Mol. Biol., 47*, 69–85.

122. Laemmli, U. K., E. Mölbert, M. K. Showe, and E. Kellenberger (1970). Form determining function of the genes required for the assembly of the head of bacteriophage T4. *J. Mol. Biol., 49*, 99–113.

123. Lederberg, E. M. (1951). Genetic studies of lysogenicity in *Escherichia coli. Genetics, 36*, 560.

124. Lederberg, J., and E. L. Tatum (1946). Gene recombination in *E. coli. Nature, 158*, 558.

125. Lederberg, J., and E. L. Tatum (1946). Novel genotypes in mixed cultures of biochemical mutants of bacteria. *Cold Spring Harbor Symp. Quant. Biol., 11*, 113–114.

126. Levine, M. (1969). Phage morphogenesis. *Ann. Rev. Genet., 3*, 323–342.

127. Linney, E. A., M. N. Hayashi, and M. Hayashi (1972). Gene A of ϕX174. I. Isolation and identification of its products. *Virology, 50*, 381–387.

128. Luftig, R. B., and C. Ganz (1972). Bacteriophage T4 head morphogenesis. II. Studies on the maturation of gene 49-defective head intermediates. *J. Virol., 9*, 377–389.

129. Luftig, R. B., W. B. Wood, and R. Okinaka (1971). Bacteriophage T4 head morphogenesis. On the nature of gene 49-defective heads and their role as intermediates. *J. Mol. Biol., 57*, 555–573.

130. Luria, S. E. (1953). *General Virology*. John Wiley & Sons, Inc., New York, p. 15.

131. Lwoff, A., and A. Gutmann (1950). Recherches sur un *Bacillus megatherium* lysogène. *Ann. Inst. Pasteur, 78*, 711–739. [English translation in (170).]

132. Lwoff, A., L. Siminovitch, and N.Kjeldgaard (1950). Induction de la production de bactériophages chez une bactérie lysogène. *Ann. Inst. Pasteur, 79*, 815–858.

133. MacHattie, L. A., D. A. Ritchie, C. A. Thomas, Jr., and C. C. Richardson (1967). Terminal repetition in permuted T2 bacteriophage DNA molecules. *J. Mol. Biol., 23*, 355–364.

134. Marvin, D. A., and B. Hohn (1969). Filamentous bacterial viruses. *Bacteriol. Rev., 33*, 172–209.

135. Mathews, C. K. (1971). *Bacteriophage Biochemistry*. Van Nostrand Reinhold Company, New York.

136. McClain, W. H., G. L. Marchin, and F. C. Neidhardt (1971). Phage-induced conversion of host valyl-tRNA synthetase. In G. E. W. Wolstenholme and M. O'Connor (eds.), *Ciba Symposium on Strategy of the Viral Genome*. Churchill Livingstone, London, pp. 191–205.

406

137. Mills, D. R., F. R. Kramer, and S. Spiegelman (1973). Complete nucleotide sequence of a replicating RNA molecule. *Science, 180,* 916–927.

138. Moody, M. F. (1973). Sheath of bacteriophage T4. III. Contraction mechanism deduced from partially contracted sheaths. *J. Mol. Biol., 80,* 613–635.

139. Morse, M. L., E. M. Lederberg, and J. Lederberg (1956). Transduction in *Escherichia coli* K-12. *Genetics, 41,* 142–156.

140. Nirenberg, M. W., T. Caskey, R. Marshall, R. Brimacombe, D. Kellogg, B. Doctor, D. Hatfield, J. Levin, F. Rottman, S. Pestka, M. Wilcox, and F. Anderson (1966). The RNA code and protein synthesis. *Cold Spring Harbor Symp. Quant. Biol., 31,* 11–24.

141. Nomura, M., B. D. Hall, and S. Spiegelman (1960). Characterization of RNA synthesized in *Escherichia coli* after bacteriophage T2 infection. *J. Mol. Biol., 2,* 306–326.

142. Okazaki, R., T. Okazaki, K. Sakabe, K. Sugimoto, and A. Sugino (1968). Mechanism of DNA chain growth. I. Possible discontinuity and unusual secondary structure of newly synthesized chains. *Proc. Natl. Acad. Sci. U.S., 59,* 598–605.

143. Pereira da Silva, L., and F. Jacob (1968). Étude génétique d'une mutation modifiant la sensibilité a l'immunité chez le bactériophage lambda. *Ann. Inst. Pasteur, 115,* 145–158.

144. Piechowski, M. M., and M. Susman (1967). Acridine resistance in phage T4D. *Genetics, 56,* 133–148.

145. Pratt, D. (1969). Genetics of single-stranded DNA bacteriophages. *Ann. Rev. Genet., 3,* 343–362.

146. Ptashne, M. (1967). Specific binding of the λ phage repressor to λ DNA. *Nature, 214,* 232–234.

147. Ptashne, M. (1971). Repressor and its action. In A. D. Hershey (ed.), *The Bacteriophage Lambda.* Cold Spring Harbor Laboratory, Cold Spring Harbor, N.Y., pp. 221–237.

148. Ptashne, M., and N. Hopkins (1968). The operators controlled by the λ phage repressor. *Proc. Natl. Acad. Sci. U.S., 60,* 1282–1287.

149. Ray, P., N. K. Sinha, H. R. Warner, and D. P. Snustad (1972). Genetic location of a mutant of bacteriophage T4 deficient in the ability to induce endonuclease II. *J. Virol., 9,* 184–186.

150. Revel, H. R., and S. E. Luria (1970). DNA glucosylation in T-even phages: Genetic determination and role in phage host interaction. *Ann. Rev. Genet., 4,* 177–192.

151. Ris, H., and B. L. Chandler (1963). The ultrastructure of genetic systems in prokaryotes and eukaryotes. *Cold Spring Harbor Symp. Quant. Biol., 28,* 1–8.

152. Riva, S., A. Cascino, and E. P. Geiduschek (1970). Coupling of late transcription to viral replication in bacteriophage T4 development. *J. Mol. Biol., 54,* 85–102.

153. Riva, S., A. Cascino, and E. P. Geiduschek (1970). Uncoupling of late transcription from DNA replication in bacteriophage T4 development. *J. Mol. Biol., 54,* 103–119.

154. Rothman, J. L. (1965). Transduction studies on the relation between prophage and host chromosome. *J. Mol. Biol., 12,* 892–912.

155. Sarabhai, A., A. O. W. Stretton, S. Brenner, and A. Bolle (1964). Co-linearity of the gene with the polypeptide chain. *Nature, 201,* 13–17.

156. Schlesinger, M. (1932). Über die Bindung des Bakteriophagen an homologe Bakterien. II. Quantitative Untersuchungen über die Bindungsgeschwindigkeit und die Sättigung. Berechnung der Teilchengrösse des Bakteriophagen aus deren Ergebnissen. *Z. Hyg. Infektionskrankh., 114,* 149–160. [English translation in (170).]

157. Schlesinger, M. (1936). The Feulgen reaction of the bacteriophage substance. *Nature, 138,* 508–509.

158. Showe, M. K., and L. W. Black (1973). Assembly core of bacteriophage T4: an intermediate in head formation. *Nature New Biol., 242,* 70–75.

159. Siegel, P. J., and M. Schaechter (1973). The role of the host cell membrane in the replication and morphogenesis of bacteriophages. *Ann. Rev. Microbiol., 27,* 261–282.

160. Silver, S. (1965). Acriflavin resistance: a bacteriophage mutation affecting the uptake of dye by the infected bacterial cells. *Proc. Natl. Acad. Sci. U.S., 53,* 24–30.

407

161. Simon, E. H., and I. Tessman (1964). Thymidine-requiring mutants of phage T4. *Proc. Natl. Acad. Sci. U.S., 50*, 526–552.

162. Simon, L. D., and T. F. Anderson (1967). The infection of *Escherichia coli* by T2 and T4 bacteriophages as seen in the electron microscope. II. Structure and function of the baseplate. *Virology, 32*, 298–305.

163. Sinsheimer, R. L. (1959). Purification and properties of bacteriophage φX174. *J. Mol. Biol., 1*, 37–42.

164. Sinsheimer, R. L. (1959). A single-stranded deoxyribonucleic acid from bacteriophage φX174. *J. Mol. Biol., 1*, 43–53.

165. Sinsheimer, R. L., B. Starman, C. Nagler, and S. Guthrie (1962). The process of infection with bacteriophage φX174. I. Evidence for a "replicative form." *J. Mol. Biol., 4*, 142–160.

166. Smith, J. D. (1972). Genetics of transfer RNA. *Ann. Rev. Genet., 6*, 235–256.

167. Smith, S. M., and N. Symonds (1973). The unexpected location of a gene conferring abnormal radiation sensitivity on phage T4. *Nature, 241*, 395–396.

168. Stavis, R. L., and J. T. August (1970). The biochemistry of RNA bacteriophage replication. *Ann. Rev. Biochem., 39*, 527–560.

169. Steitz, J. A. (1969). Polypeptide chain initiation: nucleotide sequences of the three ribosomal binding sites in bacteriophage R17 RNA. *Nature, 224*, 957–964.

170. Stent, G. S. (ed.) (1965). *Papers on bacterial viruses.* Second Ed., Little, Brown and Company, Boston.

171. Stent, G. S. (1963). *Molecular Biology of Bacterial Viruses.* W. H. Freeman and Company, Publishers, San Francisco.

172. Stent, G. S., and C. R. Fuerst (1955). Inactivation of bacteriophages by decay of incorporated radioactive phosphorus. *J. Gen. Physiol., 38*, 441–458.

173. Stevens, A. (1972). New small polypeptides associated with DNA-dependent RNA polymerase of *Escherichia coli* after infection with bacteriophage T4. *Proc. Natl. Acad. Sci. U.S., 69*, 603–607.

174. Streisinger, G., J. Emrich, and M. M. Stahl (1967). Chromosome structure in phage T4. III. Terminal redundancy and length determination. *Proc. Natl. Acad. Sci. U.S., 57*, 292–295.

175. Streisinger, G., F. Mukai, W. J. Dreyer, B. Miller, and S. Horiuchi (1961). Mutations affecting the lysozyme of phage T4. *Cold Spring Harbor Symp. Quant. Biol., 26*, 25–30.

176. Summers, D. F., and J. V. Maizel (1968). Evidence for large precursor proteins in poliovirus synthesis. *Proc. Natl. Acad. Sci. U.S., 59*, 966–971.

177. Summers, W. C., and R. B. Siegel (1969). Control of template specificity of *E. coli* RNA polymerase by a phage-coded protein. *Nature, 223*, 1111–1113.

178. Taylor, A. L. (1963). Bacteriophage-induced mutations in *Escherichia coli. Proc. Natl. Acad. Sci. U.S., 50*, 1043–1051.

179. Temin, H., and D. Baltimore (1972). RNA-directed DNA synthesis and RNA tumor viruses. *Advan. Virus Res., 17*, 129–186.

180. Tessman, E. S. (1967). Gene function in phage S13. In J. S. Colter and W. Paranchych (eds.), *The Molecular Biology of Viruses.* Academic Press, Inc., New York, pp. 193–209.

181. Tessman, I., and D. B. Greenberg (1972). Ribonucleotide reductase genes of phage T4: Map location of the thioredoxin gene *nrd*C. *Virology, 49*, 337–338.

182. Tessman, I., and E. S. Tessman (1966). Functional units of phage S13: Identification of two genes that determine the structure of the phage coat. *Proc. Natl. Acad. Sci. U.S., 55*, 1459–1462.

183. Tessman, I., E. S. Tessman, and G. S. Stent (1957). The relative radiosensitivity of bacteriophages S13 and T2. *Virology, 4*, 209–215.

184. Thomas, C. A., Jr., and L. A. MacHattie (1964). Circular T2 DNA molecules. *Proc. Natl. Acad. Sci. U.S., 52*, 1297–1301.

185. Thomas, C. A., Jr., and I. Rubenstein (1964). The arrangement of nucleotide sequences in T2 and T5 DNA molecules. *Biophys. J., 4*, 93–106.

186. Tomizawa, J., N. Anraku, and Y. Iwama (1966). Molecular mechanisms of genetic recombination in bacteriophage. VI. A mutant defective in the joining of DNA molecules. *J. Mol. Biol., 21*, 247–253.

408

187. Travers, A. A. (1970). Positive control of transcription by a bacteriophage sigma factor. *Nature, 225,* 1009–1016.

188. Valentine, R. C., R. Ward, and M. Strand (1969). The replication cycle of RNA bacteriophages. *Advan. Virus Res., 15,* 2–59.

189. Vallée, M., and J. B. Cornett (1972). A new gene of bacteriophage T4 determining immunity against ghosts and phage in T4-infected *E. coli. Virology, 48,* 777–784.

190. Vidaver, A. K., R. K. Koski, and J. L. Van Etten (1973). Bacteriophage φ6: a lipid containing virus of *Pseudomonas phaseolicola. J. Virol., 11,* 799–805.

191. Volkin, E., and L. Astrachan (1956). Phosphorus incorporation in *Escherichia coli* ribonucleic acid after infection with bacteriophage T2. *Virology, 2,* 149–161.

192. Warner, H. R., and J. E. Barnes (1966). Evidence for a dual role for the bacteriophage T4-induced deoxycytidine triphosphate nucleotidehydrolase. *Proc. Natl. Acad. Sci. U.S., 56,* 1233–1240.

193. Warner, H. R., D. P. Snustad, S. E. Jorgensen, and J. F. Koerner (1970). Isolation of bacteriophage T4 mutants defective in the ability to degrade host nucleic acid. *J. Virol., 5,* 700–708.

194. Warner, H. R., D. P. Snustad, J. F. Koerner, and J. D. Childs (1972). Identification and genetic characterization of mutants of bacteriophage T4 defective in the ability to induce exonuclease A. *J. Virol., 9,* 399–407.

195. Watson, J. D. (1963). The involvement of RNA in the synthesis of proteins. *Science, 140,* 17–26.

196. Watson, J. D. (1972). Origin of concatemeric T7 DNA. *Nature New Biol., 239,* 197–201.

197. Watson, J. D., and F. H. C. Crick (1953). Genetic implications of the structure of deoxyribonucleic acid. *Nature, 171,* 964–969.

198. Watson, J. D., and F. H. C. Crick (1953). Molecular structure of nucleic acids. *Nature, 171,* 737–739.

199. Weissmann, C., G. Feix, and H. Slor (1968). In vitro synthesis of RNA: the nature of the intermediates. *Cold Spring Harbor Symp. Quant. Biol., 33,* 83–100.

200. Weissman, C., M. A. Billeter, H. M. Goodman, J. Hindley, and H. Weber (1973). Structure and function of phage RNA. *Ann. Rev. Biochem., 42,* 303–328.

201. Wiberg, J. S. (1966). Mutants of bacteriophage T4 unable to cause breakdown of host DNA. *Proc. Natl. Acad. Sci. U.S., 55,* 614–621.

202. Wiberg, J. S., M. L. Dirksen, R. H. Epstein, S. E. Luria, and J. M. Buchanan (1962). Early enzyme synthesis and its control in *E. coli* infected with some amber mutants of bacteriophage T4. *Proc. Natl. Acad. Sci. U.S., 48,* 293–302.

203. Wiberg, J. S., S. Mendelsohn, V. Warner, K. Hercules, C. Aldrich, and J. L. Munro (1973). SP62, a viable mutant of bacteriophage T4D defective in regulation of phage enzyme synthesis. *J. Virol., 12,* 775–792.

204. Williams, R. C., and D. Fraser (1953). The morphology of the seven T-bacteriophages. *J. Bacteriol., 66,* 458–464.

205. Wilson, J. H., J. S. Kim, and J. N. Abelson (1972). Bacteriophage T4 transfer RNA: III. Clustering of the genes for T4 transfer RNAs. *J. Mol. Biol., 71,* 547–556.

206. Wolfson, J., D. Dressler, and M. Magazin (1972). Bacteriophage T7 DNA replication: a linear replicating intermediate. *Proc. Natl. Acad. Sci. U.S., 69,* 499–504.

207. Wood, W. B., and M. Henninger (1969). Attachment of tail fibers in bacteriophage T4 assembly: some properties of the reaction in vitro and its genetic control. *J. Mol. Biol., 39,* 603–618.

208. Wu, R., and E. Taylor (1971). Nucleotide sequence analysis of DNA. II. Complete nucleotide sequence of the cohesive ends of bacteriophage λ DNA. *J. Mol. Biol., 57,* 491–511.

209. Wyatt, G. R., and S. S. Cohen (1953). The bases of the nucleic acids of some bacterial and animal viruses: the occurrence of 5-hydroxymethylcytosine. *Biochem. J., 55,* 774–782.

210. Yegian, C. D., M. Mueller, G. Selzer, V. Russo, and F. W. Stahl (1971). Properties of the DNA-delay mutants of bacteriophage T4. *Virology, 46,* 900–919.

211. Yeh, Y. C., and I. Tessman (1972). Control of pyrimidine biosynthesis by phage T4: II. In vitro complementation between ribonucleotide reductase mutants. *Virology, 47,* 767–772.

212. Yeh, Y. C., E. J. Dubovi, and I. Tessman (1969). Control of pyrimidine biosynthesis by phage T4: mutants unable to catalyze the reduction of cytidine diphosphate. *Virology, 37,* 615–623.

213. Young, E. T., II, and R. L. Sinsheimer (1964). Novel intracellular forms of lambda DNA. *J. Mol. Biol., 10,* 562–564.

214. Zamecnik, P. C. (1960). Historical and current aspects of the problem of protein synthesis. In *The Harvey Lectures,* Ser. 54, Academic Press, Inc., New York, pp. 256–281.

215. Zinder, N. D. (1965). RNA phages. *Ann. Rev. Microbiol., 19,* 455–472.

216. Zinder, N. D., and J. Lederberg (1952). Genetic exchange in *Salmonella. J. Bacteriol., 64,* 679–699.

217. Zweig, M., and D. J. Cummings (1973). Cleavage of head and tail proteins during bacteriophage T5 assembly: selective host involvement in the cleavage of a tail protein. *J. Mol. Biol., 80,* 505–518.

Author Citation Index

Abelson, J. N., 75, 178, 404, 409
Abelson, P. H., 67
Adams, J. M., 401
Adams, M. H., 32, 51, 60, 246, 278, 368
Adhya, S., 333
Adler, J., 266
Ahmad-Zadeh, C., 402
Alberts, B. M., 401
Aldrich, C., 409
Amati, P., 334
Ames, B. N., 60, 334
Ammann, J., 391, 401
Anderer, F. A., 200
Anderson, E. H., 266
Anderson, E. S., 246
Anderson, F., 407
Anderson, T. F., 12, 32, 51, 60, 200, 401, 408
Anraku, N., 408
Aposhian, H. V., 67
Arber, W., 143, 266, 278, 401, 405
Arditti, R. R., 334
Asakura, S., 220
Asteriadis, G. T., 392
Astrachan, L., 60, 409
Attardi, G., 142
August, J. T., 391, 401, 408
Avitabile, A., 333

Bacon, R. H., 354
Bailey, W. T., Jr., 51, 93
Baker, W. H., 125
Baldwin, R. L., 278, 291
Baltimore, D., 67, 405, 408
Banerjee, A. K., 391, 401
Barlow, G. H., 392

Barlow, J. L., 60
Barner, H. D., 60, 67
Barnes, J. E., 409
Barnes, S. L., 206, 220
Barnett, L., 32, 106, 113, 120, 125, 142, 193, 200, 401, 402
Barrand, P., 291
Barrell, B. G., 392, 401
Basilio, C., 114
Batcheler, M. L., 404
Bautz, E. K. F., 401
Baylor, M. B., 206
Beaudreau, G., 392
Beckwith, J. R., 402
Beguin, F., 206, 406
Belfort, M., 334
Bello, L. J., 67
Benbow, R. M., 401
Bendet, I., 32, 206
Benzer, S., 32, 106, 113, 114, 125, 142, 158, 193, 200, 266, 334, 401, 402
Berg, A., 292
Berns, K. I., 148, 157, 401
Bernstein, C., 402
Bernstein, H., 402
Bertani, G., 246, 332
Bertani, L. E., 332, 333, 334
Bessman, M. J., 60, 67, 403
Bezdek, M., 334
Billeter, M. A., 392, 409
Bishop, D. H. L., 391, 392, 402
Black, L. W., 402, 407
Blum-Emerique, L., 247
Bode, V. C., 178, 291, 332
Bode, W., 114

411

413

Subject Index